ADVANCED TEXTS IN EC

Editors

Manuel Arellano   Guido Imbens   Grayham E. Mizon
Adrian Pagan   Mark Watson

Advisory Editor

C. W. J. Granger

TIME SERIES WITH LONG MEMORY

Edited by
PETER M. ROBINSON

OXFORD
UNIVERSITY PRESS

OXFORD

UNIVERSITY PRESS

Great Clarendon Street, Oxford OX2 6DP

Oxford University Press is a department of the University of Oxford.
It furthers the University's objective of excellence in research, scholarship,
and education by publishing worldwide in

Oxford New York

Auckland Bangkok Buenos Aires Cape Town Chennai
Dar es Salaam Delhi Hong Kong Istanbul Karachi Kolkata
Kuala Lumpur Madrid Melbourne Mexico City Mumbai Nairobi
São Paulo Shanghai Taipei Tokyo Toronto

Oxford is a registered trade mark of Oxford University Press
in the UK and in certain other countries

Published in the United States
by Oxford University Press Inc., New York

British Library Cataloguing in Publication Data

Data available

Library of Congress Cataloging in Publication Data
Robinson, Peter M.
Time series with long memory/Peter M. Robinson.
p. cm – (Advanced texts in econometrics)
1. Econometric models. 2. Mathematical statistics. I. Title. II. Series.
HB141 .R62 2003 330′.01′51932–dc21 2002035565

ISBN 0-19-925729-9 (hbk.)
ISBN 0-19-925730-2 (pbk.)

1 3 5 7 9 10 8 6 4 2

Typeset by Newgen Imaging Systems (P) Ltd., Chennai, India
Printed in Great Britain
on acid-free paper by
Biddles Ltd., Guildford & King's Lynn

To
Wendy and Rosa

Contents

Introduction

Dependence structure across time occupies a key role in the modelling of macroeconomic and financial data. Increasing interest has been shown in capturing the possibility of 'long memory' (also sometimes referred to as 'long-range dependence', or 'strong dependence' or 'persistence'). By long-memory time series I mean not only stationary ones but, *a fortiori*, nonstationary ones in which dependence falls off even more slowly across time. This volume collects together some articles on long memory which I hope have particular relevance or potential to modelling and statistical inference in an econometric context. By no means all of these articles were written by econometricians, however, or first appeared in econometric journals. Indeed, the long-memory literature spans the range from abstract probability theory, through mathematical statistics, to empirical studies in many areas of the sciences, besides economics. It can thus be relatively inaccessible to a researcher wishing to incorporate long memory in modelling or investigate it in data analysis, or one embarking on theoretical research.

Statistical inference on long memory has predominantly developed in a large sample framework. Superficially, this seems reasonable in that long memory can only be reliably detected for data covering a suitably long stretch of time. But in addition, as in other areas of statistics and econometrics, finite sample theory is usually mathematically intractable even in simple models, and requires very stringent assumptions. Large sample theory aims to provide rules of inference that become more reliable as sample size increases, under relatively mild conditions. However, large sample theory can pose considerably more difficult mathematical problems in a long-memory setting than in a short-memory stationary one. In short-memory series, autocovariances decay rapidly, perhaps exponentially, and the spectral density is at least bounded, and possibly very smooth. In long-memory stationary series, autocovariances are not summable, and the spectral density is unbounded. These singular mathematical properties sometimes lead to distinctive asymptotic behaviour of statistics of interest, perhaps affecting rates of convergence or leading to nonstandard limiting distributional behaviour. In other situations, asymptotic results have the same character as ones earlier developed for short-memory series, but proofs can significantly differ, and be harder. In this respect, the distinctive character of long memory has not always been appreciated, asymptotic results occasionally being claimed without adequate proofs, or even on the basis of inappropriate heuristics. Thus, in addition to covering a range of issues, indicating how ideas arose, and reflecting valuable contributions to modelling and to the estimation of parameters of interest, I will include some articles which illustrate how some of the difficulties in asymptotic statistical theory have been resolved. A number of the articles are quite challenging technically, and few are of the more 'popular', easy-to-read variety. Though some work on long memory from a Bayesian viewpoint has developed, the literature is very heavily frequentist, as reflected in the selection of articles.

The mainstream econometric time series literature over the past 15 years and more demonstrates considerable interest in long memory, in its focus on unit root series. We exclude articles on unit root series, however, these being covered only insofar as they are special cases of nonstationary fractional series. Indeed, despite the undoubted usefulness and simplicity of unit root models, and the excellent research that has been done in this area, one can envisage empirical researchers showing greater willingness in the future to consider the more flexible, fractional, classes. Unit root models essentially assume a known degree of memory, namely the order of differencing (in this case unity) which reduces the series to short-memory stationarity and invertibility. The point of fractional differencing models is not merely to allow for this memory measure to be fractional, but to allow it to be unknown, and estimable from data; statistical inference based on an incorrect order of differencing, be it unity or (as in case of short-memory stationarity) zero, are liable to be invalid, even in large samples. The bulk of the literature on estimating differencing order, both in parametric and semiparametric models, has focused on the stationary case. Work on nonstationary fractional models is developing, and some is included in this volume, but while significant new considerations do arise relative to the stationary case, many of the ideas are similar.

Several survey-type articles on long memory have been written, for example Taqqu (1986), Hampel (1987), Beran (1992), Robinson (1994), and Baillie (1996). None of these is now very up-to-date, indeed research on long memory has grown considerably in recent years. It is essential that I begin with an overview, relating the included articles and placing their contributions in perspective, so I have decided to do this in the context of a kind of survey article (Chapter 1) which sets the scene by describing the basic issues and problems, and refers to a much wider literature. I hope this will be particularly helpful to readers with little or no previous knowledge of long memory.

While a little of my earlier research touched on it, I only started to take a serious interest in long memory when I was asked to give an invited talk at the 1990 World Congress of the Econometric Society. I thought this a good excuse to learn something new so I prepared a survey on long memory (later published as Robinson 1994). My impression of the subject, circa-1990, was that on the one hand much impressive work had been done at the probabilistic end of the spectrum, and on the other, enterprising empirical work had been carried out in various substantive fields, including economics and finance, but that with some notable exceptions a large gap existed 'in between': useful, rigorous statistical theory was not well developed compared to the state of the art in short-memory time series and unit roots. Much subsequent research, including my own, has been directed to filling this gap, and the choice of articles in the present volume reflects this concern, and partly explains the over-representation of my own articles. Overall, with such a large literature, any choice is bound to be rather personal, and I hope that the relatively broad account in Chapter 1 will partly compensate, though topics such as prediction, structural change and Bayesian analysis are not discussed, and the list of references there is very far from a comprehensive bibliography; a broad long-memory time series methodology is developing. Chapter 1 contains sections on parametric modelling, semiparametric modelling, stochastic volatility models, nonstationary series, and regression and cointegration, respectively, which cover all the subsequent chapters and are ordered accordingly. At least one of the chapters connected with each of these sections includes some empirical work. Aside from Chapter 1, all articles have been previously published, except for Chapter 14.

I thank the Leverhulme Trust for their research support through a Leverhulme Trust Personal Research Professorship, and the Economic and Social Research Council for their research support through ESRC Grant R000238212. I thank Fabrizio Iacone for careful proofreading and Sue Kirkbride for secretarial help.

References

Baillie, R. T. (1996), 'Long memory processes and fractional integration in econometrics', *Journal of Econometrics*, 73, 5–59.

Beran, R. (1992), 'Statistical methods for data with long-range dependence', *Statistical Science*, 7, 404–27.

—— (1994), *Statistics for Long-Memory Processes*. New York: Chapman and Hall.

Hampel, F. (1987), 'Data analysis and self-similar processes', invited paper, ISI, Tokyo.

Robinson, P. M. (1994), 'Time series with strong dependence', in C. A. Sims, (ed.), *Advances in Econometrics: Sixth World Congress*, Vol. 1, Cambridge: Cambridge University Press, pp. 47–95.

Taqqu, M. S. (1986), 'A bibliographical guide to self-similar processes and long-range dependence', in E. Eberlein and M. Taqqu (eds), *Dependence in Probability and Statistics*, Boston: Birkhauser, pp. 137–62.

1
Long-memory Time Series

1.1 Introduction

Long memory has usually been described in terms of autocovariance or spectral density structure, in case of covariance stationary time series. Let $x_t, t = 0, \pm 1, \ldots$, be a time series indexed by time, t. It is covariance stationary, meaning that $E(x_t) = \mu$ and $\mathrm{Cov}(x_t, x_{t+j}) = \gamma(j)$ do not depend on t. If x_t has absolutely continuous spectral distribution function, then it has a spectral density, given formally by

$$f(\lambda) = \frac{1}{2\pi} \sum_{j=-\infty}^{\infty} \gamma(j) e^{-ij\lambda}, \quad -\pi \le \lambda \le \pi; \tag{1.1}$$

$f(\lambda)$ is a non-negative, even function, periodic of period 2π when extended beyond the 'Nyqvist' range $[-\pi, \pi]$. It is then common to say that x_t has long memory if

$$f(0) = \frac{1}{2\pi} \sum_{j=-\infty}^{\infty} \gamma(j) = \infty, \tag{1.2}$$

so that $f(\lambda)$ has a 'pole' at frequency zero. The opposite situation of a zero at $\lambda = 0$,

$$f(0) = \frac{1}{2\pi} \sum_{j=-\infty}^{\infty} \gamma(j) = 0, \tag{1.3}$$

is sometimes referred to as 'negative dependence' or 'anti-persistence'. The former term is natural because the second equality in (1.3) can only hold if the, positive, variance $\gamma(0)$ is balanced by predominantly negative autocovariances $\gamma(j), j \ne 0$. We then say that x_t has 'short memory' if

$$0 < f(0) < \infty. \tag{1.4}$$

These descriptions face the criticism that, consistent with (1.4), there is the possibility that $f(\lambda)$ has one or more poles or zeros at frequencies $\lambda \in (0, \pi]$, indicative of notable cyclic behaviour. We shall later refer to the modelling of such

This research was supported by a Leverhulme Trust Personal Research Professorship and ESRC Grant R000238212. I thank Fabrizio Iacone for careful checking of bibliographic details and for locating typographical errors. © P. M. Robinson 2003.

phenomena, but the bulk of interest has focused on zero frequency. Going back to the 1960s, experience of nonparametric spectral estimation for many economic time series has suggested very marked peakedness around zero frequency, see Adelman (1965), to lend support to (1.2). However, empirical evidence of long memory in various fields, such as astronomy, chemistry, agriculture, and geophysics, dates from much earlier times, see for example Newcomb (1886), Student (1927), Fairfield Smith (1938), Jeffreys (1939), and Hurst (1951). One aspect of interest was variation in the sample mean, $\bar{x} = n^{-1} \sum_{t=1}^{n} x_t$. If $f(\lambda)$ is continuous and positive at $\lambda = 0$, Féjèr's theorem indicates that

$$\text{Var}(\bar{x}) = \frac{1}{n} \sum_{j=1-n}^{n-1} \left(1 - \frac{|j|}{n}\right) \gamma(j) \tag{1.5}$$

$$\sim \frac{2\pi f(0)}{n}, \quad \text{as } n \to \infty, \tag{1.6}$$

where '$\sim$' indicates that the ratio of left and right sides tends to 1. But the empirical basis for this n^{-1} rate has been questioned, even by early experimenters, for example Fairfield Smith (1938) fitted a law $n^{-\alpha}$, $0 < \alpha < 1$ to spatial agricultural data. For future convenience, we change notation to $d = (1-\alpha)/2$, later explaining why d is referred to as the 'differencing' parameter. Fairfield Smith's law for the variance of the sample mean is thus n^{2d-1}, which from (1.5) is easily seen to arise if

$$\gamma(j) \sim c_1 j^{2d-1}, \quad \text{as } j \to \infty, \tag{1.7}$$

for $c_1 > 0$. Under additional conditions (see Yong 1974), (1.7) is equivalent to a corresponding power law for $f(\lambda)$ near zero frequency,

$$f(\lambda) \sim c_2 |\lambda|^{-2d}, \quad \text{as } \lambda \to 0, \tag{1.8}$$

for $c_2 > 0$.

The behaviour of the sample mean under such circumstances, and the form and behaviour of the best linear unbiased estimate of the population mean, was discussed by Adenstedt (1974) (included as Chapter 2 of this volume). Adenstedt anticipated the practical usefulness of (1.8) in the long-memory range $0 < d < \frac{1}{2}$, but also treated the anti-persistent case $-\frac{1}{2} < d < 0$, where (1.3) holds, as it does also for $d = -1$, which arises if a short-memory process (see (1.4)) is first-differenced; Vitale (1973) had earlier discussed similar issues in this case. The sample mean tends to be highly inefficient under anti-persistence, but for long memory Samarov and Taqqu (1988) found it to have remarkably good efficiency.

Macroeconomic series can be regarded as aggregates across many micro-units, and explanations of how long-memory behaviour might arise in macroeconomics has focused on random-parameter short-memory modelling of micro-series. Consider the random-parameter autoregressive model of order 1 ($AR(1)$),

$$X_t(\omega) = A(\omega)X_{t-1}(\omega) + \varepsilon_t(\omega), \tag{1.9}$$

where ω indexes micro-units, the $\varepsilon_t(\omega)$ are independent and homoscedastic with zero mean across ω and t, and $A(\omega)$ is a random variable with support $(-1, 1)$ or $[0, 1)$. Then, conditional on ω, $X_t(\omega)$ is a stationary $AR(1)$ sequence. Robinson (1978a) showed that the 'unconditional autovariance' which we again denote by $\gamma(j)$, is given by

$$\gamma(j) = \mathrm{Cov}\{X_t(\omega), X_{t+j}(\omega)\} = \sum_{u=0}^{\infty} E\{A(\omega)^{j+2u}\}, \qquad (1.10)$$

and that the 'unconditional spectrum' $f(\lambda)$ (1.1) at $\lambda = 0$ is proportional to $E\{(1 - A(\omega))^{-2}\}$, and thus infinite, as under long memory (1.2), if $A(\omega)$ has a density with a zero at 1 of order less than or equal to 1. One class with this property considered by Robinson (1978a) was the (possibly translated) Beta distribution, for which Granger (1980) explicitly derived the corresponding power law behaviour of the spectral density of cross-sectional aggregates $x_t = N^{-1/2} \sum_{i=1}^{N} X_t(\omega_i)$, where the ω_i are independent drawings: clearly $\mathrm{Cov}(x_t, x_{t+u})$ is $\gamma(j)$, (1.10), due to the independence properties. Indeed, if $A(\omega)$ has a Beta $(c, 2 - 2d)$ distribution on $(0, 1)$, for $c > 0, 0 < d < \frac{1}{2}$, $E\{A(\omega)^k\}$ decays like k^{2d-2}, so (1.10) decays like j^{2d-1}, as in (1.7). Intuitively, a sufficient density of individuals with close-to-unit-root behaviour produces the aggregate long memory. A similar idea was earlier employed by Mandelbrot (1971) in computer generation of long-memory time series. For further developments, for more general models than (1.9), see e.g. Robinson (1986, 1987), Goncalves and Gourieroux (1988), Lippi and Zaffaroni (1997).

The rest of the paper deals with various approaches to modelling long memory, for various kinds of data, and with relevant statistical inference. The following section provides background to estimation of parametric models. Semiparametric inference is discussed in Section 1.3. Section 1.4 describes some long-memory stochastic volatility models. Section 1.5 concerns the extension of parametric and semiparametric inference to nonstationary series. In Section 1.6 we review regression models and cointegration.

1.2 Parametric Modelling and Inference

Much interest in the possibility of long-memory or anti-persistence focuses on the parameter d, which concisely describes long-run memory properties. In practice d is typically regarded as unknown, and so its estimation is of interest. Indeed, the discussion of the previous section indicates that an estimate of d is needed even to estimate the variance of the sample mean; this was pursued by Beran (1989), for example. In order to estimate d we need to consider the modelling of dependence in more detail.

The simplest possible realistic model for a covariance stationary series is a parametric one that expresses $\gamma(j)$ for all j, or $f(\lambda)$ for all λ, as a parametric function of just two parameters, d and an unknown scale factor. Perhaps the earliest such model is 'fractional noise', which arises from considerations of self-similarity. A continuous time stochastic process $\{y(t); -\infty < t < \infty\}$ is self-similar with 'self-similarity parameter' $H \in (0, 1)$ if, for any $a > 0$, $\{y(at); -\infty < t < \infty\}$ has the same distribution as $\{a^H y(t); -\infty < t < \infty\}$. If the differences $x_t = y(t) - y(t - 1)$, for integer t, are covariance stationary, we have

$$\gamma(j) = \frac{\gamma(0)}{2} \{|j + 1|^{2H} - 2|j|^{2H} + |j - 1|^{2H}\}. \tag{1.11}$$

As $j \to \infty$, (1.11) decays like j^{2H-2}, so on taking $H = d + \frac{1}{2}$ we have again the asymptotic law (1.7); $\gamma(0)$ is the unknown scale parameter in this model. The formula for the spectral density was derived by Sinai (1976); it is complicated, but satisfies (1.8).

'Fractional noise' was extensively studied by Mandelbrot and Van Ness (1968), Hipel and McLeod (1978) and others, but, perhaps because it extends less naturally to richer stationary series, and nonstationary series, and due to its unpleasant spectral form (see the later discussion of Whittle estimates) it has received less attention in recent years than another two-parameter model, the 'fractional differencing' model proposed by Adenstedt (1974), in Chapter 2 of this volume,

$$f(\lambda) = \frac{\sigma^2}{2\pi} \left|1 - e^{i\lambda}\right|^{-2d}, \quad -\pi \le \lambda \le \pi. \tag{1.12}$$

For $d = 0$, (1.12) is just the spectral density of a white noise series (with variance σ^2), while for $0 < |d| < \frac{1}{2}$ both properties (1.7) and (1.8) hold, Adenstedt (1974) giving a formula for $\gamma(j)$ under (1.12), as well as other properties. Note that $d < \frac{1}{2}$ is necessary for integrability of $f(\lambda)$, that is for x_t to have finite variance, and this restriction is sometimes called the stationarity condition on d. Another mathematically important restriction is that of invertibility, $d > -\frac{1}{2}$. We shall discuss statistical inference on models such as (1.12).

Granger (1966) identified the 'typical spectral shape of an economic variable' as not only having a pole or singularity at zero frequency, but then decaying monotonically. Both the 'fractional differencing' and 'fractional noise' models have this simple property. However, even if monotonicity holds, as it may, at least approximately, in case of deseasonalized economic series, the notion that the entire autocorrelation structure can be explained by a single parameter, d, is highly restrictive. While the value of d determines the long-run or low-frequency behaviour of $f(\lambda)$, greater flexibility in modelling short-run, high-frequency behaviour may be desired. We referred to (1.12) as a 'fractional differencing' model because it is the spectral density of x_t generated by

$$(1 - L)^d x_t = e_t, \tag{1.13}$$

where $\{e_t\}$ is a sequence of uncorrelated variables with zero mean and variance σ^2, L represents the lag operator, $Lx_t = x_{t-1}$ and formally,

$$(1 - L)^d = \sum_{j=0}^{\infty} \frac{\Gamma(j - d)}{\Gamma(-d)\Gamma(j + 1)} L^j. \tag{1.14}$$

With $d = 1$ (and an initial condition such as $x_0 = 0$), (1.13) would describe a random walk model. Box and Jenkins (1971) stressed the vector model

$$(1 - L)^d a(L) x_t = b(L) e_t. \tag{1.15}$$

Here d is an integer, $a(L)$ and $b(L)$ are the polynomials

$$a(L) = 1 - \sum_{j=1}^{p} a_j L^j, \qquad b(L) = 1 + \sum_{j=1}^{q} b_j L^j, \tag{1.16}$$

all of whose zeros are outside the unit circle, to ensure stationarity and invertibility, and $a(L)$ and $b(L)$ have no zero in common, to ensure unambiguity of the autoregressive (AR) order p and the moving average (MA) order q. Granger and Joyeux (1980) (Chapter 3 of this volume) considered instead fractional $d \in (-\frac{1}{2}, \frac{1}{2})$ in (1.15), giving a fractional autoregressive integrated moving average model of orders p, d, q (often abbreviated as FARIMA(p, d, q) or ARFIMA(p, d, q)). It has spectral density

$$f(\lambda) = \frac{\sigma^2}{2\pi} \left| 1 - e^{i\lambda} \right|^{-2d} \left| \frac{b(e^{i\lambda})}{a(e^{i\lambda})} \right|^2, \quad -\pi \le \lambda \le \pi. \tag{1.17}$$

Much of the discussion of Granger and Joyeux (1980) concerned the simple FARIMA$(0, d, 0)$ case (1.12) of Adenstedt (1974), but they also considered estimation of d, prediction, and computer generation of long-memory series. Methods in the latter category include the aggregation approach of Mandelbrot (1971) and the Cholesky decomposition approach of Hipel and McLeod (1978); an elegant, more recent, one is due to Davies and Harte (1987), involving use of the fast Fourier transform.

Hosking (1981) provided further discussion of FARIMA(p, d, q) processes, again much of it based on Adenstedt's (1974) model (1.12), but he also gave results for the general case (1.15), especially the FARIMA$(1, d, 0)$. Further information on FARIMA(p, d, q) models was given by Sowell (1992), Chung (1994), and others.

An early proposal for estimating d, or H, used the adjusted rescaled range (R/S) statistic

$$R/S = \frac{\max_{1 \le j \le n} \sum_{t=1}^{j} (x_t - \bar{x}) - \min_{1 \le j \le n} \sum_{t=1}^{j} (x_t - \bar{x})}{\left\{ (1/n) \sum_{t=1}^{n} (x_t - \bar{x})^2 \right\}^{1/2}} \tag{1.18}$$

of Hurst (1951), Mandelbrot and Wallis (1969). Asymptotic statistical behaviour of the R/S statistic was studied early on by Mandelbrot (1975), Mandelbrot and Taqqu (1979), and it was considered in an economic context by Mandelbrot (1972). However, while it behaves well with respect to long-tailed distributions, its limit distribution is nonstandard and difficult to use in statistical inference, while it has no known optimal efficiency properties with respect to any known family of distributions. The continued popularity of R/S, for example in the finance literature, may rest in part on an inadequate appreciation of rival procedures.

While long-memory series do have distinctive features, there is no over-riding reason why traditional approaches to parametric estimation in time series should be abandoned in favour of rather special approaches like R/S. Indeed, if x_t is assumed Gaussian, the Gaussian maximum likelihood estimate (MLE) might be expected to have optimal asymptotic statistical properties, and unlike R/S, can be tailored to the particular parametric model assumed, be it (1.11), (1.12), (1.17), or whatever. The literature on the Gaussian MLE developed first with short-memory processes in mind (e.g. see Whittle 1951; Hannan 1973), and it may be helpful to provide some background to this.

One important finding was that the Gaussian likelihood can be replaced by various approximations without affecting first order limit distributional behaviour. In particular, under suitable conditions, estimates maximizing such approximations, and called 'Whittle estimates', are all $\sqrt{n}$-consistent and have the same limit normal distribution as the Gaussian MLE. Such approximations arise naturally in that treatment of pre-sample values is always an issue with time series models, but computational considerations are also an important factor. One particular Whittle estimate which seems usually particularly advantageous in the latter respect is the discrete-frequency form. Suppose the parametric spectral density has form $f(\lambda; \theta, \sigma^2) = (\sigma^2/2\pi)h(\lambda; \theta)$, where θ is an r-dimensional unknown parameter vector and σ^2 is a scalar as in (1.12). If σ^2 is regarded as varying freely from θ, and $\int_{-\pi}^{\pi} \log h(\lambda; \theta) \, d\lambda = 0$ for all admissible values of θ, then we have what might be called a 'standard parameterization'. For example, we have a standard parameterization in (1.12) with $\theta = d$, and in (1.17) with θ determining the $a_j, 1 \le j \le p$ and $b_j, 1 \le j \le q$. Define also the periodogram

$$I(\lambda) = \frac{1}{2\pi n} \left| \sum_{t=1}^{n} x_t e^{it\lambda} \right|^2 \qquad (1.19)$$

and the Fourier frequencies $\lambda_j = 2\pi j n$. Denoting by θ_0 the true value of θ, then the discrete frequency Whittle estimate of θ_0 minimizes the following approximation to a constant minus the Gaussian log likelihood,

$$\sum_{j=1}^{n-1} \frac{I(\lambda_j)}{h(\lambda_j; \theta)}. \qquad (1.20)$$

This estimate was stressed by Hannan (1973). From the viewpoint of the short-memory models under discussion at that time, it had the advantages of using directly the form of h, which is readily written down in case of autoregressive moving average (ARMA) models, Bloomfeld's (1972) spectral model, and others; by contrast, autocovariances, partial autocovariances, AR coefficients and MA coefficients, which variously occur in other types of Whittle estimate, tend to be more complicated except in special cases, indeed for (1.17) the form of autocovariances, for example, can depend on the question of multiplicity of zeros of

$a(L)$. Another advantage of (1.20) is that it makes direct use of the fast Fourier transform, which enables the periodograms $I(\lambda_j)$ to be rapidly computed even when n is very large. A third advantage is that mean-correction of x_t is dealt with simply by omission of the frequency $\lambda_0 = 0$.

Another important characteristic of Whittle estimates of θ_0, first established in case of short-memory series, is that while they are only asymptotically efficient when x_t is Gaussian, their limit distribution (in case of 'standard parameterizations') is unchanged by many departures from Gaussianity. Thus the same, relatively convenient, rules of statistical inference can be used without worrying too much about the question of Gaussianity. In particular, Hannan (1973) established this, for several Whittle forms in case x_t has a linear representation in homoscedastic stationary martingale differences having finite variance.

Hannan established first consistency under only ergodicity of x_t, so that long memory was actually included here. However, for his central limit result, with $\sqrt{n}$-convergence, which is crucial for developing statistical inference, his conditions excluded long memory, and clearly (1.20) appears easier to handle technically in the presence of a smooth h than of one with a singularity. Hannan's work was further developed in the short-memory direction, to cover 'nonstandard parameterizations' and multiple time series. One such treatment, of Robinson (1978b), has a central limit theorem that hints at how a modest degree of long memory might be covered. He reduced the problem to a central limit theorem for finitely many sample autocovariances, whose asymptotic normality had been shown by Hannan (1976) to rest crucially on square integrability of the spectral density; note that (1.12) and (1.13) are square integrable only for $d < \frac{1}{4}$. In fact for some forms of Whittle estimate, Yajima (1985) established the central limit theorem, again with $\sqrt{n}$-rate, in case of model (1.12) with $0 < d < \frac{1}{4}$.

The major breakthrough in justifying Whittle estimation in long-memory models was provided by Fox and Taqqu (1986) (Chapter 4 of this volume). Their objective function was not (1.20) but the continuous frequency form

$$\int_{-\pi}^{\pi} \frac{I(\lambda)}{h(\lambda; \theta)} \, d\lambda, \tag{1.21}$$

but Fox and Taqqu's insight applies to (1.20) also. The periodogram $I(\lambda)$ is an asymptotically unbiased estimate of the spectral density at continuity points and so $I(\lambda)$ can be expected to blow up as $\lambda \to 0$. However, since $h(\lambda; \theta)$ also blows up as $\lambda \to 0$ and appears in the denominator, some 'compensation' can be expected. More precisely, limiting distributional behaviour depends on the 'score' (the derivative in θ of (1.21)) at θ_0 being asymptotically normal; this, like (1.20), is a quadratic form in x_t, and Fox and Taqqu (1987) gave general conditions for such quadratic forms to be asymptotically normal, which then apply to Whittle estimates with long memory such that $0 < d < \frac{1}{2}$.

Fox and Taqqu (1986) assumed Gaussianity of x_t, as did Dahlhaus (1989), who also considered the actual Gaussian MLE and discrete-frequency Whittle (1.20), and established asymptotic efficiency. With reference to (1.21) Giraitis and Surgailis (1990) relaxed Gaussianity to a linear process in independent and identically distributed (i.i.d.) innovations, thus providing a partial extension of Hannan's (1973) work to long memory. Heyde and Gay (1993), Hosoya (1996) considered multivariate extensions. Overall, the bulk of this asymptotic theory has not directly concerned the discrete frequency form (1.20), and has focused mainly on the continuous frequency form (1.21), though the former benefits from the neat form of the spectral density in case of the popular FARIMA(p, d, q) class (1.17); on evaluating the integral in (1.21), we have a quadratic form involving the Fourier coefficients of $h(\lambda; \theta)^{-1}$, which are generally rather complicated for long-memory models. Also, in (1.21) and the Gaussian MLE, correction for an unknown mean must be explicitly carried out, not dealt with merely by dropping zero frequency; Monte Carlo results of Cheung and Diebold (1994) compare this approach with sample mean-correction in Gaussian MLE.

We briefly refer to other estimates that have been considered. Whittle estimation of the models (1.11), (1.12) and (1.17) requires numerical optimization, but Kashyap and Eom (1988) proposed a closed-form estimate of d in (1.12) by a log periodogram regression (across λ_j, $j = 1, \ldots, n - 1$). This idea does not extend nicely to FARIMA(p, d, q) models with $p > 0$ or $q > 0$, but it does to

$$f(\lambda) = \frac{1}{2\pi}\left|1 - e^{i\lambda}\right|^{-2d} \exp\left\{\sum_{k=1}^{p-1} \beta_k \cos((k-1)\lambda)\right\}, \quad -\pi \le \lambda \le \pi \quad (1.22)$$

(see Robinson 1994*a*) which combines (1.12) with Bloomfield's (1972) short-memory exponential model; Moulines and Soulier (1999) have recently provided asymptotic theory for log peridogram regression estimation of (1.22). They assumed Gaussianity, and ironically, for technical reasons, this is harder to avoid when a nonlinear function of the periodogram, such as the log, is involved, than in Whittle estimation, which was originally motivated by Gaussianity. Whittle estimation is also feasible with (1.22), indeed Robinson (1994*a*) noted that it can be reparameterized as

$$f(\lambda) = \frac{1}{2\pi} \exp\left\{\sum_{k=1}^{p-1} \theta_k \cos\{(k-1)\lambda\} - 2d \sum_{k=p-1}^{\infty} \frac{\cos(k\lambda)}{k}\right\}, \quad (1.23)$$

taking $\theta_1 = \beta_1$, $\theta_k = \beta_k - 2/(k-1)$, $2 \le k \le p-1$, from which it can be deduced that the limiting covariance matrix of Whittle estimates is desirably diagonal.

Generalized method of moments (GMM) has become extremely popular in econometrics, and its use has been proposed in estimating long-memory models, either in the time domain or the frequency domain. However, GMM objective functions seem in general to be less computationally attractive than (1.20), require stronger regularity conditions in asymptotic theory, and do not deal so nicely with an unknown mean. Also, unless a suitable weighting is employed they will be less efficient than Whittle estimates in the Gaussian case, have a relatively cumbersome limiting covariance matrix, and are not even asymptotically normal under $d > \frac{1}{4}$. It must be acknowledged, however, that finite-sample properties are very important in practice, and while there seems no intuitive reason why GMM estimates, for example, should be superior in this respect, it also cannot be asserted that any particular form of Whittle, such as (1.20), is always better in finite samples, or that some other, non-Whittle, estimate might not be preferable in some cases. Indeed, $\sqrt{n}$-consistency and asymptotic normality of Whittle estimates cannot even be taken for granted, having been shown not to hold over some or all of the range $d \in (0, \frac{1}{2})$ for certain nonlinear functions x_t of a underlying Gaussian long memory process (see Fox and Taqqu 1987; Giraitis and Taqqu 1999).

Nonstandard limit distributional behaviour for Whittle estimates can also arise even under Gaussianity, in certain models. As observed in Section 1.1, a spectral pole (or zero) could arise at a non-zero frequency, to explain a form of cyclic behaviour. In particular Gray, Zhang, and Woodward (1989) proposed the 'Gegenbauer' model

$$f(\lambda) = \frac{\sigma^2}{2\pi} \left| 1 - 2e^{i\lambda} \cos\omega + e^{2i\lambda} \right|^{-2d} \left| \frac{b(e^{i\lambda})}{a(e^{i\lambda})} \right|^2, \quad -\pi \le \lambda \le \pi, \quad (1.24)$$

for $\omega \in (0, \pi]$. To compare with (1.12), $f(\lambda)$ has a pole at frequency ω if $d > 0$. If ω is known then our previous discussion of estimation and asymptotic theory continues to apply, see Hosoya (1996, 1997). If ω is unknown, then Whittle procedures can be adapted, but it seems that such estimates of ω (but not of the other parameters) will be n-consistent with a nonstandard limit distribution. This was claimed by Chung (1996a,b) albeit with incomplete proof, while Giraitis, Hidalgo, and Robinson (2001) established n-consistency for an estimate of ω that even lacks a limit distribution.

1.3 Semiparametric Modelling and Inference

In the use of parametric FARIMA(p, d, q) models, correct specification of p and q is important. Under-specification of p or q leads to inconsistent estimation of AR and MA coefficients, but also of d, as does over-specification of both, due to a loss of identifiability. Order-determination procedures developed for short-memory models, such as AIC, can be adapted to FARIMA models (indeed see Beran, Bhansali and Ocker (1998) for the FARIMA($p, d, 0$) case) but there is no guarantee that the underlying model belongs to the finite-parameter class proposed. It seems especially unfortunate that an attempt to seriously model short-run features can lead to inconsistent estimation of long-run properties, if the latter happen to be the aspect of most interest.

The asymptotic behaviour (1.7) and (1.8) indicates that short-run modelling is almost irrelevant at very low frequencies and very long lags, where d dominates. It thus appears that estimates of d can be based on information arising from only one or other of these domains, and that such estimates should have validity across a wide range of short-memory behaviour. As this robustness requires estimates to essentially be based on only a vanishingly small fraction of the data as sample size increases, one expects slower rates of convergence than for estimates based on a correct finite-parameter model. However, in very long series, such as arise in finance, the degrees of freedom available may be sufficient to provide adequate precision. These estimates are usually referred to as 'semiparametric', though their slow convergence rates make them more akin to 'nonparametric' estimates in other areas of statistics, indeed some are closely related to the smoothed nonparametric spectrum estimates familiar from short-memory time-series analysis.

Before presenting some individual semiparametric estimates, it is worth stressing that not just point estimation is of interest, but also interval estimation and hypothesis testing. Perhaps the test of most interest to practitioners is a test of long memory, or rather, a test of short memory $d = 0$ against long-memory alternatives $d > 0$, or anti-persistent alternatives $d < 0$, or both, $d \neq 0$. What is then needed is a statistic with a distribution that can be satisfactorily approximated, and computed, under $d = 0$, and that has good power. In a parametric context, tests of $d = 0$—perhaps of Wald, Lagrange multiplier or likelihood-ratio type—can be based on Whittle functions such as (1.20) and the FARIMA(p, d, q) family. (Strictly speaking, much of the limit distribution theory for Whittle estimation primarily concerned with stationary long memory, $0 < d < \frac{1}{2}$, does not cover $d = 0$, or $d < 0$, but other earlier short-memory theory, such as Hannan's (1973), can provide null limit theory for testing $d = 0$.) The test statistic is based on assumed p and q, but the null limit distribution considered on this basis is generally invalid if p and q are misspecified, as discussed earlier; this can lead, for example, to mistaking unaccounted-for short-memory behaviour for long memory, and rejecting the null too often. Lo (1991) (Chapter 5 of this volume) observed the invalidity of tests for $d = 0$ based on asymptotic theory of Mandelbrot (1975) for the R/S statistic (1.18) in the presence of unanticipated short-memory autocorrelation. He proposed a corrected statistic (using smoothed nonparametric spectral estimation at frequency zero) and developed its limit distribution under $d = 0$ in the presence of a wide range of short-memory dependence (described by mixing conditions), and tested stock returns for long memory.

Lo's paper is perhaps especially notable as an early, rigorous treatment of asymptotic theory in a semiparametric context. However, the null limit theory for his modified R/S statistic is nonstandard. In principle, any number of statistics has sensitivity to long memory. Some have the character of 'method-of-moments' estimates, minimizing a 'distance' between population and sample properties. In the frequency domain, Robinson (1994b) proposed an 'averaged periodogram' estimate of d, employing what would be a consistent estimate of $f(0)$ under $d = 0$. He established consistency, requiring finiteness of only second moments and allowing for the presence of an unknown slowly varying factor $L(\lambda)$ in $f(\lambda)$, so that (1.8) is relaxed to

$$f(\lambda) \sim L(\lambda)|\lambda|^{-2d}, \quad \text{as } \lambda \to 0. \tag{1.25}$$

Delgado and Robinson (1996) proposed data-dependent choices of the bandwidth number (analogous to the one discussed later in relation to log periodogram estimation, for example) that is required in the estimation, and Lobato and Robinson

(1996) established limit distribution theory, which is complicated: the estimate is asymptotically normal for $0 \leq d < \frac{1}{4}$, but non-normal for $d \geq \frac{1}{4}$. Lobato (1997) extended Robinson's (1994b) consistency result to the averaged cross-periodogram of bivariate series. A number of other semiparametric estimates of d share this latter property, which is due to $f(\lambda)$ not being square-integrable for $d \geq \frac{1}{4}$, for example the time-domain 'variance-type' estimate of Teverovsky and Taqqu (1997).

We might then turn to the traditional statistical practice of regression. In the time domain, the asymptotic rule (1.7) suggests two approaches, nonlinearly regressing sample autocovariances on cj^{2d-1}, and ordinary least squares (OLS) regression of logged sample autocovariances on $\log j$ and an intercept. These were proposed by Robinson (1994a), the second proposal then being studied by Hall, Koul, and Turlach (1997). However, the limit distributional properties of these estimates are as complicated as those for the averaged periodogram estimate, intuitively because OLS is a very *ad hoc* procedure in this setting, the implied 'disturbances' in the 'regression model' being far from uncorrelated or homoscedastic.

Nice results can only be expected from OLS if the disturbances are suitably 'whitened'. At least for short-memory series, the (Toeplitz) covariance matrix of $x_1, \ldots, x_n$ is approximately diagonalized by a unitary transformation, such that normalized periodograms $u_j = \log\{I(\lambda_j)/f(\lambda_j)\}$ (cf. (1.20)), sufficiently resemble a zero-mean, uncorrelated, homoscedastic sequence. For long-memory series, (1.8) suggests consideration of

$$\log I(\lambda_j) \simeq \log c - 2d \log \lambda_j + u_j, \tag{1.26}$$

for a positive constant c and λ_j close to zero. This idea was pursued by Geweke and Porter-Hudak (1983) (Chapter 6 of this volume) though they instead employed a narrow band version of the 'fractional differencing' model (1.12), specifically replacing $\log \lambda_j$ by $\log |1 - e^{i\lambda_j}|$. They performed OLS regression over $j = 1, \ldots, m$, where m, a bandwidth or smoothing number, is much less than n but is regarded as increasing slowly with n in asymptotic theory. Geweke and Porter-Hudak's approach was anticipated by a remark of Granger and Joyeux (1980).

Geweke and Porter-Hudak argued, in effect, that as $n \to \infty$ their estimate $\tilde{d}$ satisfies

$$m^{1/2}(\tilde{d} - d) \to_d N\left(0, \frac{\pi^2}{24}\right), \tag{1.27}$$

giving rise to extremely simple inferential procedures. However the heuristics underlying their argument are slightly defective, and they, and some subsequent authors, did not come close to providing a rigorous proof of (1.27). A difficulty with their heuristics is that for long-memory (and anti-persistent) series the u_j are not actually asymptotically uncorrelated or homoscedastic for fixed j with $n \to \infty$, as shown by Künsch (1986), and elaborated upon by Hurvich and Beltrao (1993), Robinson (1995a). As Robinson (1995a) showed, this in itself invalidates Geweke and Porter-Hudak's (1983) argument. Even for increasing j, the approximation of the u_j by an uncorrelated, homoscedastic sequence is not very good, and this, and the nonlinearly involved periodogram, makes a proof of (1.27) non-trivial.

Robinson (1995*a*) established (1.27), explicitly in case of the approximation (1.26) rather than Geweke and Porter-Hudak's version, though indicating that the same result holds there. Robinson's result applies to the range $|d| < \frac{1}{2}$, providing simple interval estimates as well as a simple test of short memory, $d = 0$. His treatment actually covered multiple time series, possibly involving differing memory parameters, and tests for equality of these, and more efficient estimates using the equality were also given. Robinson's treatment assumed Gaussianity, but Velasco (2000) gave an extension to linear processes x_t. Both authors employed Künsch's (1986) suggestion of trimming out the lowest λ_j to avoid the anomalous behaviour of periodograms there, but Hurvich, Deo, and Brodsky (1998) showed that this was unnecessary for (1.27) to hold, under suitable conditions. They also addressed the important issue of choice of the bandwidth, m, providing optimal asymptotic minimum mean-squared error theory. When $f(\lambda)\lambda^{2d}$ is twice differentiable at $\lambda = 0$, the optimal bandwidth is of order $n^{4/5}$, but the multiplying constant depends on unknown population quantities. Hurvich and Deo (1999) proposed a consistent estimate of this constant, and hence a feasible, data-dependent choice of m. Previously, Hurvich and Beltrao (1994) had related mean squared error to integrated mean squared error in spectral density estimation, and thence proposed cross-validation procedures for choosing both m and the trimming constant.

'Log-periodogram estimation' has been greatly used empirically, deservedly so in view of its nice asymptotic properties and strong intuitive appeal. However, in view of the limited information it employs there is a concern about precision, and it is worth asking at least whether the information can be used more efficiently. Robinson (1995*a*) showed that indeed the asymptotic variance in (1.27) can be reduced by 'pooling' adjacent periodograms, prior to logging. A proposal of Künsch (1987), however, leads to an alternative frequency-domain estimate that does even better. He suggested a narrow-band discrete-frequency Whittle estimate (cf. (1.20)). Essentially this involves Whittle estimation of the 'model' $f(\lambda) = C\lambda^{-2d}$ over frequencies $\lambda = \lambda_j$, $j = 1, \ldots, m$, where m plays a similar role as in log periodogram estimation. Then C can be eliminated by a side calculation (much as the innovation variance is eliminated in getting (1.20)), and d is estimated by $\hat{d}$ which minimizes

$$\log\left\{\frac{1}{m}\sum_{j=1}^{m}\lambda_j^{2d}I(\lambda_j)\right\} - \frac{2d}{m}\sum_{j=1}^{m}\log\lambda_j. \tag{1.28}$$

There is no closed-form solution but (1.28) is easy to handle numerically. Robinson (1995*b*) (Chapter 7 of this volume) established that

$$m^{1/2}(\hat{d} - d) \to_d N\left(0, \tfrac{1}{4}\right). \tag{1.29}$$

Using the same m sequence, $\hat{d}$ is then more efficient than the log periodogram estimate $\tilde{d}$ (cf. (1.27)), while the pooled log periodogram estimate of Robinson (1995*a*) has asymptotic variance that converges to $\frac{1}{4}$ from above as the degree of pooling increases. The estimate $\hat{d}$ is only implicitly defined, but it is nevertheless easy to locate, and the linear involvement of the periodogram in (1.28) makes it possible to establish (1.29) under simpler and milder conditions than needed for (1.27), Robinson employing a linear process for x_t in martingale difference innovations. This, and the coverage of all $d \in \left(-\frac{1}{2}, \frac{1}{2}\right)$, may have implications also for further development of the asymptotic theory of parametric Whittle estimates discussed in the previous section. Another feature of the asymptotic theory of Robinson (1995a), and that of Robinson (1995b), is the purely local nature of the assumptions on $f(\lambda)$; and the way in which the theory fits in with earlier work on smoothed nonparametric spectral estimation for short-memory series; (1.8) is relaxed to

$$f(\lambda) = C|\lambda|^{-2d}(1 + O|\lambda|^\beta), \quad \text{as } \lambda \to 0, \tag{1.30}$$

where $\beta \in (0, 2]$ is analogous to the local smoothness parameter involved in the spectral estimation work, and no smoothness, or even boundedness, is imposed on f away from zero frequency. The parameter β also enters into rules for optimal choice of m; see Henry and Robinson (1996). A nice feature of the 'Gaussian semiparametric' or 'local Whittle' estimate $\hat{d}$ is that it extends naturally to multivariate series; see Lobato (1999). If only a test for short memory is desired, Lobato and Robinson (1998) provided a Lagrange multiplier one based on (1.28) that avoids estimation of d.

Work has proceeded on refinements to the semiparametric estimates $\tilde{d}$ and $\hat{d}$, and their asymptotic theory. Monte Carlo simulations have indicated bias in $\tilde{d}$, and Hurvich and Beltrao (1994), Hurvich and Deo (1999) have proposed bias-reduced estimates. Andrews and Guggenberger (2000), Robinson and Henry (2000) have developed estimates that can further reduce the bias, and have smaller asymptotic minimum mean squared error, using respectively an extended regression and higher-order kernels, Robinson and Henry (2000) at the same time introducing a unified M-estimate class that includes $\tilde{d}$ and $\hat{d}$ as special cases. Bias reduction, and a rule for bandwidth choice, also results from Giraitis and Robinson's (2000) development of an Edgeworth expansion for a modified version of $\hat{d}$. Moulines and Soulier (1999, 2000) and Hurvich and Brodsky (2001) considered a broadband version of $\tilde{d}$ originally proposed by Janacek (1982), effectively extending the regression in (1.26) over all Fourier frequencies after including cosinusoidal terms, corresponding to the model (2.12) with p, now a bandwidth number, increasing slowly with n. They showed that if $f(\lambda)\lambda^{2d}$ is analytic over all frequencies, an asymptotic mean squared error of order $(\log n)/n$ can thereby be obtained, which is not achievable by the refinements to $\tilde{d}$ and $\hat{d}$ we have discussed, though the latter require only local-to-zero assumptions on $f(\lambda)$. An alternative semiparametric estimate with nice properties was proposed by Parzen (1986) and studied by Hidalgo (2001).

1.4 Stochastic Volatility Models

We have so far presented 'long memory' as purely a second-order property of a time series, referring to autocovariances or spectral structure. These do not completely describe non-Gaussian processes, where 'memory' might usefully take on a rather different meaning. In particular, passing a process through a nonlinear filter can change asymptotic autocovariance structure. As Rosenblatt (1961) indicated, if x_t is a stationary long-memory Gaussian process satisfying (1.7), then x_t^2 has autocovariance decaying like j^{4d-2}, so has 'long memory' only when $\frac{1}{4} \leq d < \frac{1}{2}$, and even here, since $4d - 2 < 2d - 1$, x_t^2 has 'less memory' than x_t.

Many financial time series suggest a reverse kind of behaviour. Asset returns, or logged asset returns, frequently exhibit little autocorrelation, as is consistent with the efficient markets hypothesis, whereas their squares are noticeably correlated. Engle (1982) proposed to model this phenomenon by the autoregressive conditionally heteroscedastic model of order p (ARCH(p)), such that

$$x_t = \varepsilon_t \sigma_t, \tag{1.31}$$

where σ_t is the square root of

$$\sigma_t^2 \stackrel{\text{def}}{=} E(x_t^2|x_{t-1}, x_{t-2}, \ldots) = \alpha_0 + \sum_{j=1}^{p} \alpha_j x_{t-j}^2, \tag{1.32}$$

where $\alpha_0 > 0, \alpha_j \geq 0, 1 \leq j \leq p$, and ε_t is a sequence of i.i.d. random variables (possibly Gaussian). Then, under suitable conditions on the α_j, it follows that the x_t are martingale differences (and thus uncorrelated), whereas the x_t^2 have an $AR(p)$ representation, in terms of martingale difference (but not conditionally heteroscedastic) innovations. Engle's model was extended by Bollerslev (1986) to the generalized autoregressive conditionally heteroscedastic model of index p, q (GARCH(p, q)) which implies that the x_t^2 have an ARMA(max(p, q), q) representation in a similar sense.

Both ARCH and GARCH models have found considerable use in finance. However, they imply that the autocorrelations of the squares x_t^2 either eventually cut off completely or decay exponentially, whereas empirical evidence of slower decay perhaps consistent with long memory, has accumulated, see for example Whistler (1990), Ding, Granger, and Engle (1993). In fact Robinson (1991) (Chapter 8 of this volume) had already suggested ARCH-type models capable of explaining greater autocorrelation in squares, so that (1.32) is extended to

$$\sigma_t^2 = \alpha_0 + \sum_{j=1}^{\infty} \alpha_j x_{t-j}^2, \tag{1.33}$$

or replaced by

$$\sigma_t^2 = \left(\alpha_0 + \sum_{j=1}^{\infty} \alpha_j x_{t-j}\right)^2. \tag{1.34}$$

For both models, and related situations, Robinson (1991) developed Lagrange multiplier or score tests of 'no-ARCH' (which is consistent with $\alpha_j = 0, j \geq 1$) against general parameterizations in (1.33) and (1.34); such tests should be better at detecting autocorrelation in x_t^2 that falls off more slowly than ones based on the ARCH(p), (1.32), say.

So far as (1.33) is concerned, we can formally rewrite it as

$$x_t^2 - \sum_{j=1}^{\infty} \alpha_j x_{t-j}^2 = \alpha_0 + v_t, \tag{1.35}$$

where the $v_t = x_t^2 - \sigma_t^2$ are martingale differences. In section 5 of Robinson (1991) the possibility of using for α_j in (1.35) the AR weights from the FARIMA(0, d, 0) model (see (1.12)) was considered, taking $\alpha_0 = 0$, and Whistler (1990) applied this version of his test to test $d = 0$ in exchange rate series. The FARIMA(0, d, 0) case of (1.31) was further considered by Ding and Granger (1996), along with other possibilities, but sufficient conditions of Giraitis, Kokoszka, and Leipus (2000) for existence of a covariance stationary solution of (1.35) rule out long memory, though they do permit strong autocorrelation in x_t^2 that very closely approaches it, and Giraitis and Robinson (2001) have established asymptotic properties of Whittle estimates based on squares for this model. For FARIMA(p, d, q) AR weights α_j in (1.35), x_t^2 is not covariance stationary when $d > 0$, $\alpha_0 > 0$, and Baillie, Bollerslev, and Mikkelsen (1996) called this FIGARCH, a model that has since been widely applied in finance. Gaussian ML has been used to estimate it, but there seems at the time of writing to be no rigorous asymptotic theory available for this, though work is proceeding in this direction. Indeed, until recently rigorous asymptotic theory had only been given for this approach to estimating GARCH(1, 1) and ARCH(p) models within the GARCH(p, q) class.

So far as model (1.34) is concerned, Giraitis, Robinson, and Surgailis (2000) have shown that if the weights α_j decay like j^{d-1}, $0 < d < \frac{1}{2}$, then any integral power x_t^k, such as the square, has long-memory autocorrelation, satisfying (1.7) irrespective of k. This model also has the advantage over (1.33) of avoiding the non-negativity constraints on the α_j, and an ability to explain leverage, but at present lacks an asymptotic theory for parametric estimation.

Another approach to modelling autocorrelation in squares, and other nonlinear functions, alongside possible lack of autocorrelation in x_t, expresses σ_t^2 in (1.32) directly in terms of past ε_t, rather than past x_t, leading to a nonlinear MA form. Nelson (1991) proposed the exponential GARCH (EGARCH) model, where in (1.32) we take

$$\ln \sigma_t^2 = \alpha_0 + \sum_{j=1}^{\infty} \alpha_j g(\varepsilon_{t-j}), \tag{1.36}$$

where g is a user-chosen nonlinear function, for example Nelson stressed $g(z) = \theta z + \gamma(|z| - E|z|)$, which is useful in describing a leverage effect. Nelson pointed out the potential for choosing the α_j to imply long memory in σ_t^2, but stressed short memory, ARMA, weights α_j. Robinson and Zaffaroni (1997) proposed nonlinear MA models, such as

$$x_t = \varepsilon_t \left(\alpha_0 + \sum_{j=1}^{\infty} \alpha_j \varepsilon_{t-j} \right), \tag{1.37}$$

where the ε_t are an i.i.d. sequence. They showed the ability to choose the α_j such that x_t^2 has long-memory autocorrelation, and proposed use of Whittle estimation based on the x_t^2.

A closely related model to (1.37), proposed by Robinson and Zaffaroni (1998), replaces the first factor ε_t by η_t, where the η_t are i.i.d. and independent of the ε_t, again long-memory potential was shown. This model is a special case of

$$x_t = \eta_t h(\varepsilon_{t-1}, \varepsilon_{t-2}, \ldots), \tag{1.38}$$

of which the short-memory stochastic volatility model of Taylor (1986) is also a special case. Long-memory versions of Taylor's model were studied by Breidt, Crato, and de Lima (1998) (Chapter 9 of this volume), Harvey (1998), choosing

$$h(\varepsilon_{t-1}, \varepsilon_{t-2}, \ldots) = \exp\left(\alpha_0 + \sum_{j=1}^{\infty} \alpha_j \varepsilon_{t-j}\right), \tag{1.39}$$

where the α_j are MA weights in the FARIMA(p, d, q). They considered Whittle estimation based on squares, Breidt *et al.* discussing its consistency, and applying the model to stock price data. Note that asymptotic theory for ML estimates of models such as (1.37), (1.38) and (1.39) is considerably more difficult to derive, indeed it is hard to write down the likelihood, given, say, Gaussian assumptions on ε_t and η_t. To ease mathematical tractability in view of the nonlinearity in (1.39), Gaussianity of ε_t was indeed stressed by Breidt *et al.* and Harvey. In that case, we can write the exponent of h in (1.39) as $\alpha_0 + z_t$, where z_t is a stationary Gaussian, possibly long-memory, process, and likewise the second factor in (1.37). These models are all covered by modelling x_t as a general nonlinear function of a vector unobservable Gaussian process ξ_t. From an asymptotic expansion for the covariance of functions of multivariate normal vectors, Robinson (2001) indicated how long memory in nonlinear functions of x_t depends on the long memory in ξ_t and the nature of the nonlinearity involved, with application also to cyclic behaviour, cross-sectional and temporal aggregation, and multivariate models; see also Andersen and Bollerslev (1997). The allowance for quite general nonlinearity means that relatively little generality is lost by the Gaussianity assumption on ξ_t, while the scope for studying autocorrelation structure of functions such as $|x_t|$ can avoid the assumption of a finite fourth moment in x_t, which has been controversial.

1.5 Nonstationary Long Memory

Unit root models have been a major focus of econometrics during the past 15 years or so. Prior to this, modelling of economic time series typically involved a combination of short memory, $I(0)$, series satisfying (1.4), and ones that are nonstochastic, either in the sense of sequences such as dummy variables or polynomial time trends, or of conditioning on predetermined economic variables. Unit root modelling starts from the random walk model, that is (1.13) for $t \geq 1$ with $d = 1$, e_t white noise and $x_0 = 0$, and then generalizes e_t to be a more general $I(0)$ process, modelled either parametrically or nonparametrically; x_t is then said to be an $I(1)$ process. Unit root models, often with the involvement also of nonstationary time trends, have been successfully used in macroeconometrics, frequently in connection with cointegration analysis.

A key preliminary step is the testing of the unit root hypothesis. Many such tests have been proposed, often directed against $I(0)$ alternatives, and using classical Wald, Lagrange multiplier and likelihood-ratio procedures, see for example, Dickey and Fuller (1979, 1981), Phillips (1987). In many other situations, these lead to a null χ^2 limit distribution, a non-central local χ^2 limit distribution, Pitman efficiency, and a considerable degree of scope for robustness to the precise implementation of the test statistics, for example to the estimate of the asymptotic variance matrix that is employed. The unit root tests against $I(0)$ alternatives lose such properties, for example the null limit distribution is nonstandard.

The nonstandard behaviour arises essentially because the unit root is nested unsmoothly in an AR system: in the $AR(1)$ case, the process is stationary with exponentially decaying autocovariance structure when the AR coefficient α lies between -1 and 1, has unit root nonstationarity at $\alpha = 1$, and is 'explosive' for $|\alpha| > 1$. Moreover, the tests directed against AR alternatives seem not to have very good powers against fractional alternatives, as Monte Carlo investigation of Diebold and Rudebusch (1991) suggests.

There is any number of models that can nest a unit root, and the fractional class turns out to have the 'smooth' properties that lead classically to the standard, optimal asymptotic behaviour referred to earlier. Robinson (1994c) (Chapter 10 of this volume) considered the model

$$\varphi(L)x_t = u_t, \quad t \geq 1, \tag{1.40}$$

$$x_t = 0, \quad t \leq 0, \tag{1.41}$$

where u_t is an $I(0)$ process with parametric autocorrelation and

$$\varphi(L) = (1 - L)^{d_1}(1 + L)^{d_2} \prod_{j=3}^{h}(1 - 2\cos\omega_j L + L^2)^{d_j}, \tag{1.42}$$

in which the ω_j are given distinct real numbers in $(0, \pi)$, and the d_j, $1 \leq j \leq h$, are arbitrary real numbers. The initial condition (1.41) avoids an unbounded variance, the main interest being in nonstationary x_t. Robinson proposed tests for specified values of the d_j against, fractional, alternatives in the class (1.42). Thus, for example, in the simplest case the unit root hypothesis $d = 1$ can be tested, but against fractional alternatives $(1 - L)^d$ for $d > 1$, $d < 1$ or $d \neq 1$. Other null d may be of interest, e.g. $d = \frac{1}{2}$, this being the boundary between stationarity and nonstationarity in the fractional domain. The region $d \in [\frac{1}{2}, 1)$ is referred to as mean-reverting, MA coefficients of x_t decaying, albeit more slowly than under stationarity, $d < \frac{1}{2}$. The model (1.40)–(1.42) also covers seasonal and cyclical components (cf. the Gegenbauer model (1.24)) as well as stationary and overdifferenced ones. Robinson showed that his Lagrange multiplier tests enjoy the classical large-sample properties of such tests, described above.

An intuitive explanation of this outcome is that, unlike in unit root tests against AR alternatives, the test statistics are based on the null differenced x_t, which are $I(0)$ under the null hypothesis. This suggests that estimates of memory parameters d_j in (1.40)–(1.42) and of parameters describing u_t, such as Whittle estimates, will also continue to possess the kind of standard asymptotic properties—$\sqrt{n}$-consistency and asymptotic normality—under nonstationarity as we encountered in the stationary circumstances of Section 1.2. Indeed, Beran (1995), in case $\varphi(L) = (1 - L)^d$ and u_t white noise, indicated this, though the initial consistency proof he provides, an essential preliminary to asymptotic distribution theory for his implicitly defined estimate, appears to assume that the estimate lies in a neighbourhood of the true d, itself a consequence of consistency; over a suitably wide range of d-values, the objective function does not converge uniformly. Velasco and Robinson (2000) adopted a somewhat different approach, employing instead the model

$$(1 - L)^s x_t = v_t, \quad t \geq 1, \tag{1.43}$$

$$x_t = 0, \quad t \leq 0, \tag{1.44}$$

$$(1 - L)^{d-s} v_t = u_t, \quad t = 0, \pm 1, \ldots, \tag{1.45}$$

where s is the integer part of $d + \frac{1}{2}$ and u_t is a parametric $I(0)$ process such as white noise; note that v_t is a stationary $I(d - s)$ process, invertible also unless $d = -\frac{1}{2}$. The difference between the two definitions of nonstationary $I(d)$ processes, in (1.40) and (1.41) on the one hand and (1.43)–(1.45) on the other, was discussed by Marinucci and Robinson (1999); this entails, for example, convergence to different forms of fractional Brownian motion. Velasco and Robinson considered a version of discrete-frequency Whittle estimation (cf. (1.20)) but nonstationarity tends to bias periodogram ordinates, and to sufficiently reduce this they in general (for $d \geq \frac{3}{4}$ and with (1.45) modified so that v_t has an unknown mean) found it necessary to suitably 'taper' the data (e.g. see Brillinger 1975: ch. 3) and then, in order to overcome the undesirable dependence this produces between neighbouring periodograms, to use only Fourier frequencies λ_j, such that j is a multiple of p: p is the 'order' of the taper, such that $p \geq [d + \frac{1}{2}] + 1$ is required for asymptotic normality of the estimates, with $\sqrt{n}$ rate of convergence; since d is unknown a large p can be chosen for safety's sake, but the asymptotic variance is inflated by a factor varying directly with p. The theory is invariant to an additive polynomial trend of degree up to p.

The use of tapering in a nonstationary fractional setting, and a similar model to (1.43)–(1.45), originated in Hurvich and Ray (1995) (Chapter 11 of this volume). In (1.43), they required $d = 1$, while in (1.45) they allowed $-\infty < d < \frac{1}{2}$, so that (unlike Beran (1995) and Velasco and Robinson (2000)) they covered nonstationarity only up to $d < \frac{3}{2}$ (though this probably fits many applications) and on the other hand covered any degree of noninvertibility. Their concern, however, was not with asymptotic theory for parameter estimates. Hurvich and Ray found that asymptotic bias in $I(\lambda_j)$, for fixed j as $n \to \infty$, could be notably reduced by use of a cosine bell taper, leading them to recommend use of tapering (and omission of frequency λ_1) in the log periodogram estimation of d discussed in Section 1.3, in case nonstationarity is feared. Velasco (1999a) then established limit distribution theory, analogous to that described in Section 3, for log periodogram estimates in case $d \geq \frac{1}{2}$, in a semiparametric version of (1.43)–(1.45) (so u_t has nonparametric autocorrelation), using a general class of tapers; Velasco (1999b) established analogous results for local Whittle estimates, cf. (1.28). Again, there is invariance to polynomial trends, but tapering, imposed for asymptotic normality when $d \geq \frac{3}{4}$ and for consistency when $d > 1$, entails skipping frequencies and/or an efficiency loss. However, Hurvich and Chen (2000) proposed a taper, applied to first differences when $d < \frac{3}{2}$, that, with no skipping, loses less efficiency. On the other hand, Phillips (e.g. 1999, and various papers with co-workers) in the context of the model (1.40), (1.41) with $\phi(L) = (1 - L)^d$, found that untapered log periodogram and local Whittle estimates are inconsistent when $d > 1$, and not asymptotically normal when $d \geq \frac{3}{4}$. The position at present is thus that, despite its drawbacks, tapering is a wise precaution when nonstationarity is believed possible, while recent work has found it also useful in theoretical refinements even under stationarity, see for example Giraitis and Robinson (2000).

1.6 Inference on Regression and Cointegration Models

Section 1.1 provided some preparatory discussion of regression, insofar as we considered estimation of the mean of a stationary long-memory series. Rates of convergence and efficiency were discussed, but not limit distribution theory, and here some warning is in order. For short-memory series x_t, the sample mean $\bar{x} = n^{-1} \sum_{t=1}^{n} x_t$ is asymptotically normal over a wide range of dependence conditions, including not only linear processes but also various kinds of 'mixing' processes, the latter covering some forms of nonlinearity. However, Rosenblatt (1961) pointed out that if $x_t = u_t^2$, where u_t is a Gaussian long-memory process with differencing parameter d (e.g. the FARIMA$(0, d, 0)$ (1.12)), then $\bar{x}$ is not asymptotically normal when $d > \frac{1}{4}$. Taqqu (1975) described the limit distribution here as the 'Rosenblatt distribution', and considerably developed Rosenblatt's work, modelling x_t as a quite general function of u_t. The limit distribution of $\bar{x}$ is then governed not only by d but by the lowest-degree non-vanishing term in the Hermite expansion of x_t (unlike for short memory u_t, when all terms effectively contribute). In particular, when a linear term is present, there is asymptotic normality.

Taqqu's approach has been employed and extended in connection with many statistics, but while the possible consequences of non-normality must thus be borne in mind it is not clear how to cope with this in designing useful rules of inference, and we instead focus on linear, but not necessarily Gaussian, processes, where central limit theory can be anticipated (and under conditions which are in some respects milder). Early contributions here are due to Eicker (1967) (Chapter 12 of this volume) and Ibragimov and Linnik (1971; pp. 358–60).

Consider the model

$$x_t = \mu + \sum_{j=-\infty}^{\infty} \alpha_j \varepsilon_{t-j}, \quad \sum_{j=-\infty}^{\infty} \alpha_j^2 < \infty, \tag{1.46}$$

where the ε_t are i.i.d. with zero mean and finite variance. The square summability of the α_j is consistent with the long-memory properties (1.7) and (1.8), for $d < \frac{1}{2}$. Then if also

$$\sum_{j=-\infty}^{\infty} (\alpha_{j-1} + \cdots + \alpha_{j-n})^2 \to \infty, \quad \text{as } n \to \infty, \tag{1.47}$$

we have

$$\{V(\bar{x})\}^{-(1/2)}(\bar{x} - \mu) \to_d N(0, 1). \tag{1.48}$$

Condition (1.47) is merely equivalent to the essential divergence of the norming factor $V(\bar{x})$ in (1.48), and does not apply to 'noninvertible' processes; under (1.7) $V(\bar{x})$ increases like n^{2d-1} for $|d| < \frac{1}{2}$. Both Eicker (1967) and Ibragimov and Linnik (1971) covered (1.46) but Eicker also treated multiple linear regression with quite general nonstochastic errors. Thus, for the model

$$y_t = \beta' z_t + x_t, \tag{1.49}$$

with y_t, and the $r \times 1$ nonstochastic vector z_t observed, the unobserved x_t again satisfies (1.46), (1.47) with $\mu = 0$. Eicker extended (1.48) to the OLS estimate $\hat{\beta}$ of the $r \times 1$ vector β. Under stronger conditions, Eicker established 'autocorrelation-consistent' variance estimates of slope elements of β, of a different, convolution, form from the kernel spectral density type that has later become popular in econometrics. As a historical note, section 3 of Eicker's paper discussed independent, heteroscedastic x_t in (1.49) and the 'heteroskedasticity-consistent' variance estimates also later popularized in econometrics (though those earlier appeared in Eicker (1963)).

Eicker's conditions were later relaxed by Hannan (1979), but neither of them, nor Ibragimov and Linnik, explicitly discussed the impact of the asymptotic behaviour (1.7) or (1.8) for x_t, let alone parametric models like (1.12). Under a model for x_t like (1.8), with $0 < d < \frac{1}{2}$, Yajima (1988, 1991) discussed both limit distribution theory and (extending work of Grenander 1954 and Grenander and Rosenblatt 1957, for short memory x_t) efficiency of OLS in (1.49), stressing particular regressors such as polynomial time trends and indicating how the asymptotic variance simplifies as a function of d. These can be implemented given a better-than-log n-consistent (see Robinson 1994b) estimate of d, so that semiparametric modelling of x_t (cf. (1.7) or (1.8)) suffices. With a rather different treatment of nonstationary regressors, Dahlhaus (1995) presented estimates that achieve the same asymptotic efficiency as the generalized least squares (GLS) estimate; they involve d, and Dahlhaus then showed that the same efficiency can be achieved when d is estimated; d can take 'noninvertible' values, but x_t is assumed Gaussian. Deo

(1997) provided further developments along these lines, while Deo and Hurvich (1998) discussed the case of a linear time trend and $I(d)$ x_t, with $d < \frac{3}{2}$, so that the errors can be nonstationary.

Time series regression in econometrics can involve stochastic regressors, and this can significantly affect the theory. Much of the nonstochastic regression theory stresses circumstances in which the limiting 'spectral distribution function' of regressors z_t in 'Grenander's conditions' has discrete jumps, or at least a jump at zero frequency where the spectral pole of x_t is located (as is true of polynomial time trends). For stationary stochastic z_t, the existence of a spectral density may on the other hand be plausible, and in the event of sufficiently strong long memory in both z_t and x_t, specifically if z_t is $I(c)$ with $c + d \geq \frac{1}{2}$, OLS can have a nonstandard limit distribution with convergence rate slower than the usual $\sqrt{n}$. However, Robinson and Hidalgo (1997) (Chapter 13 of this volume) proposed weighted (in both the time and frequency domains) estimates that are consistent and asymptotically normal, at rate $\sqrt{n}$. A special case of these are infeasible GLS estimates, allowing arbitrarily strong stationary long memory in both z_t and x_t, so that GLS has the advantage over OLS not only of efficiency but also, interestingly, of faster convergence rate and asymptotic normality; intuitively, the explanation is that in the frequency domain $f(\lambda)^{-1}$ is involved in GLS, and this has a zero, not a pole, under long memory, avoiding integrability problems. Robinson and Hidalgo also showed that the same asymptotic theory can hold when a parametric model for $f(\lambda)$ is estimated, and gave an extension to nonlinear regression. For the same setup, Hidalgo and Robinson (2002) extended Hannan's (1963) idea of adapting to nonparametric autocorrelation in x_t. In a single-regressor version of this model Choy and Taniguchi (2001) discussed various estimates under different combinations of long memory in x_t and z_t.

This stationary stochastic regressor theory assumed z_t and x_t are uncorrelated, for otherwise estimates will be asymptotically biased (indeed for technical reasons independence was actually assumed). If (1.49) represents a cointegrating relation between y_t and z_t, x_t has lower integration order than z_t, but there is no reason in general to suppose it is uncorrelated with z_t. The cointegration literature has usually assumed y_t and z_t are $I(1)$ while x_t is $I(0)$ (e.g. see Engle and Granger 1987), and here OLS is still consistent due to the asymptotic dominance of x_t by z_t. This outcome extends to more general, fractional, nonstationary z_t, and x_t that are stationary or even less-nonstationary-than-z_t. However, it does not apply to stationary z_t, even when it has more long memory than x_t, due to simultaneous equations bias. The stationary case may be of interest in financial applications, and here Robinson (1994b) showed consistency of a narrow-band least squares (NBLS) estimate, in the frequency domain, where the number of (low) Fourier frequencies used increases more slowly than n.

Robinson and Marinucci (1997) (Chapter 14 of this volume) extended this result, making Robinson's z_t a vector and giving a rate of convergence, but mainly focused on comparisons between OLS and NBLS in nonstationary circumstances. Fractional nonstationarity of form (1.40), (1.41) in z_t was considered, while x_t can be either nonstationary or stationary, possibly with long memory. They found that for some combinations of memory parameters NBLS is asymptotically equivalent to OLS (despite losing high-frequency information), but for other combinations NBLS is superior, simultaneous equations bias not preventing consistency of OLS but possibly slowing convergence. Limit distributions are nonstandard but for the most part can be characterized by applying functional limit theory of Akonom and Gourieroux (1987), Marinucci and Robinson (2000); see also Sowell (1990), Chan and Terrin (1995). Improved versions of Robinson and Marinucci's (1997) results are in Robinson and Marinucci (2001) for scalar z_t, while Robinson and Marinucci (2000) studied interaction between stochastic and nonstochastic components in z_t. The approach in these papers indicates the dominating role of low frequencies in cointegration, no parametric modelling of dependence structure being required. It thus has the usual advantages of 'semiparametric' modelling, while both OLS and NBLS are computationally simple to implement. On the other hand, in cointegration involving $I(1)$ y_t and $I(0)$ x_t, OLS has been improved upon, for example by Phillips (1991a,b) by estimates that have a mixed normal limit distribution, with the effect that Wald statistics for testing β have classical χ^2 asymptotics. Jeganathan (1999), Robinson and Hualde (2000) have provided extensions of these results to parametric fractional models, gaining convergence rates that for some parameter combinations are faster than those of OLS and NBLS; some early empirical study of fractional cointegration is in Cheung and Lai (1993), for example.

References

Adelman, I. (1965), 'Long cycles: fact or artefact?', *American Economic Review*, 55, 444–63.

Adenstedt, R. K. (1974), 'On large-sample estimation for the mean of a stationary random sequence', *Annals of Statistics*, 2, 1095–107.

Akonom, J. and Gourieroux, C. (1987), 'A functional central limit theorem for fractional processes', preprint.

Andersen, T. G. and Bollerslev, T. (1997), 'Heterogeneous information arrivals and return volatility dynamics: increasing the long run in high frequency returns', *Journal of Finance*, 52, 975–1005.

Andrews, D. W. K. and Guggenberger, K. (2000), 'Bias-reduced log-periodogram estimator for the long memory parameter', preprint.

Baillie, R. T., Bollerslev, T., and Mikkelsen, H. O. (1996), 'Fractionally integrated generalized autoregressive conditional heteroskedasticity, *Journal of Econometrics*, 74, 3–30.

Beran, J. (1989), 'A test of location for data with slowly decaying serial correlations', *Biometrika*, 76, 261–9.

—— (1995), 'Maximum likelihood estimation of the differencing parameter for invertible short- and long-memory ARIMA models', *Journal of the Royal Statistical Society, Series B*, 57, 659–72.

Beran, J., Bhansali, R., and Ocker, D. (1998), 'On unified model selection for stationary and nonstationary short- and long-memory autoregressive processes', *Biometrika*, 85, 921–34.

Bloomfield, P. (1972), 'An exponential model for the spectrum of a scalar time series', *Biometrika*, 60, 217–26.

Bollerslev, T. (1986), 'Generalized autoregressive conditional heteroskedasticity', *Journal of Econometrics*, 31, 307–27.

Box, G. E. P. and Jenkins, G. M. (1971), *Time Series Analysis, Forecasting and Control*, San Francisco: Holden-Day.

Breidt, F. J., Crato, N., and de Lima, P. (1998), 'The detection and estimation of long memory in stochastic volatility', *Journal of Econometrics*, 83, 325–34.

Brillinger, D. R. (1975), *Time Series Data Analysis and Theory*, San Francisco: Holden Day.

Chan, N. H. and Terrin, N. (1995), 'Inference for unstable long-memory processes with applications to fractional unit root autoregressions', *Annals of Statistics*, 23, 1662–83.

Cheung, Y. W. and Lai, K. S. (1993), 'A fractional cointegration analysis of purchasing power parity', *Journal of Business and Economic Statistics*, 11, 103–12.

—— and Diebold, F. X. (1994), 'On maximum likelihood estimation of the differencing parameter of fractionally integrated noise with unknown mean', *Journal of Econometrics*, 62, 301–16.

Choy, K. and Taniguchi, M. (2001), 'Stochastic regression model with dependent disturbances', *Journal of Time Series Analysis*, 22, 175–96.

Chung, C.-F. (1994), 'A note on calculating the autocovariance of the fractionally integrated ARMA models', *Economics Letters*, 45, 293–7.

—— (1996a), 'Estimating a generalized long memory process', *Journal of Econometrics*, 73, 237–59.

—— (1996b), 'Generalized fractionally integrated autoregressive moving average process', *Journal of Time Series Analysis*, 17, 111–40.

Dahlhaus, R. (1989), 'Efficient parameter estimation for self-similar processes', *Annals of Statistics*, 17, 1749–66.

—— (1995), 'Efficient location and regression estimation for long range dependent regression models', *Annals of Statistics*, 23, 1029–47.

Davies, R. B. and Harte, D. S. (1987), 'Tests for Hurst effect', *Biometrika*, 74, 95–101.

Delgado, M. A. and Robinson, P. M. (1996), 'Optimal spectral bandwidth for long memory', *Statistica Sinica*, 6, 97–112.

Deo, R. S. (1997), 'Asymptotic theory for certain regression models with long memory errors', *Journal of Time Series Analysis*, 18, 385–93.

—— and Hurvich, C. M. (1998), 'Linear trend with fractionally integrated errors', *Journal of Time Series Analysis*, 19, 379–97.

Dickey, D. A. and Fuller, W. A. (1979), 'Distribution of the estimators for autoregressive time series with a unit root', *Journal of the American Statistical Association*, 74, 427–31.

—— (1981), 'Likelihood ratio statistics for autoregressive time series with a unit root', *Econometrica*, 49, 1057–72.

Diebold, F. X. and Rudebusch, G. D. (1991), 'On the power of Dickey–Fuller tests against fractional alternatives', *Economics Letters*, 35, 155–60.

Ding, Z. and Granger, C. W. J. (1996), 'Modelling volatility persistence of speculative returns: A new approach', *Journal of Econometrics*, 73, 185–215.

——, —— and Engle, R. F. (1993), 'A long memory property of stock market returns and a new model', *Journal of Empirical Finance*, 1, 83–106.

Eicker, F. (1963), 'Asymptotic normality and consistency of the least squares estimator for families of linear regressions', *Annals of Mathematical Statistics*, 34, 447–56.

—— (1967), 'Limit theorems for regressions with unequal and dependent errors'. In *Proceedings of the Fifth Berkeley Symposium on Mathematical Statistics and Probability*, Vol. 1, Berkeley: University of California Press, pp. 59–82.

Engle, R. F. (1982), 'Autoregressive conditional heteroscedasticity with estimates of the variance of United Kingdom inflation', *Econometrica*, 50, 987–1007.

—— and Granger, C. W. J. (1987), 'Cointegration and error correction: representation, estimation and testing', *Econometrica*, 55, 251–76.

Fairfield Smith, H. (1938), 'An empirical law describing heterogeneity in the yields of agricultural crops', *Journal of Agricultural Science*, 28, 1–23.

Fox, R. and Taqqu, M. S. (1986), 'Large sample properties of parameter estimates for strongly dependent stationary Gaussian time series', *Annals of Statistics*, 14, 517–32.

—— (1987), 'Central limit theorems for quadratic forms in random variables having long-range dependence', *Probability Theory and Related Fields*, 74, 213–40.

Geweke, J. and Porter-Hudak, S. (1983), 'The estimation and application of long-memory time series models', *Journal of Time Series Analysis*, 4, 221–38.

Giraitis, L., Hidalgo, F. J., and Robinson, P. M. (2001), 'Gaussian estimation of parametric spectral density with unknown pole', *Annals of Statistics*, 29, 987–1023.

——, Kokoszka, P., and Leipus, R. (2000), 'Stationary ARCH models: dependence structure and central limit theorem', *Econometric Theory*, 16, 3–22.

—— and Robinson, P. M. (2000), 'Edgeworth expansions for semiparametric Whittle estimation of long memory', forthcoming, *Annals of Statistics*.

—— and Robinson, P. M. (2001), 'Whittle estimation of ARCH models', *Econometric Theory*, 17, 608–31.

—— Robinson, P. M. and Surgailis, D. (2000), 'A model for long memory conditional heteroscedasticity', *Annals of Applied Probability*, 10, 1002–24.

—— and Surgailis, D. (1990), 'A central limit theorem for quadratic forms in strongly dependent random variables and its application to asymptotical normality of Whittle's estimate', *Probability Theory and Related Fields*, 86, 87–104.

—— and Taqqu, M. S. (1999), 'Whittle estimator for finite-variance non-Gaussian time series with long memory', *Annals of Statistics*, 27, 178–203.

Goncalves, E. and Gourieroux, C. (1988), 'Agrégation de processus autorégressifs d'ordre 1', *Annales d'Économie et de Statistique*, 12, 127–49.

Granger, C. W. J. (1966), 'The typical spectral shape of an economic variable', *Econometrica*, 34, 150–61.

—— (1980), 'Long memory relationships and the aggregation of dynamic models', *Journal of Econometrics*, 14, 227–38.

—— and Joyeux, R. (1980), 'An introduction to long-memory time series models and fractional differencing', *Journal of Time Series Analysis*, 1, 15–29.

Gray, H. L., Zhang, N. F. and Woodward, W. A. (1989), 'On generalized fractional processes', *Journal of Time Series Analysis*, 10, 233–57.

Grenander, U. (1954), 'On the estimation of regression coefficients in the case of an autocorrelated disturbance', *Annals of Mathematical Statistics*, 25, 252–72.

—— and Rosenblatt, M. (1957), *Statistical Analysis of Stationary Time Series*, New York: Wiley.

Hall, P., Koul, H. L., and Turlach, B. A. (1997), 'Note on convergence rates of semiparametric estimators of dependence index', *Annals of Statistics*, 25, 1725–39.

Hannan, E. J. (1963), 'Regression for Time Series', in *Time Series Analysis* (M. Rosenblatt, ed.) New York: Wiley, pp. 17–32.

—— (1973), 'The asymptotic theory of linear time series models', *Journal of Applied Probability*, 10, 130–45.

—— (1976), 'The asymptotic distribution of serial covariances', *Annals of Statistics*, 4, 396–9.

—— (1979), 'The central limit theorem for time series regression', *Stochastic Processes and their Applications*, 9, 281–9.

Harvey, A. (1998), 'Long memory in stochastic volatility', in *Forecasting Volatility in the Financial Markets* (J. Knight and S. Satchell, eds) Oxford: Butterworth-Heinemann.

Henry, M. and Robinson, P. M. (1996), 'Bandwidth choice in Gaussian semiparametric estimation of long-range dependence', in *Athens Conference on Applied Probability and Time Series Analysis*, Vol. II: Time Series Analysis (P. M. Robinson and M. Rosenblatt, eds). In Memory of E. J. Hannan, New York: Springer-Verlag, pp. 220–32.

Heyde, C. and Gay, G. (1993), 'Smoothed periodogram asymptotics and estimation for processes and fields with long-range dependence', *Stochastic Processes and their Applications*, 45, 169–87.

Hidalgo, F. J. (2001), 'Semiparametric estimation of the location of the pole', preprint.

—— and Robinson, P. M. (2002), 'Adapting to unknown disturbance autocorrelation with long memory', *Econometrica*, 70, 1545–81.

Hipel, K. W. and McLeod, A. J. (1978), 'Preservation of the adjusted rescaled range parts 1, 2 and 3', *Water Resources Research*, 14, 491–518.

Hosking, J. R. M. (1981), 'Fractional differencing', *Biometrika*, 68, 165–76.

Hosoya, Y. (1996), 'The quasi-likelihood approach to statistical inference on multiple time series with long-range dependence', *Journal of Econometrics*, 73, 217–36.

—— (1997), 'Limit theory with long-range dependence and statistical inference of related models', *Annals of Statistics*, 25, 105–37.

Hurst, H. (1951), 'Long term storage capacity of reservoirs', *Transactions of the American Society of Civil Engineers*, 116, 770–99.

Hurvich, C. M. and Beltrao, K. I. (1993), 'Asymptotics for the low-frequency estimates of the periodogram of a long memory time series', *Journal of Time Series Analysis*, 14, 455–72.

—— and Beltrao, K. I. (1994), 'Automatic semiparametric estimation of the memory parameter of a long-memory time series', *Journal of Time Series Analysis*, 15, 285–302.

—— and Brodsky, J. (2001), 'Broadband semiparametric estimation of the memory parameter of a long-memory time series using fractional exponential model', *Journal of Time Series Analysis*, 22, 221–49.

—— and Chen, W. W. (2000), 'An efficient taper for potentially overdifferenced long-memory time series', *Journal of Time Series Analysis*, 21, 155–80.

—— and Deo, R. S. (1999), 'Plug-in selection of the number of frequencies in regression estimates of the memory parameter of a long-memory time series', *Journal of Time Series Analysis*, 20, 331–41.

——, Deo, R., and Brodsky, J. (1998), 'The mean squared error of Geweke and Porter-Hudak's estimates of the memory parameter of a long memory time series', *Journal of Time Series Analysis*, 19, 19–46.

—— and Ray, B. K. (1995), 'Estimation of the memory parameter for nonstationary or noninvertible fractionally integrated processes', *Journal of Time Series Analysis*, 16, 17–41.

Ibragimov, I. A. and Linnik, Yu. V. (1971), *Independent and Stationary Sequences of Random Variables*, Groningen: Wolters-Noordhoff.

Janacek, G. J. (1982), 'Determining the degree of differencing for time series via log spectrum', *Journal of Time Series Analysis*, 3, 177–83.

Jeffreys, H. (1939), *Theory of Probability*, Oxford: Clarendon Press.

Jeganathan, P. (1999), 'On asymptotic inference in cointegrated time series with fractionally integrated errors', *Econometric Theory*, 15, 583–621.

—— (2001), 'Correction to "On asymptotic inference in cointegrated time series with fractionally integrated errors"', preprint.

Kashyap, R. and Eom, K. (1988), 'Estimation in long-memory time series model', *Journal of Time Series Analysis*, 9, 35–41.

Künsch, H. R. (1986), 'Discrimination between monotonic trends and long-range dependence', *Journal of Applied Probability*, 23, 1025–30.

—— (1987), 'Statistical aspects of self-similar processes', *Proceedings of the First World Congress of the Bernoulli Society*, VNU Science Press, 1, 67–74.

Lippi, M. and Zaffaroni, P. (1998), 'Aggregation and simple dynamics: exact asymptotic results', preprint.

Lo, A. W. (1991), 'Long-term memory in stock market prices', *Econometrica*, 59, 1279–313.

Lobato, I. G. (1997), 'Consistency of the averaged cross-periodogram in long memory series', *Journal of Time Series Analysis*, 18, 137–56.

—— (1999), 'A semiparametric two-step estimator in a multivariate long memory model', *Journal of Econometrics*, 90, 129–53.

—— and Robinson, P. M. (1996), 'Averaged periodogram estimation of long memory', *Journal of Econometrics*, 73, 303–24.

—— and Robinson, P. M. (1998), 'A nonparametric test for $I(0)$', *Review of Economic Studies*, 65, 475–95.

Mandelbrot, B. B. (1971), 'A fast fractional Gaussian noise generator', *Water Resources Research*, 7, 543–53.

—— (1972), 'Statistical methodology for non-periodic cycles: from the covariance to R/S analysis', *Annals of Economic and Social Measurement*, 1, 259–90.

—— (1975), 'Limit theorems on the self-normalized range for weakly and strongly dependent processes', *Zeitschrift für Wahrscheinlichkeitstheorie und verwandte Gebiete*, 31, 271–85.

—— and Taqqu, M. S. (1979), 'Robust R/S analysis of long-run serial correlations', *Bulletin of the International Statistical Institute*, 48, 2, 69–104.

—— and Van Ness, J. W. (1968), 'Fractional Brownian motions, fractional noises and applications', *SIAM Review*, 10, 422–37.

—— and Wallis, T. R. (1969), 'Robustness of the rescaled range R/S in the measurement of noncyclic long run statistical dependence', *Water Resources Research*, 5, 967–88.

Marinucci, D. and Robinson, P. M. (1999), 'Alternative forms of fractional Brownian motion', *Journal of Statistical Planning and Inference*, 80, 111–22.

—— (2000), 'Weak convergence of multivariate fractional processes', *Stochastic Processes and their Applications*, 86, 103–20.

Moulines, E. and Soulier, P. (1999), 'Broadband log periodogram regression of time series with long range dependence', *Annals of Statistics*, 27, 1415–39.

—— and —— (2000), 'Data driven order selection for projection estimates of the spectral density of time series with long range dependence', *Journal of Time Series Analysis*, 21, 193–218.

Nelson, D. (1991), 'Conditional heteroskedasticity in asset returns: a new approach', *Econometrica*, 59, 347–70.

Newcomb, S. (1886), 'A generalized theory of the combination of observations so as to obtain the best result', *American Journal of Mathematics*, 8, 343–66.

Parzen, E. (1986), 'Quantile spectral analysis and long-memory time series', *Journal of Applied Probability*, 23A, 41–54.

Phillips, P. C. B. (1987), 'Time series regression with a unit root', *Econometrica*, 55, 277–301.

—— (1991*a*), 'Optimal inference in cointegrated systems', *Econometrica*, 59, 283–306.

—— (1991*b*), 'Spectral regression for cointegrated time series', in *Nonparametric and Semiparametric Methods in Econometrics and Statistics* (W. A. Barnett, J. Powell, and G. Tauchen, eds) Cambridge: Cambridge University Press, pp. 413–35.

—— (1999), 'Unit root log periodogram regression', preprint.

Robinson, P. M. (1978*a*), 'Statistical inference for a random coefficient autoregressive model', *Scandinavian Journal of Statistics*, 5, 163–8.

—— (1978*b*), 'Alternative models for stationary stochastic processes', *Stochastic Processes and their Applications*, 8, 151–2.

—— (1986), 'On a model for a time series of cross-sections', *Journal of Applied Probability*, 23A, 113–25.

—— (1987), 'Stochastic differential equation models for panel data', in *Differential Equations and Applications (II)* (I. Dimovski and J. Stoyanov, eds) Rousse, pp. 879–85.

—— (1991), 'Testing for strong serial correlation and dynamic conditional heteroskedasticity in multiple regression', *Journal of Econometrics*, 47, 67–84.

—— (1994*a*), 'Time series with strong dependence', in *Advances in Econometrics* (C. A. Sims, ed.) Vol. 1, Cambridge: Cambridge University Press, pp. 47–95.

—— (1994*b*), 'Semiparametric analysis of long-memory time series', *Annals of Statistics*, 22, 515–39.

—— (1994*c*), 'Efficient tests of nonstationary hypotheses', *Journal of the American Statistical Association*, 89, 1420–37.

—— (1995*a*), 'Log-periodogram regression of time series with long range dependence', *Annals of Statistics*, 23, 1048–72.

—— (1995*b*), 'Gaussian semiparametric estimation of long-range dependence', *Annals of Statistics*, 23, 1630–61.

—— (2001), 'The memory of stochastic volatility models', *Journal of Econometrics*, 101, 195–218.

—— and Henry, M. (2000), 'Higher-order kernel semiparametric M-estimation of long memory', forthcoming, *Journal of Econometrics*.

—— and Hidalgo, J. F. (1997), 'Time series regression with long-range dependence', *Annals of Statistics*, 25, 77–104.

—— and Hualde, J. (2000), 'Cointegration in fractional systems with unknown integration orders', preprint.

—— and Marinucci, D. (1997), 'Semiparametric frequency-domain analysis of fractional cointegration', preprint.

—— and —— (2000), 'The averaged periodogram for nonstationary vector time series', *Statistical Inference for Stochastic Processes*, 3, 149–60.

—— and —— (2001), 'Narrow-band analysis of nonstationary processes', *Annals of Statistic*, 29, 947–86.

—— and Zaffaroni, P. (1997), 'Modelling nonlinearity and long memory in time series', *Fields Institute Communications*, 11, 161–70.

—— and —— (1998), 'Nonlinear time series with long memory: a model for stochastic volatility', *Journal of Statistical Planning and Inference*, 68, 359–71.

Rosenblatt, M. (1961), 'Independence and dependence', in *Proceedings of the 4th Berkeley Symposium on Mathematical Statistics and Probability*, Berkeley: University of California Press, pp. 411–43.

Samarov, A. and Taqqu, M. S. (1988), 'On the efficiency of the sample mean in long memory noise', *Journal of Time Series Analysis*, 9, 191–200.

Sinai, Y. G. (1976), 'Self-similar probability distributions', *Theory of Probability and its Applications*, 21, 64–80.

Sowell, F. B. (1990), 'The fractional unit root distribution', *Econometrica*, 58, 495–505.

—— (1992), 'Maximum likelihood estimation of stationary univatiate fractionally integrated time series models', *Journal of Econometrics*, 53, 165–88.

Student (1927), 'Errors of routine analysis', *Biometrika*, 19, 151–69.

Taqqu, M. S. (1975), 'Weak convergence to fractional Brownian motion and to the Rosenblatt process', *Zeitschrift für Wahrscheinlichkeitstheorie und verwandte Gabiete*, 31, 287–302.

Taylor, S. J. (1986), *Modelling Financial Time Series*, Chichester, UK.

Teverovsky, V. and Taqqu, M. S. (1997), 'Testing for long-range dependence in the presence of shifting means or a slowly declining trend, using a variance-type estimate', *Journal of Time Series Analysis*, 18, 279–304.

Velasco, C. (1999a), 'Non-stationary log-periodogram regression', *Journal of Econometrics*, 91, 325–71.

—— (1999b), 'Gaussian semiparametric estimation of nonstationary time series', *Journal of Time Series Analysis*, 20, 87–127.

—— (2000), 'Non-Gaussian log-periodogram regression', *Econometric Theory*, 16, 44–79.

——, and Robinson P. M. (2000), 'Whittle pseudo-maximum likelihood estimation for nonstationary time series', *Journal of the American Statistical Association*, 95, 1229–43.

Vitale, R. A. (1973), 'An asymptotically efficient estimate in time series analysis', *Quarterly Journal of Applied Mathematics*, 421–40.

Whistler, D. (1990), 'Semiparametric models of daily and intra-daily exchange rate volatility', Ph.D. thesis, University of London.

Whittle, R. (1951), *Hypothesis Testing in Time Series Analysis*, Uppsala: Almqvist.

Yajima, Y. (1985), 'On estimation of long-memory time series models', *Australian Journal of Statistics*, 27, 303–20.

—— (1988), 'On estimation of a regression model with long-memory stationary errors', *Annals of Statistics*, 16, 791–807.

—— (1991), 'Asymptotic properties of the LSE in a regression model with long memory stationary errors', *Annals of Statistics*, 19, 158–77.

Yong, C. H. (1974), *Asymptotic Behaviour of Trigonometric Series*, Hong Kong: Chinese University.

2

On Large-sample Estimation for the Mean of a Stationary Random Sequence

ROLF K. ADENSTEDT

2.1 Introduction

The following simple stochastic model is widely applied. A (possibly complex-valued) random sequence $X_t = m + Y_t, t = 0, \pm 1, \ldots$ is the sum of a constant mean value m and a disturbance. The disturbance $\{Y_t\}$ is a zero-mean, wide-sense stationary random sequence with a spectral density function $f(\lambda), -\pi \leq \lambda \leq \pi$, so that its covariance function is

$$R(\tau) = EY_{t+\tau}\bar{Y}_t = \int_{-\pi}^{\pi} e^{i\tau\lambda} f(\lambda)\, d\lambda, \quad \tau = 0, \pm 1, \ldots. \tag{2.1}$$

We consider in this paper aspects of the estimation of the mean m for this model by unbiased linear estimators

$$\hat{m} = c_0 X_0 + c_1 X_1 + \cdots + c_n X_n, \quad c_0 + c_1 + \cdots + c_n = 1, \tag{2.2}$$

formed from a large number of observed values $X_0, X_1, \ldots, X_n$. Of particular interest is the best linear unbiased estimator (BLUE) $\hat{m}_{\mathrm{BLU}}$, that is, the estimator (2.2) having minimum variance

$$\operatorname{Var} \hat{m} = E|\hat{m} - m|^2 = \sum_{j,k=0}^{n} c_j \bar{c}_k R(j - k).$$

Since, typically, calculation of $\hat{m}_{\mathrm{BLU}}$ or its variance is difficult, adequate approximations are needed in terms of more easily calculated estimators.

There is a substantial literature comparing the BLUE with other estimators, especially the least squares estimator (LSE) $\hat{m}_{\mathrm{LS}} = (n + 1)^{-1}(X_0 + \cdots + X_n)$. A good survey of results appears in Anderson (1971). Watson (1967, 1972) and Zyskind (1967) consider coincidence of the BLUE and LSE for coefficients in general linear regression models with a fixed sample size. Kruskal (1968) treats this problem from a coordinate-free Hilbert space viewpoint.

This work was previously published as Adenstedt, R. K. (1974), 'On large-sample estimation for the mean of a stationary random sequence', *Annals of Statistics*, 2, 1095–1107. Reproduced here by kind permission of the Institute of Mathematical Statistics.

The author is grateful to the editor and the referees for their helpful comments on the paper.

A general Hilbert space approach to linear estimation problems is given by Parzen (1961) and in the infinite sample size case by Rozanov (1971). Cleveland (1971) compares the BLUE to an arbitrary linear estimator using Mahalanobis distance.

Studies have also been made with special cases of our model. Fishman (1972) gives closed form expressions for $\hat{m}_{\mathrm{BLU}}$ and its variance when $\{Y_t\}$ is autoregressive of finite order. For order two he compares the BLUE and LSE with regard to small and large-sample behaviour. The first-order case is treated in detail by Chipman *et al.* (1968).

We are concerned here with asymptotic or large-sample results and will compare $\hat{m}_{\mathrm{BLU}}$ to competitors other than $\hat{m}_{\mathrm{LS}}$. Grenander (1950) considered asymptotic efficiency of the LSE relative to the BLUE in estimating means and (1954) extended the results to more general regressions. In terms of our model he showed that, as $n \to \infty$, both $\hat{m}_{\mathrm{BLU}}$ and $\hat{m}_{\mathrm{LS}}$ have asymptotic variance $2\pi f(0)/n$ as long as $f(\lambda)$ is positive and continuous. More recently Vitale (1973) considered the case when $f(\lambda)$ is continuous and positive except at the origin, where $\lambda^{-2} f(\lambda) \to L > 0$ as $\lambda \to 0$. Then $\mathrm{Var}\,\hat{m}_{\mathrm{BLU}} \sim 24\pi L/n^3$ and the estimator (2.2) with coefficients $c_k = 6k(n-k)/[n(n^2-1)]$ is asymptotically efficient, while $\hat{m}_{\mathrm{LS}}$ is not.

Related results have appeared in the applied mathematics literature on smoothing. Greville (1966) and Trench (1971) consider the minimum variance linear estimation of a polynomial $m(t)$ of fixed degree in the model $X_t = m(t) + Y_t$. Trench gives an algorithm for computing the BLUE when $\{y_t\}$ has a continuous spectral density $f(\lambda)$; while Greville gives a closed form for the BLUE when $f(\lambda) = (\sin^2 \lambda/2)^\alpha$ with α an integer. Both authors are concerned, however, with stability of minimum variance smoothing formulas, rather than asymptotic properties thereof. Trench (1973) also considers estimation of a linear functional of $m(t)$.

The main results of this paper appear in Sections 2.4 and 2.5. Sections 2.2 and 2.3 contain preliminaries. In Section 2.4 (Theorem 1) we find that, for a large class of spectral densities, the asymptotic form of $\mathrm{Var}\,\hat{m}_{\mathrm{BLU}}$ is determined solely by the behaviour of $f(\lambda)$ near $\lambda = 0$, and that asymptotically efficient estimators based only on this behaviour are available. In Section 2.5 we apply this result to the case of a spectral density $f(\lambda)$ which behaves like λ^ν at the origin, where $\nu > -1$ is any constant. We find (Theorem 3) that the $\mathrm{Var}\,\hat{m}_{\mathrm{BLU}} = O(n^{-\nu-1})$ as $n \to \infty$, and prescribe an asymptotically efficient estimator based only on the value of ν. Thereby we generalize the results of Grenander (1954) and Vitale (1973) described above.

We show in Section 2.6 that the generating functions associated with the estimators of Section 2.5 are expressible in terms of Gegenbauer polynomials, then use known properties of these polynomials to strengthen slightly the conclusions of Section 2.5.

In Sections 2.7 and 2.8 we consider cases of over- and under-estimating ν when $f(\lambda) = O(\lambda^\nu)$ as $\lambda \to 0$ and an estimator as prescribed in Section 2.5 is used. We find (Theorem 6) that the optimum rate of decrease of $\mathrm{Var}\,\hat{m}_{\mathrm{BLU}}$ is still attained if too large a value of ν is assumed, but that (Theorem 7) Var $\hat{m}_{BLU}$ decreases at a slower rate if ν is assumed to be smaller than the true value.

2.2 Uniqueness of the BLUE

From (2.1) it is not difficult to see that the covariance matrix $R = \{R(j - k); j, k = 0, 1, \ldots, n\}$ of $Y_0, \ldots, Y_n$ is positive definite as long as $R(0) = \int_{-\pi}^{\pi} f(\lambda)d\lambda > 0$, which we always assume. Then, as is well known, $\hat{m}_{BLU}$ has a unique representation (2.2). In fact, the row vector c of coefficients in $\hat{m}_{BLU}$ is given by

$$c = (c_0 \cdots c_n) = (u^T R^{-1} u)^{-1} u^T R^{-1}, \quad u = \begin{pmatrix} 1 \\ \vdots \\ 1 \end{pmatrix}, \qquad (2.3)$$

with

$$\text{Var } \hat{m}_{BLU} = (u^T R^{-1} u)^{-1} > 0. \qquad (2.4)$$

From (2.3) and (2.4) we note that $cR = u^T \text{Var } \hat{m}_{BLU}$. Since R is positive definite, this relation together with $cu = 1$ uniquely determines the coefficient vector c of $\hat{m}_{BLU}$, as well as Var $\hat{m}_{BLU}$.

2.3 Preliminaries, Notation, and Definitions

It is convenient to reformulate the problem of finding the BLUE in a concise notation that easily lends itself to analysis. For a fixed spectral density $f(\lambda)$, we define an inner product by

$$(\psi_1, \psi_2)_f = \int_{-\pi}^{\pi} \psi_1(e^{i\lambda}) \overline{\psi_2(e^{i\lambda})} f(\lambda) \, d\lambda \qquad (2.5)$$

on the space of those (complex-valued) functions $\psi(z)$ for which the associated norm $\|\psi\|_f$ is finite. Using (2.1), we may then write the variance of an estimator (2.2) as

$$\text{Var } (\hat{m}, f) = \int_{-\pi}^{\pi} |p_n(e^{i\lambda})|^2 f(\lambda) \, d\lambda = \|p_n\|_f^2, \qquad (2.6)$$

where

$$p_n(z) = c_0 + c_1 z + \cdots + c_n z^n. \qquad (2.7)$$

We employ the notation Var $(\hat{m}, f)$ to indicate dependence on $f(\lambda)$. The problem of finding $\hat{m}_{BLU}$ may now be stated as

$$\|p_n\|_f^2 = \min, \quad p_n(1) = 1. \qquad (2.8)$$

Definition 1 The polynomial (2.7) that solves (2.8) will be called the *optimum polynomial* (of degree n) for $f(\lambda)$. The minimum itself will be denoted by $\sigma_n^2(f)$.

Thus $\|p_n\|_f^2 = \sigma_n^2(f) = \text{Var } \hat{m}_{BLU}$ for the optimal polynomial. Since the optimal $p_n(z)$ is determined by $\hat{m}_{BLU}$ (and conversely), it is unique.

The square norm $\|\psi\|_f^2$ is non-decreasing in $f(\lambda)$ for a fixed function $\psi(z)$. Consequently, we have the important property that the minimum variance $\sigma_n^2(f)$ is a non-decreasing functional of $f(\lambda)$, that is, $\sigma_n^2(f) \geq \sigma_n^2(g)$ when $f(\lambda) \geq g(\lambda)$.

Clearly, $\sigma_n^2(f)$ is non-increasing in n. In the sequel we restrict attention to spectral densities for which this minimum variance does not decrease too rapidly.

Definition 2 We call $\sigma_n^2(f)$ slowly decreasing if

$$\lim_{n \to \infty} \sigma_{n+\nu}^2(f)/\sigma_n^2(f) = 1, \quad \nu = 1, 2, \dots. \tag{2.9}$$

We also define several classes of functions to be considered in the following sections. All such functions are assumed to be (Lebesgue) measurable on $[-\pi, \pi]$.

Definition 3 Let $g(\lambda)$ be defined on $[-\pi, \pi]$. We say that this function belongs to the class

L_0:   if $g(\lambda)$ is non-negative, integrable over $[-\pi, \pi]$, and continuous at $\lambda = 0$;

L_0^+:   if $g(\lambda) \in L_0$ and has a positive lower bound;

B_0:   if $g(\lambda)$ is non-negative, bounded, and continuous at $\lambda = 0$;

B_0^+:   if $g(\lambda) \in B_0$ and has a positive lower bound.

Furthermore, suppose that $g(\lambda)$ has the form

$$g(\lambda) = h(\lambda)|\lambda - \lambda_1|^{\alpha(1)}|\lambda - \lambda_2|^{\alpha(2)} \cdots |\lambda - \lambda_r|^{\alpha(r)}, \tag{2.10}$$

where r is a natural number, $\lambda_1, \lambda_2, \dots, \lambda_r$ are constants in $[-\pi, \pi]$, and $\alpha(1), \alpha(2), \dots, \alpha(r)$ are non-negative constants. We say that $g(\lambda)$ is in ZL_0^+ or ZB_0^+ according to whether $h(\lambda)$ is in L_0^+ or B_0^+, respectively.

2.4 Asymptotic Behaviour of the Minimum Variance

To relate large-sample behaviour of Var $\hat{m}_{\mathrm{BLU}}$ to behaviour of the spectral density near the origin, we study the ratio $\sigma_n^2(fg)/\sigma_n^2(f)$, where $\sigma_n^2(f)$ satisfies (2.9) and $g(\lambda)$ is suitably 'nice'. In essence we are now taking the spectral density of $\{Y_t\}$ to be $f(\lambda)g(\lambda)$ rather than $f(\lambda)$. The purpose of this section is to prove

Theorem 1 Let $f(\lambda)$ be a spectral density with slowly decreasing $\sigma_n^2(f)$, and let $g(\lambda)$ be in the class ZB_0^+. Then

$$\lim_{n \to \infty} \sigma_n^2(fg)/\sigma_n^2(f) = g(0). \tag{2.11}$$

Moreover, if $g(0) > 0$, then the BLUE $\hat{m}_f$ calculated under the hypothesis that the spectral density is $f(\lambda)$ is asymptotically efficient with respect to $f(\lambda)g(\lambda)$ in the sense that

$$\lim_{n \to \infty} \sigma_n^2(fg)/\mathrm{Var}\,(\hat{m}_f, fg) = 1. \tag{2.12}$$

Before proceeding with the proof, we obtain some preliminary results. In the remainder of this section, we adhere to the following notation of the theorem: $f(\lambda)$

will denote a spectral density with slowly decreasing $\sigma_n^2(f)$ and $\hat{m}_f$ will denote the BLUE calculated with respect to $f(\lambda)$. Moreover, the optimal polynomials for $f(\lambda)$ will always be denoted by $p_n(z)$.

Lemma 1 If $t(\lambda)$ is a non-negative trigonometric polynomial, then

$$\lim_{n \to \infty} \inf \sigma_n^2(ft)/\sigma_n^2(f) \geq t(0).$$

Proof The result is trivial if $t(0) = 0$; therefore we take $t(0) > 0$. By the Féjer–Riesz representation theorem, there is a polynomial $p(z)$ of the same degree, say ν, as $t(\lambda)$ for which $t(\lambda) = |p(e^{i\lambda})|^2$. Letting $q_0(z), q_1(z), \ldots$ be the optimal polynomials for $f(\lambda)t(\lambda)$, we define the polynomial $r(z) = p(z)q_n(z)/p(1)$ of degree $n + \nu$, and note that $r(1) = 1$. Therefore,

$$t(0)\sigma_{n+\nu}^2(f) \leq t(0)\|r\|_f^2 = \|q_n\|_{ft}^2 = \sigma_n^2(ft).$$

The proof is completed upon dividing the extreme inequality by $\sigma_n^2(f)$ and taking lim inf, with use of (2.9). ∎

Lemma 2 $K_n(\lambda) = |p_n(e^{i\lambda})|^2 f(\lambda)/\sigma_n^2(f)$ satisfies, for any $0 < \delta \leq \pi$,

$$\lim_{n \to \infty} \int_{\delta \leq |\lambda| \leq \pi} K_n(\lambda) \, d\lambda = 0. \tag{2.13}$$

Proof Let $g(\lambda) = 2$ for $0 \leq |\lambda| < \delta$ and $g(\lambda) = 1$ for $\delta \leq |\lambda| \leq \pi$. Then the integral in (2.13) is just $2 - \|p_n\|_{fg}^2/\sigma_n^2(f)$, and it suffices to show that

$$\lim_{n \to \infty} \|p_n\|_{fg}^2/\sigma_n^2(f) = 2. \tag{2.14}$$

Now clearly we can choose a non-negative trigonometric polynomial $t(\lambda)$ with the properties: $t(\lambda) \leq g(\lambda)$ and $t(0) = 2$. Then $\|p_n\|_{fg}^2 \geq \sigma_n^2(fg) \geq \sigma_n^2(ft)$. Dividing by $\sigma_n^2(f)$ and taking lim inf, with use of Lemma 1 we obtain $\lim\inf_{n \to \infty} \|p_n\|_{fg}^2/\sigma_n^2(f) \geq t(0) = 2$. But clearly also $\|p_n\|_{fg}^2/\sigma_n^2(f) \leq 2$. Thus (2.14) is established and the proof is complete. ∎

Lemma 3 If $g(\lambda)$ is in B_0, then

$$\lim_{n \to \infty} \text{Var}\,(\hat{m}_f, fg)/\sigma_n^2(f) = g(0).$$

Proof With $K_n(\lambda)$ defined as in Lemma 2, the assertion is that $\int_{-\pi}^{\pi} K_n(\lambda)g(\lambda) \, d\lambda \to g(0)$ as $n \to \infty$. This follows from a standard integral kernel argument since $K_n(\lambda)$ is non-negative and satisfies $\int_{-\pi}^{\pi} K_n(\lambda) \, d\lambda = 1$ as well as (2.13). ∎

Lemma 4 If $g(\lambda)$ is in B_0, then

$$\lim_{n \to \infty} \sup \sigma_n^2(fg)/\sigma_n^2(f) \leq g(0). \tag{2.15}$$

Proof The result is immediate from Lemma 3, since $\sigma_n^2(fg) \leq \text{Var}\,(\hat{m}_f, fg)$. ∎

Lemma 5 Let $g(\lambda)$ be in ZL_0^+. Then for any $\varepsilon > 0$ there is a trigonometric polynomial $t(\lambda)$ with the properties: $0 \leq t(\lambda) \leq g(\lambda)$ and $t(0) \geq g(0) - \varepsilon$.

Proof It suffices to prove the result for each of the factors on the right side of (2.10), for the product of the 'approximants' will yield an approximant for $g(\lambda)$ in the required sense. We consider the individual factors.

(a) $g(\lambda) \in L_0^+$. For $\varepsilon < g(0)$, we choose $0 < \delta \leq \pi$ such that $g(\lambda) \geq g(0) - \varepsilon$ for $|\lambda| \leq \delta$. Then we may use $t(\lambda) = [g(0) - \varepsilon][\frac{1}{2}(1 + \cos \lambda)]^N$, where the integer N is chosen so large that $[\frac{1}{2}(1 + \cos \delta)]^N \leq \inf g(\lambda)/[g(0) - \varepsilon]$. For $\varepsilon \geq g(0)$ we use $t(\lambda) \equiv 0$.

(b) $g(\lambda) = |\lambda - \lambda_0|^{2\nu}$, ν a positive integer. The result follows from (a) by writing $g(\lambda)$ as the product of a positive and continuous function and the trigonometric polynomial $[1 - \cos(\lambda - \lambda_0)]^\nu$.

(c) $g(\lambda) = |\lambda - \lambda_0|^\alpha$, $\alpha > 0$ not an even integer. If $\lambda_0 = 0$ we may use $t(\lambda) \equiv 0$, so we take $\lambda_0 \neq 0$. For any integer $\nu > \alpha/2$ and constant $0 < \delta < |\lambda_0|$, $h(\lambda) = [\max\{\delta, |\lambda - \lambda_0|\}]^{-2\nu+\alpha}$ is positive and continuous. Thus $g_1(\lambda) = h(\lambda)|\lambda - \lambda_0|^{2\nu}$ may be approximated according to (a) and (b). But $g_1(\lambda) \leq g(\lambda)$ and $g_1(0) = g(0)$, so the approximant for $g_1(\lambda)$ also serves for $g(\lambda)$. ∎

With these preliminary results, the proof of the theorem stated at the beginning of this section is quite short.

Proof of Theorem 1 Any function $g(\lambda)$ in ZB_0^+ is in ZL_0^+ and in B_0, hence satisfies (2.15). To prove Theorem 1 it suffices thus to show that

$$\lim_{n \to \infty} \inf \sigma_n^2(fg)/\sigma_n^2(f) \geq g(0). \tag{2.16}$$

For any $\varepsilon > 0$, choosing a trigonometric polynomial $t(\lambda)$ with the properties in Lemma 1, we have $\sigma_n^2(fg) \geq \sigma_n^2(ft)$, and so with use of Lemma 1, the left side of (2.16) is seen to be bounded below by $\lim \inf_{n \to \infty} \sigma_n^2(ft)/\sigma_n^2(f) \geq t(0) \geq g(0) - \varepsilon$. Letting $\varepsilon \to 0$, we obtain (2.16). The second assertion (2.12) follows immediately from (2.11) and Lemma 3. ∎

2.5 Application to Certain Spectral Densities

Theorem 1 enables us to obtain the rate of decrease of Var $\hat{m}_{\text{BLU}}$ and asymptotically efficient estimators for large classes of spectral densities, in particular for many spectral densities characterized by a zero (or infinity) of fixed finite order at the origin. As representatives of such spectral densities we take

$$f_\alpha(\lambda) = (2\pi)^{-1}|1 - e^{i\lambda}|^{2\alpha} = 2^{2\alpha-1}\pi^{-1}(\sin^2 \lambda/2)^\alpha. \tag{2.17}$$

Here, α (not necessarily an integer) is a constant, with $\alpha > -\frac{1}{2}$ for integrability. The main problem is in obtaining the BLUE for $f_\alpha(\lambda)$.

Lemma 6 The covariance function corresponding to the spectral density (2.17) is

$$R_\alpha(\tau) = (-1)^\tau \frac{\Gamma(2\alpha + 1)}{\Gamma(\alpha + \tau + 1)\Gamma(\alpha - \tau + 1)}, \quad \tau = 0, \pm 1, \ldots. \quad (2.18)$$

We take $1/\Gamma(z) = 0$ for z a non-positive integer.

Proof Since $f_\alpha(\lambda)$ is even, by making the variable change $\lambda = \pi - 2\theta$ in $\int_{-\pi}^\pi e^{i\tau\lambda} f_\alpha(\lambda)\, d\lambda$ we derive

$$R_\alpha(\tau) = (-1)^\tau 2^{2\alpha+1}\pi^{-1} \int_0^{\pi/2} \cos 2\tau\theta \cos^{2\alpha}\theta\, d\theta.$$

Since the above integral has value $\pi\Gamma(2\alpha + 1)/[2^{2\alpha+1}\Gamma(\alpha + \tau + 1)\Gamma(\alpha - \tau + 1)]$ as given by Erdélyi (1953: vol. 1, p. 12), we obtain (2.17). ■

The next two combinatorial lemmas are presented in a more general form, needed in Section 2.7, than presently required. $B(p, q) = \Gamma(p)\Gamma(q)/\Gamma(p + q)$ represents the beta function.

Lemma 7 For any constants $a, b > -1$,

$$\sum_{k=0}^n \binom{n}{k} B(a + k + 1, b + n - k + 1) = B(a + 1, b + 1). \quad (2.19)$$

Proof With both sides expressed in terms of beta integrals, the identity is obvious. ■

Lemma 8 For $\alpha > -\frac{1}{2}$ and β a non-negative integer,

$$S_j(n, \alpha, \beta)$$

$$\equiv (-1)^j \sum_{k=0}^n (-1)^k \binom{n}{k} \frac{\Gamma(\alpha + \beta + k + 1)\Gamma(\alpha + \beta + n - k + 1)}{\Gamma(\alpha + k - j + 1)\Gamma(\alpha - k + j + 1)}$$

$$= (-1)^\beta n! \sum_{\nu=0}^{2\beta} (-1)^\nu$$

$$\times \binom{j + \beta}{\nu}\binom{n - j + \beta}{2\beta - \nu} \frac{\Gamma^2(\alpha + \beta + 1)}{\Gamma(\alpha + \beta - \nu + 1)\Gamma(\alpha - \beta + \nu + 1)} \quad (2.20)$$

for $j = 0, 1, \ldots, n$, with $\binom{N}{\nu}$ taken as zero for $\nu > N$ and $\nu < 0$.

Proof With use of $\Gamma(z + 1) = z\Gamma(z)$ repeatedly and of Leibnitz's rule, the result follows upon taking the derivative $\partial^{n+2\beta}/\partial x^{j+\beta}\partial y^{n-j+\beta}$ of the identity

$$\sum_{k=0}^{n}(-1)^k\binom{n}{k}x^{\alpha+\beta+k}y^{\alpha+\beta+n-k} = x^{\alpha+\beta}y^{\alpha+\beta}(y-x)^n$$

and then setting $x = y = 1$. ∎

Theorem 2 The BLUE $\hat{m}_\alpha$ calculated with respect to the spectral density (2.17) has coefficients

$$c_k = c_k(n, \alpha)$$
$$= \binom{n}{k}\frac{B(\alpha + k + 1, \alpha + n - k + 1)}{B(\alpha + 1, \alpha + 1)}, \quad k = 0, 1, \ldots, n, \qquad (2.21)$$

and variance

$$\sigma_n^2(f_\alpha) = \mathrm{Var}\,(\hat{m}_\alpha, f_\alpha) = B(n + 1, 2\alpha + 1)/B(\alpha + 1, \alpha + 1). \qquad (2.22)$$

Proof As noted at the end of Section 2.2 it suffices to show that

$$\sum_{k=0}^{n}c_k(n, \alpha) = 1 \qquad (2.23)$$

and

$$\sum_{k=0}^{n}c_k(n, \alpha)R_\alpha(k - j) = \sigma_n^2(f_\alpha), \quad j = 0, 1, \ldots, n, \qquad (2.24)$$

with $\sigma_n^2(f_\alpha)$ given by (2.22). (2.23) follows directly from Lemma 7 with $a = b = \alpha$. As for (2.24), using (2.18), (2.21), and (2.22) we may reduce the assertion to (2.20) with $\beta = 0$. ∎

A straightforward application of Stirling's formula for the gamma function yields, from (2.22), a

Corollary 1 As $n \to \infty$,

$$\sigma_n^2(f_\alpha) \sim n^{-2\alpha-1}\Gamma(2\alpha + 1)/B(\alpha + 1, \alpha + 1). \qquad (2.25)$$

From (2.25) we conclude that $\sigma_n^2(f_\alpha)$ is slowly decreasing, and may immediately apply Theorem 1. We state the result in terms of the original model in Section 2.1.

Theorem 3 Let the disturbance $\{Y_t\}$ have spectral density function $f(\lambda) = f_\alpha(\lambda)g(\lambda)$, where $\alpha > -\frac{1}{2}$ and $g(\lambda) \in ZB_0^+$ (Definition 3) with $g(0) > 0$. Then

$$\mathrm{Var}\,\hat{m}_{\mathrm{BLU}} \sim n^{-2\alpha-1}\Gamma(2\alpha + 1)g(0)/B(\alpha + 1, \alpha + 1), \quad n \to \infty.$$

Moreover, the estimator $\hat{m}_\alpha = \sum_{k=0}^n c_k(n, \alpha) X_k$ with coefficients given in (2.21) is asymptotically efficient in the sense that

$$\text{Var } \hat{m}_{\text{BLU}}/\text{Var } \hat{m}_\alpha \to 1, \quad n \to \infty.$$

For $\alpha = 0$ and $\alpha = 1$ the theorem gives slightly strengthened versions of results of Grenander (1954) and Vitale (1973), respectively. The theorem establishes a conjecture of Vitale for integral α.

2.6 The Optimal Polynomials for $f_\alpha(\lambda)$

We denote by $C_n^{(a)}(x)$ the Gegenbauer (or ultraspherical) polynomials, that is, the polynomials orthogonal on $[-1, 1]$ with respect to the weight function $(1 - x^2)^{a-1/2}$ and with the standardization $C_n^{(a)}(1) = \Gamma(n + 2a)/[n!\Gamma(2a)]$ for $a \neq 0$. We shall relate the optimal polynomials for $f_\alpha(\lambda)$ to the $C_n^{(a)}(x)$ and use the relation to establish a lemma required in Sections 2.7 and 2.8. This lemma will also enable us to strengthen Theorem 3 slightly. Without further reference we make use of well-known properties of the $C_n^{(a)}(x)$, as given by Erdélyi (1953: vol. 2).

Lemma 9 The optimal polynomials

$$P_{n,\alpha}(z) = \sum_{k=0}^n c_k(n, \alpha) z^k \tag{2.26}$$

for $f_\alpha(\lambda)$ satisfy

$$p_{n,\alpha}(e^{2i\theta}) = e^{in\theta} C_n^{(\alpha+1)}(\cos\theta)/C_n^{(\alpha+1)}(1). \tag{2.27}$$

Proof In terms of the coefficients (2.21), it is known that

$$C_n^{(\alpha+1)}(\cos\theta)/C_n^{(\alpha+1)}(1) = \sum_{k=0}^n c_k(n, \alpha)\cos(n - 2k)\theta.$$

The right side, however, is easily seen to be $e^{-in\theta} p_{n,\alpha}(e^{2i\theta})$, since the $c_k(n, \alpha) = c_{n-k}(n, \alpha)$ and are real.

With use of the relation between the Gegenbauer polynomials and the hypergeometric function $F(a, b; c; x)$, and a quadratic transformation on the hypergeometric function, we can rewrite (2.27) as

$$p_{n,\alpha}(e^{2i\theta}) = e^{in\theta} F(-n/2, n/2 + \alpha + 1; \alpha + 3/2; \sin^2\theta).$$

This result is given essentially by Greville (1996), in a result attributed to Sheppard (1913), for α an integer and n even. ∎

Lemma 10 For $\delta > 0$, $n^{\alpha+1} \max_{\delta \leq |\lambda| \leq \pi} |p_{n,\alpha}(e^{i\lambda})|$ is bounded uniformly in n.

Proof Since $C_n^{(\alpha+1)}(1) \sim n^{2\alpha+1}/\Gamma(2\alpha + 2)$ as $n \to \infty$, and because of (2.27), it suffices to show that $n^{-\alpha} \max_{\varepsilon \le |\theta| \le \pi/2} |C_n^{(\alpha+1)}(\cos\theta)|$ is bounded uniformly in n when $\varepsilon > 0$. This follows for $-1 < \alpha < 0$ from the inequality

$$(\sin\theta)^{\alpha+1} |C_n^{(\alpha+1)}(\cos\theta)| < (n/2)^\alpha / \Gamma(\alpha+1), \quad 0 \le \theta \le \pi,$$

and for $\alpha = 0$ from $C_n^{(1)}(\cos\theta) = \sin(n+1)\theta/\sin\theta$. The desideratum may then be obtained inductively for all $\alpha > 0$ from the recurrence formula

$$2\alpha(1 - x^2)C_n^{(\alpha+1)}(x) = (2\alpha + n)C_n^{(\alpha)}(x) - (n + 1)xC_{n+1}^{(\alpha)}(x). \quad \blacksquare$$

Theorem 4 The conclusions of Theorem 3 remain valid if the condition therein that $g(\lambda) \in ZB_0^+$ is replaced by the weaker condition that $g(\lambda) \in ZL_0^+$.

Proof Theorem 3 followed directly from Theorem 1. In the proof of Theorem 1 boundedness of $g(\lambda)$ entered only through Lemmas 3 and 4, and these lemmas would be true for unbounded $g(\lambda)$ (in L_0) if (2.13) could be replaced by the stronger condition that

$$\lim_{n \to \infty} \max_{\delta \le |\lambda| \le \pi} K_n(\lambda) = 0, \quad \delta > 0. \tag{2.28}$$

Since our current $K_n(\lambda) = |p_{n,\alpha}(e^{i\lambda})|^2 f_\alpha(\lambda)/\sigma_n^2(f_\alpha)$ satisfies (2.28), as is seen from Lemma 10, and (2.25), the result follows. $\blacksquare$

2.7 Overestimating the Zero Order

In applying the estimator $\hat{m}_\alpha$ of Theorem 3 for a spectral density that vanishes at the origin, we may guess incorrectly the true order α of the zero. In this section we consider certain cases when our estimate is too high; the opposite case is studied in Section 2.8.

We consider initially the following situation: the true spectral density is $f_\alpha(\lambda)$ but we use the estimator $\hat{m}_{\alpha+\beta}$ having coefficients $c_k(n, \alpha + \beta)$. β is restricted to the non-negative integers; the case of nonintegral β remains open. We define the *asymptotic efficiency* (if it exists) of $\hat{m}_{\alpha+\beta}$ relative to $\hat{m}_\alpha$ for the spectral density $f_\alpha(\lambda)$ by

$$e(\alpha, \beta) = \lim_{n \to \infty} \sigma_n^2(f_\alpha)/\mathrm{Var}\,(\hat{m}_{\alpha+\beta}, f_\alpha).$$

With use of (2.18) and (2.21) we may write

$$\mathrm{Var}\,(\hat{m}_{\alpha+\beta}, f_\alpha) = \sum_{j,k=0}^{n} c_j(n, \alpha + \beta)c_k(n, \alpha + \beta)R_\alpha(j - k)$$

$$= \frac{\Gamma(2\alpha + 1)}{B(\alpha + \beta + 1, \alpha + \beta + 1)\Gamma(2\alpha + 2\beta + n + 2)}$$

$$\times \sum_{j=0}^{n} c_j(n, \alpha + \beta)S_j(n, \alpha, \beta), \tag{2.29}$$

where the $S_j(n, \alpha, \beta)$ are given in (2.20). We introduce the notation $\langle x \rangle_r = x(x-1)\cdots(x-r+1)$ for r a non-negative integer, with $\langle x \rangle_0 = 1$. Writing

$$\binom{j+\beta}{\nu}\binom{n-j+\beta}{2\beta-\nu} = \sum_{r=0}^{\nu}\sum_{s=0}^{2\beta-\nu} a_{r,s}\langle j \rangle_r \langle n-j \rangle_s$$

and employing Lemma 7, we find that for $\nu = 0, 1, \ldots, 2\beta$ and $n \geq 2\beta$,

$$\sum_{j=0}^{n}\binom{n}{j}\binom{j+\beta}{\nu}\binom{n-j+\beta}{2\beta-\nu} B(\alpha+\beta+j+1, \alpha+\beta+n-j+1)$$

$$= \sum_{r,s} a_{r,s}\langle n \rangle_{r+s} \sum_{j=0}^{n-r-s}\binom{n-r-s}{j}$$

$$\times B(\alpha+\beta+j+r+1, \alpha+\beta+n-j-r+1)$$

$$\sim a_{\nu,2\beta-\nu}n^{2\beta} B(\alpha+\beta+\nu+1, \alpha+3\beta-\nu+1)$$

as $n \to \infty$. Since $a_{\nu,2\beta-\nu} = [\nu!(2\beta-\nu)!]^{-1}$, with (2.20) and (2.21) we therefore obtain

$$\lim_{n\to\infty}(n^{2\beta}n!)^{-1}\sum_{j=0}^{n}c_j(n, \alpha+\beta)S_j(n, \alpha, \beta)$$

$$= S_\beta(2\beta, \alpha, \beta)\Gamma(2\alpha+2\beta+2)/[(2\beta)!\Gamma(2\alpha+4\beta+2)]. \qquad (2.30)$$

Referring to the proof of Lemma 8, we see however that

$$S_\beta(2\beta, \alpha, \beta) = (-1)^\beta \frac{\partial^{4\beta}}{\partial x^{2\beta}\partial y^{2\beta}} x^{\alpha+\beta}y^{\alpha+\beta}(y-x)^{2\beta}\Big|_{x=y=1}$$

is a polynomial of degree 2β in α that vanishes for $\alpha = -1, -2, \ldots, -2\beta$ and has value $[(2\beta)!]^2\binom{2\beta}{\beta}$ at $\alpha = 0$, so that

$$S_\beta(2\beta, \alpha, \beta) = (2\beta)!\binom{2\beta}{\beta}\Gamma(\alpha+2\beta+1)/\Gamma(\alpha+1). \qquad (2.31)$$

Combining (2.29), (2.30) and (2.31), we arrive at an asymptotic expression for $\text{Var}(\hat{m}_{\alpha+\beta}, f_\alpha)$. Using (2.25) and asymptotic expressions for those gamma functions with argument involving n, we arrive after a bit of algebra at

Theorem 5 For $\alpha > -\frac{1}{2}$ and $\beta \geq 0$ an integer,

$$e(\alpha, \beta) = \frac{\Gamma(2\alpha+2)\Gamma^2(\alpha+\beta+1)\Gamma(2\alpha+4\beta+2)}{\binom{2\beta}{\beta}\Gamma(\alpha+1)\Gamma(\alpha+2\beta+1)\Gamma^2(2\alpha+2\beta+2)}. \qquad (2.32)$$

The efficiency of $\hat{m}_{\alpha+\beta}$ for a more general spectral density $f_\alpha(\lambda)g(\lambda)$ may be treated by an integral kernel argument as used in Theorem 4.

Theorem 6 Let $\{Y_t\}$ have spectral density $f(\lambda) = f_\alpha(\lambda)g(\lambda)$, where $g(\lambda) \in ZL_0^+$, $g(0) > 0$, and $\alpha > -\frac{1}{2}$. Then for any integer $\beta \geq 0$,

$$\mathrm{Var}\,(\hat{m}_{\alpha+\beta}, f) \sim g(0)\sigma_n^2(f_\alpha)/e(\alpha, \beta), \quad n \to \infty \qquad (2.33)$$

and $\hat{m}_{\alpha+\beta}$ has the asymptotic efficiency $e(\alpha, \beta)$ in (2.32).

Proof (2.33) states that $\int_{-\pi}^{\pi} K_n(\lambda)g(\lambda)\,d\lambda \to g(0)$ as $n \to \infty$, where now

$$K_n(\lambda) = e(\alpha, \beta)|p_{n,\alpha+\beta}(e^{i\lambda})|^2 f_\alpha(\lambda)/\sigma_n^2(f_\alpha).$$

Since $\beta \geq 0$, from (2.25) and Lemma 10 we conclude that $K_n(\lambda)$ satisfies (2.28). Also $K_n(\lambda) \geq 0$, and $\int_{-\pi}^{\pi} K_n(\lambda)\,d\lambda \to 1$ as $n \to \infty$ by definition of $e(\alpha, \beta)$. Thus (2.33) follows. Since $\sigma_n^2(f) \sim g(0)\sigma_n^2(f_\alpha), n \to \infty$, by Theorem 4, the second assertion is also clear.

Since $e(\alpha, \beta) > 0$, we see that overestimation still yields an estimator whose variance decreases at the optimal rate. From (2.32) we observe: (a) $e(\alpha, \beta)$ decreases to $1/\binom{2\beta}{\beta}$ as $\alpha \to \infty$, and (b) $e(\alpha, \beta) \to 1$ as $\alpha \to -\frac{1}{2}$. We note that $e(0, 1) = \frac{5}{6}$, as obtained by Vitale (1973). ∎

2.8 Underestimating the Zero Order

We consider now the case when the estimator $\hat{m}_\alpha$, $\alpha =$ integer, is used but the true spectral density has a higher order zero at the origin. The results stand in marked contrast to those of the previous section. We shall prove, after some preliminaries, the following theorem.

Theorem 7 Let α be a non-negative integer and let the true spectral density be $f(\lambda) = f_{\alpha+1}(\lambda)g(\lambda)$, where $g(\lambda) \in L_0$. Then

$$\lim_{n\to\infty} n^{2\alpha+2}\mathrm{Var}\,(\hat{m}_\alpha, f) = [(2\alpha + 1)!/\alpha!]^2\pi^{-1} \int_{-\pi}^{\pi} g(\lambda)\,d\lambda. \qquad (2.34)$$

Thus, if $f(\lambda) = f_{\alpha+\beta+1}(\lambda)g(\lambda)$ with $\beta \geq 0$, $g(\lambda) \in ZL_0^+$ and $g(0) > 0$, we see (using Theorem 4) that the efficiency of $\hat{m}_\alpha$ is $O(n^{-2\beta-1})$; $\hat{m}_\alpha$ is far from efficient. (2.34) is given by Vitale (1973) for $\alpha = 0$.

In the following, α remains a fixed integer and we suppress dependence on it in some cases. We note that then $R_{\alpha+1}(\tau) = (-1)^\tau \binom{2\alpha+2}{\alpha+1+\tau}$, with $R_{\alpha+1}(\tau) = 0$ for $|\tau| \geq \alpha + 2$. We may also now write

$$c_k(n, \alpha) = (2\alpha + 1)\binom{2\alpha}{\alpha}\frac{(k+1)\cdots(k+\alpha)(n-k+1)\cdots(n-k+\alpha)}{(n+1)(n+2)\cdots(n+2\alpha+1)},$$

which defines a polynomial of degree 2α in k, with zeroes at $k = -1, -2, \ldots, -\alpha$ and $n + 1, n + 2, \ldots, n + \alpha$. Thus the $c_k(n, \alpha)$ satisfy the difference equation

$$\sum_{k=-\alpha-1-\nu}^{\alpha+1-\nu} c_k(n, \alpha) R_{\alpha+1}(k + \nu) = 0, \quad \nu = 0, \pm 1, \ldots. \quad (2.35)$$

Lemma 11 Define, for $\nu = 0, \pm 1, \ldots$

$$S_\nu(n) = \sum_{k=0}^{n} c_k(n, \alpha) R_{\alpha+1}(k + \nu). \quad (2.36)$$

Then $S_\nu(n) = 0$ for $\nu \geq \alpha + 2$ and $-n + 1 \leq \nu \leq -1$, while

$$\lim_{n\to\infty} n^{\alpha+1} S_\nu(n) = (-1)^\nu [(2\alpha + 1)!/\alpha!] \binom{\alpha + 1}{\nu}, \quad 0 \leq \nu \leq \alpha + 1. \quad (2.37)$$

Proof For $\nu \geq \alpha + 2$, $S_\nu(n)$ involves only the $R_{\alpha+1}(\tau)$ for $\tau \geq \alpha + 2$, which all vanish. For $-n + 1 \leq \nu \leq -1$ we may write $S_\nu(n)$ as the left side of (2.35) by adding terms with $c_k(n, \alpha) = 0$ and deleting terms with $R_{\alpha+1}(k + \nu) = 0$. Finally, suppose that $0 \leq \nu \leq \alpha + 1$. Note that

$$\lim_{n\to\infty} n^{\alpha+1} c_k(n, \alpha) = (2\alpha + 1) \binom{2\alpha}{\alpha} (k + \alpha)!/k!. \quad (2.38)$$

Since, for $n \geq \alpha + 1 - \nu$, the upper summation limit in (2.36) can be taken as $\alpha + 1 - \nu$, as $n \to \infty$ we have

$$(-1)^\nu (\alpha!)^2 \nu!(\alpha + 1 - \nu)! n^{\alpha+1} S_\nu(n)/[(2\alpha + 1)!(2\alpha + 2)!]$$

$$\to \sum_{k=0}^{\alpha+1-\nu} (-1)^k \binom{\alpha + 1 - \nu}{k} B(k + \alpha + 1, \nu + 1) = B(\alpha + 1, \alpha + 2)$$

with use of (2.38) and the beta integral, and (2.37) follows. ∎

Lemma 12 Theorem 7 holds if $g(\lambda)$ is a non-negative trigonometric polynomial.

Proof Setting $g(\lambda) = \sum_{\nu=-\gamma}^{\gamma} b_\nu e^{i\nu\lambda}$ we easily find that $f(\lambda) = f_{\alpha+1}(\lambda) g(\lambda)$ has covariance function $R(\tau) = \sum_\nu b_\nu R_{\alpha+1}(\tau + \nu)$. Consequently,

$$n^{2\alpha+2} \text{Var}(\hat{m}_\alpha, f) = \sum_{\nu=-\gamma}^{\gamma} b_\nu n^{2\alpha+2} W_\nu(n),$$

where

$$W_\nu(n) = \sum_{k=0}^{n} c_k(n, \alpha) S_{\nu-k}(n)$$

and the $S_\nu(n)$ are defined in (2.36). Now, since $c_k(n, \alpha) = c_{n-k}(n, \alpha)$ and $R_{\alpha+1}(\tau) = R_{\alpha+1}(-\tau)$, we find that $S_{-\nu}(n) = S_{-n+\nu}(n)$ and $W_{-\nu}(n) = W_\nu(n)$. Since $c_k(n, \alpha) = 0$ for $-\alpha \leq k \leq -1$, from Lemma 11 and (2.38) we find that, for $1 \leq \nu \leq \gamma$ and $n \geq \nu$,

$$n^{2\alpha+2} W_\nu(n) = n^{2\alpha+2} \sum_{k=0}^{\alpha+1} c_{\nu-k}(n, \alpha) S_k(n)$$

$$\rightarrow A \sum_{k=0}^{\alpha+1} (-1)^k \binom{\alpha+1}{k} (\nu - k + 1)(\nu - k + 2) \cdots (\nu - k + \alpha) = 0$$

as $n \rightarrow \infty$, where A depends only on α. The last sum vanishes since $(\nu - k + 1)$ $(\nu - k + 2) \cdots (\nu - k + \alpha)$ is a polynomial of degree α in k. Since $S_{-n}(n) = S_0(n)$, from Lemma 11 and (2.38) we also obtain, as $n \rightarrow \infty$,

$$n^{2\alpha+2} W_0(n) = 2n^{2\alpha+2} c_0(n, \alpha) S_0(n) \rightarrow 2[(2\alpha + 1)!/\alpha!]^2.$$

The result then follows because $2b_0 = \pi^{-1} \int_{-\pi}^{\pi} g(\lambda) \, d\lambda$. ∎

Proof of Theorem 7 For convenience, we write $I(g) = \int_{-\pi}^{\pi} g(\lambda) \, d\lambda$ and define

$$\psi_n(\lambda) = [\alpha!/(2\alpha + 1)!]^2 n^{2\alpha+2} |p_{n,\alpha}(e^{i\lambda})|^2 f_{\alpha+1}(\lambda).$$

We wish to show that $I(\psi_n g) \rightarrow \pi^{-1} I(g)$ as $n \rightarrow \infty$. We let $t_\nu(\lambda)$, $\nu = 0, 1, \ldots$, denote the Féjer means of the Fourier series for $g(\lambda)$, and note that $I(t_\nu) = I(g)$. Then

$$|\pi^{-1} I(g) - I(\psi_n g)|$$

$$\leq |\pi^{-1} I(t_\nu) - I(\psi_n t_\nu)| + \int_{-\delta}^{\delta} \psi_n(\lambda) |t_\nu(\lambda) - g(\lambda)| \, d\lambda$$

$$+ \int_{\delta \leq |\lambda| \leq \pi} \psi_n(\lambda) |t_\nu(\lambda) - g(\lambda)| \, d\lambda \qquad (2.39)$$

for any $0 < \delta \leq \pi$. The proof will be complete if we show that each term on the right side approaches zero as first $n \rightarrow \infty$, then $\nu \rightarrow \infty$, then $\delta \rightarrow 0$.

By Lemma 12, the first term tends to zero as $\nu \rightarrow \infty$ for each fixed ν. By Lemma 10, $\psi_n(\lambda)$ is bounded by a constant C_δ independent of n for $|\lambda| \geq \delta > 0$. Thus the last term in (2.39) is bounded uniformly in n by $C_\delta I(|t_\nu - g|)$, which for fixed δ vanishes as $\nu \rightarrow \infty$ by a property of the Féjer means. From Lemma 12, $I(\psi_n) \rightarrow 2$ as $n \rightarrow \infty$. Therefore, as $n \rightarrow \infty$, the second term on the right side of (2.39) is bounded by $2 \sup_{|\lambda| \leq \delta} |t_\nu(\lambda) - g(\lambda)|$. That this expression approaches zero as $\nu \rightarrow \infty$, then $\delta \rightarrow 0$, follows from the fact that the $t_\nu(\lambda)$ are continuous at $\lambda = 0$ uniformly in ν, as is easily proved using properties of Féjer's kernel. ∎

2.9 Remarks

The results of this paper may easily be extended to the estimation of m in the model $X_t = m e^{i\mu t} + Y_t$, where μ is a fixed constant in $[-\pi, \pi]$. With $\tilde{X}_t = X_t e^{-i\mu t}$ and $\tilde{Y}_t = Y_t e^{-i\mu t}$, this new model reduces to our original: $\tilde{X}_t = m + \tilde{Y}_t$. If $\{Y_t\}$ has spectral density $f(\lambda)$ then $\{\tilde{Y}_t\}$ has spectral density $\tilde{f}(\lambda) = f(\lambda + \mu)$, with definition of $f(\lambda)$ extended outside $[-\pi, \pi]$ by periodicity. In the new model it is thus behaviour of $f(\lambda)$ near $\lambda = \mu$ that is pertinent.

The types of 'multipliers' $g(\lambda)$ allowed in Theorems 1 and 3, and defined in Section 2.3, may seem overly complicated. However, we want to stress that it is only continuity of and behaviour near $\lambda = 0$ of the spectral density that are important. It does not matter if there are discontinuities or isolated zeroes away from the origin.

In Sections 2.5 through 2.8 we have dealt with spectra behaving like $\lambda^{2\alpha}$ near $\lambda = 0$. We should remark that spectra corresponding to $\alpha > 1$ are unlikely to occur in practice, except possibly through difference filtering. However, spectra with $-\frac{1}{2} < \alpha \leq 1$ would seem to be practically useful.

References

Anderson, T. W. (1971), *The Statistical Analysis of Time Series*. Wiley, New York.

Chipman, J. S. *et al.* (1968), 'Efficiency of the sample mean when residuals follow a first-order stationary Markoff process', *Journal of the American Statistical Association*, 63, 1237–46.

Cleveland, W. S. (1971), 'Projection with the wrong inner product and its application to regression with correlated errors and linear filtering of time series', *Annals of Mathematics and Statistics*, 42, 616–24.

Erdélyi, A. *et al.* (1953), *Higher Transcendental Functions*, 1 and 2. McGraw-Hill, New York.

Fishman, G. S. (1972), 'Estimating the mean of observations from a wide-sense stationary autoregressive sequence', *Journal of the American Statistical Association*, 67, 402–6.

Grenander, U. (1950), 'Stochastic processes and statistical inference', *Arkiv für Matematik*, 1, 195–277.

—— (1954), 'On the estimation of regression coefficients in the case of an autocorrelated disturbance', *Annals of Mathematics and Statistics*, 25, 252–72.

Greville, T. N. E. (1966), 'On stability of linear smoothing formulas', *SIAM Journal of Numerical Analysis*, 3, 157–70.

Kruskal, W. (1968), 'When are Gauss–Markov and least squares estimators identical? A coordinate-free approach', *Annals of Mathematics and Statistics*, 39, 70–75.

Parzen, E. (1961), 'An approach to time series analysis', *Annals of Mathematics and Statistics*, 32, 951–89.

Rozanov, J. A. (1971), *Infinite-Dimensional Gaussian Distributions*. American Mathematical Society, Providence.

Sheppard, W. F. (1913), 'Reduction of error by means of negligible differences', *Proceedings of the Fifth International Congress of Mathematicians*, 2, 348–84, Cambridge.

Trench, W. F. (1971), 'Discrete minimum variance smoothing of a polynomial plus random noise', *Journal of Mathematical Analysis and Applications*, 35, 630–645.

—— (1973), 'Strong discrete minimum variance estimation of a polynomial plus random noise', *Journal of Mathematical Analysis and Applications*, 41, 20–33.

Vitale, R. A. (1973), 'An asymptotically efficient estimate in time series analysis', *Quarterly Applications of Mathematics*, 30, 421–40.

Watson, G. S. (1967), 'Linear least squares regression', *Annals of Mathematics and Statistics*, 38, 1679–99.

—— (1972), 'Prediction and the efficiency of least squares', *Biometrika*, 59, 91–98.

Zyskind, G. (1967), 'On canonical forms, non-negative covariance matrices and best and simple least squares linear estimators in linear models', *Annals of Mathematics and Statistics*, 38, 1092–109.

3

An Introduction to Long-memory Time Series Models and Fractional Differencing

C. W. J. GRANGER AND ROSELYNE JOYEUX

3.1 On Differencing Time Series

It has become standard practice for time series analysts to consider differencing their series 'to achieve stationarity'. By this they mean that one differences to achieve a form of the series that can be identified as an ARMA model. If a series does need differencing to achieve this, it means that strictly the original, undifferenced series has infinite variance. There clearly can be problems when a variable with infinite variance is regressed on another such variable, using least squares techniques, as illustrated by Granger and Newbold (1974). A good recent survey of this topic is by Plosser and Schwert (1978). This has led time series analysts to suggest that econometricians should at least consider differencing their variables when building models. However, econometricians have been somewhat reluctant to accept this advice, believing that they may lose something of importance. Phrases such as differencing 'zapping out the low frequency components' are used. At first sight the two viewpoints appear irreconcilable, but it will be seen that by considering a general enough class of models, both sides of the controversy can be correct.

Suppose that x_t is a series that, when differenced d times, gives the series y_t, which has an ARMA representation. x_t will then be called an integrated series, with parameter d, and denoted $x_t \sim I(d)$. If y_t has spectrum $f(\omega)$, then x_t does not strictly possess a spectrum, but from filtering considerations the spectrum of x_t can be thought of as

$$f_x(\omega) = |1 - z|^{-2d} f(\omega), \quad \omega \neq 0, \tag{3.1}$$

where $z = e^{-i\omega}$. This follows by noting that differencing a series once multiplies its spectrum by $|1-z|^2 = 2(1-\cos \omega)$. If y_t is strictly ARMA, then $\lim_{\omega \to 0} f(\omega) = c$, where c is a constant. c is taken to be positive, as, if $c = 0$, this may be thought to be an indication that the series has been over-differenced. It follows that

$$f_x(\omega) = c\omega^{-2d} \quad \text{for } \omega \text{ small.}$$

This chapter was previously published as Granger, C. W. J. and Joyeux, R. (1980), 'An introduction to long-memory time series models and fractional differencing', *Journal of Time Series Analysis*, 1, 15–29. Copyright ©1980 C. W. J. Granger and R. Joyeux, reproduced by kind permission of Blackwell Publishers Ltd.

Now consider the case where $f_x(\omega)$ is given by (3.1), but d is a fraction $0 < d < 1$. This corresponds to a filter $a(B) = (1 - B)^d$ which when applied to x_t results in an ARMA series. It will be shown that if $\frac{1}{2} \leqslant d < 1$, the x_t has infinite variance, and so the ordinary Box–Jenkins identification procedure will suggest that differencing is in order, but if x_t is differenced, the spectrum becomes

$$f_{\Delta x}(\omega) = [2(1 - \cos \omega)]^{2(1-d)} f(\omega),$$

so that $f_{\Delta x}(0) = 0$, and an ARMA model with invertible moving average compon-ent is no longer completely appropriate. Thus, in this case, the time series analysts will suggest differencing to get finite variance, but if the series is differenced, its zero frequency component will be removed and the econometrician's fears are realized. It seems that neither differencing nor not differencing is appropriate with data having the spectrum (3.1) with fractional d. In later sections, properties of these series are discussed and some open questions mentioned.

It should be pointed out that if a series has spectrum of the form (3.1), with fractional d, then it is possible to select a model of the usual ARMA(p, d, q) type, with integer d, which will closely approximate this spectrum at all frequencies *except* those near zero. Thus, models using fractional d will not necessarily provide clearly superior shortrun forecasts, but they may give better longer-run forecasts where modelling the low frequencies properly is vital. It will be seen that fractional d models have special long-memory properties which can give them extra potential in long-run forecasting situations. It is this possibility that makes consideration of single series of this class of interest. The generalized differencing and the solution of the difference or not controversy, together with the chance of obtaining superior relationships between series, is a further reason for believing that these models may be of importance. A discussion of how fractional integrated models may arise is given in Granger (1980), and is summarized below.

Long-memory models have been much considered by workers in the field of water resources. Good recent surveys are those by Lawrance and Kottegoda (1977) and Hipel and McLeod (1978). Many aspects of the models were first investigated by Mandelbrot (e.g. 1968, 1971) in a series of papers. However, the fundamental reasoning underlying the long-memory models is quite different in these previous papers than that utilized here. The models that arise are not identical in details, and the statistical techniques used both differ and sometimes have different aims. However, it should be emphasized that many of the results to be reported have close parallels in this previous literature. The results presented below are closer in form to the classical time series approach and are, hopefully, easier to interpret. It should also be realized that this paper represents just those results achieved at the start of a much more detailed, and wider-ranging, investigation.

3.2 Time Series Properties

Consider a series x_t with spectrum

$$f(\omega) = \alpha(1 - \cos \omega)^{-d}, \tag{3.2}$$

where α is a positive constant. This is a series which, if differenced d times, will produce white noise. However, we now consider the case $-1 < d < 1$, but $d \neq 0$, so that 'fractional differencing' may be required. It will be assumed that x_t is derived from a linear filter applied to zero-mean white noise and that x_t has zero mean.

The autocovariances, if they exist, will be given by

$$\mu_\tau = \int_0^{2\pi} \cos \tau \omega f(\omega) \, d\omega$$

$$= \int_0^\pi \alpha 2^{-d} \cos \tau \omega (\sin \omega/2)^{-2d} \, d\omega$$

by noting that

$$(1 - \cos \omega) = 2(\sin \omega/2)^2.$$

Using the standard formula (Gradshteyn and Ryzhik 1965: p. 372, eqn (3.631.8))

$$\int_0^\pi \sin^{v-1} x \cos ax \, dx = \frac{\pi \cos a\pi/2}{2^{v-1} \cdot v \cdot B((v+a+1)/2, (v-a+1)/2)}$$

and some algebra gives

$$\mu_\tau = \alpha \cdot 2^{1+d} \sin(\pi d) \frac{\Gamma(\tau + d)}{\Gamma(\tau + 1 - d)} \cdot \Gamma(1 - 2d) \quad \text{provided } -1 < d < \tfrac{1}{2}, d \neq 0.$$

It will be seen later that, if $d \geqslant \tfrac{1}{2}$, then μ_0, the variance, is infinite. It follows that the autocorrelations are given by

$$\rho_\tau = \frac{\Gamma(1-d)}{\Gamma(d)} \cdot \frac{\Gamma(\tau+d)}{\Gamma(\tau+1-d)}. \tag{3.3}$$

Using the standard approximation derived from Sheppard's formula, that for j large, $\Gamma(j+a)/\Gamma(j+b)$ is well approximated by j^{a-b}, it follows that

$$\rho_\tau \simeq A(d)\tau^{2d-1} \tag{3.4}$$

for τ large, and $d < \tfrac{1}{2}, d \neq 0$.

Note that for a stationary ARMA model

$$\rho_\tau \approx A\theta^\tau, \quad |\theta| < 1$$

for τ large, and these values tend to zero exponentially and thus quicker than ρ_τ given by (3.4). This illustrates the 'long-memory' aspect of series with spectrum (3.1) or (3.2).

The infinite moving average representation of x_t will be denoted by

$$x_t = \sum_{j=0}^\infty b_j \varepsilon_{t-j} = b(B)\varepsilon_t,$$

using the backward operator, B. It follows that the spectrum of x_t is

$$f(\omega) = ab(z)b(\bar{z}),$$

where $z = e^{-i\omega}$ and $a = \sigma_\varepsilon^2/2$. Thus, as

$$f(\omega) = \alpha(1 - z)^{-d}(1 - \bar{z})^{-d},$$

we can take $\alpha = a$ and

$$b(z) = (1 - z)^{-d}.$$

Such a filter $b(B)$ will be called an integrating filter of order d. Using the standard binomial expansion

$$(1 - z)^{-d} = 1 + \sum_{k=1}^{\infty} \frac{\Gamma(k + d)z^k}{\Gamma(d)\Gamma(k + 1)},$$

it follows that

$$b_j = \frac{\Gamma(j + d)}{\Gamma(d)\Gamma(j + 1)} \quad j \geq 1 \tag{3.5}$$

$$\simeq Aj^{d-1} \tag{3.6}$$

for j large and an appropriate constant A.

Consider now an MA(∞) model with b_j, $j \geq 1$, given exactly by (3.6), that is,

$$y_t = A \sum_{j=1}^{\infty} j^{d-1}\varepsilon_{t-j} + \varepsilon_t$$

so that $b_0 = 1$. This series has variance

$$V(y) = A^2\sigma_\varepsilon^2 \left(1 + \sum_{j=1}^{\infty} j^{2(d-1)}\right).$$

From the theory of infinite series, it is known that

$$\sum_{j=1}^{\infty} j^{-s} \quad \text{converges for } s > 1$$

but otherwise diverges. Since it is easily shown that the variance of x_t and that of y_t differ only by a finite quantity, it follows that the variance of x_t is finite provided $d < \frac{1}{2}$, but is infinite if $d \geq \frac{1}{2}$.

The AR(∞) representation of x_t is

$$\sum_{j=0}^{\infty} a_j x_{t-j} = \varepsilon_t, \quad a_0 = 1,$$

that is,

$$a(B)x_t = \varepsilon_t,$$

which gives spectrum

$$f(\omega) = \frac{\alpha}{a(z)a(\bar{z})}$$

so that, comparing with (3.2),

$$a(z) = (1 - z)^d.$$

Hence

$$a_j = \frac{\Gamma(j - d)}{\Gamma(1 - d)\Gamma(j + 1)}, \quad j \geqslant 1 \tag{3.7}$$

and, for j large

$$|a_j| \simeq A j^{-(1+d)}. \tag{3.8}$$

From (3.6) and (3.8), it is seen that b_j and $|a_j|$ tend to zero slower than exponential. It follows that no ARMA(p, q) model, with finite p and q would provide an adequate approximation for large j. From (3.5) and (3.7) it can be noted that a_j is positive and b_j negative if d is negative, and a_j is negative and b_j positive if d is positive.

The case $d = 0$ has been excluded throughout this section but this is just the white noise case, so that ρ_j, b_j and a_j all are zero for $j > 0$.

If a series is generated by the more general model, compared to (3.2),

$$x_t = (1 - B)^{-d} a'(B)\varepsilon_t \tag{3.9}$$

where $0 < a'(0) < \infty$ and ε_t is white noise, results (3.4), (3.6) and (3.8) continue to hold for j large.

A filter of the form

$$a(B) = (1 - B)^d \tag{3.10}$$

which, using the previously introduced phrase, is an integrating filter of order $-d$, can also be called a fractional differencing operator. This is easily seen by taking $d = \frac{1}{2}$, as then applying $a(B)$ twice corresponds to an ordinary, full difference and thus applying it once gives a half, or fractional, difference. The idea of half differencing should not be confused with that of differencing over a half sampling interval, as the two concepts are quite unrelated.

It is not clear at this time if integrated models with non-integer d occur in practice and only extensive empirical research can resolve this issue. However, some aggregation results presented in Granger (1980) do suggest that these models may be expected to be relevant for actual economic variables. It is proved there, for example, that if x_{jt}, $j = 1, \ldots, n$ are set of independent series, each generated by an AR(1) model, so that

$$x_{jt} = \alpha_j x_{j,t-1} + \varepsilon_{jt}, \quad j = 1, \ldots, N,$$

where the ε_{jt} are independent, zero-mean white noise and if the α_j's are values independently drawn from a beta distribution on $(0, 1)$, where

$$dF(\alpha) = \frac{2}{B(p,q)}\alpha^{2p-1}(1 - \alpha^2)^{q-1}\,d\alpha, \quad 0 \leqslant \alpha \leqslant 1 \text{ and } p > 0, q > 0,$$

then if

$$\bar{x} = \sum_{j-1}^{N} x_{j,t} \text{ for } N \text{ large, } \bar{x} \sim I(1 - q/2).$$

The shape of the distribution from which the α's are drawn is only critical near 1 for this result to hold.

A more general result arises from considering x_{jt} generated by

$$x_{jt} = \alpha_j x_{j,t-1} + y_{j,t} + \beta_j W_t + \varepsilon_{jt},$$

where the series $y_{j,t}$, W_1 and ε_{jt} are all independent of each other for all j, ε_{jt} are white noise with variances σ_j^2, $y_{j,t}$ has spectrum $f_y(\omega, \theta_j)$ and is at least potentially observable for each micro-component. It is assumed that there is no feedback in the system and the various parameters α, θ_j, β, and σ^2 are all assumed to be drawn from populations and the distribution function for the α's independent are generated by an AR(1) model, plus an independent causal series $y_{j,t}$ and a common factor causal series W_1. With these assumptions, it is shown that

(i)
$$\bar{x}_t \sim I(d_x),$$

where d_x is the largest of the three terms $1 - q/2 + d_y$, $1 - q + d_w$ and $1 - q/2$, where $y_t \sim I(d_y)$, $W_t \sim I(d_w)$, and

(ii) if a transfer function model of the form

$$\bar{x}_t = a_1(B)\bar{y}_t + a_2(B)W_t + e_t$$

is fitted, then both $a_1(B)$ and $a_2(B)$ are integrating filters of order $1 - q$.

In Granger (1980) it was shown that integrated models may arise from micro-feedback models and also from large-scale dynamic econometric models that are not too sparse. Thus, at the very least, it seems that integrated series can occur from realistic aggregation situations, and so do deserve further consideration.

There are a number of ways in which data can be generated to have the long-memory properties, at least to a good order of approximation. Mandelbrot (1971) has come down heavily in favour of utilizing aggregates of the form just discussed and has conducted simulation studies to show that series with the appropriate properties are achieved. An alternative technique, which appears to be less efficient but to have easier interpretation, has been proposed by Hipel and McLeod (1978). Suppose that you are interested in generating a series with autocorrelations ρ_τ, $\tau = 0, 1, \ldots$, with $\rho_0 = 1$. Define the $N \times N$ correlation matrix

$$C_N = [\rho_{|i-j|}],$$

and let this have Cholesky decomposition

$$C_N = M \cdot M^{\mathrm{T}},$$

where T denotes transpose and

$$M = [m_{ij}] \text{ is an } N \times N \text{ lower triangular matrix.}$$

Then it is easily shown that if $e_t, t = 1, \ldots, N$ are terms in a Gaussian white noise series with zero mean and unit variance, then the series

$$y_t = \sum_{i=1}^{t} m_{ti} e_i \qquad (3.11)$$

will have the autocorrelations ρ_τ. The generating process is seen to be non-stationary and is expensive for large N values. However, by using ρ_τ given by (3.3), a series with long-memory properties is generated.

An obvious alternative is to use a long autoregressive representation, of the form

$$x_t + \sum_{j=1}^{m} a_{j,m} x_{t-j} = \varepsilon_t, \qquad (3.12)$$

where ε_t is white noise and the $a_{j,m}$ are generated by solving the first m Walker–Yule equations with the theoretical values of ρ_τ given by (3.3), that is

$$\begin{bmatrix} a_{1,m} \\ a_{2,m} \\ \vdots \\ a_{m,m} \end{bmatrix} = -C_m^{-1} \begin{bmatrix} \rho_1 \\ \rho_2 \\ \vdots \\ \rho_m \end{bmatrix} \qquad (3.13)$$

and

$$a_{j,m} = 0, \quad j > m.$$

Clearly, m will have to be fairly large for a reasonable approximation to be achieved. The obvious problem with this technique is what starting-up values to use for the y's. If the starting-up values do not belong to a long-memory process, such as using a set of zeros, the long-memory property of the model means that it will take a long time to forget this incorrect starting-up procedure. To generate data, we decided to combine the Hipel–McLeod and the autoregressive methods. The Hipel–McLeod method was used to generate N observations which are then used as start-up values for the autoregressive equation, taking $m = N$, and then n values for y_{jt} are generated. The methods just described are appropriate only for $-1 < d < \frac{1}{2}, d \neq 0$. To generate data y_t with $\frac{1}{2} \leqslant d < 1$, x_t is first formed for $-\frac{1}{2} < d < 0$ and then y_t generated by

$$y_t = x_t + y_{t-1}.$$

We have had a little experience with this generating procedure, with $d = 0.25$ and $d = 0.45$. Using $N = 50$ and $n = 400$ or 1000, the estimated autocorrelations did not compare well with the theoretical ones, but using $N = 100$ and $n = 100$ or 400 the estimated and theoretical autocorrelations matched closely for $d = 0.25$ but were less good for $d = 0.45$. Clearly, more study is required to determine the comparative advantages of alternative generation methods and the properties of the series produced.

3.3 Forecasting and Estimation of d

The obvious approach to forecasting models with spectra given by (3.2) is to find an ARMA or ARIMA model which has a spectrum approximating this form. Unfortunately, this is a rather difficult problem as functions of the form (3.2), when expressed in $z = e^{-iw}$, are not analytic in z and so standard approximation theory using rational functions does not apply. Whilst realizing that much deeper study is required, in the initial stages of our investigations we have taken a very simple viewpoint, and used $AR(m)$ models of the type discussed above, in equations (3.12) and (3.13). In practice, one would expect series to possibly have spectral shapes of form (3.2) at low frequencies but to have different shapes at other frequencies. These other shapes, which can perhaps be thought of as being generated by short-memory ARMA models, will be important for short-term forecasts but will be of less relevance for long-term forecasts. Thus, the $AR(m)$ model may be useful for forecasting 10 or 20 steps ahead, say, in a series with no seasonal. If m is taken to be 50, for example, a rather high order model is being used, but it should be noted that the parameters of the model depend only on d, and so the model is seen to be highly parsimonious. The method will, nevertheless, require quite large amounts of data. To use this autoregressive forecasting model, the value of d is required.

There are a variety of ways of estimating the essential parameter model d. The water resource engineers use a particular re-scaled range variable which has little intuitive appeal (see e.g. Lawrance and Kottegoda 1977). Other techniques could be based on estimates of the logarithm of the spectrum at low frequencies. At this time we are taking a very pragmatic approach, by using $AR_d(m)$ models, with a grid of d values from -0.9 to $+0.4$, excluding $d = 0$, forming ten-step forecasts and then estimating the mean squared errors for each of the models with different d-values, together with white noise random walk models. Some initial results that have been obtained are discussed in the following section. The method we have used is clearly arbitrary and sub-optimal, as the following theory shows. Suppose that the observed series x_t has two uncorrelated components

$$x_t = y_t + z_t,$$

where y_t is a 'pure' long-memory series, having spectrum given exactly by (3.2), and z_t is a stationary standard short-memory model that can be represented by an $ARMA(p, q)$ model with small p and q and with all roots not near unity. For large

τ, the autocovariances μ_τ^2 of z_t will be negligible, and so

$$\rho_\tau^x \simeq A\rho_\tau^y,$$

where ρ_τ^y is given by (3.3) and

$$A = \frac{\text{Var } y}{\text{Var } x}. \tag{3.14}$$

To derive an autoregressive model of order m appropriate for long-run forecasting, the coefficients a_{jm} in (3.12) should be solved from the new Walker–Yule equations

$$\boldsymbol{a}_m = \begin{bmatrix} 1 & A\rho_1 & A\rho_2 & \cdots & A\rho_{m-1} \\ A\rho_1 & 1 & A\rho_1 & \cdots & A\rho_{m-2} \\ \vdots & A\rho_1 & 1 & \ddots & \\ A\rho_{m-1} & & & & 1 \end{bmatrix}^{-1} \boldsymbol{\rho}_m \cdot A$$

where $\boldsymbol{a}_m = (a_{1m}, a_{2m}, \ldots, a_{mm})$ and $\boldsymbol{\rho}_m^1 = (\rho_1, \rho_2, \ldots, \rho_m)$ which may be written

$$\boldsymbol{a}_m = [(1 - A)\boldsymbol{I}_m + A\boldsymbol{C}_m]^{-1} \boldsymbol{\rho}_m \cdot A, \tag{3.15}$$

where $\boldsymbol{I}_m$ is the $m \times m$ unit matrix and $\boldsymbol{C}_m$ is the autocovariance matrix as introduced in the previous section. Obviously, if $A = 1$, (3.15) becomes identical to (3.13). Equation (3.15) may be rewritten

$$\boldsymbol{a}_m = \boldsymbol{C}_m^{-1}\left[\boldsymbol{I}_m + D\boldsymbol{C}_m^{-1}\right]^{-1}\boldsymbol{\rho}_m, \tag{3.16}$$

where

$$D = \frac{1 - A}{A}$$

and this can be expanded as

$$\boldsymbol{a}_m = \boldsymbol{C}_m^{-1}\left[\boldsymbol{I}_m - D\boldsymbol{C}_m^{-1} + \frac{D^2}{2}\boldsymbol{C}_m^{-2} + \cdots\right]\boldsymbol{\rho}_m.$$

Thus, the zero-order approximation, assuming D is small, is

$$\boldsymbol{a}_m^{(0)} = \boldsymbol{C}_m^{-1}\boldsymbol{\rho}_m,$$

which is identical to (3.13). The first-order approximation is

$$\boldsymbol{a}_m^{(1)} = \left[\boldsymbol{C}_m^{-1} - D\boldsymbol{C}_m^{-2}\right]\boldsymbol{\rho}_m, \quad \text{etc.}$$

There now effectively become two parameters to estimate, d and D. As stated before, in our preliminary investigation just zero-order approximations were used, and the relevance of this with real data needs investigation.

Table 3.1. Forecasting properties of
long-memory models

d	$V(d)$	$S_N(d)$	$R_{10}(d)$
(1) *Ten-step forecasts* ($N = 10$)			
−0.9	1.81	1.81	0.369E–05
−0.8	1.648	1.648	0.227E–04
−0.7	1.504	1.504	0.745E–04
−0.6	1.380	1.380	0.182E–03
−0.5	1.273	1.273	0.365E–03
−0.4	1.183	1.182	0.611E–03
−0.3	1.109	1.108	0.840E–03
−0.2	1.052	1.051	0.868E–03
−0.1	1.014	1.014	0.485E–03
0.1	1.019	1.017	0.223E–02
0.2	1.099	1.078	0.185E–01
0.3	1.316	1.204	0.857E–01
0.4	2.070	1.425	0.312E–00
0.5		1.791	
0.6		2.376	
0.7		3.289	
0.8		4.691	
0.9		6.816	
(2) *Twenty-step forecasts* ($N = 20$)			
−0.9	1.81	1.81	0.460E–06
−0.8	1.648	1.648	0.331E–05
−0.7	1.504	1.504	0.127E–04
−0.6	1.380	1.380	0.362E–04
−0.5	1.273	1.273	0.823E–04
−0.4	1.183	1.183	0.164E–03
−0.3	1.109	1.109	0.263E–03
−0.2	1.052	1.052	0.315E–03
−0.1	1.014	1.014	0.204E–03
0.1	1.019	1.018	0.126E–02
0.2	1.099	1.085	0.121E–01
0.3	1.316	1.232	0.645E–01
0.4	2.070	1.510	0.270E–00
0.5		2.016	
0.6		2.913	
0.7		4.487	
0.8		7.222	
0.9		11.938	

The techniques discussed in this section only apply for the range $-1 < d <$
0.5, $d \neq 0$. If d lies in the region $0.5 \leqslant d < 1$ a number of approaches could be
taken, for instance, one could first difference and then get a series with $-1 < d <$
0, or one could apply the fractional differencing operator $(1 - B)^{0.5}$, and then get

a series with $0 < d < 0.5$. The first of these two suggestions is much the easier, but the second may provide a better estimate of the original d. Clearly, much more investigation is required.

As an indication of the forecasting potential of the long-memory models, the Table 3.1 shows the following quantities:

$$V(d) = \text{variance of } y_t = (1 - B)^{-d}\varepsilon_t,$$

where ε_t is white noise with unit variance. y_t is thus a pure long-memory series, with spectrum given by (3.2) for all frequencies and with $\alpha = \frac{1}{2}\pi$,

$$S_N(d) = \sum_{j=0}^{N-1} b_j^2(d),$$

which is the variance of the N-step forecast error using the optimal forecast for y_{n+N}, using $y_{n-j}, j \geq 0$, and where b_j are the theoretical moving average coefficients given by (3.5), and

$$R_N^2(d) = \frac{V(d) - S_N(d)}{V(d)}$$

which is a measure of the N-step forecastability of y_t. Clearly, this measure only applies for series with finite variance, and so V and R^2 are only defined for $d < 0.5$. The table shows these quantities for both ten and twenty-step forecasts. It is seen that variance decreases as d goes from -0.9 to -0.1 and then increases again as d goes from 0.1 to 0.4. It might be noted that $d = -1$, which corresponds to a differenced white noise, has $V(-1) = 2$ and, of course, $R_N(-1) = 0$, $N > 1$. For d negative the amount of forecastability is low, but as d approaches 0.5 the results are much more impressive. For example, with $d = 0.4$, which corresponds to a finite variance model and would presumably be identified as low-order ARMA by the standard Box–Jenkins techniques, the table shows that ten-step forecast error variance is 30 per cent less than forecasts using just the mean and twenty-step forecast error variance is 27 per cent less than simple short-memory forecasting models would produce. It is clear that the long-memory models would be of greatest practical importance if the real world corresponded to d values around 0.5. A slightly curious feature of the table is that $R_N(d)$ does not quite increase monotonically as d goes from -0.9 to $+0.4$, as there is a slight dip at $d = -0.1$.

3.4 Practical Experience

To this point in time we have had only limited experience with the techniques discussed above, but it has been encouraging.

Using the method described in section three, series of length 400 were generated with $d = 0.25$ and $d = 0.45$, using an AR(100) approximation with an initial 100 terms being generated by the Hipel–McLeod moving average model for use as start-up values.

The following two tables show the theoretical and estimated autocorrelations for levels and differences together with some estimated partial autocorrelations. These allow ARIMA models to be identified and estimated by standard techniques.

$d = 0.25$	Level			Differences
Lag	Est. Autocorr.	Theor. Autocorr.	Est. Partial	Est. Autocorr.
1	0.41	0.33	0.41	−0.42
2	0.31	0.24	0.17	−0.02
3	0.24	0.19	0.08	−0.09
4	0.27	0.17	0.15	0.03
5	0.27	0.15	0.11	0.01
6	0.26	0.14	0.07	−0.04
7	0.29	0.13	0.13	0.11
8	0.19	0.12	−0.03	−0.08
9	0.18	0.11	0.01	0.02
10	0.14	0.11	−0.01	−0.08
11	0.20	0.10	0.07	0.10
12	0.15	0.10	−0.03	−0.08
13	0.18	0.09	0.06	0.05
14	0.15	0.09	0.00	−0.02
15	0.15	0.09	0.02	−0.01

The approximate standard error for small lags is 0.05. For levels, the first 96 estimated autocorrelations are all non-negative and the first 22 are more than twice the standard error. If an AR(2) model is identified, it is estimated to be

$$x_t = 0.34x_{t-1} + 0.17x_{t-2} - 0.03 + \varepsilon_t$$

$$(6.85) \qquad (3.45) \qquad (-0.05)$$

(brackets show t-values) and the estimated residuals pass the usual simple white noise tests. An alternative model might be identified as ARIMA $(0, 1, 1)$, but this will have similar long-run forecasting properties as a random walk.

For levels, the first thirty-four autocorrelations are non-negative and the next sixty-five are negative but very small. The first sixteen are greater than twice the standard error. The results suggest that the generating mechanism has not done a good job of reproducing the larger lag autocorrelations. A relevant ARIMA model

$d = 0.45$

Lag	Levels			Differences	
	Est. autocorr.	Theor. autocorr.	Est. partial corr.	Est. autocorr.	Est. partial
1	0.67	0.82	0.67	−0.38	−0.38
2	0.59	0.76	0.25	−0.03	−0.20
3	0.53	0.74	0.12	−0.07	−0.19
4	0.51	0.71	0.13	0.06	−0.06
5	0.45	0.70	0.01	−0.07	−0.11
6	0.44	0.69	0.07	0.02	−0.07
7	0.41	0.68	0.04	−0.05	−0.11
8	0.41	0.67	0.08	0.02	−0.09
9	0.40	0.66	0.06	0.02	−0.04
10	0.39	0.65	0.03	0.01	−0.02
11	0.37	0.65	0.01	0.05	−0.06
12	0.31	0.64	−0.07	−0.08	−0.04
13	0.31	0.63	0.02	0.02	−0.01
14	0.28	0.63	−0.01	−0.02	−0.04
15	0.27	0.63	0.01	0.02	−0.03

for this data might be IMA(1, 1) and the following model was estimated:

$$(1 - B)x_t = \varepsilon_t - 0.61\varepsilon_{t-1} + 0.008.$$

$$(15.5) \qquad (0.4)$$

The grid-search method was applied to each series, that is $AR_d(50)$ models, with parameters depending just on different d-values, were used to forecast ten-steps ahead. The various mean-squared (ten-step) forecast errors (MSE) were found to be:

(i) $d = 0.25$

Selected d	Resulting MSE
$d = 0.1$	1.29
$d = 0.2$	1.36
$d = 0.25$	1.30
$d = 0.3$	1.263
$d = 0.4$	1.264

MSE (random walk) = 2.45
MSE (mean) = 1.45
MSE (estimated AR(2)) = 1.36

MSE (random walk) is the ten-step forecast error mean-squared error if one had assumed the series were a random walk. MSE (mean) is that resulting from forecasts made by a model of mean plus white noise, where estimate of the mean is continually updated. The grid method 'estimates' d to be about 0.3, which is near the true value of 0.25, and produces forecasts that have a ten-step error variance which is somewhat better than the fitted AR(2) model and is considerably better than using a random walk or IMA(1, 1) model. The theoretical MSE with $d = 0.25$ is 1.13, which suggests that the forecasting method used is not optimal. The results are biased in favour of the AR(2) model, as its parameters are estimated over the same data set from which the mean-squared forecast errors were estimated.

(ii) $d = 0.45$

Selected d	Resulting MSE
$d = 0.1$	2.27
$d = 0.2$	2.15
$d = 0.3$	2.06
$d = 0.4$	2.02
$d = 0.45$	2.01

MSE (random walk) $= 2.84$
MSE (mean) $= 2.42$

The theoretical achievable MSE using $d = 0.45$ is 1.58. Once more, the grid search has apparently selected the correct d (although values of d greater than 0.45 were not considered), the forecast method used was not optimal but the ten-step MSE achieved was about thirty per cent better than if a random walk or IMA(1, 1) model had been used.

The grid search procedure has been used by us on just one economic series so far, the US monthly index of consumer food prices, not seasonally adjusted, for the period January 1947 to June 1978. The ordinary identification of this series is fairly interesting. The correlogram of the raw series clearly indicates that first differencing is required. For the first differenced series, the autocorrelations up to lag 72 are all positive and contain fifteen values greater than twice the standard error. The first twenty-four autocorrelations are:

Lag	1	2	3	4	5	6	7	8	9	10	11	12
r_k	0.36	0.22	0.17	0.11	0.14	0.22	0.17	0.08	0.17	0.21	0.19	0.24

Lag	13	14	15	16	17	18	19	20	21	22	23	24
r_k	0.18	0.07	0.05	0.10	0.07	0.11	0.05	0.07	0.05	0.08	0.15	0.11

The standard errors are 0.05 and 0.07 for the two rows. The partial auto-correlations are generally small except for the first, but the sixth, ninth, and tenth are greater than twice the standard error. The parsimonious model probably identified by standard procedures is thus ARIMA(1, 1, 0).

Using first differenced series, plus extra d differencing, the grid gives the following ten-step forecasting mean square forecast errors for the series in level form:

d	−0.9	−0.8	−0.7	−0.6	−0.5	−0.4	−0.3	−0.2	−0.1
MSE	139	128	115	101	86.1	72.1	58.4	46.1	35.7

d	0.1	0.2	0.3	0.35	0.4	0.45
MSE	21.6	17.8	16.1	16.03	16.4	17.32

The variance of the whole series is 1204.6 and the ten-step forecast MSE using just a random walk model is 27.6. The evidence thus suggests that the original series should be differenced approximately 1.35 times and that substantially superior ten-step forecasts then result. The ten-step forecast MSE using an ARIMA$(1, 1, 0)$ model is 19.85, which should be compared to the minimum grid value of 16.03. The actual model fitted was

$$(1 - B)x_t = .37(1 - B)x_{t-1} + .246t + \varepsilon_t.$$
$$(7.63) \qquad\qquad (4.77)$$

There is clearly plenty of further work required on such questions as how best to estimate d, how best to form forecasts for integrated models and the properties of these estimates and forecasts. It is clear that the techniques we have used in this paper are by no means optimal but hopefully they do illustrate the potential of using long-memory models and will provoke further interest in these models. It is planned to investigate the above questions and also to find if these models appear to occur, and can be used to improve long-term forecasts, in actual economic data.

Appendix 3.1 The $d = 0$ Case

The $d = 0$ case can be considered from a number of different viewpoints, which lead to different models. Some of these viewpoints are:

(i) If $f(\omega) = \alpha(1 - \cos \omega)^{-d}$ then simply taking $d = 0$ gives the usual stationary case with $f(0) = c$, where c is a positive but finite constant. This corresponds to taking $d = 0$ in an ARIMA(p, d, q) model.

(ii) If one considers aggregates of the form

$$x_{jt} = \alpha_j x_{j,t-1} + \beta_j e_t$$

$$z_t = \sum_{j=1}^{N} x_{jt}$$

then approximately

$$z_t = \left[\int \frac{\beta}{1 - \alpha B} \, dF(\alpha, \beta) \right] e_t$$

that is,

$$z_t = \bar{\beta} \log(1 - B)e_t$$

if α, β are independent and α is rectangular on $(0, 1)$.

This corresponds to a series with spectrum proportional to $\log(1-z)\log(1-\bar{z})$, which takes the form $(\log \omega)^2$ for small ω and so is infinite at $\omega = 0$. The moving average form corresponding to this model has $b_j \simeq A/j$ for j large, which is the same as equation (3.6) with $d = 0$. The autocovariances take the form

$$\mu_\tau \simeq \sigma_e^2 \frac{\log \tau}{\tau} \quad \text{for large } \tau.$$

This type of model can be thought of as arising from applying filters of the form $[(1 - B)^d - 1]/d$ to white noise, and then letting $d \to 0$.

(iii) By looking at equations (3.4) and (3.8), one could ask what models correspond to autoregressive equations with

$$|a_j| \simeq \frac{A}{j} \quad \text{for large } j$$

or have autocovariances of form

$$\mu_\tau \simeq \frac{A}{\tau}$$

for large τ, which arose in Section 3.2 from a particular aggregation.

The relationships between these various viewpoints and the relevance for forecastings need further investigation.

References

Gradshteyn, I. S. and Ryzhik, I. M. (1965), *Tables of Integrals, Series and Products* (4th edn), Academic Press.

Granger, C. W. J. (1980), 'Long memory relationships and the aggregation of dynamic models', *Journal of Econometrics*, 14, 227–38.

—— and Newbold, P. (1974), 'Spurious regressions in economics', *Journal of Econometrics*, 2, 111–20.

Hipel, K. W. and McLeod, A. J. (1978), 'Preservation of the rescaled adjusted range Parts 1, 2, and 3', *Water Resources Research*, 14, 491–518.

Lawrance, A. J. and Kottegoda, N. T. (1977), 'Stochastic modelling of riverflow Time Series', *Journal of the Royal Statistical Society*, A, 140, 1–47.

Mandelbrot, B. B. and Van Ness, J. W. (1968), 'Fractional Brownian motions, fractional noises and applications', *SIAM Review*, 10, 422–37.

—— (1971), 'A fast fractional Gaussian noise generator', *Water Resources Research*, 7, 543–53.

Plosser, C. I. and Schwert, G. W. (1978), 'Money, income and sunspots: measuring economic relationships and the effects of differencing', *Journal of Monetary Economics*, 4, 637–60.

Rosenblatt, M. (1976), 'Fractional integrals of stochastic processes and the central limit theorem', *Journal of Applied Probability*, 13, 723–32.

4

Large-sample Properties of Parameter Estimates for Strongly Dependent Stationary Gaussian Time Series

ROBERT FOX AND MURAD S. TAQQU

4.1 Introduction

Let $X_j, j \geq 1$, be a stationary Gaussian sequence with mean μ and spectral density $\sigma^2 f(x, \theta)$, $-\pi \leq x \leq \pi$, where $\mu, \sigma^2 > 0$ and the vector $\theta \in E \subset R^p$ are unknown parameters. Denote the covariance by $\sigma^2 r_k(\theta)$, so that

$$E(X_j - \mu)(X_{j+k} - \mu) = \sigma^2 r_k(\theta) = \sigma^2 \int_{-\pi}^{\pi} e^{ikx} f(x, \theta) \, dx.$$

(We are not assuming that σ^2 is the variance of X_j.) Let $R_N(\theta)$ be the $N \times N$ matrix with j, kth entry $r_{j-k}(\theta)$. Thus $\sigma^2 R_N(\theta)$ is the covariance matrix of $X_1, \ldots, X_N$. Our object is to estimate θ and σ^2 based on the observations $X_1, \ldots, X_N$.

We are interested in *strongly dependent* sequences X_j, that is, in sequences $f(x, \theta) \sim |x|^{-\alpha(\theta)} L_\theta(x)$ as $x \to 0$, where $0 < \alpha(\theta) < 1$ and $L_\theta(x)$ varies slowly at 0. These sequences have covariances that decrease too slowly to permit the normalized partial sums

$$S_{[Nt]} = \frac{\sum_{j=1}^{[NT]} X_j}{\sqrt{N}}, \quad t \geq 0$$

to converge weakly to Brownian motion. Because of this fact, strongly dependent sequences play an important role in the theory of self-similar stochastic processes. Two examples, fractional Gaussian noise and fractional ARMA's, are described at the end of this section. We will estimate simultaneously all the unknown parameters symbolized by the vector θ, and not just the exponent $\alpha(\theta)$ in isolation.

This work was previously published as Fox, R. and Taqqu, M. S. (1986), 'Large-sample properties of parameter estimates for strongly dependent stationary Gaussian time series,' *Annals of Statistics* **14**, 517–32. Reproduced here by kind permission of the Institute of Mathematical Statistics.

We would like to thank Hans Kuensch for suggesting the proof of Lemma 5. We also thank the Associate Editor and the referees of *Annals of Statistics* for suggestions which helped us improve the presentation of the original paper.

This research was supported by the National Science Foundation grant ECS-80-15585 at Cornell University.

A number of approaches to parameter estimation for strongly dependent sequences have been considered in the literature. These include the R/S technique, periodogram estimation, and maximum likelihood estimation. Theoretical properties of the R/S estimates have been investigated by Mandelbrot (1975) and Mandelbrot and Taqqu (1979). Periodogram estimation has been considered by Mohr (1981), Graf (1983), and Geweke and Porter-Hudak (1983). Hipel and McLeod (1978) have discussed computational considerations involved in the application of maximum likelihood estimation. See also Todini and O'Connell (1979).

The study of maximum likelihood estimation for strongly dependent sequences is a special case of the problem of maximum likelihood estimation for dependent observations. Sweeting (1980) has given conditions under which the maximum likelihood estimator is consistent and asymptotically normally distributed. Basawa and Prakasa Rao (1980) and Basawa and Scott (1983) survey theorems and examples in this area. In order to apply these results it would be necessary to study the second derivatives of $R_N^{-1}(\theta)$. To avoid this difficulty we will follow the approach suggested by Whittle (1951). This involves maximizing

$$\frac{1}{\sigma} \exp\left\{ -\frac{\mathbf{Z}' A_N(\theta)\mathbf{Z}}{2N\sigma^2} \right\}. \tag{4.1}$$

Here $\mathbf{Z} = (X_1 - \bar{X}_N, \ldots, X_N - \bar{X}_N)'$, $\bar{X}_N = (1/N)\sum_{j=1}^N X_j$, and $A_N(\theta)$ is the $N \times N$ matrix with entries $[A_N(\theta)]_{jk} = a_{j-k}(\theta)$, where

$$a_j(\theta) = \frac{1}{(2\pi)^2} \int_{-\pi}^{\pi} e^{ijx}[f(x,\theta)]^{-1}dx. \tag{4.2}$$

For the approximation of the inverse of $(R(\theta))_{i\geq0,j\geq0}$ by $(A_N(\theta))_{i\geq0,j\geq0}$, see Bleher (1981) and also Beran and Kuensch (1985). Notice that, by Parseval's relation, the doubly infinite matrix $A(\theta)$ with entries $a_{j-k}(\theta)$, $-\infty < j, k < \infty$, is the inverse of the doubly infinite matrix $R(\theta)$ with entries $r_{j-k}(\theta)$, $-\infty < j, k < \infty$.

Thus we define estimators $\bar{\theta}_N$ and $\bar{\sigma}_N^2$ to be those values of θ and σ^2 which maximize $(1/\sigma)\exp\{-(1/2N\sigma^2)\mathbf{Z}' A_N(\theta)\mathbf{Z}\}$. This is equivalent to choosing $\bar{\theta}_N$ to minimize

$$\sigma_N^2(\theta) = \frac{\mathbf{Z}' A_N(\theta)\mathbf{Z}}{N} \tag{4.3}$$

and then setting $\bar{\sigma}_N^2 = \sigma_N^2(\bar{\theta}_N)$. It will be convenient to use the fact that

$$\sigma_N^2(\theta) = \frac{1}{2\pi} \int_{-\pi}^{\pi} [f(x,\theta)]^{-1}I_N(x)\,dx, \tag{4.4}$$

where

$$I_N(x) = \frac{\left| \sum_{j=1}^N e^{ijx}(X_j - \bar{X}_N) \right|^2}{2\pi N}. \tag{4.5}$$

Walker (1964) showed that $\bar{\theta}_N$ is consistent and asymptotically normal in many cases in which the sequence $\{X_j\}$ is weakly dependent (and not necessarily

Gaussian). Hannan (1973) improved these results and was able to use them to prove consistency and asymptotic normality of the maximum likelihood estimator. Dunsmuir and Hannan (1976) gave extensions to the case of vector-valued observations.

We show that for strongly dependent Gaussian sequences $\{X_j\}$ the estimator $\bar{\theta}_N$ is consistent and asymptotically normal. The conditions under which this result holds are given in Section 4.2. Our results apply to fractional Gaussian noise and fractional ARMA's.

Fractional Gaussian noise was introduced by Mandelbrot and Van Ness (1968) and has been widely used to model strongly dependent geophysical phenomena. It is a stationary Gaussian sequence with mean 0 and covariance

$$r_k = EX_j X_{j+k} = \frac{C}{2}\left\{|k+1|^{2H} - 2|k|^{2H} + |k-1|^{2H}\right\},$$

where H is a parameter satisfying $\frac{1}{2} < H < 1$ and $C > 0$. This covariance satisfies

$$r_k \sim CH(2H-1)k^{2H-2} \quad \text{as } k \to \infty.$$

The spectral density $f(x, H)$ of fractional Gaussian noise is given by

$$f(x, H) = CF(H) f_0(x, H), \tag{4.6}$$

where

$$f_0(x, H) = (1 - \cos x) \sum_{k=-\infty}^{\infty} |x + 2k\pi|^{-1-2H}, \quad -\pi \leq x \leq \pi, \tag{4.7}$$

and

$$F(H) = \left\{\int_{-\infty}^{\infty} (1 - \cos x)|x|^{-1-2H} \, dx\right\}^{-1} \tag{4.8}$$

(see Sinai 1976). As $x \to 0$ we have

$$f(x, H) \sim \frac{CF(H)}{2} |x|^{1-2H}.$$

Fractional Gaussian noise is the unique Gaussian sequence with the property that $S_{mN} = \sum_{j=1}^{mN} X_j$ has the same distribution as $m^H S_N$ for all $m, N \geq 1$. Further properties are discussed in Mandelbrot and Taqqu (1979).

Another example to which our results apply is fractional ARMA. To define it, let $g(x, \xi) = \sum_{j=0}^{p} \xi_j x^j$ and $h(x, \phi) = \sum_{j=0}^{q} \phi_j x^j$, where $\xi = (\xi_0, \ldots, \xi_r)$ and $\phi = (\phi_0, \ldots, \phi_q)$. Suppose that $g(x, \xi)$ and $h(x, \phi)$ have no zeros on the unit circle and no zeros in common. For $0 < d < \frac{1}{2}$, define the spectral density

$$f(x, d, \xi, \phi) = C|e^{ix} - 1|^{-2d} \left|\frac{g(e^{ix}, \xi)}{h(e^{ix}, \phi)}\right|^2, \quad -\pi \leq x \leq \pi. \tag{4.9}$$

A Gaussian sequence with mean 0 and spectral density $f(x, d, \xi, \phi)$ is called a fractional ARMA process. Heuristically, it is the sequence which, when differenced d times, yields an ARMA process with spectral density

$$C \left| \frac{g(e^{ix}, \xi)}{h(e^{ix}, \phi)} \right|^2.$$

Granger and Joyeux (1980) and Hosking (1981) have proposed the use of fractional ARMA's to model strongly dependent phenomena, since

$$f(x, d, \xi, \phi) \sim C \left| \frac{g(1, \xi)}{h(1, \phi)} \right|^2 |x|^{-2d} \quad \text{as } x \to 0.$$

4.2 Statements of the Theorems

Let X_j, $j \geq 1$ be a stationary Gaussian sequence with mean μ and spectral density $\sigma^2 f(x, \theta)$ where $\mu, \sigma^2 > 0$ and $\theta \in E$ are unknown parameters. The set $E \subset R^p$ is assumed to be compact. Let σ_0^2 and θ_0 be the true values of the parameters. We assume that θ_0 is in the interior of E.

If θ and θ' are distinct elements of E, we suppose that the set $\{x : f(x, \theta) \neq f(x, \theta')\}$ has positive Lebesgue measure, so that different θ's correspond to different dependence structures. Assume the functions $f(x, \theta)$ are normalized so that

$$\int_{-\pi}^{\pi} \log f(x, \theta) \, dx = 0, \quad \theta \in E.$$

Let $f^{-1}(x, \theta) = 1/f(x, \theta)$.

Remark The condition $\int_{-\pi}^{\pi} \log f(x, \theta) \, dx > -\infty$ guarantees that the sequence $\{X_j\}$ admits a backward expansion

$$X_j = \sigma \sum_{k=0}^{\infty} b_k(\theta) \varepsilon_{j-k},$$

where ε_j, $j \geq 1$ are independent standard normal random variables. The first coefficient $b_0(\theta)$ is the one-step prediction standard deviation of the sequence $Y_j = X_j/\sigma$. It is given by

$$b_0(\theta) = 2\pi \exp \left\{ \frac{1}{2\pi} \int_{-\pi}^{\pi} \log f(x, \theta) \, dx \right\}.$$

(See Hannan 1970: p. 137.) If $\int_{-\pi}^{\pi} \log f(x, \theta) \, dx = 0$ for all θ, it follows that $b_0(\theta) = 2\pi$ and so $\{Y_j\}$ has one-step prediction standard deviation independent of θ.

We will refer to the following conditions.

Conditions A We say that $f(x, \theta)$ satisfies conditions A.1–A.6 if there exists $0 < \alpha(\theta) < 1$ such that for each $\delta > 0$

(A.1) $g(\theta) = \int_{-\pi}^{\pi} \log f(x, \theta)\, dx$ can be differentiated twice under the integral sign.

(A.2) $f(x, \theta)$ is continuous at all (x, θ), $x \neq 0$, $f^{-1}(x, \theta)$ is continuous at all (x, θ), and

$$f(x, \theta) = O(|x|^{-\alpha(\theta)-\delta}) \quad \text{as } x \to 0.$$

(A.3) $\partial/\partial\theta_j\, f^{-1}(x, \theta)$ and $\partial^2/\partial\theta_j\, \partial\theta_k\, f^{-1}(x, \theta)$ are continuous at all (x, θ),

$$\frac{\partial}{\partial\theta_j} f^{-1}(x, \theta) = O(|x|^{\alpha(\theta)-\delta}) \quad \text{as } x \to 0, \quad 1 \leq j \leq p,$$

and

$$\frac{\partial^2}{\partial\theta_j\, \partial\theta_k} f^{-1}(x, \theta) = O(|x|^{\alpha(\theta)-\delta}) \quad \text{as } x \to 0, \quad 1 \leq j, k \leq p.$$

(A.4) $\partial/\partial x f(x, \theta)$ is continuous at all (x, θ), $x \neq 0$, and

$$\frac{\partial}{\partial x} f(x, \theta) = O(|x|^{-\alpha(\theta)-1-\delta}) \quad \text{as } x \to 0.$$

(A.5) $\partial^2/\partial x\, \partial\theta_j\, f^{-1}(x, \theta)$ is continuous at all (x, θ), $x \neq 0$, and

$$\frac{\partial^2}{\partial x\, \partial\theta_j} f^{-1}(x, \theta) = O(|x|^{\alpha(\theta)-1-\delta}) \quad \text{as } x \to 0, \quad 1 \leq j \leq p.$$

(A.6) $\partial^3/\partial^2 x\, \partial\theta_j\, f^{-1}(x, \theta)$ is continuous at all (x, θ), $x \neq 0$, and

$$\frac{\partial^3}{\partial^2 x\, \partial\theta_j} f^{-1}(x, \theta) = O(|x|^{\alpha(\theta)-2-\delta}) \quad \text{as } x \to 0, \quad 1 \leq j \leq p.$$

Remark The constants that appear in the $O(\cdot)$ conditions may depend on the parameters θ and δ. If $L(x)$ varies slowly as $x \to 0$, then $L(x) = O(|x|^{-\delta})$ as $x \to 0$ for every $\delta > 0$ (see Feller 1971: p. 277). Thus (A.1)–(A.6) will be satisfied if the indicated continuity holds, $f(x, \theta)$ varies regularly as $x \to 0$ with exponent $-\alpha(\theta)$, $\partial/\partial x/f(x, \theta)$ with exponent $-\alpha(\theta) - 1$, $\partial/\partial\theta_j$, $f^{-1}(x, \theta)$, and $\partial^2/\partial\theta_j\, \partial\theta_k\, f^{-1}(x, \theta)$ with exponent $\alpha(\theta)$, $\partial^2/\partial x \partial\theta_j\, f^{-1}(x, \theta)$ with exponent $\alpha(\theta) - 1$, and $\partial^3/\partial^2 x\, \partial\theta_j\, f^{-1}(x, \theta)$ with exponent $\alpha(\theta) - 2$.

Definition of the Estimator Consider the quadratic form $\sigma_N^2(\theta)$ given in (4.3). Let θ_N be a value of θ which minimizes $\sigma_N^2(\theta)$. Put $\bar{\sigma}_N^2 = \sigma_N^2(\bar{\theta}_N)$. The following theorem establishes the strong consistency of the estimators $\bar{\theta}_N$ and $\bar{\sigma}_N^2$.

Theorem 1 If $f(x, \theta)$ satisfies conditions (A.2) and (A.4), then with probability 1

$$\lim_{N \to \infty} \bar{\theta}_N = \theta_0 \quad \text{and} \quad \lim_{n \to \infty} \bar{\sigma}_N^2 = \sigma_0^2.$$

To state the next theorem, let $W(\theta)$ be the $p \times p$ matrix with j, kth entry

$$w_{jk}(\theta) = \int_{-\pi}^{\pi} f(x, \theta) \frac{\partial^2}{\partial \theta_j \, \partial \theta_k} f^{-1}(x, \theta) \, \mathrm{d}x.$$

Theorem 2 If conditions (A.1)–(A.6) are satisfied then the random vector $\sqrt{N}(\bar{\theta}_N - \theta_0)$ tends in distribution to a normal random vector with mean 0 and covariance matrix $4\pi W^{-1}(\theta_0)$.

Theorems 1 and 2 are proven in Section 4.3.

Remark 1 As in Hannan (1973), it can be shown that the results hold if $\sigma_N^2(\theta)$ in (4.4) is replaced by

$$\sum_k f^{-1}\left(\frac{2\pi k}{N}, \theta\right) I_N\left(\frac{2\pi k}{N}\right),$$

where $-N/2 < k \le [N/2]$. This last expression may be useful for computational purposes.

Remark 2 $\bar{\theta}_N - \bar{\theta}_0$ is asymptotically of the order of $1/\sqrt{N}$. Geweke and Porter-Hudak (1983) have obtained asymptotic results for an estimator resulting from a regression based on the periodogram estimates. This estimator converges to the true value of the parameter at a slower speed than $1/\sqrt{N}$.

Remark 3 The sample mean $\overline{X}_N$ converges to $\mu = E X_j$ at a slower speed than $\bar{\theta}_N$ converges to θ_0 because (up to a slowly varying function in the normalization) $N^{1/2-\alpha/2}(\overline{X}_N - \mu)$ converges to a normal distribution (see Taqqu 1975).

Applications Theorems 1 and 2 can be applied to fractional Gaussian noise and to fractional ARMA. In order to apply them to fractional Gaussian noise, restrict the parameter H to a compact subset of $(\frac{1}{2}, 1)$ and choose the normalization constant $CF(H)$ in (4.6) as

$$CF(H) = \exp\left[-\frac{1}{2\pi} \int_{-\pi}^{\pi} \log\left\{(1 - \cos x) \sum_{k=-\infty}^{\infty} |x + 2k\pi|^{-1-2H}\right\} \, \mathrm{d}x\right],$$
$$(4.10)$$

so that

$$\int_{-\pi}^{\pi} \log f(x, H) \, \mathrm{d}x = 0.$$

Similarly, Theorems 1 and 2 can be applied to a fractional ARMA process by restricting the parameter (d, ξ, ϕ) to a compact set and choosing C in (4.9) as

$$C = C(d, \xi, \phi) = \exp\left[-\frac{1}{2\pi} \int_{-\pi}^{\pi} \log\left\{|e^{ix} - 1|^{-2d} \left|\frac{g(e^{ix}, \xi)}{h(e^{ix}, \phi)}\right|^2\right\} \, \mathrm{d}x\right].$$
$$(4.11)$$

Theorem 3 The conclusions of Theorems 1 and 2 hold if $X_j - \mu$ is fractional Gaussian noise with $\frac{1}{2} < H < 1$ or a fractional ARMA process with $0 < d < \frac{1}{2}$.

Theorem 3 is proven in Section 4.4.

Remark We have supposed that the mean μ of the sequence X_j is unknown. If it is known, merely replace the periodogram $I_N(x)$ in (4.5) by

$$\tilde{I}_N(x) = \frac{\left| \sum_{j=1}^{N} e^{ijx}(X_j - \mu) \right|^2}{2\pi N}.$$

Corollary 1 When $\mu = EX_j$ is known, Theorems 1, 2, and 3 hold if $I_N(x)$ is replaced by $\tilde{I}_N(x)$.

4.3 Proofs of Theorems 1 and 2

Retain the assumptions and definitions made in Section 2 prior to the statement of Thorem 1. Introduce $r_k(\theta) = \int_{-\pi}^{\pi} e^{ikx} f(x, \theta) \, dx$, so that $E(X_j - \mu)(X_{j+k} - \mu) = \sigma_0^2 r_k(\theta_0)$. Adopt the convention that functions defined in $[-\pi, \pi]$ are extended to $[-2\pi, 2\pi]$ in such a way as to have period 2π.

Lemma 1 Let $g(x, \theta)$ be a continuous function on $[-\pi, \pi] \times E$. If (A.2) and (A.4) hold, then with probability 1

$$\lim_{N \to \infty} \int_{-\pi}^{\pi} g(x, \theta) I_N(x) \, dx = \sigma_0^2 \int_{-\pi}^{\pi} g(x, \theta) f(x, \theta_0) \, dx$$

uniformly in θ.

Proof Note that $I_N(x)$ has Fourier coefficients

$$\int_{-\pi}^{\pi} e^{ikx} I_N(x) \, dx = \begin{cases} W(k, N), & |k| < N, \\ 0, & |k| > N, \end{cases}$$

where

$$W(k, N) = \frac{1}{N} \sum_{j=1}^{N-k} (X_j - \overline{X}_N)(X_{j+k} - \overline{X}_N)$$

$$= \frac{1}{N} \sum_{j=1}^{N-k} \{X_j - \mu - (\overline{X}_N - \mu)\}\{X_{j+k} - \mu - (\overline{X}_N - \mu)\}$$

$$= \frac{\sum_{j=1}^{N-k} (X_j - \mu)(X_{j+k} - \mu)}{N} + \frac{N-k}{N}(\overline{X}_N - \mu)^2$$

$$- (\overline{X}_N - \mu)\frac{\sum_{j=1}^{N-k}(X_j - \mu)}{N} - (\overline{X}_N - \mu)\frac{\sum_{j=k+1}^{N}(X_j - \mu)}{N}.$$

The sequence $\{X_j\}$ is ergodic since it is Gaussian with spectral density $f(x, \theta_0)$ that satisfies $\int_{-\pi}^{\pi} f(x, \theta_0) \, dx > -\infty$. Therefore $(\overline{X}_N - \mu)$ tends to 0 as $N \to \infty$, as do the last three terms on the right-hand side. Hence

$$\lim_{N \to \infty} W(k, N) = \lim_{N \to \infty} \frac{1}{N} \sum_{j=1}^{N-k} (X_j - \mu)(X_{j+k} - \mu)$$

$$= \sigma_0^2 r_k(\theta_0).$$

This means that the proof of Lemma 1 can be carried out exactly as that of Theorem 1 of Hannan (1973). ∎

Proof of Theorem 1   The proof uses Lemma 1 and the fact that $\int_{-\pi}^{\pi} \log f(x, \theta) \, dx = 0$ for all θ. Proceed as in the proof of Theorem 1 of Hannan (1973). ∎

To establish Theorem 2 we use the following four lemmas.

Lemma 2   Let $\{b_N\}$ be a sequence of constants tending to ∞. Let $\partial/\partial\theta \, \sigma_N^2(\theta)$ be the random vector with jth component equal to $\partial/\partial\theta_j \, \sigma_N^2(\theta)$. If (A.2)–(A.4) hold and Y is a random vector such that $b_N \, \partial/\partial\theta \, \sigma_N^2(\theta_0)$ tends to Y in distribution as $N \to \infty$, then $b_N(\overline{\theta}_N - \theta_0)$ tends to $(-2\pi/\sigma_0^2) W^{-1}(\theta_0) Y$ in distribution as $N \to \infty$.

Proof   Let $\partial^2/\partial\theta^2 \, \sigma_N^2(\theta)$ be the $p \times p$ random matrix with j, kth entry $\partial^2/\partial\theta_j \, \partial\theta_k \, \sigma_N^2(\theta)$. According to the mean value theorem

$$\frac{\partial}{\partial\theta} \sigma_N^2(\overline{\theta}_N) = \frac{\partial}{\partial\theta} \sigma_N^2(\theta_0) + \left[\frac{\partial^2}{\partial\theta^2} \sigma_N^2(\theta_N^*) \right] (\overline{\theta}_N - \theta_0),$$

where $|\theta_N^* - \theta_0| < |\overline{\theta}_N - \theta_0|$. Since θ_0 is in the interior of E, Theorem 1 implies that $\overline{\theta}_N$ is in the interior of E for large N. Since $\overline{\theta}_N$ minimizes $\sigma_N^2(\theta)$, it follows that $\partial/\partial\theta \, \sigma_N^2(\overline{\theta}_N) = 0$ for large N. Thus for large N

$$\frac{\partial}{\partial\theta} \sigma_N^2(\theta_0) = \left[-\frac{\partial^2}{\partial\theta^2} \sigma_N^2(\theta_N^*) \right] (\overline{\theta}_N - \theta_0).$$

Because

$$\frac{\partial^2}{\partial\theta_j \, \partial\theta_k} \sigma_N^2(\theta) = \frac{1}{2\pi} \int_{-\pi}^{\pi} \frac{\partial}{\partial\theta_j \, \partial\theta_k} f^{-1}(x, \theta) I_N(x) \, dx,$$

it follows from Lemma 1 and Theorem 1 that with probability 1

$$\frac{\partial^2}{\partial\theta_j \, \partial\theta_k} \sigma_N^2(\theta_N^*) \to \frac{\sigma_0^2}{2\pi} w_{jk}(\theta_0).$$

Therefore

$$-b_N \frac{\sigma_0^2}{2\pi} W(\theta_0)(\overline{\theta}_N - \overline{\theta}_0)$$

tends in distribution to Y, completing the proof of Lemma 2. ∎

Lemma 3 If (A.1), (A.2), and (A.3) hold then

$$w_{jk}(\theta) = \int_{-\pi}^{\pi} \left(\frac{\partial}{\partial \theta_j} f^{-1}(x, \theta) \right) \left(\frac{\partial}{\partial \theta_k} f^{-1}(x, \theta) \right) f^2(x, \theta) \, dx.$$

Proof If the right-hand side is denoted J, then by (A.1),

$$0 = \int_{-\pi}^{\pi} \frac{\partial^2}{\partial \theta_j \, \partial \theta_k} \log f(x, \theta) \, dx = -J + \int_{-\pi}^{\pi} f^{-1}(x, \theta) \frac{\partial^2}{\partial \theta_j \, \partial \theta_k} f(x, \theta) \, dx.$$

Hence

$$\begin{aligned}
w_{jk}(\theta) &= \int_{-\pi}^{\pi} f(x, \theta) \frac{\partial^2}{\partial \theta_j \, \partial \theta_k} f^{-1}(x, \theta) \, dx \\
&= 2J - \int_{-\pi}^{\pi} f^{-1}(x, \theta) \frac{\partial^2}{\partial \theta_j \, \partial \theta_k} f(x, \theta) \, dx \\
&= J. \blacksquare
\end{aligned}$$

Lemma 4 If conditions (A.2) and (A.4) hold, then for every $\delta > 0$

$$r_k(\theta) = O(k^{\alpha(\theta)-1+\delta}) \quad \text{as } k \to \infty.$$

Proof Fix θ and put $f(x) = f(x, \theta)$. Since f is periodic,

$$\begin{aligned}
2|r_k(\theta)| &= \left| \int_{-\pi}^{\pi} e^{ikx} \left[f(x) - f\left(x + \frac{\pi}{k}\right) \right] dx \right| \\
&\leq \int_{-\pi}^{\pi} \left| f(x) - f\left(x + \frac{\pi}{k}\right) \right| dx \\
&= \int_{-\pi}^{-2\pi/k} + \int_{-2\pi/k}^{\pi/k} + \int_{\pi/k}^{\pi}.
\end{aligned}$$

Conditions (A.2) and (A.4) imply that there is a constant $C = C(\theta, \delta)$ such that

$$f(x) \leq C|x|^{-\alpha(\theta)-\delta}$$

and

$$\left| \frac{\partial}{\partial x} f(x) \right| \leq C|x|^{-\alpha(\theta)-1-\delta}$$

for x bounded away from $\pm 2\pi$, say $|x| \leq 2\pi - 1$. (Since f is periodic, f need not be continuous at 2π.) By the mean value theorem

$$\begin{aligned}
\int_{\pi}^{2\pi/k} \left| f(x) - f\left(x + \frac{\pi}{k}\right) \right| dx &\leq C \frac{\pi}{k} \int_{-\pi}^{-2\pi/k} \left| x + \frac{\pi}{k} \right|^{-\alpha(\theta)-1-\delta} dx \\
&= C \frac{\pi}{k} \int_{-\pi+\pi/k}^{-\pi/k} |x|^{-\alpha(\theta)-1-\delta} dx = O(k^{\alpha(\theta)-1+\delta})
\end{aligned}$$

as $k \to \infty$. A similar argument shows that

$$\int_{\pi/k}^{\pi} \left| f(x) - f\left(x + \frac{\pi}{k}\right) \right| dx = O(k^{\alpha(\theta)-1+\delta}).$$

We also have

$$\int_{-2\pi/k}^{\pi/k} \left| f(x) - f\left(x + \frac{\pi}{k}\right) \right| dx \leq \int_{-2\pi/k}^{\pi/k} f(x)\, dx + \int_{-2\pi/k}^{\pi/k} f\left(x + \frac{\pi}{k}\right) dx$$

$$\leq C \int_{-2\pi/k}^{\pi/k} |x|^{-\alpha(\theta)-\delta}\, dx$$

$$+ C \int_{-2\pi/k}^{\pi/k} \left| x + \frac{\pi}{k} \right|^{-\alpha(\theta)-\delta}\, dx$$

$$= 2C \int_{-2\pi/k}^{\pi/k} |x|^{-\alpha(\theta)-\delta}\, dx = O(k^{\alpha(\theta)-1+\delta}).$$

This completes the proof of Lemma 4. ∎

Lemma 5 If conditions (A.3), (A.5), and (A.6) hold, then for every $\delta > 0$ and every $1 \leq j \leq p$,

$$\int_{-\pi}^{\pi} e^{ikx} \left[\frac{\partial}{\partial\theta_j} f^{-1}(x, \theta) \right] dx = 0(k^{-\alpha(\theta)-1+\delta}) \quad \text{as } k \to \infty.$$

Proof Since $\partial/\partial\theta_j f^{-1}(x, \theta)$ is symmetric, integration by parts yields

$$e_k(\theta) := \int_{-\pi}^{\pi} e^{ikx} \frac{\partial}{\partial\theta_j} f^{-1}(x, \theta)\, dx$$

$$= -\frac{1}{ik} \int_{-\pi}^{\pi} e^{ikx} \frac{\partial^2}{\partial x \partial\theta_j} f^{-1}(x, \theta)\, dx.$$

The argument in Lemma 4 can now be applied since

$$\frac{\partial^2}{\partial x\, \partial\theta_j} f^{-1}(x, \theta) = O(|x|^{\alpha(\theta)-1-\delta}) \quad \text{with } 0 < -(\alpha(\theta) - 1) < 1.$$

We thus get

$$e_k(\theta) = \frac{1}{k} O(k^{-(\alpha(\theta)-1)-1+\delta})$$

$$= O(k^{-\alpha(\theta)-1+\delta}) \quad \text{as } k \to \infty. \blacksquare$$

The proof of Theorem 2 uses the following result which is a consequence of Theorem 4 of Fox and Taqqu (1983).

Proposition 1 Let $f(x)$ and $g(x)$ be symmetric real-valued functions whose sets of discontinuities have Lebesgue measure 0. Suppose that there exist $\alpha < 1$ and $\beta < 1$ such that $\alpha + \beta < \frac{1}{2}$ and such that for each $\delta > 0$

$$f(x) = O(|x|^{-\alpha-\delta}) \quad \text{as } x \to 0$$

and

$$g(x) = O(|x|^{-\beta-\delta}) \quad \text{as } x \to 0.$$

If $\{X_j\}$ is a stationary, mean 0, Gaussian sequence with the spectral density $f(x)$, then

$$\sqrt{N} \left\{ \int_{-\pi}^{\pi} I_N(x) g(x) \, dx - E \int_{-\pi}^{\pi} I_N(x) g(x) \, dx \right\}$$

tends in distribution to a normal random variable with mean 0 and variance

$$4\pi \int_{-\pi}^{\pi} [f(x)g(x)]^2 \, dx.$$

Remark Proposition 1 is established by showing that the cumulants of order greater than two tend to zero as $N \to \infty$. Non-Gaussian limits may occur if the $\{X_j\}$ are non-Gaussian (see Fox and Taqqu 1985).

Proof of Theorem 2 Let α_0 denote $\alpha(\theta_0)$. Define $m_N = E\partial/\partial\theta \, \sigma_N^2(\theta_0)$ and let $m_{N,j} = E\partial/\partial\theta_j \, \sigma_N^2(\theta_0)$ be the jth coordinate of m_N. Let $c_1, \ldots, c_p$ be fixed constants and consider the random variable

$$Y_N = \sum_{j=1}^{p} c_j \left[\frac{\partial}{\partial\theta_j} \sigma_N^2(\theta_0) - m_{N,j} \right]$$

$$= \frac{1}{2\pi} \int_{-\pi}^{\pi} \left[\sum_{j=1}^{p} c_j \frac{\partial}{\partial\theta_j} f^{-1}(x, \theta_0) \right] I_N(x) \, dx - \sum_{j=1}^{p} c_j m_{N,j}.$$

Under condition (A.3) the function in brackets in $O(|x|^{\alpha_0-\delta})$ as $x \to 0$ for every $\delta > 0$. Apply Proposition 1 with

$$\alpha = \alpha_0,$$

$$\beta = -\alpha_0,$$

$$f(x) = \sigma_0^2 f(x, \theta_0),$$

and

$$g(x) = \sum_{j=1}^{p} c_j \frac{\partial}{\partial\theta_j} f^{-1}(x, \theta_0),$$

and conclude that $\sqrt{N}Y_N$ tends in distribution as $N \rightarrow \infty$ to a normal random variable with mean 0 and variance s^2 given by

$$s^2 = \frac{\sigma_0^4}{\pi} \int_{-\pi}^{\pi} f^2(x, \theta_0) \left[\sum_{j=1}^{p} c_j \frac{\partial}{\partial \theta_j} f^{-1}(x, \theta_0) \right]^2 dx$$

$$= \sum_{j=1}^{p} \sum_{k=1}^{p} c_j c_k \frac{\sigma_0^4}{\pi} \int_{-\pi}^{\pi} f^2(x, \theta_0) \left(\frac{\partial}{\partial \theta_j} f^{-1}(x, \theta_0) \right) \left(\frac{\partial}{\partial \theta_k} f^{-1}(x, \theta_0) \right) dx.$$

An application of Lemma 3 yields

$$s^2 = \sum_{j=1}^{p} \sum_{k=1}^{p} c_j c_k \frac{\sigma_0^4}{\pi} w_{ij}(\theta_0).$$

Since $c_1, \ldots, c_p$ were arbitrary, we have shown that $\sqrt{N}(\partial/\partial \theta \, \sigma_N^2(\theta_0) - m_N)$ tends in distribution to a normal random vector with mean 0 and covariance matrix $\sigma_0^4/\pi \, W(\theta_0)$. Therefore, Theorem 2 will follow from Lemma 2 if we show that under the conditions of Theorem 2

$$\lim_{N \rightarrow \infty} \sqrt{N} m_{N,l} = 0, \quad l = 1, \ldots, p.$$

To prove this, define

$$\mu_{N,l} = E \left\{ \frac{1}{2\pi} \int_{-\pi}^{\pi} \frac{\partial}{\partial \theta_l} f^{-1}(x, \theta_0) \tilde{I}_N(x) \, dx \right\},$$

where

$$\tilde{I}_N(x) = \frac{\left| \sum_{j=1}^{N} e^{ijx}(X_j - \mu) \right|^2}{2\pi N}.$$

It follows from lemma 8.1 of Fox and Taqqu (1983) that $\lim_{N \rightarrow \infty} \sqrt{N}(m_{N,l} - \mu_{N,l}) = 0$. Thus it suffices to show

$$\lim_{N \rightarrow \infty} \sqrt{N} \mu_{N,l} = 0, \quad l = 1, \ldots, p. \tag{4.12}$$

We have

$$\mu_{N,l} = \frac{1}{(2\pi)^2 N} \sum_{j=1}^{N} \sum_{k=1}^{N} e_{j-k}(\theta_0)(X_j - \mu)(X_k - \mu),$$

where

$$e_k = e_k(\theta) = \int_{-\pi}^{\pi} e^{ikx} \frac{\partial}{\partial \theta_l} f^{-1}(x, \theta) \, dx.$$

Set also $r_k = r_k(\theta_0)$. Then

$$\mu_{N,l} = \frac{\sigma_0^2}{(2\pi)^2 N} \sum_{j=1}^{N} \sum_{k=1}^{N} e_{j-k} r_{j-k}.$$

Note that $e_k r_k$ is the kth Fourier coefficient of the convolution

$$h(x) = \int_{-\pi}^{\pi} \left(\frac{\partial}{\partial \theta_l} f^{-1}(y, \theta_0) \right) f(y - x, \theta_0) \, \mathrm{d}y.$$

Note also that

$$h(0) = \int_{-\pi}^{\pi} \left(\frac{\partial}{\partial \theta_l} f^{-1}(y, \theta_0) \right) f(y, \theta_0) \, \mathrm{d}y$$

$$= -\int_{-\pi}^{\pi} \frac{\partial}{\partial \theta_l} \log f(y, \theta_0) \, \mathrm{d}y = 0,$$

where we have used (A.1).

To prove (4.12), observe that by Lemmas 4 and 5, there is a $0 < \delta < \frac{1}{4}$ such that as $k \to \infty$,

$$e_k r_k = O(k^{-2+2\delta}). \tag{4.13}$$

Observe also that $e_k r_k$ is the kth Fourier coefficient of h, so that

$$\sum_{k=-\infty}^{\infty} e_k r_k = h(0) = 0. \tag{4.14}$$

We have

$$\frac{(2\pi)^2}{\sigma_0^2} \sqrt{N} \mu_{N,l} = \sqrt{N} \frac{\sum_{j=1}^{N} \sum_{j=1}^{N} e_{j-k} r_{j-k}}{N}$$

$$= \sqrt{N} \sum_{|k|<N} e_k r_k \left(1 - \frac{k}{N} \right)$$

$$= N^{1/2} \sum_{|k|<N} e_k r_k - N^{-1/2} \sum_{|k|<N} k e_k r_k.$$

Because of (4.14), the first term equals $-N^{1/2} \sum_{|k|\geq N} e_k r_k$, which is $O(N^{-1/2+2\delta})$ by (4.13). The second term is also $O(N^{-1/2+2\delta})$ by (4.13). These terms tend to zero as $N \to \infty$, establishing (4.12). This completes the proof of Theorem 2. ∎

Remark Conditions (A.4), (A.5), and (A.6) were used in the proofs of Lemmas 4 and 5 to show that $r_k = O(k^{\alpha(\theta)-1+\delta})$ and $e_k = O(k^{-\alpha(\theta)-1+\delta})$ as $k \to \infty$. In specific cases, a Tauberian theorem may be applied to $f(x, \theta)$ and $\partial/\partial \theta_j f^{-1}(x, \theta)$ to yield such estimates on r_k and e_k.

4.4 Proof of Theorem 3

In order to verify that conditions A are satisfied for fractional Gaussian noise and fractional ARMA's, it is convenient to check the following conditions which are stronger than conditions A.

Conditions B We say that $f(x, \theta)$ satisfies conditions (B.1)–(B.4) if there is a continuous function $0 < \alpha(\theta) < 1$ and constants $C(\delta)$, $C_0(\delta)$ such that for each $\delta > 0$:

(B.1) $f(x, \theta)$ is continuous at all (x, θ), $x \neq 0$ and

$$f(x, \theta) \geq C_0(\delta)|x|^{-\alpha(\theta)+\delta}.$$

(B.2) $f(x, \theta) \leq C(\delta)|x|^{-\alpha(\theta)-\delta}.$

(B.3) $\partial/\partial\theta_j \, f(x, \theta)$ and $\partial^2/\partial\theta_j \, \partial\theta_k f(x, \theta)$ are continuous at all (x, θ), $x \neq 0$,

$$\left| \frac{\partial}{\partial\theta_j} f(x, \theta) \right| \leq C(\delta)|x|^{-\alpha(\theta)-\delta}, \quad 1 \leq j \leq p,$$

and

$$\left| \frac{\partial^2}{\partial\theta_j \, \partial\theta_k} f(x, \theta) \right| \leq C(\delta)|x|^{-\alpha(\theta)-\delta}, \quad 1 \leq j, \; k \leq p.$$

(B.4) $\partial/\partial x f(x, \theta)$, $\partial^2/\partial x \, \partial\theta_j f(x, \theta)$, and $\partial^3/\partial x^2 \, \partial\theta_j f(x, \theta)$ are continuous at all (x, θ), $x \neq 0$,

$$\left| \frac{\partial}{\partial x} f(x, \theta) \right| \leq C(\delta)|x|^{-\alpha(\theta)-1-\delta},$$

$$\left| \frac{\partial^2}{\partial x \, \partial\theta_j} f(x, \theta) \right| \leq C(\delta)|x|^{-\alpha(\theta)-1-\delta}, \quad 1 \leq j \leq p,$$

and

$$\left| \frac{\partial^3}{\partial x^2 \, \partial\theta_j} f(x, \theta) \right| \leq C(\delta)|x|^{-\alpha(\theta)-2-\delta}, \quad 1 \leq j \leq p.$$

Note that conditions B do not involve the function $f^{-1}(x, \theta)$. The constants $C(\delta)$ and $C_0(\delta)$ which appear in conditions B are required to be independent of θ.

Lemma 6 If f satisfies conditions (B.1)–(B.4), then f satisfies conditions (A.1)–(A.6).

Proof Suppose that f satisfies conditions B. It is easily seen that conditions (A.2)–(A.6) are satisfied. For example

$$\left| \frac{\partial}{\partial\theta_j} f^{-1}(x, \theta) \right| = \left| \frac{\partial}{\partial\theta_j} f(x, \theta) \right| \bigg/ f^2(x, \theta) \leq \frac{C(\delta)}{C_0^2(\delta)} |x|^{\alpha(\theta)-3\delta}.$$

This implies that $\partial/\partial\theta_j f^{-1}(x, \theta)$ is continuous and that $\partial/\partial\theta_j f^{-1}(x, \theta) = O(|x|^{\alpha(\theta)-3\delta})$ as $x \to 0$.

We check that condition A.1 is satisfied. Let v_j be the jth unit vector in R^p, that is, the vector with jth component equals 1 and all other components equal 0. Then we have

$$\frac{\int_{-\pi}^{\pi} \log f(x, \theta + \varepsilon v_j) \, dx - \int_{-\pi}^{\pi} \log f(x, \theta) \, dx}{\varepsilon}$$

$$= \int_{-\pi}^{\pi} \frac{\log f(x, \theta + \varepsilon v_j) - \log f(x, \theta)}{\varepsilon} \, dx.$$

By the mean value theorem this integrand is majorized for each $x \neq 0$ by

$$\left| \frac{\partial}{\partial \theta_j} \log f(x, \theta^*(x)) \right| = \left| \frac{\partial}{\partial \theta_j} f(x, \theta^*(x)) \right| \Big/ f(x, \theta^*(x)),$$

where $|\theta^*(x) - \theta| < |\varepsilon|$. Under conditions B.1 and B.3 this quotient is at most $C(\delta)/C_0(\delta)|x|^{\alpha_m - \alpha_M - 2\delta}$, where

$$\alpha_m = \min_{\theta \in E} \alpha(\theta)$$

and

$$\alpha_M = \max_{\theta \in E} \alpha(\theta).$$

Since $\alpha_m - \alpha_M > -1$ we can choose δ so that $\alpha_m - \alpha_M - 2\delta > -1$, and thus

$$\int_{-\pi}^{\pi} |x|^{\alpha_m - \alpha_M - 2\delta} \, dx < \infty.$$

Hence the dominated convergence theorem implies that $\int_{-\pi}^{\pi} \log f(x, \theta) \, dx$ can be differentiated under the integral sign. A similar argument shows that a second differentiation under the integral sign can also be performed. ∎

Proof of Theorem 3 For fractional Gaussian noise, $f(x, H) = CF(H) f_0(x, H)$, where $f_0(x, H)$ is defined in (4.7) and $CF(H)$ is defined in (4.10) as

$$CF(H) = \exp \left\{ -\frac{1}{2\pi} \int_{-\pi}^{\pi} \log f_0(x, H) \, dx \right\}.$$

According to Lemma 6 it suffices to show that $f(x, H)$ satisfies conditions B.1–B.4 with $\alpha(H) = 2H - 1$. We will show that $f_0(x, H)$ satisfies conditions B with $\alpha(H) = 2H - 1$. Then Lemma 6 implies that $CF(H)$ is twice continuously differentiable, which means that $f(x, H)$ satisfied conditions B.

Note that

$$f_0(x, H) = (1 - \cos x) \left[|x|^{-1-2H} + f_1(x, H) \right],$$

where

$$f_1(x, H) = \sum_{k \neq 0} |x + 2k\pi|^{-1-2H}.$$

Since $1 - \cos x \sim |x|^2/2$ as $x \to 0$, conditions B will hold for $f_0(x, H)$ if $f_1(x, H)$ is three times continuously differentiable at all (x, H). A standard theorem on differentiation of series (e.g. Rudin 1964: th. 7.17), shows that this is indeed the case. Thus conditions B are satisfied for fractional Gaussian noise.

It is even simpler to verify conditions B for a fractional ARMA process because the divergent term is already factored out in that case. ■

References

Basawa, I. V. and Prakasa Rao, B. L. S. (1980), *Statistical Inference for Stochastic Processes*. Academic, London.

—— and Scott, D. J. (1983), 'Asymptotic optimal inference for nonergodic models', *Springer Lecture Notes in Statistics* 17. Springer, New York.

Beran, J. and Kuensch, H. (1985), 'Location estimators for processes with long range dependence'. Preprint.

Bleher, P. M. (1981), 'Inversion of Toeplitz matrices', *Transactions of the Moscow Mathematical Society*, (Issue 2), 201–29.

Dunsmuir, W. and Hannan, E. J. (1976), 'Vector linear time series models', *Advances in Applied Probability*, 8, 339–64.

Feller, W. (1971). *An Introduction to Probability Theory and Its Applications*, 2. Wiley, New York.

Fox, R. and Taqqu, M. S. (1983), 'Central limit theorems for quadratic forms in random variables having long-range dependence'. Technical Report 590, School of Operations Research and Industrial Engineering, Cornell University.

—— and Taqqu, M. S. (1985), 'Noncentral limit theorems for quadratic forms in random variables having long-range dependence', *Annals of Probability*, 13, 428–46.

Geweke, J. and Porter-Hudak, S. (1983), 'The estimation and application of long memory time series models', *Journal of Time Series Analysis*, 4, 221–38.

Graf, H.-P. (1980), Long-range correlations and estimation of the self-similarity parameter. Ph.D. dissertation, Swiss Federal Institute of Technology.

Granger, C. W. J. and Joyeux, R. (1980), 'An introduction to long memory time series models and fractional differencing', *Journal of Time Series Analysis*, 1, 15–30.

Hannan, E. J. (1970), *Multiple Time Series*. Wiley, New York.

—— (1973), 'The asymptotic theory of linear time series models', *Journal of Applied Probability*, 10, 130–45.

Hipel, I. and McLeod, A. (1978), 'Preservation of the rescaled adjusted range 1', *Water Resources Research*, 14, 491–508.

Hosking, J. R. M. (1981). 'Fractional differencing', *Biometrika*, 68, 165–76.

Mandelbrot, B. B. (1975), 'Limit theorems for the self-normalized range', *Zeitschrift für Wahrscheinlichkeitstheorie and Verwandte Gebiete*, 31, 271–85.

—— and Taqqu, M. S. (1979), 'Robust R/S analysis of long run serial correlation', Proceedings of the 42nd session of the International Statistical Institute, Manila. *Bulletin of the ISI* 48 (Book 2), 69–104.

——— and Van Ness, J. W. (1968), 'Fractional Brownian motions, fractional noises and applications', *SIAM Reviews*, 10, 422–37.

Mohr, D. (1981), Modeling data as a fractional Gaussian noise. Ph.D. dissertation, Princeton University.

Rudin, W. (1964), *Principles of Mathematical Analysis*. McGraw-Hill, New York.

Sinai, Ya. G. (1976). 'Self-similar probability distributions', *Theory of Probability and its Applications*, 21, 64–80.

Sweeting, T. J. (1980), 'Uniform asymptotic normality of the maximum likelihood estimator', *Annals of Statistics*, 8, 1375–81.

Taqqu, M. S. (1975), 'Weak convergence to fractional Brownian motion and to the Rosenblatt process', *Zeitschrift für Wahrscheinlichkeitstheorie and Verwandte Gebiete*, 31, 287–302.

Todini, E. and O'Connell, P. E. (1979), *Hydrological Simulation of Lake Nasser* 1. Institute of Hydrology, Wallingford, UK.

Walker, A. M. (1964), 'Asymptotic properties of least squares estimator of parameters of the spectrum of a stationary non-deterministic time series', *Journal of the Australian Mathematical Society*, 4, 363–84.

Whittle, P. (1951), *Hypothesis Testing in Time Series Analysis*. Hafner, New York.

5
Long-term Memory in Stock Market Prices

ANDREW W. LO

5.1 Introduction

That economic time series can exhibit long-range dependence has been a hypothesis of many early theories of the trade and business cycles. Such theories were often motivated by the distinct but nonperiodic cyclical patterns that typified plots of economic aggregates over time, cycles of many periods, some that seem nearly as long as the entire span of the sample. In the frequency domain such time series are said to have power at low frequencies. So common was this particular feature of the data that Granger (1966) considered it the 'typical spectral shape of an economic variable'. It has also been called the 'Joseph effect' by Mandelbrot and Wallis (1968), a playful but not inappropriate biblical reference to the Old Testament prophet who foretold of the seven years of plenty followed by the seven years of famine that Egypt was to experience. Indeed, Nature's predilection towards long-range dependence has been well-documented in hydrology, meteorology, and geophysics, and to the extent that the ultimate sources of uncertainty in economics are natural phenomena like rainfall or earthquakes, we might also expect to find long-term memory in economic time series.[1]

The presence of long-memory components in asset returns has important implications for many of the paradigms used in modern financial economics. For example, optimal consumption/savings and portfolio decisions may become extremely sensitive to the investment horizon if stock returns were long-range dependent. Problems also arise in the pricing of derivative securities (such as options and futures) with martingale methods, since the class of continuous time stochastic processes most commonly employed is inconsistent with long-term memory (e.g. see

This work was previously published as Lo, A.W. (1991), 'Long-term memory in stock market prices', *Econometrica* 59, 1279–313. Copyright © the Econometric Society. Reproduced by kind permission.

I am grateful to Joseph Haubrich for stimulating my interest in this area and for many enlightening discussions, and to Lars Hansen for extensive comments and suggestions. I also thank David Aldous, Buzz Brock, John Campbell, Yin-Wong Cheung, John Heaton, Blake LeBaron, Bruce Lehmann, Nancy Lo, Craig MacKinlay, Whitney Newey, Pierre Perron, Jim Poterba, Murad Taqqu, Jean-Luc Vila, Lian Wang, Jeff Wooldridge, two referees, and participants at various seminars for useful comments. Sunghwan Shin provided excellent research assistance. Research support from the Batterymarch Fellowship, the International Financial Service Research Center at M.I.T., the National Science Foundation (Grant Nos. SES-8520054, SES-8821583), the John M. Olin Fellowship at the NBER, the Rodney L. White Fellowship at the Wharton School, and the University of Pennsylvania Research Foundation is gratefully acknowledged.

Maheswaran 1990; Maheswaran and Sims 1990; Sims 1984). Traditional tests of the capital asset pricing model and the arbitrage pricing theory are no longer valid since the usual forms of statistical inference do not apply to time series exhibiting such persistence. And the conclusions of more recent tests of 'efficient' markets hypotheses or stock market rationality also hang precariously on the presence or absence of long-term memory.[2]

Among the first to have considered the possibility and implications of persistent statistical dependence in asset returns was Mandelbrot (1971). Since then, several empirical studies have lent further support to Mandelbrot's findings. For example, Greene and Fielitz (1977) claim to have found long-range dependence in the daily returns of many securities listed on the New York Stock Exchange. More recent investigations have uncovered anomalous behaviour in long-horizon stock returns;[3] alternately attributed to speculative fads and to time-varying conditional expected returns, these long-run swings may be further evidence of the Joseph effect.

In this paper I develop a test for such forms of long-range dependence using a simple generalization of a statistic first proposed by the English hydrologist Harold Edwin Hurst (1951). This statistic, called the 'rescaled range' or 'range over standard deviation' or 'R/S' statistic, has been refined by Mandelbrot (1972, 1975) and others in several important ways (e.g. see Mandelbrot and Taqqu (1979) and Mandelbrot and Wallis (1968, 1969a–1969c)). However, such refinements were not designed to distinguish between short-range and long-range dependence (in a sense to be made precise below), a severe shortcoming in applications of R/S analysis to recent stock returns data since Lo and MacKinlay (1988, 1990) show that such data display substantial short-range dependence. Therefore, to be of current interest, any empirical investigation of long-term memory in stock returns must first account for the presence of higher frequency autocorrelation.

By modifying the rescaled range appropriately, I construct a test statistic that is robust to short-range dependence and derive its limiting distribution under both short-range and long-range dependence. Contrary to the findings of Greene and Fielitz (1977) and others, when this statistic is applied to daily and monthly stock return indexes over several different sample periods and sub-periods, there is no evidence of long-range dependence once the effects of short-range dependence are accounted for. Monte Carlo experiments indicate that the modified R/S test has reasonable power against at least two particular models of long-range dependence, suggesting that the time series behaviour of stock returns may be adequately captured by more conventional models of short-range dependence.

The particular notions of short-term and long-term memory are defined in Section 5.2 and some illustrative examples are given. The test statistic is presented in Section 5.3 and its limiting distributions under the null and alternative hypotheses are derived via functional central limit theory. In Section 5.4 the empirical results are reported, and Monte Carlo simulations that illustrate the size and power of the test in finite samples are presented in Section 5.5. I conclude in Section 5.6.

5.2 Long-range Versus Short-range Dependence

To develop a method for detecting long-term memory, the distinction between long-range and short-range statistical dependence must be made precise. One of the most widely used concepts of short-range dependence is the notion of 'strong-mixing' due to Rosenblatt (1956), a measure of the decline in statistical dependence between events separated by successively longer spans of time. Heuristically, a time series is strong-mixing if the maximal dependence between events at any two dates becomes trivially small as the time span between those two dates increases. By controlling the rate at which the dependence between past and future events declines, it is possible to extend the usual laws of large numbers and central limit theorems to dependent sequences of random variables. I adopt strong-mixing as an operational definition of short-range dependence in the null hypothesis of Section 5.2.1. In Section 5.2.2, I give examples of alternatives to short-range dependence such as the class of fractionally-differenced processes proposed by Granger and Joyeux (1980), Hosking (1981), and Mandelbrot and Van Ness (1968).

5.2.1 The null hypothesis

Let P_t denote the price of an asset at time t and define $X_t \equiv \log P_t - \log P_{t-1}$ to be the continuously compounded single-period return of that asset from $t-1$ to t. With little loss in generality, let any dividend payments be reinvested in the asset so that X_t is indeed the total return of the asset between $t-1$ and t.[4] It is assumed throughout that

$$X_t = \mu + \varepsilon_t, \tag{5.1}$$

where μ is an arbitrary but fixed parameter and ε_t is a zero mean random variable. Let this stochastic process $\{X_t, (\omega)\}$ be defined on the probability space $(\Omega, \mathcal{F}, P)$ and define

$$\alpha(\mathcal{A}, \mathcal{B}) \equiv \sup_{\{A \in \mathcal{A}, B \in \mathcal{B}\}} |P(A \cap B) - P(A)P(B)|, \quad \mathcal{A} \subset \mathcal{F}, \ \mathcal{B} \subset \mathcal{F}. \tag{5.2}$$

The quantity $\alpha(\mathcal{A}, \mathcal{B})$ is a measure of the dependence between the two σ-fields $\mathcal{A}$ and $\mathcal{B}$ in $\mathcal{F}$. Denote by $\mathcal{B}_s^t$ the Borel σ-field generated by $\{X_s(\omega), \dots, X_t(\omega)\}$, that is, $\mathcal{B}_s^t \equiv \sigma(X_s(\omega), \dots, X_t(\omega)) \subset \mathcal{F}$. Define the coefficients α_k as

$$\alpha_k \equiv \sup_j \alpha\left(\mathcal{B}_{-\infty}^j, \mathcal{B}_{j+k}^\infty\right). \tag{5.3}$$

Then $\{X_t(\omega)\}$ is said to be *strong-mixing* if $\lim_{k \to \infty} \alpha_k = 0$.[5] Such mixing conditions have been used extensively in the recent literature to relax the assumptions that ensure the consistency and asymptotic normality of various econometric estimators (e.g. see Chan and Wei 1988; Phillips 1987; White 1980; White and Domowitz 1984). As Phillips (1987) observes, these conditions are satisfied by a great many stochastic processes, including all Gaussian finite-order stationary ARMA models. Moreover, the inclusion of a moment condition also allows for heterogeneously distributed sequences, an especially important extension in view of the apparent instabilities of financial time series.

In addition to strong mixing, several other conditions are required as part of the null hypothesis in order to develop a sampling theory for the test statistic proposed in Section 5.3. In particular, the null hypothesis is composed of the following four conditions on ε_t:

(A1) $E[\varepsilon_t] = 0$ for all t;

(A2) $\sup_t E\left[|\varepsilon_t|^\beta\right] < \infty$ for some $\beta > 2$;

(A3) $0 < \sigma^2 = \lim_{n\to\infty} E\left[\frac{1}{n}\left(\sum_{j=1}^n \varepsilon_j\right)^2\right] < \infty$;

(A4) $\{\varepsilon_t\}$ is strong-mixing with mixing coefficients α_k that satisfy $\sum_{k=1}^\infty \alpha_k^{1-(2/\beta)} < \infty$.

Condition (A1) is standard. Conditions (A2) through (A4) are restrictions on the maximal degree of dependence and heterogeneity allowable while still permitting some form of the law of large numbers and the (functional) central limit theorem to obtain. Although (A2) rules out infinite variance marginal distributions of ε_t such as those in the stable family with characteristic exponent less than 2, the disturbances may still exhibit unconditional leptokurtosis via time-varying conditional moments (e.g. conditional heteroscedasticity). Moreover, since there is a trade-off between (A2) and (A4), the uniform bound on the moments can be relaxed if the mixing coefficients decline faster than (A4) requires.[6] For example, if ε_t is required to have finite absolute moments of all orders (corresponding to $\beta \to \infty$), then α_k must decline faster than $1/k$. However, if ε_t is restricted to have finite moments only up to order 4, then α_k must decline faster than $1/k^2$. These conditions are discussed at greater length by Phillips (1987).

Of course, it is too much to hope that *all* forms of short-memory processes are captured by (A1)–(A4). For example, if ε_t were the first difference of a stationary process, its spectral density at frequency zero vanishes, violating (A3). Yet such a process certainly need not be long-range dependent. A more subtle example is given by Ibragimov and Rozanov (1978)—a stationary Gaussian process with spectral density function

$$f(\omega) = \exp\left\{\sum_{k=1}^\infty \frac{\cos k\omega}{k \log k + 1}\right\}, \tag{5.4}$$

which is strong-mixing but has unbounded spectral density at the origin. The stochastic process with $1/f(\omega)$ for its spectral density is also strong-mixing, but $1/f(\omega)$ vanishes at the origin. Although neither process is long-range dependent, they both violate (A3). Unfortunately, a general characterization of the implications of such processes for the behaviour of the test statistic proposed in Section 5.3 is currently unavailable. Therefore, a rejection of the null hypothesis does not necessarily imply that long-range dependence is present but merely that, if the rejection is not a type I error, the stochastic process does not satisfy all four conditions simultaneously. Whether or not the composite null $(A1)$–$(A4)$ is a useful one must therefore depend on the particular application at hand.

In particular, although mixing conditions have been widely used in the recent literature, several other sets of assumptions might have served equally well as our short-range dependent null hypothesis. For example, if $\{\varepsilon_t\}$ is assumed to be stationary and ergodic, the moment condition $(A2)$ can be relaxed and more temporal dependence than $(A4)$ is allowable (see Hall and Heyde 1980). Whether or not the assumption of stationarity is a restrictive one for financial time series is still an open question. There is ample evidence of changing variances in stock returns over periods longer than 5 years, but unstable volatilities can be a symptom of *conditional* heteroscedasticity which can manifest itself in stationary time series. Since the empirical evidence regarding changing conditional moments in asset returns is mixed, allowing for nonstationarities in our null hypothesis may still have value. Moreover, $(A1)$–$(A4)$ may be weakened further, allowing for still more temporal dependence and heterogeneity, hence widening the class of processes contained in our null hypothesis.[7]

Note, however, that conditions $(A1)$–$(A4)$ are satisfied by many of the recently proposed stochastic models of persistence, such as those of Campbell and Mankiw (1987), Fama and French (1988), and Poterba and Summers (1988). Therefore, since such models of longer-term correlations are contained in our null, the kinds of long-range dependence that $(A1)$–$(A4)$ were designed to exclude are quite different. Although the distinction between dependence in the short run and the long run may appear to be a matter of degree, strongly dependent processes behave so differently from weakly dependent time series that the dichotomy proposed in our null seems most natural. For example, the spectral densities at frequency zero of strongly dependent processes are either unbounded or zero whereas they are nonzero and finite for processes in our null. The partial sums of strongly dependent processes do not converge in distribution at the same rate as weakly dependent series. And graphically, their behaviour is marked by cyclical patterns of all kinds, some that are virtually indistinguishable from trends.

5.2.2 Long-range dependent alternatives

In contrast to the short-term memory of 'weakly dependent' (i.e., mixing) processes, natural phenomena often display long-term memory in the form of nonperiodic cycles. This has led several authors to develop stochastic models that exhibit dependence even over very long time spans, such as the fractionally-integrated time series models of Granger (1980), Granger and Joyeux (1980), Hosking (1981), and Mandelbrot and Van Ness (1968). These stochastic processes are not strong-mixing, and have autocorrelation functions that decay at much slower rates than those of weakly dependent processes. For example, let X_t satisfy the following difference equation:

$$(1 - L)^d X_t = \varepsilon_t, \quad \varepsilon_t \sim WN(0, \sigma_\varepsilon^2), \tag{5.5}$$

where L is the lag operator and ε_t is white noise. Granger and Joyeux (1980) and Hosking (1981) show that when the quantity $(1 - L)^d$ is extended to noninteger powers of d in the mathematically natural way, the result is a well-defined time series that is said to be 'fractionally-differenced' of order d (or, equivalently, 'fractionally-integrated' of order $-d$). Briefly, this involves expanding the expression $(1 - L)^d$ via the binomial theorem for noninteger powers:

$$(1 - L)^d = \sum_{k=0}^{\infty} (-1)^k \binom{d}{k} L^k, \tag{5.6}$$

$$\binom{d}{k} \equiv \frac{d(d - 1)(d - 2) \cdots (d - k + 1)}{k!},$$

and then applying the expansion to X_t:

$$(1 - L)^d X_t = \sum_{k=0}^{\infty} (-1)^k \binom{d}{k} L^k X_t = \sum_{k=0}^{\infty} A_k X_{t-k} = \varepsilon_t, \tag{5.7}$$

where the autoregressive coefficients A_k are often re-expressed in terms of the gamma function:

$$A_k = (-1)^k \binom{d}{k} = \frac{\Gamma(k - d)}{\Gamma(-d)\Gamma(k + 1)}. \tag{5.8}$$

X_t may also be viewed mechanically as an infinite-order MA process since

$$X_t = (1 - L)^{-d}\varepsilon_t = B(L)\varepsilon_t, \qquad B_k = \frac{\Gamma(k + d)}{\Gamma(d)\Gamma(k + 1)}. \tag{5.9}$$

It is not obvious that such a definition of fractional-differencing might yield a useful stochastic process, but Granger (1980), Granger and Joyeux (1980), and Hosking (1981) show that the characteristics of fractionally-differenced time series are interesting indeed. For example, it may be shown that X_t is stationary and invertible for $d \in (-\frac{1}{2}, \frac{1}{2})$ (see Hosking 1981), and exhibits a unique kind of dependence that is positive or negative depending on whether d is positive or negative, that is, the autocorrelation coefficients of X_t are of the same sign as d. So slowly do the autocorrelations decay that when d is positive their sum diverges to infinity, and collapses to zero when d is negative.[8] To develop a sense of this long-range dependence, compare the autocorrelations of a fractionally-differenced X_t with those of a stationary AR(1) in Table 5.1. Although both the AR(1) and the fractionally-differenced ($d = \frac{1}{3}$) series have first-order autocorrelations of 0.500, at lag 25 the AR(1) autocorrelation is 2.98×10^{-8} whereas the fractionally-differenced series has autocorrelation 0.173, declining only to 0.109 at lag 100.

In fact, the defining characteristic of long-range dependent processes has been taken by many to be this slow decay of the autocovariance function. Therefore, more generally, long-range dependent processes may be defined to be those

Table 5.1. Comparison of autocorrelation functions of fractionally differenced time series $(1 - L)^d X_t = \varepsilon_t$, for $d = \frac{1}{3}, -\frac{1}{3}$ with that of an AR(1)$X_t = \rho X_{t-1} + \varepsilon_t$, $\rho = 0.5$. The variance of ε_t was chosen to yield a unit variance for X_t in all three cases

Lag k	$\rho(k)$ $\left[d = \frac{1}{3}\right]$	$\rho(k)$ $\left[d = -\frac{1}{3}\right]$	$\rho(k)$ [AR(1), $\rho = 0.5$]
1	0.500	−0.250	0.500
2	0.400	−0.071	0.250
3	0.350	−0.036	0.125
4	0.318	−0.022	0.063
5	0.295	−0.015	0.031
10	0.235	−0.005	0.001
25	0.173	−0.001	2.98×10^{-8}
50	0.137	-3.24×10^{-4}	8.88×10^{-16}
100	0.109	-1.02×10^{-4}	7.89×10^{-31}

processes with autocovariance functions γ_k such that

$$\gamma_k \sim \begin{cases} k^\nu L(k) & \text{for } \nu \in (-1, 0) \text{ or} \\ -k^\nu L(k) & \text{for } \nu \in (-2, -1), \end{cases} \quad \text{as } k \to \infty, \qquad (5.10)$$

where $L(k)$ is any slowly varying function at infinity.[9] This is the definition I shall adopt in the analysis to follow. As an example, the autocovariance function of the fractionally-differenced process (5.5) is

$$\gamma_k = \frac{\sigma_\varepsilon^2 \Gamma(1 - 2d)\Gamma(k + d)}{\Gamma(d)\Gamma(1 - d)\Gamma(k + 1 - d)} \sim ck^{2d-1} \quad \text{as } k \to \infty, \qquad (5.11)$$

where $d \in (-\frac{1}{2}, \frac{1}{2})$ and c is some constant. Depending on whether d is negative or positive, the spectral density of (5.5) at frequency zero, given by

$$f(\lambda) \cong (1 - e^{-i\lambda})^{-d}(1 - e^{i\lambda})^{-d}\sigma_\varepsilon^2 \sim \sigma_\varepsilon^2 \lambda^{-2d} \quad \text{as } \lambda \to 0, \qquad (5.12)$$

will either be zero or infinite; thus such processes violate condition (A3).[10] Furthermore, the results of Helson and Sarason (1967) show that these processes are not strong-mixing; hence they also violate condition (A4) of our null hypothesis.[11]

5.3 The Rescaled Range Statistic

To detect long-range or 'strong' dependence, Mandelbrot has suggested using the range over standard deviation or R/S statistic, also called the 'rescaled range', which was developed by Hurst (1951) in his studies of river discharges. The R/S statistic is the range of partial sums of deviations of a time series from its mean, rescaled by its standard deviation. Specifically, consider a sample of returns $X_1, X_2, \ldots, X_n$ and let $\overline{X}_n$ denote the sample mean $(1/n) \sum_j X_j$. Then the classical rescaled range statistic, denoted by $\tilde{Q}_n$, is defined as

$$\tilde{Q}_n \equiv \frac{1}{S_n} \left[\underset{1 \leqslant k \leqslant n}{\mathrm{Max}} \sum_{j=1}^{k} (X_j - \overline{X}_n) - \underset{1 \leqslant k \leqslant n}{\mathrm{Min}} \sum_{j=1}^{k} (X_j - \overline{X}_n) \right], \qquad (5.13)$$

where S_n is the usual (maximum likelihood) standard deviation estimator:

$$S_n \equiv \left[\frac{1}{n} \sum_j (X_j - \overline{X}_n)^2 \right]^{1/2}. \qquad (5.14)$$

The first term in brackets in (5.13) is the maximum (over k) of the partial sums of the first k deviations of X_j from the sample mean. Since the sum of all n deviations of the X_j's from their mean is zero, this maximum is always nonnegative. The second term in (5.13) is the minimum (over k) of this same sequence of partial sums; hence it is always nonpositive. The difference of the two quantities, called the 'range' for obvious reasons, is therefore always nonnegative; hence $\tilde{Q}_n \geqslant 0$.[12]

In several seminal papers Mandelbrot, Taqqu, and Wallis demonstrate the superiority of R/S analysis to more conventional methods of determining long-range dependence, such as analyzing autocorrelations, variance ratios, and spectral decompositions. For example, Mandelbrot and Wallis (1969a) show by Monte Carlo simulation that the R/S statistic can detect long-range dependence in highly non-Gaussian time series with large skewness and kurtosis. In fact, Mandelbrot (1972, 1975) reports the almost-sure convergence of the R/S statistic for stochastic processes with infinite variances, a distinct advantage over autocorrelations and variance ratios which need not be well-defined for such processes. Further aspects of the R/S statistic's robustness are derived in Mandelbrot and Taqqu (1979). Mandelbrot (1972) also argues that, unlike spectral analysis which detects periodic cycles, R/S analysis can detect *nonperiodic* cycles, cycles with periods equal to or greater than the sample period.

Although these claims may all be contested to some degree, it is a well-established fact that long-range dependence can indeed be detected by the 'classical' R/S statistic. However, perhaps the most important shortcoming of the rescaled range is its sensitivity to short-range dependence, implying that any incompatibility between the data and the predicted behaviour of the R/S statistic under the null hypothesis need not come from long-term memory, but may merely be a symptom of short-term memory.

To see this, first observe that under a simple i.i.d. null hypothesis, it is well-known (and is a special case of Theorem 1 below) that as n increases without bound, the rescaled range converges in distribution to a well-defined random variable V when properly normalized, that is,

$$\frac{1}{\sqrt{n}} \tilde{Q}_n \Rightarrow V, \qquad (5.15)$$

where '$\Rightarrow$' denotes weak convergence and V is the range of a Brownian bridge on the unit interval.[13]

Now suppose, instead, that $\{X_j\}$ were short-range dependent—for example, let X_j be a stationary AR(1):[14]

$$\varepsilon_t = \rho\varepsilon_{t-1} + \eta_t, \quad \eta_t \sim WN(0, \sigma_\eta^2), \quad |\rho| \in (0, 1). \quad (5.16)$$

Although $\{\varepsilon_t\}$ is short-range dependent, it yields a $\tilde{Q}_n$ that does not satisfy (5.15). In fact, it may readily be shown that for (5.16) the limiting distribution of $\tilde{Q}_n/\sqrt{n}$ is ξV where $\xi \equiv \sqrt{(1 + \rho)/(1 - \rho)}$ (see Proposition 1 below). For some portfolios of common stock, $\hat{\rho}$ is as large as 50 per cent, implying that the mean of $\tilde{Q}_n/\sqrt{n}$ may be biased upward by 73 per cent! Since the mean of V is $\sqrt{\pi/2} \approx 1.25$, the mean of the classical rescaled range would be 2.16 for such an AR(1) process. Using the critical values of V reported in Table 5.2, it is evident that a value of 2.16 would yield a rejection of the null hypothesis at any conventional significance level.

This should come as no surprise since the values in Table 5.2 correspond to the distribution of V, not ξV. Now by taking into account the 'short-term' auto-correlations of the X_j's—by dividing Q_n by ξ for example—convergence to V may be restored. But this requires knowledge of ξ which, in turn, requires knowledge of ρ. Moreover, if X_j follows a short-range dependent process other than an AR(1), the expression for ξ will change, as Proposition 1 below shows. Therefore, correcting for short-range dependence on a case-by-case basis is impractical. Ideally, we would like to correct for short-term memory without taking too strong a position on what form it takes. This is precisely what the modified rescaled range of Section 5.3.1 does—its limiting distribution is invariant to many forms of short-range dependence, and yet it is still sensitive to the presence of long-range dependence.

Although aware of the effects of short-range dependence on the rescaled range, Mandelbrot (1972, 1975) did not correct for this bias since his focus was the relation of the R/S statistic's logarithm to the logarithm of the sample size as the sample size increases without bound. For short-range dependent time series such as strong-mixing processes, the ratio $\log \tilde{Q}_n / \log n$ approaches $\frac{1}{2}$ in the limit, but converges to quantities greater or less than $\frac{1}{2}$ according to whether there is positive or negative long-range dependence. The limit of this ratio is often denoted by H and is called the 'Hurst' coefficient. For example, the fractionally-differenced process (5.1) satisfies the simple relation: $H = d + \frac{1}{2}$.

Table 5.2. Fractiles of the distribution $F_V(v)$

$P(V < v)$	0.005	0.025	0.050	0.100	0.200	0.300	0.400	0.500
v	0.721	0.809	0.861	0.927	1.018	1.090	1.157	1.223
$P(V < v)$	0.543	0.600	0.700	0.800	0.900	0.950	0.975	0.995
v	$\sqrt{\pi/2}$	1.294	1.374	1.473	1.620	1.747	1.862	2.098

Mandelbrot and Wallis (1969a) suggest estimating the Hurst coefficient by plotting the logarithm of $\tilde{Q}_n$ against the logarithm of the sample size n. Beyond some large n, the slope of such a plot should settle down to H. However, although $H = \frac{1}{2}$ across general classes of short-range dependent processes, the finite-sample properties of the *estimated* Hurst coefficient are not invariant to the form of short-range dependence. In particular, Davies and Harte (1987) show that even though the Hurst coefficient of a stationary Gaussian AR(1) is precisely $\frac{1}{2}$, the 5 per cent Mandelbrot regression test rejects this null hypothesis 47 per cent of the time for an autoregressive parameter of 0.3. Additional Monte Carlo evidence is reported in Section 5.5.

5.3.1 The modified R/S statistic

To distinguish between long-range and short-range dependence, the R/S statistic must be modified so that its statistical behaviour is invariant over a general class of short memory processes, but deviates for long-memory processes. This is accomplished by the following statistic Q_n:

$$Q_n \equiv \frac{1}{\hat{\sigma}_n(q)} \left[\underset{1 \leqslant k \leqslant n}{\text{Max}} \sum_{j=1}^{k} (X_j - \overline{X}_n) - \underset{1 \leqslant k \leqslant n}{\text{Min}} \sum_{j=1}^{k} (X_j - \overline{X}_n) \right], \qquad (5.17)$$

where

$$\hat{\sigma}_n^2(q) \equiv \frac{1}{n} \sum_{j=1}^{n} (X_j - \overline{X}_n)^2$$

$$+ \frac{2}{n} \sum_{j=1}^{q} \omega_j(q) \left\{ \sum_{i=j+1}^{n} (X_i - \overline{X}_n)(X_{i-j} - \overline{X}_n) \right\} \qquad (5.18)$$

$$= \hat{\sigma}_x^2 + 2 \sum_{j=1}^{q} \omega_j(q) \hat{\gamma}_j, \quad \omega_j(q) \equiv 1 - \frac{j}{q+1}, \quad q < n, \qquad (5.19)$$

and $\hat{\sigma}_x^2$ and $\hat{\gamma}_j$ are the usual sample variance and autocovariance estimators of X.

Q_n differs from $\tilde{Q}_n$ only in its denominator, which is the square root of a consistent estimator of the partial sum's variance. If $\{X_t\}$ is subject to short-range dependence, the variance of the partial sum is not simply the sum of the variances of the individual terms, but also includes the autocovariances. Therefore, the estimator $\hat{\sigma}_n(q)$ involves not only sums of squared deviations of X_j, but also its weighted autocovariances up to lag q. The weights $\omega_j(q)$ are those suggested by Newey and West (1987) and always yield a positive $\hat{\sigma}_n^2(q)$, an estimator of 2π times the (unnormalized) spectral density function of X_t at frequency zero using a Bartlett window. Theorem 4.2 of Phillips (1987) demonstrates the consistency of $\hat{\sigma}_n(q)$ under the following conditions:[15]

($A2'$) $\sup_t E[|\varepsilon_t|^{2\beta}] < \infty$ for some $\beta > 2$.

($A5$) As n increases without bound, q also increases without bound such that
$q \sim o(n^{1/4})$.

By allowing q to increase with (but at a slower rate than) the number of observations n, the denominator of Q_n adjusts appropriately for general forms of short-range dependence. Of course, although the conditions ($A2'$) and ($A5$) ensure the consistency of $\hat{\sigma}^2(q)$, they provide little guidance in selecting a truncation lag q. Monte Carlo studies such as Andrews (1991) and Lo and MacKinlay (1989) have shown that when q becomes large relative to the sample size n, the finite-sample distribution of the estimator can be radically different from its asymptotic limit. However q cannot be chosen too small since the autocovariances beyond lag q may be substantial and should be included in the weighted sum. Therefore, the truncation lag must be chosen with some consideration of the data at hand. Andrews (1991) does provide a data-dependent rule for choosing q; however its minimax optimality is still based on an asymptotic mean-squared error criterion—little is known about how best to pick q in finite samples. Some Monte Carlo evidence is reported in Section 5.5.

Since there are several other consistent estimators of the spectral density function at frequency zero, conditions ($A2'$) and ($A5$) can be replaced with weaker assumptions if conditions ($A1$), ($A3$), and ($A4$) are suitably modified. If, for example, X_t, is m-dependent (so that observations spaced greater than m periods apart are independent), it is well-known that the spectral density at frequency zero may be estimated consistently with a finite number of unweighted autocovariances (e.g. see Hansen (1982: lemma 3.2)) Other weighting functions may be found in Hannan (1970: chapter V.4) and may yield better finite-sample properties for Q_n than the Bartlett window without altering the limiting null distribution derived in the next section.[16]

5.3.2 The asymptotic distribution of Q_n

To derive the limiting distribution of the modified rescaled range Q_n under our null hypothesis, consider the behaviour of the following standardized partial sum:

$$W_n(\tau) \equiv \frac{1}{\sigma \sqrt{n}} S_{[n\tau]}, \quad \tau \in [0, 1], \tag{5.20}$$

where S_k denotes the partial sum $\sum_{j=1}^{k} \varepsilon_j$ and $[n\tau]$ is the greatest integer less than or equal to $n\tau$. The sample paths of $W_n(\tau)$ are elements of the function space $\mathcal{D}[0, 1]$, the space of all real-valued functions on [0,1] that are right-continuous and possess finite left limits. Under certain conditions it may be shown that $W_n(\tau)$ converges weakly to a Brownian motion $W(\tau)$ on the unit interval, and that well-behaved functionals of $W_n(\tau)$ converge weakly to the same functionals of

Brownian motion (see Billingsley 1968 for further details). Armed with these results, the limiting distribution of the modified rescaled range may be derived in three easy steps, summarized in the following theorem.[17]

Theorem 1[18] If $\{\varepsilon_t\}$ satisfies assumptions (A1), (A2$'$), (A3)–(A5), then as n increases without bound:

(a) $\displaystyle \operatorname*{Max}_{1\le k\le n} \frac{1}{\hat{\sigma}_n(q)\sqrt{n}} \sum_{j=1}^{k}(X_j - \overline{X}_n) \quad \Rightarrow \quad \operatorname*{Max}_{0\le\tau\le1} W^{\mathrm{o}}(\tau) \equiv M^{\mathrm{o}},$

(b) $\displaystyle \operatorname*{Min}_{1\le k\le n} \frac{1}{\hat{\sigma}_n(q)\sqrt{n}} \sum_{j=1}^{k}(X_j - \overline{X}_n) \quad \Rightarrow \quad \operatorname*{Min}_{0\le\tau\le1} W^{\mathrm{o}}(\tau) \equiv m^{\mathrm{o}},$

(c) $\displaystyle \frac{1}{\sqrt{n}}Q_n \Rightarrow M^{\mathrm{o}} - m^{\mathrm{o}} \equiv V.$

Parts (a) and (b) of Theorem 1 follow from lemmas A.1 and A.2 of Appendix 5.1, and Theorem 4.2 of Phillips (1987), and show that the maximum and minimum of the partial sum of deviations of X_j from its mean converge respectively to the maximum and minimum of the celebrated Brownian bridge $W^{\mathrm{o}}(\tau)$ on the unit interval, also called 'pinned' or 'tied-down' Brownian motion because $W^{\mathrm{o}}(0) = W^{\mathrm{o}}(1) = 0$. That the limit of the partial sums is a Brownian bridge is not surprising since the summands are deviations from the mean and must therefore sum to zero at $k = n$. Part (c) of the theorem follows immediately from Lemma A.2 and is the key result, allowing us to perform large sample statistical inference once the distribution function for the range of the Brownian bridge is obtained. This distribution function is implicitly contained in Feller (1951), and is given explicitly by Kennedy (1976) and Siddiqui (1976) as[19]

$$F_V(v) = 1 + 2\sum_{k=1}^{\infty}(1 - 4k^2v^2)e^{-2(kv)^2}. \tag{5.21}$$

Critical values for tests of any significance level are easily obtained from this simple expression (5.21) for F_V. The values most commonly used are reported in Table 5.2. The moments of V may also be readily computed from (5.21); a simple calculation shows that $E[V] = \sqrt{\pi/2}$ and $E[V^2] = \pi^2/6$, thus the mean and standard deviation of V are approximately 1.25 and 0.27, respectively. Plots of F_V and f_V are given in Fig. 5.1, along with Gaussian distribution and density functions (with the same mean and variance as V) for comparison. The distribution of V is positively skewed and most of its mass falls between $\frac{3}{4}$ and 2.

5.3.3 The relation between Q_n and $\tilde{Q}_n$

Since Q_n and $\tilde{Q}_n$ differ solely in how the range is normalized, the limiting behaviour of our modified R/S statistic and Mandelbrot's original will only coincide when $\hat{\sigma}_n(q)$ and s_n are asymptotically equivalent. From the definitions of $\hat{\sigma}_n(q)$ and s_n, it is apparent that the two will generally converge in probability to different limits in the presence of autocorrelation. Therefore, under the weakly dependent null

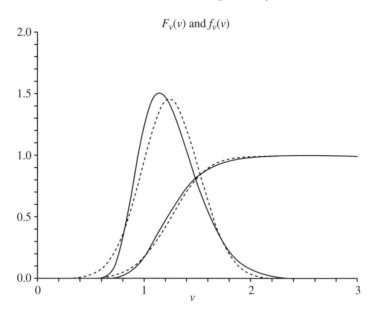

$F_v(v)$ and $f_v(v)$

Fig. 5.1. Distribution and density function of the range V of a Brownian bridge. Dashed curves are the normal distribution and density functions with mean and variance equal to those of V ($\sqrt{\pi/2}$ and $\pi^2/6$, respectively).

hypothesis the statistic $\tilde{Q}_n/\sqrt{n}$ will converge to the range V of a Brownian bridge multiplied by some constant. More formally, we have the almost trivial result:

Proposition 1 If $\lim_{n\to\infty} E\left[\sum_{j=1}^n \varepsilon_j^2/n\right]$ is finite and positive, then under assumptions $(A1)$–$(A4)$, $\tilde{Q}_n/\sqrt{n} \Rightarrow \xi V$ where

$$\xi^2 \equiv \frac{\lim_{n\to\infty} E\left[(1/n)\left(\sum_{j=1}^n \varepsilon_j\right)^2\right]}{\lim_{n\to\infty} E\left[(1/n)\sum_{j=1}^n \varepsilon_j^2\right]}. \tag{5.22}$$

Therefore, normalizing the range by s_n in place of $\hat{\sigma}_n(q)$ changes the limiting distribution of the rescaled range by the multiplicative constant ξ. This result was used above to derive the limiting distribution of $\tilde{Q}_n$ in the AR(1) case, and closed-form expressions for ξ for general stationary ARMA(p, q) processes may readily be obtained using (5.22).

Since it is robust to many forms of heterogeneity and weak dependence, tests based on the modified R/S statistic Q_n cover a broader set of null hypotheses than those using $\tilde{Q}_n$. More to the point, the modified rescaled range is able to distinguish between short-range and long-range dependence—the classical rescaled range cannot. Whereas an extreme value for Q_n indicates the likelihood of long-range dependence, a rejection based on the $\tilde{Q}_n$ statistic is also consistent with short-range dependence in the data. Of course, it is always possible to tabulate the limiting distribution of the classical R/S statistic under a particular model of short-range dependence, but this obviously suffers from the drawback of specificity. The modified rescaled range converges weakly to the range of a Brownian bridge under *general* forms of weak dependence.

Despite its sensitivity to short-range dependence, the classical R/S statistic may still be used to test for independently and identically distributed X_t's. Indeed, the AR(1) example of Section 5.3 and the results of Davies and Harte (1987) suggest that such a test may have considerable power against non-i.i.d. alternatives. However, since there is already a growing consensus among financial economists that stock market prices are not independently and identically distributed, this null hypothesis is of less immediate interest. For example, it is now well-known that aggregate stock market returns exhibit significant serial dependence for short-horizon holding periods and are therefore not independently distributed.

5.3.4 The behaviour of Q_n under long-memory alternatives

To complete the analysis of the modified rescaled range, its behaviour under long-range dependent alternatives remains to be investigated. Although this depends of course on the specific alternative at hand, surprisingly general results are available based on the following result from Taqqu (1975).

Theorem 2 (Taqqu) Let $\{\varepsilon_t\}$ be a zero-mean stationary Gaussian stochastic process such that

$$\sigma_n^2 \equiv \text{Var}[S_n] \sim n^{2H} L(n), \qquad (5.23)$$

where S_n is the partial sum $\sum_{j=1}^{n} \varepsilon_j$, $H \in (0, 1)$, and $L(n)$ is a slowly varying function at infinity. Define the following function on $\mathcal{D}[0, 1]$:

$$W_n(\tau) \equiv \frac{1}{\sigma_n} S_{[n\tau]}, \qquad \tau \in (0, 1).$$

Then $W_n(\tau) \Rightarrow W_H(\tau)$, where $W_H(\tau)$ is a fractional Brownian motion of order H on $[0, 1]$.

Theorem 2 is a functional central limit theorem for strongly dependent processes, and is only a special case of Taqqu's (1975) considerably more general results. In contrast to the usual functional central limit theorem in which properly normalized partial sums converge to a standard Brownian motion, Theorem 2 states that long-range dependent partial sums converge weakly to a *fractional* Brownian motion, first defined by Mandelbrot and Van Ness (1968) as the following stochastic integral:

$$W_H(\tau) \equiv \frac{1}{\Gamma(H + \frac{1}{2})} \int_0^\tau (\tau - x)^{H-(1/2)} \, dW(x). \qquad (5.24)$$

Observe that when $H = \frac{1}{2}$, $W_H(\tau)$ reduces to a standard Brownian motion. In that case, there is no long-range dependence, the variance of the partial sums grows at rate n, and the spectral density at frequency zero is finite and positive. If

$H \in (\frac{1}{2}, 1)$ ($H \in (0, \frac{1}{2})$), there is positive (negative) long-range dependence, the variance grows faster (slower) than n, hence the spectral density at frequency zero is infinite (zero).

In a fashion analogous to Theorem 1, the behaviour of Q_n under long-range dependent alternatives may now be derived in several steps using Lemmas A.2, A.3, and Theorem 2:

Theorem 3 Let $\{\varepsilon_t\}$ be a zero-mean stationary Gaussian stochastic process with autocovariance function γ_k such that

$$\gamma_k \sim \begin{cases} k^{2H-2}L(k) & \text{for } H \in (\frac{1}{2}, 1) \text{ or} \\ -k^{2H-2}L(k) & \text{for } H \in (0, \frac{1}{2}) \end{cases} \quad \text{as} \quad k \to \infty, \qquad (5.25)$$

where $L(k)$ is a slowly varying function at infinity. Then as n and q increase without bound such that $(q/n) \to 0$, we have:

(a) $\displaystyle \underset{1 \leqslant k \leqslant n}{\text{Max}} \frac{1}{\sigma_n} \sum_{j=1}^{k} (X_j - \overline{X}_n) \implies \underset{0 \leqslant \tau \leqslant 1}{\text{Max}} \ W_H^o(\tau) \equiv M_H^o,$

(b) $\displaystyle \underset{1 \leqslant k \leqslant n}{\text{Min}} \frac{1}{\sigma_n} \sum_{j=1}^{k} (X_j - \overline{X}_n) \implies \underset{0 \leqslant \tau \leqslant 1}{\text{Min}} \ W_H^o(\tau) \equiv m_H^o,$

(c) $\displaystyle R_n \equiv \frac{\hat{\sigma}_n(q)\sqrt{n}}{\sigma_n} \cdot \frac{1}{\sqrt{n}} Q_n \implies M_H^o - m_H^o \equiv V_H,$

(d) $\displaystyle a_n \equiv \frac{\sigma_n}{\hat{\sigma}_n(q)\sqrt{n}} \overset{p}{\to} \begin{cases} \infty & \text{for } H \in (\frac{1}{2}, 1), \\ 0 & \text{for } H \in (0, \frac{1}{2}), \end{cases}$

(e) $\displaystyle \frac{1}{\sqrt{n}} Q_n = a_n R_n \overset{p}{\to} \begin{cases} \infty & \text{for } H \in (\frac{1}{2}, 1), \\ 0 & \text{for } H \in (0, \frac{1}{2}), \end{cases}$

where $\hat{\sigma}_n(q)$ is defined in (5.18), σ_n is defined in Theorem 2, and $W_H^o(\tau) \equiv W_H(\tau) - \tau W_H(1)$.[20]

Theorem 3 shows that the modified rescaled range test is consistent against a class of long-range dependent stationary Gaussian alternatives. In the presence of positive strong dependence, the R/S statistic diverges in probability to infinity; in the presence of negative strong dependence, it converges in probability to zero. In either case, the probability of rejecting the null hypothesis approaches unity for all stationary Gaussian stochastic processes satisfying (5.25), a broad set of alternatives that includes all fractionally-differenced Gaussian ARIMA(p, d, q) models with $d \in (-\frac{1}{2}, \frac{1}{2})$.

From (a) and (b) of Theorem 3 it is apparent that the normalized *population* rescaled, $R_n/\sqrt{n}$, converges to zero in probability. Therefore whether or not $Q_n/\sqrt{n}$ approaches zero or infinity in the limit depends entirely on the limiting behaviour of the ratio $\sigma_n / \hat{\sigma}_n(q)$. That is,

$$\frac{Q_n}{\sqrt{n}} = \frac{\sigma_n}{\hat{\sigma}_n(q)} \frac{R_n}{\sqrt{n}} \tag{5.26}$$

so that if the ratio $\sigma_n / \hat{\sigma}_n(q)$ diverges fast enough to overcompensate for the convergence of $R_n / \sqrt{n}$ to zero, then the test will reject in the upper tail, otherwise it will reject in the lower tail. This is determined by whether d lies in the interval $(0, \frac{1}{2})$ or $(-\frac{1}{2}, 0)$. When $d = 0$, the ratio $\sigma_n / \hat{\sigma}_n(q)$ converges to unity in probability and, as expected, the normalized R/S statistic converges in distribution to the range of the standard Brownian bridge.

Of course, if one is interested exclusively in fractionally-differenced alternatives, a more efficient means of detecting long-range dependence might be to estimate the fractional differencing parameter directly. In such cases, the approaches taken by Geweke and Porter-Hudak (1983), Sowell (1990b), and Yajima (1985, 1988) may be preferable. The modified R/S test is perhaps most useful for detecting departures into a broader class of alternative hypotheses, a kind of 'portmanteau' test statistic that may complement a comprehensive analysis of long-range dependence.

5.4 R/S Analysis for Stock Market Returns

The importance of long-range dependence in asset markets was first considered by Mandelbrot (1971). More recently, the evidence uncovered by Fama and French (1988), Lo and MacKinlay (1988), and Poterba and Summers (1988) may be symptomatic of a long-range dependent component in stock market prices. In particular, Lo and MacKinlay (1988) show that the ratios of k-week stock return variances to k times the variance of one-week returns generally exceed unity when k is small (2–32). In contrast, Poterba and Summers (1988) find that this same variance ratio falls below one when k is much larger (96 and greater).

To see that such a phenomenon can easily be generated by long-range dependence, denote by X_t the time-t return on a stock and let it be the sum of two components X_{at} and X_{bt} where

$$(1 - L)^d X_{at} = \varepsilon_t, \quad (1 - \rho L) X_{bt} = \eta_t,$$

and assign the values $(-0.2, 0.25, 1, 1.1)$ to the parameters $(d, \rho, \sigma_\varepsilon^2, \sigma_\eta^2)$. Let the ratio of the k-period return variance to k times the variance of X_t be denoted by $VR(k)$. Then a simple calculation will show that for the parameter values chosen:

$$VR(2) = 1.04, \quad VR(10) = 10.4,$$
$$VR(3) = 1.06, \quad VR(50) = 0.97,$$
$$VR(4) = 1.07, \quad VR(100) = 0.95,$$
$$VR(5) = 1.06, \quad VR(250) = 0.92.$$

The intuition for this pattern of variance ratios comes from observing that $VR(k)$ is a weighted sum of the first $k-1$ autocorrelation coefficients of X_t with linearly declining weights (see Lo and MacKinlay 1988). When k is small the autocorrelation of X_t is dominated by the positively autocorrelated AR(1) component X_{bt}. But since the autocorrelations of X_{bt} decay rapidly relative to those of X_{at}, as k grows the influence of the long-memory component eventually outweighs that of the AR(1), ultimately driving the variance ratio below unity.

5.4.1 The evidence for weekly and monthly returns

Greene and Fielitz (1977) were perhaps the first to apply R/S analysis to common stock returns. More recent applications include Booth and Kaen (1979) (gold prices), Booth, Kaen, and Koveos (1982) (foreign exchange rates), and Helms, Kaen, and Rosenman (1984) (futures contracts). These and earlier applications of R/S analysis by Mandelbrot and Wallis (1969a) have three features in common: (i) They provide no sampling theory with which to judge the statistical significance of their empirical results; (ii) they use the $\tilde{Q}_n$ which is not robust to short-range dependence; and (iii) they do not focus on the R/S statistic itself, but rather on the regression of its logarithm on (sub)sample sizes. The shortcomings of (i) and (ii) are apparent from the discussion in the preceding sections. As for (iii), Davies and Harte (1987) show such regression tests to be significantly biased toward rejection even for a stationary AR(1) process with an autoregressive parameter of 0.3.

To test for long-term memory in stock returns, I use data from the Center for Research in Security Prices (CRSP) monthly and daily returns files. Tests are performed for the value- and equal-weighted CRSP indexes. Daily observations for the returns indexes are available from 3 July 1962 to 31 December 1987 yielding a sample size of 6409 observations. Monthly indexes are each composed of 744 observations from 30 January 1926 to 31 December 1987. The following statistic is computed for the various returns indexes:

$$V_n(q) \equiv \frac{1}{\sqrt{n}} Q_n \overset{a}{\sim} V,$$

where the distribution F_V of V is given in (5.21). Using the values in Table 5.2 a test of the null hypothesis may be performed at the 95 per cent level of confidence by accepting or rejecting according to whether V_n is or is not contained in the interval $[0.809, 1.862]$ which assigns equal probability to each tail.

$V_n(q)$ is written as a function of q to emphasize the dependence of the modified rescaled range on the truncation lag. To check the sensitivity of the statistic to the lag length, $V_n(q)$ is computed for several different values of q. The normalized classical Hurst-Mandelbrot rescaled range $\tilde{V}_n$ is also computed for comparison, where

$$\tilde{V}_n \equiv \frac{1}{\sqrt{n}} \tilde{Q}_n \overset{a}{\sim} \xi V.$$

Table 5.3 reports results for the daily equal- and value-weighted returns indexes. The panel labelled 'Equal-weighted' contains the $V_n(q)$ and $\tilde{V}_n$ statistics for

Table 5.3. R/S analysis of daily equal- and value-weighted CRSP stock returns indexes from 3 July 1962 to 31 December 1987 using the classical rescaled range $\tilde{V}_n$ and the modified rescaled range $V_n(q)$. Entries in the %-bias columns are computed as $[(\tilde{V}_n/V_n(q)) - 1]\,100$, and are estimates of the bias of the classical R/S statistic in the presence of short-term dependence

Time period	Sample size	$\tilde{V}_n$	$V_n(90)$	%-Bias	$V_n(1980)$	%-Bias	$V_n(270)$	%-Bias	$V_n(360)$	%-Bias
Equal-weighted										
620703–871231	6409	2.63*	1.46	79.9	1.45	81.1	1.50	75.2	1.50	75.4
620703–750428	3204	3.18*	1.61	97.0	1.57	102.0	1.63	95.2	1.62	96.8
750429–871231	3205	1.45	0.92	57.2	0.97	49.0	1.05	38.5	1.14	27.3
620703–681217	1602	2.40*	1.39	72.2	1.46	64.7	1.72	39.7	1.78	34.8
681219–750428	1602	2.03*	1.07	90.7	1.10	84.9	1.19	70.6	1.23	65.3
750428–810828	1602	1.35	0.89	51.6	1.23	9.5	1.49	−9.2	1.71	−21.0
810831–871231	1603	1.79	1.15	55.8	1.10	62.4	1.18	51.6	1.27	41.4
Value-weighted										
620703–871231	6409	1.55	1.29	20.8	1.26	22.9	1.30	19.1	1.33	16.8
620703–750428	3204	1.97*	1.43	37.3	1.39	41.4	1.43	37.5	1.45	35.5
750429–871231	3205	1.29	1.22	5.8	1.24	4.1	1.32	−2.3	1.42	−9.4
620703–681217	1602	1.67	1.43	16.8	1.45	15.3	1.62	3.4	1.69	−1.3
681219–750428	1602	1.85	1.34	38.2	1.34	38.2	1.40	31.7	1.45	27.1
750428–810828	1602	1.08	1.12	−3.7	1.26	−14.7	1.34	−19.4	1.42	−24.2
810831–871231	1603	1.50	1.38	8.8	1.37	9.2	1.50	−0.3	1.63	−8.0

*Significance at the 5% level.

the equal-weighted index for the entire sample period (the first row), two equally-partitioned sub-samples (the next two rows), and four equally-partitioned sub-samples (the next four rows). The modified rescaled range is computed with q-values of 90, 180, 270, and 360 days. The columns labelled '%-Bias' report the estimated bias of the original rescaled range $\tilde{V}_n$, and is $100 \cdot (\hat{\xi} - 1)$ where $\hat{\xi} = \hat{\sigma}_n(q)/s_n = \tilde{V}_n/V_n$.

Although Table 5.3 shows that the classical R/S statistic $\tilde{V}_n$ is statistically significant at the 5 per cent level for the daily equal-weighted CRSP returns index, the modified R/S statistic V_n is not. While $\tilde{V}_n$ is 2.63 for the entire sample period the modified R/S statistic is 1.46 with a truncation lag of 90 days, and 1.50 with a truncation lag of 360 days. The importance of normalizing by $\hat{\sigma}(q)$ is clear—dividing by s_n imparts a potential upward bias of 80 per cent!

The statistical insignificance of the modified R/S statistics indicates that the data are consistent with the short-memory null hypothesis. The stability of the $V_n(q)$ across truncation lags q also supports the hypothesis that there is little dependence in daily stock returns beyond one or two months. For example, using 90 lags yields a V_n of 1.46 whereas 270 and 360 lags both yield 1.50, virtually the same point estimate. The results are robust to the sample period—none of the sub-period $V_n(q)$'s are significant. The classical rescaled range is significant only in the first half of the sample for the value-weighted index, and is insignificant when the entire sample is used.

Table 5.4 reports similar results for monthly returns indexes with four values of q employed: 3, 6, 9, and 12 months. None of the modified R/S statistics are statistically significant at the 5 per cent level in any sample period or sub-period for either index. The percentage bias is generally lower for monthly data, although it still ranges from -0.2 to 25.3 per cent.

To develop further intuition for these results, Fig. 5.2 contains the autocorrelograms of the daily and monthly equal-weighted returns indexes, where the maximum lag is 360 for daily returns and 12 for monthly. For both indexes only the lowest order autocorrelation coefficients are statistically significant. For comparison, alongside each of the index's autocorrelogram is the autocorrelogram of the fractionally-differenced process (5.1) with $d = 0.25$ and the variance of the disturbance chosen to yield a first-order autocorrelation of $\frac{1}{3}$. Although the general shapes of the fractionally-differenced autocorrelograms seem consistent with the data, closer inspection reveals that the index autocorrelations decay much more rapidly. Therefore, although short-term correlations are large enough to drive $\tilde{Q}_n$ and Q_n apart, there is little evidence of long-range dependence in Q_n itself.

Additional results are available for weekly and annual stock returns data but since they are so similar to those reported here, I have omitted them to conserve space. Although the annual data spans 115 years (1872–1986), neither the classical nor the modified R/S statistics are statistically significant over this time span.

The evidence in Tables 5.3 and 5.4 shows that the null hypothesis of short-range dependence cannot be rejected by the data—there is little support for long-term memory in US stock returns. With adjustments for autocorrelation at lags up to one calendar year, estimates of the modified rescaled range are consistent with

Table 5.4. R/S Analysis of monthly equal- and value-weighted CRSP Stock Returns Indexes from 30 January 1926 to 31 December 1987 using the classical rescaled range $\tilde{V}_n$ and the modified rescaled range $V_n(q)$

Time Period	Sample Size	$\tilde{V}_n$	$V_n(3)$	%-Bias	$V_n(6)$	%-Bias	$V_n(9)$	%-Bias	$V_n(12)$	%-Bias
Equal-weighted										
260130–871231	744	1.17	1.07	9.1	1.10	6.6	1.09	7.2	1.06	10.4
260130–561231	372	1.32	1.21	9.4	1.26	5.1	1.24	7.1	1.18	12.1
570131–871231	372	1.37	1.26	8.4	1.23	11.1	1.27	7.6	1.30	5.2
260130–410630	186	1.42	1.31	8.3	1.40	1.6	1.39	2.6	1.32	8.0
410731–561231	186	1.60	1.42	13.1	1.34	20.0	1.28	25.3	1.28	25.1
570131–720630	186	1.20	1.04	15.9	0.99	21.9	1.03	17.4	1.07	12.3
720731–871231	186	1.57	1.51	3.8	1.51	4.3	1.55	1.2	1.57	-0.2
Value-weighted										
260130–871231	744	1.33	1.27	4.5	1.26	5.5	1.22	8.4	1.19	11.1
260130–561231	372	1.57	1.51	4.5	1.51	4.3	1.44	9.5	1.38	14.5
570131–871231	372	1.28	1.22	4.4	1.18	7.9	1.21	5.6	1.24	2.7
260130–410630	186	1.57	1.52	3.2	1.55	1.0	1.49	5.5	1.42	10.6
410731–561231	186	1.26	1.18	6.4	1.11	12.9	1.07	17.1	1.08	16.1
570131–720630	186	1.05	0.96	9.3	0.92	14.7	0.95	10.9	1.01	4.7
720731–871231	186	1.51	1.48	1.6	1.45	4.0	1.47	2.4	1.49	1.1

Entries in the %-Bias columns are computed as $[(\tilde{V}_n/V_n(q)) - 1] \cdot 100$, and are estimates of the Bias of the classical R/S statistic in the presence of short-term dependence.

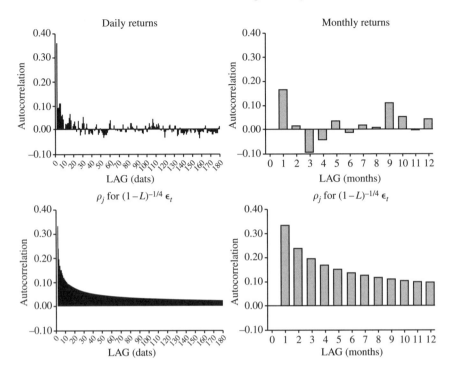

Fig. 5.2. Autocorrelograms of equally-weighted CRSP daily and monthly stock returns indexes and fractionally-differenced process with $d = 1/4$. The sample period for the daily index is July 1962–December 1987, and is January 1926–December 1987 for the monthly index.

the null hypothesis of weak dependence. This reinforces Kandel and Stambaugh's (1989) contention that the long-run predictability of stock returns uncovered by Fama and French (1988) and Poterba and Summers (1988) may not be 'long-run' in the time series sense, but may be the result of more conventional models of short-range dependence.[21] Of course, since our inferences rely solely on asymptotic distribution theory, we must check our approximations before dismissing the possibility of long-range dependence altogether. The finite-sample size and power of the modified rescaled range test are considered in the next sections.

5.5 Size and Power

To explore the possibility that the inability to reject the null hypothesis of short-range dependence is merely a symptom of low power, and to check the quality of Section 5.3's asymptotic approximations for various sizes, I perform several illustrative Monte Carlo experiments. Section 5.5.1 reports the empirical size of the test statistic under two Gaussian null hypotheses: i.i.d. and AR(1) disturbances. Section 5.5.2 presents power results against the fractionally-differenced process (5.1) for $d = \frac{1}{3}$ and $-\frac{1}{3}$.

5.5.1 The size of the R/S test

Table 5.5a contains simulation results for the modified R/S statistic with sample sizes of 100, 250, 500, 750, and 1000 under the null hypothesis of independently and identically distributed Gaussian errors. All simulations were performed on an IBM 4381 in double precision using the random generator G05DDF from the Numerical Algorithms Group Fortran Library Mark 12. For each sample size the statistic $V_n(q)$ is computed with $q = 0, 5, 10, 25, 50$, and with q chosen by Andrews' (1991) data-dependent formula:

$$ q = [k_n], \qquad k_n \equiv \left(\frac{3n}{2}\right)^{1/3} \cdot \left(\frac{2\hat{\rho}}{1 - \hat{\rho}^2}\right)^{2/3}, $$

where $[k_n]$ denotes the greatest integer less than or equal to k_n, and $\hat{\rho}$ is the estimated first-order autocorrelation coefficient of the data.[22] (Note that this is an optimal truncation lag only for an AR(1) data-generating process—a different expression obtains if, for example, the data-generating process were assumed to be an ARMA(1,1). See Andrews (1991) for further details.) In this case, the entry reported in the column labelled 'q' is the mean of the q's chosen, with the population standard deviation reported in parentheses below the mean. When $q = 0$, $V_n(q)$ is identical to Mandelbrot's classical R/S statistic $\tilde{V}_n$.

The entries in the last three columns of Table 5.5a show that the classical R/S statistic tends to reject too frequently—even for sample sizes of 1000 the empirical size of a 5 per cent test based on $\tilde{V}_n$ is 5.9 per cent. The modified R/S statistic tends to be conservative for values of q that are not too large relative to the sample size. For example, with 100 observations and 5 lags the empirical size of the 5 per cent test using $V_n(q)$ is 2.1 per cent. However, with 50 lags this test has a rejection rate of 31 per cent! That the sampling properties worsen with the number of lags is not surprising—the imprecision with which the higher-order autocovariances are estimated can introduce considerable noise into the statistic (e.g. see, Lo and MacKinlay (1989)). But for 1000 observations and 5 lags, the size of a 5 per cent test based on $V_n(q)$ is 5.1 per cent. Andrews' procedure yields intermediate results, with sizes in between those of the classical R/S statistic and the closest of the modified R/S statistics.

Table 5.5(b) reports the results of simulations under the null hypothesis of a Gaussian AR(1) with autoregressive coefficient 0.5 (recall that such a process is weakly dependent). The last three columns confirm the example of Section 5.3 and accord well with the results of Davies and Harte (1987): tests based on the classical R/S statistic have considerable power against an AR(1) null. In samples of only 100 observations the empirical size of the 5 per cent test based on $\tilde{V}_n$ is 38 per cent and increases to 62 per cent for sample sizes of 1000. In contrast, the empirical sizes of tests based on $V_n(q)$ are much closer to their nominal values since the geometrically declining autocorrelations are taken into account by the denominator $\hat{\sigma}_n(q)$ of $V_n(q)$. When q is chosen via Andrews' procedure, this yields conservative test sizes, ranging from 2.8 per cent for a sample of 100, to 4.3 per cent for a sample of 1000.

Table 5.5a. Finite sample distribution of the modified R/S statistic under an i.i.d. null hypothesis

n	q	Min	Max	Mean	S.D.	Size 1%-test	Size 5%-test	Size 10%-test
100	0	0.534	2.284	1.144	0.263	0.029	0.095	0.153
100	5	0.649	1.913	1.179	0.207	0.002	0.021	0.050
100	10	0.710	1.877	1.223	0.175	0.000	0.003	0.012
100	25	0.858	2.296	1.383	0.186	0.001	0.014	0.039
100	50	0.918	3.119	1.694	0.360	0.137	0.313	0.414
100	0.97 (0.83)	0.557	2.164	1.150	0.247	0.019	0.070	0.127
250	0	0.496	2.527	1.183	0.270	0.021	0.075	0.133
250	5	0.580	2.283	1.196	0.243	0.008	0.041	0.089
250	10	0.654	2.048	1.211	0.221	0.003	0.021	0.054
250	25	0.757	1.905	1.264	0.176	0.000	0.001	0.006
250	50	0.877	2.206	1.372	0.169	0.000	0.005	0.020
250	0.97 (0.83)	0.497	2.442	1.185	0.263	0.017	0.064	0.120
500	0	0.518	2.510	1.201	0.267	0.015	0.061	0.117
500	5	0.589	2.357	1.207	0.252	0.008	0.047	0.094
500	10	0.630	2.227	1.215	0.240	0.004	0.032	0.073
500	25	0.677	2.051	1.240	0.210	0.000	0.008	0.029
500	50	0.709	1.922	1.285	0.176	0.000	0.001	0.005
500	0.96 (0.82)	0.549	2.510	1.202	0.263	0.014	0.057	0.112
750	0	0.558	2.699	1.207	0.270	0.014	0.061	0.120
750	5	0.597	2.711	1.212	0.260	0.009	0.049	0.101
750	10	0.615	2.553	1.217	0.251	0.006	0.039	0.087
750	25	0.677	2.279	1.235	0.228	0.001	0.017	0.052
750	50	0.758	1.971	1.266	0.198	0.000	0.002	0.015
750	0.96 (0.83)	0.558	2.670	1.208	0.268	0.013	0.058	0.117
1000	0	0.542	2.577	1.211	0.270	0.014	0.059	0.113
1000	5	0.566	2.477	1.214	0.262	0.011	0.051	0.103
1000	10	0.570	2.405	1.218	0.256	0.008	0.045	0.089
1000	25	0.616	2.203	1.231	0.237	0.003	0.025	0.061
1000	50	0.716	2.036	1.253	0.211	0.000	0.007	0.029
1000	0.96 (0.81)	0.549	2.546	1.212	0.268	0.012	0.056	0.111

Each set of rows of a given sample size n corresponds to a separate and independent Monte Carlo experiment based on 10,000 replications. A lag q of 0 corresponds to Mandelbrot's Classical R/S statistic, and a noninteger lag value indicates the mean lag (standard deviation given in parentheses) chosen via Andrews' (1991) data-dependent procedure assuming an AR(1) data-generating process. Standard errors for the empirical size may be computed using the usual normal approximation; they are 9.95×10^{-4}, 2.18×10^{-3}, and 3.00×10^{-3} for the 1%, 5%, and 10% tests, respectively.

Table 5.5b. Finite sample distribution of the modified R/S statistic under an
AR(1) null hypothesis with autoregressive coefficient 0.5

n	q	Min	Max	Mean	S.D.	Size 1%-test	Size 5%-test	Size 10%-test
100	0	0.764	3.418	1.764	0.402	0.203	0.382	0.486
100	5	0.634	1.862	1.201	0.220	0.003	0.027	0.059
100	10	0.693	1.805	1.178	0.176	0.000	0.010	0.030
100	25	0.779	2.111	1.290	0.175	0.000	0.005	0.015
100	50	0.879	3.013	1.571	0.341	0.074	0.198	0.284
100	5.61 (1.25)	0.636	1.974	1.195	0.219	0.004	0.028	0.063
250	0	0.865	3.720	1.913	0.432	0.309	0.505	0.614
250	5	0.597	2.478	1.268	0.262	0.005	0.038	0.086
250	10	0.615	2.137	1.212	0.228	0.003	0.023	0.063
250	25	0.734	1.811	1.218	0.177	0.000	0.004	0.015
250	50	0.809	2.119	1.304	0.166	0.000	0.003	0.010
250	8.07 (1.07)	0.603	2.357	1.227	0.242	0.004	0.030	0.071
500	0	0.836	4.392	1.980	0.456	0.363	0.559	0.665
500	5	0.622	2.557	1.302	0.285	0.012	0.055	0.109
500	10	0.579	2.297	1.236	0.256	0.007	0.039	0.085
500	25	0.627	1.980	1.214	0.215	0.001	0.015	0.041
500	50	0.734	1.894	1.243	0.178	0.000	0.002	0.009
500	10.40 (0.99)	0.577	2.353	1.236	0.256	0.007	0.039	0.085
750	0	0.839	4.211	2.017	0.459	0.389	0.592	0.696
750	5	0.567	2.637	1.323	0.291	0.011	0.062	0.118
750	10	0.557	2.429	1.253	0.265	0.007	0.043	0.091
750	25	0.614	2.114	1.222	0.232	0.003	0.022	0.058
750	50	0.702	1.891	1.235	0.200	0.000	0.005	0.022
750	12.03 (0.93)	0.556	2.324	1.244	0.260	0.007	0.041	0.088
1000	0	0.926	4.327	2.045	0.465	0.414	0.617	0.716
1000	5	0.625	2.768	1.340	0.296	0.014	0.065	0.125
1000	10	0.592	2.622	1.268	0.272	0.009	0.047	0.096
1000	25	0.608	2.350	1.231	0.244	0.004	0.030	0.072
1000	50	0.636	1.997	1.236	0.217	0.001	0.011	0.038
1000	13.30 (0.89)	0.590	2.548	1.252	0.265	0.008	0.043	0.090

Each set of rows of a given sample size n corresponds to a separate and independent Monte Carlo experiment based on 10,000 replications. A lag q of 0 corresponds to Mandelbrot's classical R/S statistic, and a noninteger lag value indicates the mean lag (standard deviation given in parentheses) chosen via Andrews' (1991) data-dependent procedure, assuming an AR(1) data-generating process. Standard errors for the empirical size may be computed using the usual normal approximation; they are 9.95×10^{-4}, 2.18×10^{-3}, and 3.00×10^{-3} for the 1%, 5%, and 10% tests, respectively.

5.5.2 Power against fractionally-differenced alternatives

Tables 5.6a and 5.6b report the power of the R/S tests against the Gaussian fractionally-differenced alternative:

$$(1 - L)^d \varepsilon_t = \eta_t, \quad \eta_t \text{ i.i.d. } N(0, \sigma_\eta^2),$$

with $d = \frac{1}{3}$ and $-\frac{1}{3}$, and $\sigma_\eta^2 = \Gamma^2(1 - d)/\Gamma(1 - 2d)$ so as to yield a unit variance for ε_t. For sample sizes of 100, tests based on $V_n(q)$ have very little power, but when the sample size reaches 250 the power increases dramatically. According to Table 5.6a, the power of the 5 per cent test with $q = 5$ against the $d = \frac{1}{3}$ alternative is 33.5 per cent with 250 observations, 62.8 per cent with 500 observations, and 84.6 per cent with 1000 observations. Although Andrews' automatic truncation lag procedure is generally less powerful, its power is still 63.0 per cent for a sample size of 1000. Also, the rejections are generally in the right tail of the distribution, as the entries in the 'Max' column indicate. This is not surprising in light of Theorem 3, which shows that under this alternative the modified R/S statistic diverges in probability to infinity.

For a fixed sample size, the power of the $V_n(q)$-based test declines as the number of lags is increased. This is due to the denominator $\hat{\sigma}_n(q)$, which generally increases with q since there is positive dependence when $d = \frac{1}{3}$. The increase in the denominator decreases the mean and variance of the statistic, shifting the distribution towards the left and pulling probability mass from both tails, thereby reducing the frequency of draws in the right tail's critical region, where virtually all the power is coming from.

Against the $d = -\frac{1}{3}$ alternative, Table 5.6b shows that the test seems to have somewhat higher power. However, in contrast to Table 5.6a the rejections are now coming from the left tail of the distribution, as Theorem 3 predicts. Although less powerful, tests based on Andrews' procedure still exhibit reasonable power, ranging from 33.1 per cent in samples of 100 observations to 94.5 per cent in samples of 1000.

For the larger sample sizes the power again declines as the number of lags increases, due to the denominator $\hat{\sigma}_n(q)$, which declines as q increases because the population autocorrelations are all negative when $d = -\frac{1}{3}$. The resulting increase in the mean of $V_n(q)$'s sampling distribution overwhelms the increase in its variability, leading to a lower rejection rate from the left tail.

Tables 5.6a and 5.6b show that the modified R/S statistic has reasonable power against at least two specific models of long-term memory. However, these simulations are merely illustrative—a more conclusive study would include further simulations with several other values for d, and perhaps with short-range dependence as well.[23] Moreover, since our empirical work has employed data sampled at different frequencies (implying different values of d for different sample sizes), the trade-off between the time span of the data and the frequency of observation for the test's power may be an important issue. Nevertheless, the simulation results suggest that short-range dependence may be the more significant feature of recent stock market returns.

Table 5.6a. Power of the modified R/S statistic under a Gaussian fractionally differenced alternative with differencing parameter $d = 1/3$

n	q	Min	Max	Mean	S.D.	Power 1%-test	Power 5%-test	Power 10%-test
100	0	0.729	4.047	2.025	0.513	0.429	0.600	0.680
100	5	0.635	2.089	1.361	0.242	0.001	0.014	0.065
100	10	0.686	1.746	1.237	0.171	0.000	0.005	0.015
100	25	0.723	2.148	1.208	0.156	0.000	0.002	0.008
100	50	0.823	2.803	1.399	0.328	0.035	0.096	0.154
100	4.27 (1.43)	0.650	2.330	1.411	0.257	0.003	0.040	0.104
250	0	0.938	5.563	2.678	0.709	0.774	0.878	0.918
250	5	0.713	2.924	1.699	0.364	0.153	0.335	0.442
250	10	0.705	2.304	1.475	0.277	0.006	0.091	0.186
250	25	0.681	1.852	1.264	0.175	0.000	0.003	0.012
250	50	0.756	1.971	1.208	0.140	0.000	0.002	0.007
250	6.63 (1.36)	0.711	2.596	1.619	0.317	0.067	0.240	0.360
500	0	1.061	7.243	3.336	0.929	0.924	0.967	0.980
500	5	0.731	3.726	2.055	0.491	0.450	0.628	0.709
500	10	0.692	2.944	1.750	0.384	0.197	0.385	0.494
500	25	0.623	2.164	1.429	0.258	0.001	0.045	0.123
500	50	0.687	1.763	1.271	0.178	0.000	0.003	0.009
500	8.96 (1.37)	0.709	3.201	1.809	0.384	0.242	0.447	0.557
750	0	1.228	8.059	3.799	1.052	0.972	0.990	0.995
750	5	0.838	4.280	2.313	0.565	0.620	0.766	0.830
750	10	0.769	3.421	1.955	0.447	0.370	0.557	0.655
750	25	0.734	2.478	1.569	0.310	0.042	0.195	0.304
750	50	0.722	1.925	1.359	0.223	0.000	0.005	0.036
750	10.58 (1.33)	0.798	3.324	1.942	0.421	0.363	0.559	0.657
1000	0	1.398	8.615	4.174	1.174	0.985	0.996	0.998
1000	5	0.898	4.672	2.521	0.635	0.720	0.846	0.892
1000	10	0.779	3.766	2.121	0.504	0.494	0.669	0.747
1000	25	0.641	2.734	1.686	0.354	0.135	0.322	0.431
1000	50	0.628	2.118	1.441	0.259	0.001	0.052	0.138
1000	11.87 (1.31)	0.766	3.613	2.044	0.454	0.446	0.630	0.718

The variance of the process has been normalized to unity. Each set of rows of a given sample size n corresponds to a separate and independent Monte Carlo experiment based on 10,000 replications. A lag q of 0 corresponds to Mandelbrot's classical R/S statistic, and a noninteger lag value indicates the mean lag (standard deviation given in parentheses) chosen via Andrews' (1991) data-dependent procedure assuming an AR(1) data-generating process.

Table 5.6b. Power of the modified R/S statistic under a Gaussian fractionally differenced alternative with differencing parameter $d = -1/3$

n	q	Min	Max	Mean	S.D.	Power 1%-test	Power 5%-test	Power 10%-test
100	0	0.367	1.239	0.678	0.120	0.670	0.858	0.923
100	5	0.637	1.710	1.027	0.153	0.006	0.054	0.134
100	10	0.762	2.030	1.217	0.161	0.000	0.001	0.005
100	25	0.953	2.638	1.587	0.207	0.014	0.095	0.211
100	50	1.052	3.478	2.033	0.354	0.425	0.679	0.785
100	2.94 (0.99)	0.478	1.621	0.889	0.155	0.131	0.331	0.466
250	0	0.303	1.014	0.561	0.089	0.951	0.991	0.997
250	5	0.549	1.479	0.851	0.128	0.146	0.409	0.571
250	10	0.632	1.752	1.005	0.143	0.007	0.065	0.152
250	25	0.833	1.936	1.292	0.157	0.000	0.001	0.004
250	50	0.977	2.357	1.594	0.186	0.007	0.078	0.198
250	4.20 (0.86)	0.448	1.437	0.796	0.129	0.301	0.578	0.716
500	0	0.292	0.819	0.479	0.071	0.997	1.000	1.000
500	5	0.458	1.244	0.728	0.105	0.517	0.794	0.888
500	10	0.555	1.489	0.861	0.121	0.111	0.366	0.543
500	25	0.706	1.735	1.105	0.143	0.000	0.004	0.022
500	50	0.881	2.089	1.356	0.157	0.000	0.002	0.011
500	5.45 (0.77)	0.443	1.318	0.725	0.108	0.529	0.793	0.887
750	0	0.276	0.700	0.433	0.063	1.000	1.000	1.000
750	5	0.422	1.070	0.659	0.094	0.764	0.932	0.973
750	10	0.499	1.262	0.779	0.109	0.325	0.641	0.789
750	25	0.689	1.570	1.001	0.132	0.003	0.049	0.138
750	50	0.837	1.802	1.227	0.148	0.000	0.000	0.002
750	6.34 (0.73)	0.424	1.133	0.682	0.099	0.679	0.892	0.951
1000	0	0.257	0.775	0.403	0.057	1.000	1.000	1.000
1000	5	0.401	1.149	0.613	0.085	0.895	0.978	0.993
1000	10	0.487	1.376	0.725	0.099	0.525	0.809	0.907
1000	25	0.633	1.596	0.930	0.121	0.020	0.154	0.306
1000	50	0.778	1.820	1.139	0.139	0.000	0.000	0.006
1000	7.01 (0.70)	0.412	1.235	0.651	0.092	0.789	0.945	0.978

The variance of the process has been normalized to unity. Each set of rows of a given sample size n corresponds to a separate and independent Monte Carlo experiment based on 10,000 replications. A lag q of 0 corresponds to Mandelbrot's classical R/S statistic, and a noninteger lag value indicates the mean lag (standard deviation given in parentheses) chosen via Andrews' (1991) data-dependent procedure assuming an AR(1) data-generating process.

5.6 Conclusion

Using a simple modification of the Hurst-Mandelbrot rescaled range that accounts for short-term dependence, and contrary to previous studies, I find little evidence of long-term memory in historical US stock market returns. If the source of serial correlation is lagged adjustment to new information, the absence of strong dependence in stock returns should not be surprising from an economic standpoint, given the frequency with which financial asset markets clear. Surely financial security prices must be immune to persistent informational asymmetries, especially over longer time spans. Perhaps the fluctuations of aggregate economic output are more likely to display such long-run tendencies, as Kondratiev and Kuznets have suggested, and this long memory in output may eventually manifest itself in the return to equity. But if some form of long-range dependence is indeed present in stock returns, it will not be easily detected by any of our current statistical tools, especially in view of the optimality of the R/S statistic in the Mandelbrot and Wallis (1969) sense. Direct estimation of particular parametric models may provide more positive evidence of long-term memory and is currently being pursued by several investigators.[24]

Appendix 5.1

Proofs of the theorems rely on the following three lemmas:

Lemma A.1 (Herrndorf (1984)) If $\{\varepsilon_t\}$ satisfies assumptions $(A1)$–$(A4)$ then as n increases without bound, $W_n(\tau) \Rightarrow W(\tau)$.

Lemma A.2 (Extended Continuous Mapping Theorem)[25] Let h_n and h be measurable mappings from $\mathcal{D}[0, 1]$ to itself and denote by E the set of $x \in \mathcal{D}[0, 1]$ such that $h_n(x_n) \to h(x)$ fails to hold for some sequence x_n converging to x. If $W_n(\tau) \Rightarrow W(\tau)$ and E is of Wiener-measure zero, that is, $P(W \in E) = 0$, then $h_n(W_n) \Rightarrow h(W)$.

Lemma A.3 Let $R_n \Rightarrow R$ where both R_n and R have nonnegative support, and let $P(R = 0) = P(R = \infty) = 0$. If $a_n \overset{p}{\to} \infty$, then $a_n R_n \overset{p}{\to} \infty$. If $a_n \overset{p}{\to} 0$, then $a_n R_n \overset{p}{\to} 0$.

Proof of Theorem 1 Let $S_n = \sum_{j=1}^{n} \varepsilon_j$ and define the following function $Y_n(\tau)$ on $\mathcal{D}[0, 1]$:

$$Y_n(\tau) = \frac{1}{\sigma\sqrt{n}} S_{[n\tau]}, \quad \tau \in [0, 1], \tag{A.1}$$

where $[n\tau]$ denotes the greatest integer less than or equal to $n\tau$, and σ is defined in condition $(A3)$ of the null hypothesis. By convention, set $Y_n(0) \equiv 0$. Under conditions $(A1)$, $(A2')$, $(A3)$, and $(A4)$ Herrndorf (1984) has shown that

$Y_n(\tau) \Rightarrow W(\tau)$. But consider:

$$\text{Max}_{1 \leqslant k \leqslant n} \frac{1}{\hat{\sigma}_n(q)\sqrt{n}} \sum_{j=1}^{k} (X_j - \bar{X}_n) = \text{Max}_{1 \leqslant k \leqslant n} \frac{1}{\hat{\sigma}_n(q)\sqrt{n}} \left(S_k - \frac{k}{n} S_n \right), \quad \text{(A.2a)}$$

$$= \text{Max}_{0 \leqslant \tau \leqslant 1} Z_n(\tau), \quad \text{(A.2b)}$$

where

$$Z_n(\tau) \equiv Y_n(\tau) - \frac{[n\tau]}{n} Y_n(1). \quad \text{(A.2c)}$$

Since the sequence of functions h_n that map $Y_n(\tau)$ to $Z_n(\tau)$ satisfies the conditions of Lemma A.2, where the limiting mapping h takes $Y_n(\tau)$ to $Y_n(\tau) - \tau Y_n(1)$, it may be concluded that

$$h_n(Y_n(\tau)) = Z_n(\tau) \Rightarrow h(W(\tau)) = W(\tau) - \tau W(1) = W^0(\tau). \quad \text{(A.3)}$$

If the estimator $\hat{\sigma}_n(q)$ is substituted in place of σ in the construction of $Z_n(\tau)$, then under conditions $(A2')$ and $(A5)$, theorem 4.2 of Phillips (1987) shows that $(A.3)$ still obtains. The rest of the theorem follows directly from repeated application of Lemma A.2. ∎

Proof of Theorem 2 See Davydov (1970) and Taqqu (1975). ∎

Proof of Theorem 3 Parts (a)–(c) follow directly from Theorem 2 and Lemma A.2, and part (e) follows immediately from Lemma A.3. Therefore, we need only prove (d). Let $H \in (\frac{1}{2}, 1)$ so that $\gamma(k) \sim k^{2H-2}L(k)$. This implies that

$$\text{Var}[S_n] \sim n^{2H} L(n). \quad \text{(A.4)}$$

Therefore, to show that $a_n \overset{p}{\to} \infty$, it suffices to show that

$$\frac{\hat{\sigma}^2(q)}{n^{2H-1}L(n)} \overset{p}{\to} 0. \quad \text{(A.5)}$$

Consider the population counterpart to (A.5):

$$\frac{\sigma^2(q)}{n^{2H-1}L(n)} = \frac{1}{n^{2H-1}L(n)} \left(\sigma_\varepsilon^2 + 2 \sum_{j=1}^{q} \omega_j \gamma_j \right), \quad \text{(A.6)}$$

where $\omega_j = 1 - j/(q+1)$. Since by assumption $\gamma_j \sim j^{2H-2}L(j)$, there exists some integer q_a and $M > 0$ such that for $j > q_a$, $\gamma_j < Mj^{2H-2}L(j)$. Now it is well known that a slowly-varying function satisfies the inequality $j^{-\varepsilon} < L(j) < j^\varepsilon$ for any $\varepsilon > 0$ and $j > q_b$, for some $q_b(\varepsilon)$. Choose $\varepsilon < 2 - 2H$, and observe that

$$\gamma_j < Mj^{2H-2}j^\varepsilon, \quad j > q_0 \equiv \max(q_a, q_b), \quad \text{(A.7)}$$

which implies

$$2 \sum_{j=1}^{q} \omega_j \gamma_j < 2 \sum_{j=1}^{q_0} \omega_j \gamma_j + 2M \sum_{j=q_0+1}^{q} \omega_j j^{2H-2+\varepsilon}, \qquad \text{(A.8)}$$

where, without loss of generality, we have assumed that $q > q_0$. As q increases without bound, the first sum of the right-side of (A.8) remains finite, and the second sum may be bounded by observing that its summands are positive and decreasing, hence (see, for example, Buck (1978, ch. 5.5)):

$$2M \sum_{j=q_0+1}^{q} \omega_j j^{2H-2+\varepsilon} \leqslant 2M \int_{q_0}^{q} \left(1 - \frac{x}{q+1}\right) x^{2H-2+\varepsilon} dx, \qquad \text{(A.9a)}$$

$$\sim 0(q^{2H-1+\varepsilon}), \qquad \text{(A.9b)}$$

where the asymptotic equivalence follows by direct integration. If $q \sim O(n^\delta)$ where $\delta \in (0, 1)$, a weaker condition than required by our null hypothesis, then the ratio $\sigma^2(q)/(n^{2H-1}L(n))$ is at most of order $O(n^{(2H-1+\varepsilon)(\delta-1)})$, which converges to zero. If we can now show that (A.6) and its sample counterpart are equal in probability, then we are done. This is accomplished by the following sequence of inequalities:

$$E\left|\frac{\hat\sigma^2(q)}{n^{2H-1}} - \frac{\sigma^2(q)}{n^{2H-1}}\right| = \frac{1}{n^{2H-1}} E\left|\left(\hat\sigma_\varepsilon^2 - \sigma_\varepsilon^2\right) + 2\sum_{j=1}^{q} \omega_j\left(\hat\gamma_j - \gamma_j\right)\right|, \quad \text{(A.10a)}$$

$$\leqslant \frac{E|\hat\sigma_\varepsilon^2 - \sigma_\varepsilon^2|}{n^{2H-1}} + \frac{2}{n^{2H-1}} \sum_{j=1}^{q} \omega_j E|\hat\gamma_j - \gamma_j|, \qquad \text{(A.10b)}$$

$$\leqslant \frac{E|\hat\sigma_\varepsilon^2 - \sigma_\varepsilon^2|}{n^{2H-1}} + \frac{2}{n^{2H-1}} \sum_{j=1}^{q} \omega_j \sqrt{E(\hat\gamma_j - \gamma_j)^2}. \quad \text{(A.10c)}$$

But since Hosking (1984: th. 2) provides rates of convergence for sample auto-covariances of stationary Gaussian processes satisfying (5.25), an integral evaluation similar to that in (A.9) shows that the sum in (A.10c) vanishes asymptotically when $q \sim 0(n)$. This completes the proof. Since the proof for $H \in (0, \frac{1}{2})$ is similar, it is left to the reader. ∎

Notes

1. Haubrich (1990) and Haubrich and Lo (1989) provide a less fanciful theory of long-range dependence in economic aggregates.
2. See Leroy (1989) and Merton (1987) far excellent surveys of this recent literature.

3. See, for example, Fama and French (1988), Jegadeesh (1990*a,b*), and Poterba and Summers (1988).

4. This is in fact how the stock returns data of Section 5.4 are constructed.

5. There are several other ways of measuring the degree of statistical dependence, giving rise to other notions of 'mixing'. For further details, see Eberlein and Taqqu (1986), Rosenblatt (1956), and White (1984).

6. See Herrndorf (1985). One of Mandelbrot's (1972) arguments in favour of R/S analysis is that finite second moments are not required. This is indeed the case if we are interested only in the almost sure convergence of the statistic. But since for purposes of inference the *limiting distribution* is required, a stronger moment condition is needed here.

7. Specifically, that the sequence $\{\varepsilon_t\}$ is strong-mixing may be replaced by the weaker assumption that it is a near-epoch dependent function of a strong-mixing process. See McLeish (1977) and Wooldridge and White (1988) for further details.

8. Mandelbrot and others have called the $d < 0$ case 'anti-persistence', reserving the term 'long-range dependence' for the $d > 0$ case. However, since both cases involve autocorrelations that decay much more slowly than those of more conventional time series, I call both long-range dependent.

9. A function $f(x)$ is said to be *regularly varying at infinity with index* ρ if $\lim_{t\to\infty} f(tx)/f(t) = x^\rho$ for all $x > 0$; hence regularly varying functions are functions that behave like power functions asymptotically. When $\rho = 0$, the function is said to be *slowly varying at infinity*, since it behaves like a constant for large x. An example of a function that is slowly varying at infinity is $\log x$. See Resnick (1987) for further properties of regularly varying functions.

10. This has also been advanced as a definition of long-range dependence—see, for example, Mandelbrot (1972).

11. Note, Helson and Sarason (1967) only consider the case of linear dependence; hence their conditions are sufficient to rule out strong-mixing but not necessary. For example, white noise may be approximated by a nonlinear deterministic time series (e.g. the tent map) and will have constant spectral density, but will be strongly dependent. I am grateful to Lars Hansen for pointing this out.

12. The behaviour of $\tilde{Q}_n$ may be better understood by considering its origins in hydrological studies of reservoir design. To accommodate seasonalities in riverflow, a reservoir's capacity must be chosen to allow for fluctuations in the supply of water above the dam while still maintaining a relatively constant flow of water below the dam. Since dam construction costs are immense, the importance of estimating the reservoir capacity necessary to meet long term storage needs is apparent. The range is an estimate of this quantity. If X_j is the riverflow (per unit time) above the dam and $\overline{X}_n$ is the desired riverflow below the dam, the bracketed quantity in (5.13) is the capacity of the reservoir needed to ensure this smooth flow given the pattern of flows in periods 1 through n. For example, suppose annual riverflows are assumed to be 100, 50, 100, and 50 in years 1 through 4. If a constant annual flow of 75 below the dam is

desired each year, a reservoir must have a minimum total capacity of 25 since it must store 25 units in years 1 and 3 to provide for the relatively dry years 2 and 4. Now suppose instead that the natural pattern of riverflow is 100, 100, 50, 50 in years 1 through 4. To ensure a flow of 75 below the dam in this case, the minimum capacity must increase to 50 so as to accommodate the excess storage needed in years 1 and 2 to supply water during the 'dry spell' in years 3 and 4. Seen in this context, it is clear that an increase in persistence will increase the required storage capacity as measured by the range. Indeed, it was the apparent persistence of 'dry spells' in Egypt that sparked Hurst's life-long fascination with the Nile, leading eventually to his interest in the rescaled range.

13. See Billingsley (1968) for the definition of weak convergence. I discuss the Brownian bridge and V more formally below.

14. It is implicitly assumed throughout that white noise has a Lebesgue-integrable characteristic function to avoid the pathologies of Andrews (1984).

15. Andrews (1991) has improved the rate restriction in (A5) to $o(n^{1/2})$, and it has been conjectured that $o(n)$ is sufficient.

16. For example, Andrews (1991) and Gallant (1987) both advocate the use of Parzen weights, which also yields a positive semi-definite estimator of the spectral density at frequency zero but is optimal in an asymptotic mean-square error sense.

17. Mandelbrot (1975) derives similar limit theorems for the statistic $\tilde{Q}_n$ under the more restrictive i.i.d. assumption, in which case the limiting distribution will coincide with that of Q_n. Since our null hypothesis includes weakly dependent disturbances, I extend his results via the more general functional central limit theorem of Herrndorf (1984, 1985).

18. Proofs of theorems are given in Appendix 5.1.

19. I am grateful to David Aldous and Yin-Wong Cheung for these last two references.

20. Although it is tempting to call $W_H^o(\tau)$ a 'fractional Brownian bridge', this is not the most natural definition despite the fact that it is 'tied down'. See Jonas (1983: ch. 3.3) for a discussion.

21. Moreover, several papers have suggested that these long-run results may be spurious. See, for example, Kim, Nelson, and Startz (1991), Richardson (1989), and Richardson and Stock (1990).

22. For this procedure, the Newey–West autocorrelation weights (5.19) are replaced by those suggested by Andrews (1991):

$$w_j = 1 - \left| \frac{j}{k_n} \right|.$$

23. The very fact that the modified R/S statistic yields few rejections under the null simulations of Section 5.5.1 shows that the test may have low power against *some* long-range dependent alternatives, since the pseudo-random number generator used in those simulations is, after all, a long-range dependent process. A more striking example is the 'tent' map, a particularly simple nonlinear

deterministic map (it has a correlation dimension of 1) which yields sequences that are virtually uncorrelated but long-range dependent. In particular, the tent map is given by the following recursion:

$$X_t = \begin{cases} 2X_{t-1} & \text{if } X_{t-1} < \frac{1}{2}, \\ 2(1 - X_{t-1}) & \text{if } X_{t-1} \geq \frac{1}{2}, \end{cases} \quad t = 1, \ldots, T, \quad X_0 \in (0, 1).$$

As an illustration, I performed two Monte Carlo experiments using the tent map to generate samples of 500 and 1000 observations (each with 10,000 replications) with an independent uniform (0, 1) starting value for each replication. Neither the Mandelbrot rescaled range, nor its modification with fixed or automatic truncation lags have any power against the tent map. In fact, the finite sample distributions are quite close to the null distribution. Of course, one could argue that if the dynamics and the initial condition were unknown, then even if a deterministic system *were* generating the data, the resulting time series would be short-range dependent 'for all practical purposes' and *should* be part of our null. I am grateful to Lars Hansen for suggesting this analysis.

24. See, for example, Boes *et al.* (1989), Diebold and Rudebusch (1989), Fox and Taqqu (1986), Geweke and Porter-Hudak (1983), Porter-Hudak (1990), Sowell (1990*a*,*b*), and Yajima (1985, 1988).
25. See Billingsley (1968) for a proof.

References

Andrews, D. (1984), 'Non-strong mixing autoregressive processes', *Journal of Applied Probability*, 21, 930–4.
—— (1991), 'Heteroskedasticity and autocorrelation consistent covariance matrix estimation', *Econometrica*, 59, 817–58.
Billingsley, P. (1968): *Convergence of Probability Measures*. New York: John Wiley and Sons.
Boes, D., Davis, R. and Gupta, S. (1989), 'Parameter estimation in low-order fractionally differenced ARMA processes', Working Paper, Department of Statistics, Colorado State University.
Booth, G., and Kaen, F. (1979), 'Gold and silver spot prices and market information efficiency', *Financial Review*, 14, 21–6.
——, ——, and Koveos, P. (1982), '*R/S* Analysis of foreign exchange rates under two international monetary regimes', *Journal of Monetary Economics*, 10, 407–15.
Buck, R. (1978): *Advanced Calculus*. New York: McGraw-Hill.
Campbell, J., and Mankiw, G. (1987), 'Are output fluctuations transitory?' *Quarterly Journal of Economics*, 102, 857–80.
Chan, N., and Wei, C. (1988), 'Limiting distributions of least squares estimates of unstable autoregressive processes', *Annals of Statistics*, 16, 367–401.

Davies, R., and Harte, D. (1987), 'Tests for Hurst Effect', *Biometrika*, 74, 95–101.

Davydov, Yu (1970), 'The invariance principle for stationary processes', *Theory of Probability and Its Applications*, 15, 487–89.

Diebold, F., and Rudebusch, G. (1989), 'Long memory and persistence in aggregate output', to appear in *Journal of Monetary Economics*.

Eberlein, E., and Taqqu, M. (1986), *Dependence in Probability and Statistics: A survey of Recent Results*. Boston: Birkhäuser.

Fama, E., and French, K. (1988), 'Permanent and temporary components of stock prices', *Journal of Political Economy*, 96, 246–73.

Feller, W. (1951), 'The asymptotic distribution of the range of sums of independent random variables', *Annals of Mathematical Statistics*, 22, 427–32.

Fox, R., and Taqqu, M. (1986), 'Large-sample properties of parameter estimates for strongly dependent stationary Gaussian time series', *Annals of Statistics*, 14, 517–32.

Gallant, R. (1987): *Nonlinear Statistical Models*. New York: John Wiley and Sons.

Geweke, J., and Porter-Hudak, S. (1983), 'The estimation and application of long memory time series models', *Journal of Time Series Analysis*, 4, 221–38.

Granger, C. (1966), 'The typical spectral shape of an economic variable', *Econometrica*, 34, 150–61.

—— (1980), 'Long memory relationships and the aggregation of dynamic models', *Journal of Econometrics*, 14, 227–38.

—— and Joyeux, R. (1980), 'An introduction to long-memory time series models and fractional differencing', *Journal of Time Series Analysis*, 1, 15–29.

Greene, M., and Fielitz, B. (1977), 'Long-term dependence in common stock returns', *Journal of Financial Economics*, 4, 339–49.

Hall, P., and Heyde, C. (1980), *Martingale Limit Theory and Its Application*. New York: Academic Press.

Hannan, E. (1970), *Multiple Time Series*. New York: John Wiley and Sons.

Hansen, L. (1982), 'Large sample properties of generalized method of moments estimators', *Econometrica*, 50, 1029–54.

Haubrich, J. (1990), 'Consumption and fractional-differencing: old and new anomalies', Federal Reserve Bank of Cleveland Working Paper No. 9010.

—— and Lo, A. (1989), 'The sources and nature of long-term dependence in the business cycle', NBER Working Paper No. 2951.

Helms, B., Kaen, F., and Rosenman, R. (1984), 'Memory in commodity futures contracts', *Journal of Futures Markets*, 4, 559–67.

Helson, H., and Sarason, D. (1967), 'Past and future', *Mathematica Scandinavia*, 21, 5–16.

Herrndorf, N. (1984), 'A functional central limit theorem for weakly dependent sequences of random variables', *Annals of Probability*, 12, 141–53.

—— (1985), 'A functional central limit theorem for strongly mixing sequences of random variables', *Z. Wahrscheinlichkeitstheorie verw. Gebiete*, 541–50.

Hosking, J. (1981), 'Fractional differencing', *Biometrika*, 68, 165–76.

—— (1984), 'Asymptotic distributions of the sample mean, autocovariances and autocorrelations of long-memory time series', Mathematics Research Center Technical Summary Report #2752, University of Wisconsin-Madison.

Hurst, H. (1951), 'Long term storage capacity of reservoirs', *Transactions of the American Society of Civil Engineers*, 116, 770–99.

Ibragimov, I., and Rozanov, Y. (1978), *Gaussian Random Processes*. New York: Springer-Verlag.

Jegadeesh, N. (1990a), 'Evidence of predictable behavior of security returns', *Journal of Finance*, 45, 881–98.

—— (1990b), 'Seasonality in stock price mean reversion: evidence from the US and UK', to appear in *Journal of Finance*.

Jonas, A. (1983), 'Persistent memory random processes', Doctoral Dissertation, Department of Statistics, Harvard University.

Kandel, S., and Stambaugh, R. (1989), 'Modelling expected stock returns for long and short horizons', Rodney L. White Center Working Paper No. 42-88, Wharton School, University of Pennsylvania.

Kennedy, D. (1976), 'The distribution of the maximum Brownian excursion', *Journal of Applied Probability*, 13, 371–6.

Kim, M., Nelson, C., and Startz, R. (1991), 'Mean reversion in stock prices: a reappraisal of the empirical evidence', *Review of Economic Studies*, 58, 515–28.

LeRoy, S. (1989), 'Efficient capital markets and martingales', *Journal of Economic Literature*, 27, 1583–621.

Lo, A., and MacKinlay, C. (1988), 'Stock market prices do not follow random walks: evidence from a simple specification test', *Review of Financial Studies*, 1, 41–66.

——(1989), 'The size and power of the variance ratio test in finite samples: a Monte Carlo investigation', *Journal of Econometrics*, 40, 203–38.

——(1990), 'When are contrarian profits due to stock market overreaction?' *Review of Financial Studies*, 3, 175–206.

Maheswaran, S. (1990), 'Predictable short-term variation in asset prices: theory and evidence', Working Paper, Carlson School of Management, University of Minnesota.

—— and Sims, C. (1990), 'Empirical implications of arbitrage free asset markets', Discussion Paper, Department of Economics, University of Minnesota.

Mandelbrot, B. (1971), 'When can price be arbitraged efficiently? A limit to the validity of the random walk and martingale models', *Review of Economics and Statistics*, 53, 225–36.

—— (1972), 'Statistical methodology for non-periodic cycles: from the covariance to R/S analysis', *Annals of Economic and Social Measurement*, 1, 259–90.

—— (1975), 'Limit theorems on the self-normalized range for weakly and strongly dependent processes', *Z. Wahrscheinlichkeitstheorie verw. Gebiete*, 31, 271–85.

—— and Taqqu, M. (1979), 'Robust R/S analysis of long-run serial correlation', *Bulletin of the International Statistical Institute*, 48, Book 2, 59–104.

—— and Van Ness, J. (1968), 'Fractional Brownian motion, fractional noises and applications', *S.I.A.M. Review*, 10, 422–37.

—— and Wallis, J. (1968), 'Noah, Joseph and operational hydrology', *Water Resources Research*, 4, 909–18.

—— (1969*a*), 'Computer experiments with fractional Gaussian noises. Parts 1, 2, 3', *Water Resources Research*, 5, 228–67.

—— (1969*b*), 'Some long run properties of geophysical records', *Water Resources Research*, 5, 321–40.

—— (1969*c*), 'Robustness of the rescaled range R/S in the measurement of noncyclic long run statistical dependence', *Water Resources Research*, 5, 967–88.

McLeish, D. (1977), 'On the invariance principle for nonstationary mixingales', *Annals of Probability*, 5, 616–21.

Merton, R. (1987), 'On the current state of the stock market rationality hypothesis', in *Macroeconomics and Finance: Essays in Honor of Franco Modigliani*, ed. by R. Dornbusch, F. Fischer, and J. Bossons. Cambridge: M.I.T. Press.

Newey, W., and West, K. (1987), 'A simple positive definite, heteroscedasticity and autocorrelation consistent covariance matrix', *Econometrica*, 55, 703–5.

Phillips, P. (1987), 'Time series regression with a unit root', *Econometrica*, 55, 277–301.

Porter-Hudak, S. (1990), 'An application of the seasonal fractionally differenced model to the monetary aggregates', *Journal of the American Statistical Association*, 85, 338–44.

Poterba, J., and Summers, L. (1988), 'Mean reversion in stock returns: evidence and implications', *Journal of Financial Economics*, 22, 27–60.

Resnick, S. (1987), *Extreme Values, Regular Variation, and Point Processes*. New York: Springer-Verlag.

Richardson, M. (1989), 'Temporary components of stock prices: a skeptic's view', Working Paper, Wharton School, University of Pennsylvania.

—— and Stock, J. (1990), 'Drawing inferences from statistics based on multiyear asset returns', *Journal of Financial Economics*, 25, 323–48.

Rosenblatt, M. (1956), 'A central limit theorem and a strong mixing condition', *Proceedings of the National Academy of Sciences*, 42, 43–7.

Siddiqui, M. (1976), 'The asymptotic distribution of the range and other functions of partial sums of stationary processes', *Water Resources Research*, 12, 1271–6.

Sims, C. (1984), 'Martingale-like behavior of asset prices and interest rates', Discussion Paper No. 205, Center for Economic Research, Department of Economics, University of Minnesota.

Sowell, Fallaw (1990*a*), 'The fractional unit root distribution', *Econometrica*, 58, 495–506.

—— (1990*b*), 'Maximum likelihood estimation of stationary univariate fractionally integrated time series models', Working Paper, Graduate School of Industrial Administration, Carnegie-Mellon University.

Taqqu, M. (1975), 'Weak convergence to fractional Brownian motion and to the Rosenblatt process', *Z. Wahrscheinlichkeitstheorie verw. Gebiete*, 31, 287–302.

White, H. (1980), 'A heteroscedasticity-consistent covariance matrix estimator and a direct test for heteroscedasticity', *Econometrica*, 48, 817–38.

—— (1984), *Asymptotic Theory for Econometricians*. New York: John Wiley and Sons.

—— and Domowitz, I. (1984), 'Nonlinear regression with dependent observations', Econometrica, 52, 143–62.

Wooldridge, J., and White, H. (1988), 'Some invariance principles and central limit theorems for dependent heterogeneous processes', *Econometric Theory*, 4, 210–30.

Yajima, Y. (1985), 'On estimation of long-memory time series models', *Australian Journal of Statistics*, 27, 303–20.

——(1988), 'On estimation of a regression model with long-memory stationary errors', *Annals of Statistics*, 16, 791–807.

6

The Estimation and Application of Long-memory Time Series Models

JOHN GEWEKE AND SUSAN PORTER-HUDAK

6.1 Introduction

In the most widely applied models for stationary time series the spectral density function is bounded at the frequency $\lambda = 0$ and the autocorrelation function decays exponentially. This is true in stable autoregressive moving average models, and it is characteristic of estimates obtained by nonparametric spectral methods, for example. Yet these properties do not appear widely characteristic of many time series (Hurst 1951; Granger and Joyeux 1980). The failure of these models to represent the spectral density at low frequencies adequately suggests that many-step-ahead forecasts obtained using these models could be inferior to those produced by models that permit unbounded spectral densities at $\lambda = 0$ and autocorrelation functions that do not decay exponentially. Limited empirical evidence to this effect has been reported by Granger and Joyeux, and is supported by somewhat more extensive investigations reported by Porter-Hudak (1982).

Attention has recently been given to two single-parameter models in which the spectral density function is proportional to λ^{-r}, $1 < r < 2$, for λ near 0, and the asymptotic decay of the autocorrelation function is proportional to τ^{r-1}. Because the spectral density function is unbounded at $\lambda = 0$—equivalently, the autocorrelation function is not summable—these are *long-memory models* (defined by McLeod and Hipel 1978).

The earlier model was introduced by Mandelbrot and Van Ness (1968) and Mandelbrot (1971) to formalize Hurst's empirical findings using cumulative river flow data. Let $Y_t = \int_{-\infty}^{t} (t - s)^{H-1/2} \, dB(s)$, where $B(s)$ is Brownian motion and $H \in (0, 1)$. Then $X_t = Y_t - Y_{t-1}$ is a *simple fractional Gaussian noise*. (We have added the word 'simple' to the definition given in the literature, to emphasize the difference between that model and the extension of it introduced

This chapter was previously published as Geweke, J. and Porter-Hudak, S. (1983), 'The estimation and application of long-memory time series models', *Journal of Time Series Analysis*, 4, 221–38. Copyright © 1983 V. Geweke and S. Porter-Hudak, reproduced by kind permission of Blackwell Publishers Ltd.

This work was undertaken while the first author was Associate Professor and the second was a graduate student in the Department of Economics, University of Wisconsin-Madison. Portions of this article are taken from Susan Porter-Hudak's Ph.D. thesis. Financial support from National Science Foundation grant SES-8207639 is acknowledged.

here.) Jonas (1981) has shown that its spectral density is

$$f_1(\lambda; H) = \sigma^2 (2\pi)^{-2H-2} \Gamma(2H+1) \sin(\pi H) 4 \sin^2(\lambda/2)$$

$$\times \sum_{n=-\infty}^{\infty} |n + (\lambda/2\pi)|^{-2H-1}.$$

(We use the convention that the spectral density of a time series $\{X_t\}$ is $f(\lambda) = \sum_{s=-\infty}^{\infty} R_x(s) \exp(-i\lambda s)$, where $R_x(s)$ is the autocovariance function of $\{X_t\}$.)
A little manipulation of this expression shows that

$$\lim_{\lambda \to 0} \lambda^{2H-1} f_1(\lambda; H) = (2\sigma^2/\pi) \Gamma(2H+1) \sin(\pi H).$$

The corresponding autocorrelation function is

$$\rho_1(\tau; H) = 0.5(|\tau - 1|^{2H} - 2|\tau|^{2H} + |\tau + 1|^{2H}).$$

A Taylor series expansion in $(\tau - 1)/\tau$ and $(\tau + 1)/\tau$ shows

$$\lim_{\tau \to \infty} \tau^{2-2H} \rho_1(\tau; H) = H(2H - 1),$$

and from the Schwarz inequality $\rho_1(\tau; H) \gtrless 0$ for $\tau \neq 0$ as $H \gtrless 0.5$. In empirical work interest has centered on the cases $0.5 < H < 1$.

The fractional Gaussian noise model is designed to account for the long term behaviour of the time series in question. Realistically, it seems unlikely that all the second moment properties of the series would be well described by any single-parameter function, but shorter term behaviour could be modelled by more conventional means. Toward this end, $\{X_t\}$ will be said to be a *general fractional Gaussian noise* if its spectral density is of the form $f_1(\lambda; H) f_u(\lambda)$ where $f_u(\lambda)$ is a positive continuous function bounded above and away from zero on the interval $[-\pi, \pi]$. The spectral density $f_u(\lambda)$ therefore is that of a short memory time series (McLeod and Hippel) with autoregressive representation (Rozanov 1967: pp. 77–78). This class of models is broad: it includes, for example, time series arising from $\phi(B)X_t = \theta(B)X_t'$, where $\phi(B)$ and $\theta(B)$ are invertible polynomials of finite order in the lag operator B and X_t' is a simple fractional Gaussian noise.

The second long-memory model was proposed independently by Granger and Joyeux (1980) and Hosking (1981). It can be motivated by the observation that some time series appear to have unbounded spectral densities at the frequency $\lambda = 0$, but the spectral densities of first differences of the same series appear to vanish at $\lambda = 0$. This suggests the model $(1 - B)^d X_t = \varepsilon_t$, where $d \in (-0.5, 0.5)$ and ε_t is serially uncorrelated. By this it is meant that the spectral density of $\{X_t\}$ is

$$f_2(\lambda; d) = (\sigma^2/2\pi)|1 - e^{-i\lambda}|^{-2d} = (\sigma^2/2\pi)\{4 \sin^2(\lambda/2)\}^{-d}.$$

A series with the spectral density $f_2(\lambda; d)$ will be called a *simple integrated series*. Clearly $\lim_{\lambda \to 0} \lambda^{2d} f_2(\lambda; d) = (\sigma^2/2\pi)$. The corresponding autocorrelation function (for $d \neq 0$) is

$$\rho_2(\tau; d) = \Gamma(1 - d)\Gamma(\tau + d)/\{\Gamma(d)\Gamma(\tau + 1 - d)\},$$

and $\lim_{\tau \to \infty} \tau^{1-2d} \rho_2(\tau; d) = \Gamma(1 - d)/\Gamma(d) = \pi/\sin(\pi d)$ (Granger and Joyeux, 1980: p. 17). A simple integrated series has the autoregressive representation

$$\sum_{j=0}^{\infty} a_2(j, d) X_{t-j} = \varepsilon_t, \quad a_2(j; d) = \Gamma(j - d)/\{\Gamma(-d)\Gamma(j + 1)\},$$

and moving average representation

$$X_t = \sum_{j=0}^{\infty} b_2(j; d)\varepsilon_{t-j}, \quad b_2(j; d) = \Gamma(j + d)/\{\Gamma(j + 1)\Gamma(d)\}.$$

(Hosking 1981: p. 167; Granger and Joyeux, 1980, eqn (6) is in error.) Since $\lim_{j \to \infty} j^{(1+d)} a_2(j; d) = 1/\Gamma(-d)$ and $\lim_{j \to \infty} b_2(j; d) = 1/\Gamma(d)$, $a_2(j; d)$ and $b_2(j; d)$ are each square summable; but $\sum_{j=0}^{\infty} |a_2(j; d)| > \infty$ if and only if $d > 0$, while $\sum_{j=0}^{\infty} |b_2(j; d)| < \infty$ if and only if $d < 0$.

It seems desirable to extend the integrated series model in the same way as the fractional Gaussian noise model. Hence $\{X_t\}$ will be called a *general integrated series* if its spectral density is of the form $f_2(\lambda; d) f_u(\lambda)$, where $f_u(\lambda)$ has the same characteristics as before. Whenever the process $\{\varepsilon_t\}$ is i.i.d. as well as being serially uncorrelated, we shall add the adjective *linear* to describe the process.

There are obvious similarities in simple fractional Gaussian noise and simple integrated series, to which Granger and Joyeux (1980), Hosking (1981), and Jonas (1981) have referred obliquely. In Section 6.2 it is shown that the spectral density function of a general fractional Gaussian noise with parameter H is that of a general integrated series with parameter $H - \frac{1}{2}$, and vice versa. Which model one uses therefore depends on practical considerations, to which the rest of the paper is addressed. In Section 6.3 a consistent, computationally efficient estimator $\hat{d}$ of the parameter d for general integrated series is introduced, and its asymptotic distribution is derived. (By virtue of the results of Section 6.2, $\hat{H} = \hat{d} + \frac{1}{2}$ provides a consistent estimator of H, and $\hat{H} - H$ and $\hat{d} - d$ have the same asymptotic distribution.) Results with synthetic time series are reported in Section 6.4, and with actual time series in Section 6.5. On the basis of these results it can be concluded that the estimator $\hat{d}$ is at least as reliable in finite sample as any estimator of d and H suggested to date, and is more attractive computationally.

6.2 The Equivalence of General Fractional Gaussian Noise and General Integrated Series

We provide the following characterization of the relationship of the two models.

Theorem 6.1 $\{X_t\}$ is a general integrated series with parameter $d(-\frac{1}{2} < d < \frac{1}{2})$ if, and only if, it is also a general fractional Gaussian noise with parameter $H = d+\frac{1}{2}$.

Proof Let $d = (L-2)/2$ and $H = (L-1)/2$, $1 < L < 3$.

$$f_1\{\lambda; (L-1)/2\}/f_2\{\lambda; (L-2)/2\}$$

$$= \pi^{-1}\Gamma(L)\sin\{\pi(L-1)/2\}|\sin(\lambda/2)|^L \sum_{n=-\infty}^{\infty} |n + (\lambda/2\pi)|^{-L}$$

$$= \pi(L)\sin\{\pi(L-1)/2\}|\sin(\lambda/2)/(\lambda/2)|^L \sum_{n=-\infty}^{\infty} |(\lambda/2\pi)/\{n + (\lambda/2\pi)\}|^L.$$

It suffices to show that this ratio is continuous in λ and bounded above and below by positive numbers on $[-\pi, \pi]$. The function $|\sin(\lambda/2)/(\lambda/2)|^L$ has these characteristics. Furthermore

$$1 < \sum_{n=-\infty}^{\infty} |(\lambda/2\pi)/\{n + (\lambda/2\pi)\}|^L$$

$$= 1 + \sum_{n=1}^{\infty} |(\lambda/2\pi)/\{n + (\lambda/2\pi)\}|^L + \sum_{n=1}^{\infty} |(\lambda/2\pi)/\{n - (\lambda/2\pi)\}|^L$$

$$< 1 + 2|(\lambda/2\pi)/\{1 + (\lambda/2\pi)\}|^L \zeta(L),$$

where $\zeta(L) = \sum_{n=1}^{\infty} n^{-L}$ is Riemann's zeta function; $\zeta(L) < \infty$ for $L > 1$. The function $\sum_{n=-\infty}^{\infty} |(\lambda/2\pi)/\{n + (\lambda/2\pi)\}|^L$ is therefore continuous and bounded above and below by positive numbers on $[-\pi, \pi]$ also.

The proof shows that $f_1(\lambda; (L-1)/2)/f_2(\lambda; (L-2)/2)$ is bounded uniformly in λ. It is not bounded uniformly in L, and its behaviour as $L \to 1$ and $L \to 3$ indicates potential complications in applied work. To examine the case $L \to 1$, write

$$f_1\{\lambda; (L-1)/2\}/f_2\{\lambda; (L-2)/2\}$$

$$= \sin(\lambda/2)\pi^{-1}\Gamma(L)\frac{\sin\{\pi(L-1)/2\}}{\sin(\pi L)}\sin(\pi L) \sum_{n=-\infty}^{\infty} |n + (\lambda/2\pi)|^{-L}.$$

Substitute $\sin(\pi L) = \pi/\{\Gamma(L)\Gamma(1-L)\}$ (Gradshteyn and Ryzhik 1980: 8.334.3) and $\sum_{n=-\infty}^{\infty} |n + (\lambda/2\pi)|^{-L} = \zeta(L, \lambda/2\pi) + \zeta(L, -\lambda/2\pi) - (-|\lambda|/2\pi)^{-L}$ (where $\zeta(L, q) = \sum_{n=0}^{\infty} (n+q)^{-L}$ is Riemann's general zeta function) to obtain

$$\sin(\lambda/2)\pi^{-L+1}\Gamma(L)\frac{\sin\{\pi(L-1)/2\}}{\sin(\pi L)\Gamma(L)\Gamma(1-L)}$$

$$\times \{\zeta(L, \lambda/2\pi) + \zeta(L, -\lambda/2\pi) - (-|\lambda|/2\pi)^{-L}\}.$$

Since $\lim_{L \to 1} \zeta(L, q)/\Gamma(1 - L) = -1$ (Gradshteyn and Ryzhik 1980: 9.533.1) the limiting value of this expression as $L \to 1$ is $\sin(\lambda/2)$. Hence if a general integrated series were to be described as a general fractional Gaussian noise, then as $d \to -\frac{1}{2}$ a noninvertible moving average component would be required.

As $L \to 3$ the ratio approaches zero, but the approach is uniform in λ: $\lim_{L \to 3} f_1\{\lambda; (L - 1)/2\}/[f_2\{\lambda; (L - 2)/2\} \sin\{\pi(L - 1)/2\}]$ is continuous and is bounded above and below by positive constants. This limiting case seems to pose no practical problems. ∎

6.3 A Simple Estimation Procedure for General Integrated Series

Consider the problem of estimating the parameter d in the general integrated series model. Suppose $(1 - B)^d X_t = u_t$, where u_t is a stationary linear process with spectral density function $f_u(\lambda)$ which is finite, bounded away from zero and continuous on the interval $[-\pi, \pi]$. The spectral density function of $\{X_t\}$ is $f(\lambda) = (\sigma^2/2\pi)\{4\sin^2(\lambda)\}^{-d} f_u(\lambda)$, and

$$\ln\{f(\lambda)\} = \ln\{\sigma^2 f_u(0)/2\pi\} - d \ln\{4\sin^2(\lambda/2)\} + \ln\{f_u(\lambda)/f_u(0)\}. \quad (6.1)$$

Suppose that a sample of $\{X_t\}$ of size T is available. Let $\lambda_{j,T} = 2\pi j/T$ ($j = 0, \ldots, T - 1$) denote the harmonic ordinates, and $I(\lambda_{j,T})$ denote the periodogram at these ordinates. Evaluate (6.1) at $\lambda_{j,T}$ and rearrange to obtain

$$\ln\{I(\lambda_{j,T})\} = \ln\{\sigma^2 f_u(0)/2\pi\} - d \ln\{4\sin^2(\lambda_{j,T}/2)\}$$
$$+ \ln\{f_u(\lambda_{j,T})/f_u(0)\} + \ln\{I(\lambda_{j,T})/f(\lambda_{j,T})\}. \quad (6.2)$$

The proposed estimate of d is motivated by the formal similarity of (6.2) and a simple linear regression equation: $\ln\{I(\lambda_{j,T}/2)\}$ is analogous to the dependent variable, $\ln\{4\sin^2(\lambda_{j,T})\}$ is the explanatory variable, $\ln\{I(\lambda_{j,T})/f(\lambda_{j,T})\}$ is the disturbance, the slope coefficient is $-d$, and the intercept term is $\ln\{\sigma^2 f_u(0)/2\pi\}$ plus the mean of $\ln\{I(\lambda_{j,T})/f(\lambda_{j,T})\}$. The term $\ln\{f_u(\lambda_{j,T})/f_u(0)\}$ becomes negligible as attention is confined to harmonic frequencies nearer to zero. The proposed estimator is the slope coefficient in the least squares regression of $\ln\{I(\lambda_{j,T})\}$ on a constant and $\ln\{4\sin^2(\lambda_{j,T}/2)\}$ in the sample $j = 1, \ldots, g(T)$; $g(T)$ will be described subsequently. It will be shown that when $d < 0$ the estimator is consistent, and the conventional interpretation of the standard error for the slope coefficient is appropriate asymptotically. We conjecture that the results remain true for $d \geqslant 0$, and experimental evidence to that effect will be provided subsequently.

To begin, consider (6.2) for $j = 1, \ldots, n$, for any arbitrarily chosen positive integer n. As T increases, the harmonic frequencies $\lambda_{j,T}$ all approach zero and the term $\ln\{f_u(\lambda_{j,T})/f_u(0)\}$ may be ignored. When $d < 0$ the coefficients in the moving average representation of $\{X_t\}$ are absolutely summable. Hence the random variables $Z_{j,T} = \sum_{t=1}^{T} X_t \exp(-i\lambda_{j,T})/\{f(\lambda_{j,T})\}^{1/2}$ are asymptotically i.i.d. normal (Hannan, 1973: th. 3). The terms $\ln\{I(\lambda_{j,T})/f(\lambda_{j,T})\}$ are therefore also asymptotically i.i.d., and their distribution can be derived by change of variable techniques (Porter-Hudak 1982: appendix C). The distribution is of the Gumbel type; the asymptotic mean of $\ln\{I(\lambda_{j,T})/f(\lambda_{j,T})\}$ is $-C$ (C is Euler's constant, $0.57721\ldots$) and its variance is $\pi^2/6$. Consider invoking a conventional central limit theorem to obtain the asymptotic distribution of the least squares slope coefficient in (6.2). This requires that certain conditions on the regressors be met (to be discussed shortly) and that the disturbance term be i.i.d. with finite mean and variance. This condition will generally never be met for fixed n or T, but for fixed n it may be approximated to any specified degree of accuracy by choosing T sufficiently large. Once the accuracy criterion is met the number of ordinates can be increased to $n + 1$, and T can then be further increased until the criterion of accuracy is again met. In this way a function $g(T)$ is defined, such that if $n = g(T)$ then the least squares slope estimator $\hat{d}$ of d is asymptotically normal. Clearly $g(T)$ must satisfy $\lim_{T\to\infty} g(T) = \infty$, $\lim_{T\to\infty} g(T)/T = 0$, but beyond this all that is known is that $g(T)$ exists.

Let $U(j, T) = \ln\{4\sin^2(\lambda_{j,T}/2)\}$, $\bar{U}(T, n) = n^{-1}\sum_{j=1}^{n} U(j, T)$. The asymptotic normality of $\hat{d}$ follows from the Lindberg–Lévy central limit theorem (Kendall and Stuart 1972: pp. 206–8).

$$\lim_{T\to\infty} \sum_{j=1}^{g(T)}\{U(j, T) - \bar{U}(T, g(T))\}^2 = \infty, \qquad (6.3)$$

$$\lim_{T\to\infty} \{U(1, T)\}^2 \bigg/ \left[\sum_{j=1}^{g(T)}\{U(j, T) - \bar{U}(T, g(T))\}^2\right] = 0. \qquad (6.4)$$

(In (6.4), use has been made of the fact that $\sup_{j=1,\ldots,g(T)}\{U(j, T)\}^2 = \{U(1, T)\}^2$.) In showing (6.3) and (6.4) it is simpler to work with $W(j, T) = \ln(\lambda_{j,T})$ and $\bar{W}(T, n) = n^{-1}\sum_{j=1}^{n} W(j, T)$ in lieu of $U(j, T)$ and $\bar{U}(T, n)$, respectively. The substitution is justified by the fact that for any $g(T)$ that satisfies $\lim_{T\to\infty} g(T)/T = 0$, $\lim_{T\to\infty} U(j, T)/W(j, T) = 2$ ($j = 1, \ldots, g(T)$) and

$$\lim_{T\to\infty} \left\{\sum_{j=1}^{g(T)} U(j, T) \bigg/ \sum_{j=1}^{g(T)} W(j, T)\right\}^2$$

$$= \lim_{T\to\infty} \sum_{j=1}^{g(T)}\{U(j, T)\}^2 \bigg/ \sum_{j=1}^{g(T)}\{Z(j, T)\}^2 = 4.$$

To simplify notation let $n = g(T)$. To establish (6.3) observe that

$$\sum_{j=1}^{n}\{W(j, T) - \bar{W}(j, T)\}^2 = \sum_{j=1}^{n}(\ln j)^2 - n^{-1}\left(\sum_{j=1}^{n}\ln j\right)^2. \qquad (6.5)$$

Make the substitutions

$$\sum_{j=1}^{n}(\ln j)^2 = \left(n + \tfrac{1}{2}\right)(\ln n)^2 - 2n \ln n + 2n + C_n$$

where $-(\ln 2)^2/2 - 2 < C_n < -2$ (Buck 1978: p. 252), and

$$\sum_{j=1}^{n}\ln j = \left(n + \tfrac{1}{2}\right)\ln n - n + D_n,$$

where $1 - \ln 2/2 < D_n < 1$ (Buck 1978: p. 252). Expression (6.5) then reduces to

$$n - \left(\tfrac{1}{2} + \tfrac{1}{4}n^{-1}\right)(\ln n)^2 + (1 - 2D_n - n^{-1}D_n)\ln n + C_n + 2D_n - n^{-1}D_n^2, \quad (6.6)$$

which for large n is dominated by $n \to \infty$. On the other hand, $\{W(1, T)\}^2 = (\ln 2\pi - \ln T)^2$. So long as

$$\lim_{T \to \infty}(\ln T)^2/g(T) = 0, \qquad (6.7)$$

(6.4) will be satisfied. This appears to be a weak requirement: for example, it is satisfied by $g(T) = cT^\alpha, 0 < \alpha < l$.

In inference about d from $\hat{d}$, use can be made of the known variance of the analogue of the disturbance term in (6.2): in large samples, the distribution of $\hat{d}$ is approximated by

$$N\left(d, \pi^2 \bigg/ \left[6\sum_{j=1}^{g(T)}\{U(j, T) - \bar{U}(T, g(T))\}^2\right]\right).$$

A natural, somewhat simpler procedure is to assume that $\hat{d}$ is normal with standard deviation given by the usual ordinary least squares arithmetic. This assumption is justified so long as s^2 consistently estimates $\mathrm{var}\{I(\lambda_{j,T})/f(\lambda_{j,T})\}$ in (6.2), which will be guaranteed if the least squares intercept b_0 converges in probability to the population intercept $\ln\{\sigma^2 f_u(0)/2\pi\} - C$. Since $b_0 = n^{-1}\sum_{j=1}^{n}\ln\{I(\lambda_{j,T})\} - \hat{d}\bar{U}(T, n)$, this will occur if $\mathrm{plim}\,(\hat{d} - d)\bar{U}(T, n) = 0$. As argued above, $\bar{U}(T, n)$ behaves like $n^{-1}\sum_{j=1}^{n}\ln(2\pi j/T) = \ln 2\pi - \ln T + \ln n + \ln n/2n - 1 + D_n/n$ when n is large, so $\lim \bar{U}(T, n)/\ln(n/T) = 1$. From (6.6), the asymptotic standard error of $\hat{d}$ is proportional to $n^{-1/2}$. Hence b_0 is consistent for the population intercept, and $\mathrm{plim}\,s^2 = \pi^2/6$, if $\lim_{T \to \infty}\{g(T)\}^{-1/2}\ln\{g(T)/T\} = 0$. This condition is implied by (6.7).

These results are collected in the following theorem.

Theorem 6.2 Suppose $\{X_t\}$ is a general integrated linear process, with $d < 0$. Let $I(\lambda_{j,T})$ denote the periodogram of $\{X_t\}$ at the harmonic frequencies $\lambda_{j,T} = \pi j/T$

in a sample of size T. Let $b_{1,T}$ denote the ordinary least squares estimator of β_1 in the regression equation $\ln\{I(\lambda_{j,T})\} = \beta_0 + \beta_1 \ln\{4\sin^2(\lambda_{j,T}/2)\} + u_{j,T}$, $j = 1, \ldots, n$. Then there exists a function $g(T)$ (which will have the properties $\lim_{T\to\infty} g(T) = \infty$, $\lim_{T\to\infty} g(T)/T = 0$) such that if $n = g(T)$ then plim $b_1 = -d$. If $\lim_{T\to\infty} (\ln T)^2/g(T) = 0$, then $(b_1 + d)/\{v\hat{a}r(b_1)\}^{1/2} \to^{\mathcal{D}} N(0, 1)$, where $v\hat{a}r(b_1)$ is the usual least squares estimator of $\mathrm{var}(b_1)$.

Once the parameter 'd' is estimated, the relationship $f_u(\lambda) = f(\lambda)/f_2(\lambda; d)$ may be used to estimate $f_u(\lambda)$. This can be done as follows. First, the exact finite Fourier transform of the sequence $\{X_t, t = 1, \ldots, T\}$ is computed at the harmonic ordinates. The exact, rather than the fast, Fourier transform is used because the fast Fourier transform presumes circular stationarity; this presumption seems a poor approximation for long-memory models, and in experiments along the lines of those reported in the next two sections the exact Fourier transform was a much more satisfactory (though more costly) procedure. Second, the Fourier transform of the series is multiplied by $(1 - \exp(-i\lambda))^{\hat{d}}$. Third, the exact inverse Fourier transform of this product is computed. In the final step conventional procedures—for example, ARMA models, long autoregressions, or nonparametric spectral methods—are used to model $f_u(\lambda)$. Since $\hat{d}$ is consistent for d, the estimator (or the implicit estimator) of $f_u(\lambda)$ in the last step is consistent. The conventional distribution theory for the estimator will not be applicable, because $\hat{d}$ rather than d is used in the second step; indeed, because $\mathrm{var}(\hat{d}) = O(g(T)^{-1})$ and $\lim_{T\to\infty} g(T)/T = 0$, conventional standard errors in the final step could be badly misleading.

Multi-step-ahead forecasting is straightforward. With $\hat{d}$ in hand, the autoregression coefficient estimates $(a_2(j; \hat{d})$, see Section 6.1) for the 'long memory' portion of the series can be computed. The estimated autoregressive representation for the 'short memory' portion of the series (corresponding to $f_u(\lambda)$) can also be computed. The convolution of the lag operators in these two representations provides the lag operator for the estimated autoregressive representation of $\{X_t\}$. Multi-step-ahead forecasts are then computed in the usual way.

It is clear that through suitable integer differencing the concept of integrated series can be extended, and these methods applied to cases $d \geqslant \frac{1}{2}$. The choice of 'suitable' integer involves subjective judgment and may become critical if d lies exactly between two integers and Theorem 2 is to be invoked. An alternative in such cases is to fractionally difference the original series as in steps one through three of the estimation procedure for $f_u(\lambda)$, using a prespecified value of d in lieu of $\hat{d}$, to ensure that for the transformed series $d \in (-\frac{1}{2}, \frac{1}{2})$.

6.4 Simulation Results

The results on estimation are asymptotic and restricted to the case $d < 0$. They leave open the questions of how large a sample is required for their reliability, how rapidly the number of ordinates used to estimate d, $g(T)$, should increase with sample size T, and whether they are valid for $d > 0$. Given the seemingly insurmountable difficulties in obtaining analytical answers to all of these questions, the most productive line of attack is to conduct some experiments with synthetic series. We report here the findings from a very limited simulation study; a more thorough investigation along these lines is warranted before these methods are used widely in empirical work.

The generation of synthetic data to mimic the realizations of long-memory models has proved to be a challenge in its own right, precisely because widely separated points in the generated sample must be highly correlated. In early work, short memory approximations to long-memory models were employed, with unsatisfactory results (see the discussion and citations in McLeod and Hipel 1978: p. 497). A direct procedure (McLeod and Hipel 1978) is to compute the pertinent $T \times T$ covariance matrix C for given values of the parameter H or d and sample size T, and then compute the Cholesky decomposition $C = MM'$. Given a standardized normal sequence of synthetic random variables $\underline{e} = (e_1, \ldots, e_T)'$ the synthetic long-memory series is then $M\underline{e}$. The computation of M requires storage proportional to T^2 and time proportional to T^3, and is impractical for values of T exceeding 200 or so. In a variant on this method Granger and Joyeux (1980) used the Cholesky decomposition to obtain the first 100 values of the simulated sample, and then used the truncated autoregression $\sum_{j=0}^{100} a_2(j; d)x_{t-j} = e_t$ to obtain successive values of the series.

In our simulations we employed the recursion algorithm usually attributed to Levinson (1947) and Durbin (1960), as extended by Whittle (1963); Jonas (1981) has recognized the potential of this method in the simulation of long-memory time series. The Levinson–Durbin–Whittle algorithm implicitly provides the Cholesky decomposition of a Toeplitz matrix. (The distinguishing feature of a Toeplitz matrix A is $a_{ij} = a_{i-j}$; the variance matrix of T successive realizations of a wide sense stationary process is a Toeplitz matrix.) Given the values of the autocovariance function $R(0), R(1), \ldots, R(T)$ for a wide sense stationary process $\{X_t\}$ the algorithm provides the coefficients in the linear projection of X_t on $X_{t-1}, \ldots, X_{t-p}$ and the associated residual variance, for $p = 1, \ldots, T$. Storage requirements and computation time are proportional to T^2; the advantage in computation time relative to the Cholesky decomposition stems from the exploitation of the fact that the variance matrix is a Toeplitz form. (For $T = 400$, computation time with 64-bit double precision arithmetic is less than one minute using a VAX-11/780.)

Evidence on the characteristics of series generated in this way is provided in Table 6.1. For the example reported there, $T = 265$. In each of 4000 replications the first 26 values of the sample autocovariance function were computed. The average of these values over the replications is the estimated autocovariance function of the simulated data reported in Table 6.1; standard errors for these estimates are reported parenthetically. Since at least one million pairs of generated data are involved in each average, the standard errors are quite small. In no case do the estimated means differ from the population values by more than two standard errors. These results compare quite favourably with those of Granger and Joyeux (1980: pp. 25–6), and underscore the importance of using exact methods rather than short memory approximations in the synthesis of long-memory time series.

Table 6.1. Autocovariance functions estimated from synthetic
data[a]

Lag	$d = 0.25$		$d = 0.45$	
	True population	Synthetic population estimate[b]	True population	Synthetic population estimate[c]
0	1.000	0.998	1.000	1.017
1	0.333	0.332	0.818	0.835
2	0.238	0.238	0.765	0.782
3	0.195	0.194	0.735	0.752
4	0.169	0.168	0.715	0.731
5	0.151	0.150	0.699	0.715
6	0.138	0.137	0.686	0.702
7	0.128	0.126	0.676	0.692
8	0.119	0.117	0.667	0.683
9	0.113	0.112	0.659	0.674
10	0.107	0.106	0.652	0.667
11	0.102	0.101	0.646	0.661
12	0.098	0.097	0.640	0.655
13	0.094	0.092	0.635	0.650
14	0.090	0.089	0.631	0.645
15	0.087	0.087	0.626	0.641
16	0.084	0.084	0.622	0.637
17	0.082	0.082	0.619	0.634
18	0.080	0.079	0.615	0.630
19	0.078	0.078	0.612	0.627
20	0.076	0.075	0.609	0.624
21	0.074	0.073	0.606	0.620
22	0.072	0.071	0.603	0.618
23	0.070	0.068	0.600	0.616
24	0.069	0.069	0.598	0.613
25	0.068	0.068	0.595	0.611

[a] See text for data generation method.
[b] Standard errors for these estimates range from 0.0014 to 0.0019.
[c] Standard errors for these estimates range from 0.0096 to 0.0097.

The results of the experiments conducted are provided in Table 6.2. The experiments were designed to investigate the effects of alternative T, of alternative true models, and of using the known variance $\pi^2/6$ of the disturbance of the regression equation (6.2) as an alternative to the least squares estimator s^2, all for $d > 0$. The rule $g(T) = T^\alpha$ ($\alpha = 0.5, 0.6, 0.7$) was held fixed. The fractions reported in Table 6.2 can be regarded as point estimates of the difference between one and the true size of the confidence intervals in finite sample; with 300 replications, the standard error of these point estimates is about 0.0125.

Table 6.2. Fraction of replications* in which true d was outside 95% confidence interval for d based on Theorem 2

Model α	$d = 0.2, \phi = 0$			$d = 0.35, \phi = 0$			$d = 0.44, \phi = 0$			$d = 0.25, \phi = 0.5$		
	0.5	0.6	0.7	0.5	0.6	0.7	0.5	0.6	0.7	0.5	0.6	0.7
$T = 50, \hat{\sigma}^2 = s^2$	0.123	0.073	0.047	0.107	0.070	0.070	0.133	0.100	0.080	0.136	0.216	0.390
$T = 50, \hat{\sigma}^2 = \pi^2/6$	0.063	0.057	0.030	0.033	0.027	0.040	0.067	0.057	0.067	0.070	0.113	0.356
$T = 100, \hat{\sigma}^2 = s^2$	0.090	0.067	0.063	0.087	0.073	0.053	0.077	0.083	0.063	0.103	0.183	0.446
$T = 100, \hat{\sigma}^2 = \pi^2/6$	0.057	0.060	0.053	0.053	0.037	0.040	0.047	0.057	0.037	0.083	0.130	0.473
$T = 200, \hat{\sigma}^2 = s^2$	0.087	0.117	0.090	0.063	0.043	0.053	0.073	0.100	0.057	0.090	0.156	0.566
$T = 200, \hat{\sigma}^2 = \pi^2/6$	0.067	0.080	0.100	0.023	0.037	0.053	0.050	0.057	0.043	0.050	0.117	0.526
$T = 300, \hat{\sigma}^2 = s^2$	0.057	0.057	0.047	0.070	0.060	0.053	0.060	0.090	0.053	0.073	0.103	0.500
$T = 300, \hat{\sigma}^2 = \pi^2/6$	0.067	0.060	0.047	0.037	0.033	0.050	0.050	0.053	0.060	0.053	0.070	0.513

*300 replications, $g(T) = T^\alpha$ $(1 - \phi B)(1 - B)^d X_t = \varepsilon_t$.

The results for simple integrated series, in the first three panels of Table 6.2, are better than those for the general integrated series in the fourth panel. Most markedly, confidence intervals evidently are more reliable when the true variance $\pi^2/6$ is used than when the conventional regression s^2 is employed to construct standard errors. Second, the reliability of confidence intervals is insensitive to the size of d. Finally—by comparison with these patterns—the effect of sample size on the reliability of the confidence intervals is negligible; in any event a sample size of 100 or more certainly seems adequate. Overall, the experimental results strongly suggest that the asymptotic results in Theorem 2 are adequate in samples of models size when the true variance $\pi^2/6$ is used. They also support the conjecture that those results are valid for $d > 0$ as well as for $d < 0$.

For the general integrated series considered these generalizations remain true: true variance leads to a more reliable confidence interval than s^2, and the effect of increasing sample size is small by comparison. When $g(t) = T^{0.5}$ the results are much like those for the simple integrated series, but as more ordinates are incorporated in the log periodogram regression the reliability of the confidence intervals deteriorates. This is not surprising, since as ordinates are added the contribution of the term $\ln\{f_u(\lambda_j, T)/f_u(0)\}$ in (6.2) becomes non-negligible. The results in the last panel suggest that in empirical work $g(T)$ should be kept small if d appears sensitive to choice of $g(T)$.

6.5 Forecasting with Long-memory Models

The development of long-memory models has been motivated, in part, by the hope that through providing a better description of the low frequency portion of the spectral density than is possible with conventional models, they will also provide better several-step-ahead forecasts. It should be revealing to compare the mean square error of several-step-ahead forecasts generated by these models with those of other models, using actual time series, estimated parameter values, and out-of-sample forecasts. We undertook such a comparison, using three postwar monthly price indices.

Four different models were used to produce forecasts. The first is a simple integrated series, in which the parameter d was estimated as described in Section 6.3. The second is a general integrated series of the form

$$\phi(B)(1 - B)^d X_t = \theta(B)\varepsilon_t,$$

where $\phi(B)$ and $\theta(B)$ are polynomials of finite order. The parameter d was estimated as in the first model, and the series $(1-B)^{\hat{d}} X_t$ was formed using the frequency domain methods described in Section 6.3. Beginning with this series standard ARMA modelling procedures were used to identify and estimate $\phi(B)$ and $\theta(B)$. We have discussed reasons why these two particular models might produce good several-step-ahead forecasts if one of the long-memory models is an appropriate description of the behaviour of the series. The third model is conventional ARIMA, in which only integer differencing of the series is presumed. Since parsimoniously parameterized ARIMA models are incapable of describing the low frequency behaviour of series with long-memory properties, several-step-ahead forecasts for this model should be poor if a long memory is in fact appropriate. The final model is an autoregression of order fifty, estimated by ordinary least squares. A fiftieth order autoregression can describe a long-memory model better than a parsimoniously parameterized ARIMA model. The larger the value of $d \in (-\frac{1}{2}, \frac{1}{2})$ the more closely it approximates a simple integrated series, and it may approximate a general integrated series better than a mixed fractional difference and ARMA model. In practice, the profligate parameterization of this model would presumably increase mean square error due to imperfectly estimated parameters to a greater extent than is the case with the other three models.

The estimation procedure in the first two models requires that the number of ordinates, $g(T)$, used in the periodogram regressions be specified. In the absence of any previous experience with this estimator, in each case we chose the number of ordinates for which the corresponding value of d minimized the mean square error of 20-step-ahead, in-sample forecasts. In most cases the plot of mean square error against the number of ordinates was a smooth parabola with a well defined minimum, and in all cases the number of ordinates selected was between $T^{0.55}$ and $T^{0.6}$. Forecasts were computed by calculating the autoregressive representation implied by the estimated model, truncating at the fiftieth term, and then recursively computing n-step-ahead forecasts.

The consumer price index for food ('Food CPI', January 1947–July 1978), the Wholesale Price Index ('WPF', January 1947–February 1977) and the Consumer Price Index ('CPI', January 1947–February 1976) were selected for study. This choice was motivated in part by Mandelbrot's argument (1973) that prices which are fractional Gaussian noise will remain so even with attempted arbitrage, and in part by Granger's demonstration (1980) that aggregates of first order Markov processes are (under certain assumptions about the distribution of parameters across the processes) simple integrated series. Granger and Joyeux have studied the mean square error of forecast for Food CPI from January 1947 to July 1978, using a simple integrated series model and a grid search estimator for d.

Sample autocorrelation functions for both levels and first differences of these series are provided in Table 6.3. For the differences, these functions exhibit the slow decay typical of long-memory models but not of stable autoregressive moving average models with a few parameters. The first fifty sample autocorrelation coefficients are all positive in each case; for the Food CPI and the WPI over half of these are significantly different from zero by the usual rule of thumb, and for the CPI they are all significant. From Table 6.3 it is clear that the CPI is more sluggish than the other two series. In the first two models, it was found necessary to begin with the series $(1 - B)^{1.25}\text{CPI}_t$ (formed as described in Section 6.3) whereas for the other two only a conventional first difference was required.

Further evidence on the long-memory characteristics of these series is provided in Figs 6.1–6.3, in which $\ln\{I(\lambda_{j,T})\}$ (the 'dependent variable' in eqn (6.2)) is plotted against $\ln\{4\sin^2(\lambda_{j,T})\}$ (the 'regressor' in eqn (6.2)). For each figure the periodogram ordinates were computed from the entire sample; the number of

Table 6.3. Estimated autocorrelation function for price series

Lags	Levels			Differences		
	Food CPI	WPI	CPI	Food CPI	WPI	CPI
1	0.985	0.983	0.985	0.333	0.294	0.548
2	0.970	0.966	0.969	0.244	0.317	0.538
3	0.956	0.950	0.954	0.148	0.321	0.562
4	0.942	0.933	0.938	0.103	0.156	0.499
5	0.928	0.917	0.922	0.142	0.276	0.470
6	0.915	0.901	0.906	0.218	0.292	0.496
7	0.903	0.884	0.891	0.168	0.186	0.446
8	0.891	0.867	0.875	0.080	0.209	0.465
9	0.879	0.851	0.860	0.173	0.210	0.476
10	0.867	0.834	0.844	0.204	0.085	0.473
11	0.855	0.817	0.829	0.183	0.268	0.374
12	0.842	0.800	0.815	0.250	0.245	0.390
13	0.829	0.784	0.800	0.157	0.046	0.370
14	0.817	0.767	0.786	0.091	0.232	0.300
15	0.804	0.750	0.771	0.038	0.103	0.319
16	0.792	0.733	0.757	0.093	0.072	0.319
17	0.780	0.716	0.743	0.067	0.179	0.242
18	0.768	0.699	0.729	0.109	0.128	0.289
19	0.757	0.682	0.715	0.047	0.063	0.186
20	0.747	0.665	0.702	0.055	0.164	0.240
21	0.736	0.647	0.689	0.054	0.014	0.209
22	0.725	0.630	0.677	0.071	0.007	0.183
23	0.714	0.614	0.665	0.132	0.093	0.166
24	0.702	0.597	0.653	0.123	0.062	0.154
25	0.690	0.580	0.641	0.029	0.080	0.139
26	0.679	0.564	0.629	0.028	0.114	0.154
27	0.667	0.547	0.618	0.002	0.029	0.153
28	0.655	0.529	0.608	0.022	0.128	0.125
29	0.644	0.511	0.597	0.068	0.142	0.158
30	0.632	0.493	0.587	0.101	0.155	0.169
31	0.619	0.475	0.577	0.050	0.140	0.116
32	0.607	0.458	0.567	0.074	0.167	0.206
33	0.594	0.440	0.557	0.049	0.079	0.167
34	0.582	0.424	0.548	0.049	0.084	0.124
35	0.569	0.410	0.539	0.128	0.153	0.100
36	0.557	0.395	0.529	0.152	0.091	0.178
37	0.543	0.382	0.520	0.091	0.161	0.153
38	0.531	0.368	0.512	0.073	0.123	0.153
39	0.519	0.355	0.503	0.060	0.077	0.172
40	0.507	0.342	0.495	0.108	0.123	0.189

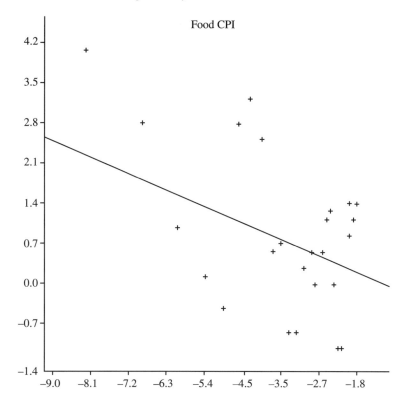

Fig. 6.1. Linear plot of log periodogram versus deterministic regressor.

ordinates was selected as just described. If the series in question were in fact a general fractional Gaussian noise or a general integrated series, then the relationship should be approximately linear, and the standard deviation of the plotted points about the line should be about $\pi/\sqrt{6} = 1.28$. These features seem characteristic of the plots for WPI and CPI (Figs 6.2 and 6.3), but more arguable for Food CPI (Fig. 6.1).

For all three series 10- and 20-step-ahead forecasts were made beginning with March 1951 and at 10-month intervals thereafter. Forecasts were made using a 50-order autoregressive equation using the representation given in Section 6.1. The parameter estimates based on the entire sample were used to produce the in-sample forecasts. For the out-of-sample forecasts parameter estimates are functions of the data preceding and including the month in which the forecast is made; the model is reestimated each time new forecasts are constructed. The mean square error of forecast is computed for the levels of the series. Results are reported in Tables 6.4–6.6.

For a given method of forecasting, it is always the case that the mean square error of the out-of-sample forecast exceeds that of the in-sample forecast, and

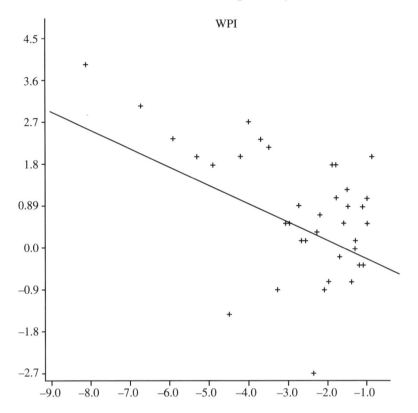

Fig. 6.2. Linear plot of log periodogram versus deterministic regressor.

of course 20-step-ahead forecasts are less accurate than 10-step-ahead forecasts. The comparatively sluggish CPI has a substantially lower mean square error of forecast (by whatever method) than the other two; the larger value of $\hat{d}$ for this series reflects the slower decay in the sample autocorrelation function reported in Table 6.3.

When comparisons across different forecasting procedures are made, the results are mixed but suggestive. The autoregression of order 50 ('AR50') provides the best out-of-sample forecasts for Food CPI and CPI, but performs miserably for WPI. Conventional ARMA models ('ARMA') generally perform badly: only for WPI did they approach the performance of the long-memory models. Mixed fractional differencing and ARMA representation ('d/ARMA') provides the best forecast for WPI, but for other series was superior only to the ARMA model. The forecasts provided by the simple integrated series model ('d') were never the best, but they were never markedly inferior to the others, either. The mean square error of 20-step-ahead forecasts for the simple integrated series model is on average 10% higher than that of the best set of forecasts (from whichever model). By contrast, for mixed integrated series and ARMA it is 31%, for ARMA 73%, and

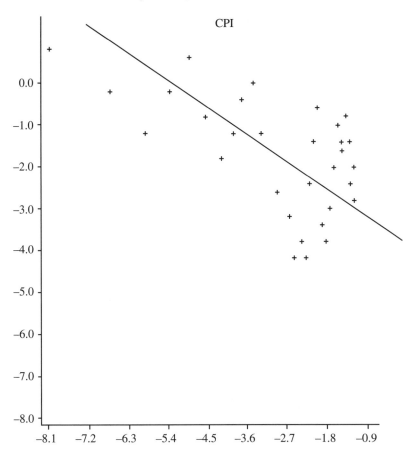

Fig. 6.3. Linear plot of log periodogram versus deterministic regressor.

Table 6.4. Comparison of mean square error of forecast
Consumer Food Price Index, 1947–1978(6)

Estimation method	Mean square error of forecast			
	In sample		Out of sample	
	10-step	20-step	10-step	20-step
d[a]	8.359	17.837	26.310	36.223
d/ARMA[b]	7.804	16.996	34.174	46.305
ARMA[c]	8.969	18.647	41.727	56.732
AR50	8.046	16.754	24.017	33.477
$\hat{d} = .35$	8.361	17.923	28.360	37.443
(Granger and Joyeux)				

[a] $\hat{d} = 0.423$, standard error $= 0.146$; $T = 378$, $g(T) = 26 \doteq T^{0.55}$.
[b] $(1 - B)^{\hat{d}} X_t = (1 - 0.188B - 0.03B^2 - 0.1B^3 - 0.15B^4)\varepsilon_t$, where X_t is the first differenced series.
[c] $(1 - 0.33B)X_t = 0.26 + \varepsilon_t$, where X_t is the first differenced series.

Table 6.5. Comparison of mean square error of forecast
Wholesale Price Index, 1947–1977

Estimation method	Mean square error of forecast			
	In sample		Out of sample	
	10-step	20-step	10-step	20-step
d[a]	7.830	16.675	10.200	22.540
d/ARMA[b]	8.258	17.259	8.773	19.763
ARMA[c]	8.739	17.736	8.832	25.195
AR50	9.010	18.030	38.212	49.242

[a] $\hat{d} = 0.417$, standard error $= 0.112$; $T = 363$, $g(T) = 39 \doteq T^{0.6}$.
[b] $(1 - B)^{\hat{d}} X_t = (1 - 0.25B - 0.15B^2)(1 + 0.14B^{12})\varepsilon_t$, where X_t is the first differenced series.
[c] $(1 - 0.18B - 0.19B^2 - 0.19B^3 - 0.16B^5)X_t = 0.12 + \varepsilon_t$, where X_t is the first differenced series.

Table 6.6. Comparison of mean square error of
forecast Consumer Price Index, 1947–1976

Estimation method	Mean square error of forecast			
	In sample		Out of sample	
	10-step	20-step	10-step	20-step
d[a]	0.838	2.140	1.895	3.325
d/ARMA[b]	0.732	1.852	2.775	4.792
ARMA[c]	1.311	2.782	3.751	7.474
AR50	1.402	2.717	1.417	3.079

[a] $\hat{d} = 0.701$, standard error $= 0.155$; $T = 351$, $g(T) = 34 \doteq T^{0.6}$.
[b] $(1 - B)^{\hat{d}} X_t = (1 - 0.5B)\varepsilon_t$, where X_t is the first differenced series.
[c] $(1 - 0.22B - 0.19B^2 - 0.26B^3 - 0.1B^4 - 0.06B^5)X_t = \varepsilon_t$, where X_t is the first differenced series.

for an autoregression of order fifty 50%. The results suggest that use of a simple integrated series model may be a reasonable, conservative procedure for producing several-step-ahead univariate forecasts. Further experimentation along these lines with other series is clearly required to confirm this judgment.

References

Buck, R. C. (1978), *Advanced Calculus*, McGraw-Hill: New York.
Durbin, J. (1960), 'The fitting of time series models', *Review of the International Statistical Institute*, 28, 233–44.

Gradshteyn, I. S. and Ryzhik, I. M. (1980), *Tables of Integrals, Series and Products*. Academic Press: New York.

Granger, C. W. G. and Joyeux, R. (1980), 'An introduction to long memory time series models and fractional differencing,' *Journal of Time Series Analysis*, 1 (1), 15–29.

—— (1980), 'Long memory relationships and the aggregation of dynamic models', *Journal of Econometrics*, 14, 227–38.

Hannan, E. J. (1973), 'Central limit theorems for time series regression', *Zeitschrift für Wahrscheinlichkeitstheorie und Verwandte Gebiete*, 26, 157–70.

Hosking, J. R. M. (1981), 'Fractional differencing', *Biometrika*, 68 (1), 165–76.

Hurst, H. E., (1951), 'Long term storage capacity of reservoirs', Transactions of the American Society of Civil Engineers, 116, 770–99.

Jonas, A. (1981), 'Long memory self similar time series models', Unpublished Harvard University manuscript.

Kendall, M. and Stuart, A. (1977), *The Advanced Theory of Statistics*, Vol. 1, 4th ed. Charles Griffin: London.

Levinson, N. (1947), 'The Wiener RMS criterion in filter design and prediction', *Journal of Mathematical Physics*, 25, 261–78.

Mandelbrot, B. B. and van Ness, J. W. (1968), 'Fractional Brownian motion, fractional noises and applications', *SIAM Review*, 10, 422–37.

—— (1971), 'A fast-fractional Gaussian noise generator', *Water Resources Research* 7, 543–53.

—— (1973), 'Statistical methodology for nonperiodic cycles: from the covariance to R/S analysis', *Review of Economic and Social Measurement*, pp. 259–90.

McLeod, A. I. and Hipel, K. W. (1978), 'Preservation of the rescaled adjusted range, I. A. reassessment of the Hurst phenomenon', *Water Resources Research*, 14(3), 491–518.

Porter-Hudak, S. (1982), 'Long-term memory modelling—a simplified spectral approach', Unpublished University of Wisconsin Ph.D. thesis.

Rozanov, Y. A. (1967), *Stationary Random Processes*. Holden-Day: San Francisco.

Whittle, P. (1963), 'On the fitting of multivariate autoregressions, and the approximate canonical factorization of spectral density matrix', *Biometrika*, 50, 129–34.

7

Gaussian Semiparametric Estimation of Long-range Dependence

P. M. ROBINSON

7.1 Introduction

Several estimates are now available for the slope of the logged spectral density of a long-range dependent covariance stationary scalar process $x_t, t = 0, \pm 1, \ldots$, which is observed at times $t = 1, \ldots, n$. Denote by γ_j the lag-j autocovariance of x_t and by $f(\lambda)$ the spectral density of x_t, such that $\gamma_j = E(x_0 - Ex_0)(x_j - Ex_0) = \int_{-\pi}^{\pi} \cos(j\lambda) f(\lambda) \, d\lambda$. It is assumed that

$$f(\lambda) \sim G\lambda^{1-2H} \quad \text{as } \lambda \to 0+, \tag{7.1}$$

for $G \in (0, \infty)$ and $H \in (0, 1)$. The parameter H is sometimes called the self-similarity parameter. In case $H = \frac{1}{2}$, $f(\lambda)$ tends to a finite positive constant at zero frequency, whereas if $H \in (\frac{1}{2}, 1)$ it tends to infinity and if $H \in (0, \frac{1}{2})$ it tends to zero. A recently published survey of relevant literature up to about 1990 is Robinson (1994a).

Finite-parameter models for $f(\lambda)$ over the full frequency band $(-\pi, \pi]$ which are consistent with the property (7.1), have been considered (such as fractional autoregressive moving average models), as have methods of estimating an unknown H and additional parameters. In particular, the asymptotic distributional properties of Gaussian parameter estimates have been derived by Fox and Taqqu (1986), Dahlhaus (1989) and Giraitis and Surgailis (1990), in case $H \in (\frac{1}{2}, 1)$ and under regularity conditions. These properties are highly desirable ones: $n^{1/2}$-consistency and asymptotic normality and, when x_t is actually Gaussian, asymptotic efficiency. However, these properties also depend on correct specification of $f(\lambda)$ over $(-\pi, \pi]$, and in the event of any misspecification, estimates will, in general, be inconsistent. In particular, misspecification of $f(\lambda)$ at high frequencies can lead to an inconsistent estimate of the parameter H, which characterizes low frequency behaviour. The spectral density of a fractional autoregressive moving average model has the rather strange mathematical property of

This work was previously published as Robinson, P. M. (1995) 'Gaussian semiparametric estimation of long-range dependence', *The Annals of Statistics*, 23, 1630–61.

This research was funded by the Economic and Social Research Council (ESRC) reference number R000233609.

I thank two referees for helpful comments, Joris Pinkse, Ignacio Lobato and Paolo Zaffaroni for carrying out the numerical work and Ignacio Lobato for correcting a number of errors.

being infinitely differentiable at all frequencies in $[-\pi, \pi]$ except at zero frequency, where (for $H > \frac{1}{2}$) it is discontinuous and unbounded.

To overcome such criticisms, 'semiparametric' estimates of H in (7.1) have been proposed which can be justified as consistent in the absence of full parametric or other global assumptions on $f(\lambda)$. One of these, due to Robinson (1994*b*), is consistent when $H \in (\frac{1}{2}, 1)$ under mild conditions which do not include Gaussianity or even loose restrictions on $f(\lambda)$ away from zero frequency (apart from integrability on $(-\pi, \pi]$, a consequence of covariance stationarity). The limiting distributional properties of this estimate are rather complicated, however (see Lobato and Robinson (1996)). One alternative semiparametric estimate of H is due to Geweke and Porter-Hudak (1983). Robinson (1995) recently established desirable asymptotic properties of modified and more efficient versions of this estimate, but under the restrictive condition of Gaussianity, which may not be necessary but seems difficult to replace by weaker but reasonably comprehensible conditions because the estimate involves nonlinear transformation of the periodogram.

This chapter discusses a semiparametric 'Gaussian' estimate of H which, unlike the estimates of Geweke and Porter-Hudak (1983) and Robinson (1995), is not defined in closed form, but which dominates these estimates in several respects. It is asymptotically more efficient. Much weaker assumptions than Gaussianity are imposed. (Giraitis and Surgailis (1990) likewise avoided Gaussianity in their study of a parametric Gaussian estimate.) Trimming out low frequency components (introduced in Robinson's (1995) estimate following a suggestion of Künsch (1986)) is avoided, as is the additional user-chosen number needed for Robinson's (1994*b*) estimate. The current estimate was suggested by Künsch (1987), but he did not establish or conjecture any statistical properties. The estimate is described in the following section. In Section 7.3 it is shown to be weakly consistent under only slight additional conditions to (7.1), with only finite second moments of x_t assumed. In Section 7.4 the estimate is shown to be asymptotically normally distributed under somewhat stronger conditions, including a fourth moment condition on x_t. Three technical lemmas are proved in Section 7.5. The proof of consistency and asymptotic normality involves some relatively unusual features. The objective function optimized by the (implicitly defined) estimate behaves in a nonuniform way over the parameter space. The mean-value theorem argument used in the central limit theorem requires not just a preliminary consistency proof for the estimate, but a rate of consistency. Throughout, we aim for minimal conditions on the spectrum away from zero frequency and on the bandwidth number involved in the semiparametric estimation, and our proofs cover simultaneously cases $0 < H < \frac{1}{2}$, $H = \frac{1}{2}$ and $1 > H > \frac{1}{2}$, whether or not H is known a priori to be consistent with an infinite or a zero spectrum at zero frequency. Section 7.6 contains Monte Carlo evidence of finite-sample performance, and comparison with an estimate similar to those of Geweke and Porter-Hudak (1983), Künsch (1986) and Robinson (1995), as well as an application to the series of Nile minima.

It has been stated in the literature that fractional autoregressive moving average processes which satisfy (7.1) also satisfy the time domain property

$$\gamma_j \sim g j^{2H-2} \quad \text{as } j \to \infty, \tag{7.2}$$

where $g < 0$ for $0 < H < \frac{1}{2}$ and $g > 0$ for $\frac{1}{2} < H < 1$. Condition (7.2) (extended to include a slowly varying function) has been stressed by Taqqu (1975), for example. With $g = 2G\Gamma(2 - 2H)\cos\pi H$, it is known that for $0 < H < \frac{1}{2}$, (7.2) implies (7.1) (Yong 1974: p. 90), whereas for $\frac{1}{2} < H < 1$, (7.1) and (7.2) are equivalent if the γ_j are quasimonotonically convergent to zero, that is, $\gamma_j \to 0$ as $j \to \infty$ and for some $C < \infty$, $\gamma_{j+1} \leq \gamma_j(1 + C/j)$ for all large enough j (Yong 1974: p. 75). In general, (7.1) does not imply (7.2).

7.2 Semiparametric Gaussian Estimate

Define the discrete Fourier transform and periodogram of x_t:

$$w(\lambda) = (2\pi n)^{-1/2} \sum_{t=1}^{n} x_t e^{it\lambda}, \qquad I(\lambda) = |w(\lambda)|^2, \qquad (7.3)$$

where correction for an unknown mean of x_t is unnecessary because the statistics (7.3) will be computed only at frequencies $\lambda_j = 2\pi j/n$ for $j = 1, \ldots, m$, where m is an integer less than $\frac{1}{2}n$. Because our estimate is not defined in closed form, it is convenient to denote by G_0 and H_0 the true parameter values, and by G and H any admissible values.

Consider the objective function (see Künsch 1987)

$$Q(G, H) = \frac{1}{m} \sum_{j=1}^{m} \left\{ \log G\lambda_j^{1-2H} + \frac{\lambda_j^{2H-1}}{G} I_j \right\}, \qquad (7.4)$$

writing $I_j = I(\lambda_j)$. Define the closed interval of admissible estimates of H_0, $\Theta = [\Delta_1, \Delta_2]$, where Δ_1 and Δ_2 are numbers picked such that $0 < \Delta_1 < \Delta_2 < 1$. We can choose Δ_1 and Δ_2 arbitrarily close to 0 and 1, respectively, or we can choose them to reflect weak prior knowledge on H_0, for example, $\Delta_1 = \frac{1}{2}$ if we are confident that $f(\lambda) \not\to 0$ as $\lambda \to 0$. Clearly the estimate

$$(\hat{G}, \hat{H}) = \arg\min_{\substack{0 < G < \infty \\ H \in \Theta}} Q(G, H)$$

exists. We can also write

$$\hat{H} = \arg\min_{H \in \Theta} R(H),$$

where

$$R(H) = \log \hat{G}(H) - (2H - 1)\frac{1}{m}\sum_1^m \log \lambda_j, \qquad \hat{G}(H) = \frac{1}{m}\sum_1^m \lambda_j^{2H-1} I_j.$$

Were we to take $m = \frac{1}{2}n$ for n even, or $m = \frac{1}{2}(n - 1)$ for n odd, $\hat{H}$ would be a Gaussian estimate of H_0 in the parametric model $f(\lambda) = G_0|\lambda|^{1-2H_0}$, $\lambda \in (-\pi, \pi]$, and the asymptotic theory of $\hat{H}$ would be effectively covered by that of Fox and Taqqu (1986) and Giraitis and Surgailis (1990) in case $H_0 \in (\frac{1}{2}, 1)$ and $\Delta_1 > \frac{1}{2}$, although these authors considered an integral form in place of the summation form (7.4). In our asymptotic theory, m tends to infinity more slowly than n, so that the proportion of the frequency band $(-\pi, \pi]$ involved in the estimation degenerates relatively slowly to zero as n increases. The derivation of such asymptotic theory is quite different from that of Fox and Taqqu (1986), Dahlhaus (1989) and Giraitis and Surgailis (1990) for parametric Gaussian estimates. Our $\hat{H}$ is only $m^{1/2}$-consistent and is thus much less efficient than these estimates when they happen to be based on a correct parametric model. Like Giraitis and Surgailis (1990), we avoid the Gaussianity assumption of Fox and Taqqu (1986) and Dahlhaus (1989), and unlike any of these authors, we allow H_0 to be $\frac{1}{2}$ or less than $\frac{1}{2}$, as well as greater than $\frac{1}{2}$, also permitting the set of admissible values of $\hat{H}$ to include ones in $(0, \frac{1}{2}]$ as well as $(\frac{1}{2}, 1)$. An integral form of (7.4) could be considered, but we prefer the discrete form (7.4), partly for its computational convenience and partly because, in case of an unknown mean of x_t, it avoids dependence on the sample mean; for parametric Gaussian estimates, Cheung and Diebold (1994) have found that the slow convergence of the sample mean when $H_0 > \frac{1}{2}$ can produce inferior finite sample behaviour in estimates of H_0 based on the integral form of objective function.

7.3 Consistency of Estimates

The following assumptions are introduced.

Assumption A1 As $\lambda \to 0+$,

$$f(\lambda) \sim G_0 \lambda^{1-2H_0},$$

where $G_0 \in (0, \infty)$ and $H_0 \in [\Delta_1, \Delta_2]$.

Assumption A2 In a neighbourhood $(0, \delta)$ of the origin, $f(\lambda)$ is differentiable and

$$\frac{\mathrm{d}}{\mathrm{d}\lambda} \log f(\lambda) = O(\lambda^{-1}) \quad \text{as } \lambda \to 0+.$$

Assumption A3 We have

$$x_t - Ex_0 = \sum_{j=0}^{\infty} \alpha_j \varepsilon_{t-j}, \quad \sum_{j=0}^{\infty} \alpha_j^2 < \infty, \tag{7.5}$$

where

$$E(\varepsilon_t | F_{t-1}) = 0, \quad E(\varepsilon_t^2 | F_{t-1}) = 1, \quad \text{a.s., } t = 0, \pm 1, \ldots,$$

in which F_t is the σ-field of events generated by ε_s, $s \le t$, and there exists a random variable ε such that $E\varepsilon^2 < \infty$ and for all $\eta > 0$ and some $K > 0$, $P(|\varepsilon_t| > \eta) \le KP(|\varepsilon| > \eta)$.

Assumption A4 As $n \to \infty$,

$$\frac{1}{m} + \frac{m}{n} \to 0.$$

Assumption A1 is just the basic model (7.1), with H_0 contained in the interval of admissible estimates $[\Delta_1, \Delta_2]$, while Assumption A2 is a regularity condition, analogous to ones imposed in the parametric case by Fox and Taqqu (1986) and Giraitis and Surgailis (1990). Assumption A3 takes the innovations in the Wold representation (7.5) to be a square-integrable martingale difference sequence that need not be strictly stationary, but satisfies a mild homogeneity restriction. Assumption A4 is minimal, because m must tend to infinity for consistency, while it must do so more slowly than n because A1 specifies $f(\lambda)$ only as $\lambda \to 0+$. The estimate of Robinson (1994b) was shown to be consistent under conditions which are in some respects weaker, but the latter estimate may be less attractive for practical use than $\hat{H}$ because it is sensitive not only to m, but to an additional user-chosen constant.

Theorem 1 Let Assumptions A1–A4 hold. Then

$$\hat{H} \to_p H_0 \quad \text{as } n \to \infty.$$

Proof For $\frac{1}{2} > \delta > 0$ let $N_\delta = \{H: |H - H_0| < \delta\}$ and $\overline{N}_\delta = (-\infty, \infty) - N_\delta$. Then for $S(H) = R(H) - R(H_0)$,

$$P(|\hat{H} - H_0| \geq \delta) = P(\hat{H} \in \overline{N}_\delta \cap \Theta)$$

$$= P\left(\inf_{\overline{N}_\delta \cap \Theta} R(H) \leq \inf_{N_\delta \cap \Theta} R(H)\right) \leq P\left(\inf_{\overline{N}_\delta \cap \Theta} S(H) \leq 0\right),$$

because $H_0 \in N_\delta \cap \Theta$. Now define $\Theta_1 = \{H: \Delta \leq H \leq \Delta_2\}$, where $\Delta = \Delta_1$ when $H_0 < \frac{1}{2} + \Delta_1$ and $H_0 \geq \Delta > H_0 - \frac{1}{2}$ otherwise. When $H_0 \geq \frac{1}{2} + \Delta_1$, define $\Theta_2 = \{H : \Delta_1 \leq H < \Delta\}$, and otherwise take Θ_2 to be empty. It follows that

$$P(|\hat{H} - H_0| \geq \delta) \leq P\left(\inf_{\overline{N}_\delta \cap \Theta_1} S(H) \leq 0\right) + P\left(\inf_{\Theta_2} S(H) \leq 0\right). \qquad (7.6)$$

It is necessary to treat Θ_1 and Θ_2 separately because of the nonuniform behaviour of $R(H)$ around $H = H_0 - \frac{1}{2}$. The set Θ_2 is empty when, for example, $\Delta_1 \geq \frac{1}{2}$ (so knowledge that $H_0 \geq \frac{1}{2}$ is used in fixing Θ) or when, on the other hand, $H_0 < \frac{1}{2}$. The first probability on the right of (7.6) is bounded by

$$P\left(\sup_{\Theta_1} |T(H)| \geq \inf_{\overline{N}_\delta \cap \Theta_1} U(H)\right) \qquad (7.7)$$

with the definitions

$$T(H) = \log\left\{\frac{\hat{G}(H)}{G(H)}\right\} - \log\left\{\frac{\hat{G}(H_0)}{G_0}\right\} - \log\left\{\frac{1}{m}\sum_1^m j^{2(H-H_0)} \middle/ \frac{m^{2(H-H_0)}}{2(H-H_0)+1}\right\}$$

$$+ 2(H - H_0)\left\{\frac{1}{m}\sum_1^m \log j - (\log m - 1)\right\},$$

$$U(H) = 2(H - H_0) - \log\{2(H - H_0) + 1\},$$

$$G(H) = G_0\frac{1}{m}\sum_1^m \lambda_j^{2(H-H_0)},$$

so that $S(H) = U(H) - T(H)$. Here, $U(H)$ is the deterministic part of $S(H)$ obtained by replacing I_j by $G\lambda_j^{1-2H}$ and Riemann sums by integrals, $T(H)$ being the remainder. Because the function $x - \log(1 + x)$ achieves a unique relative and absolute minimum on $(-1, \infty)$ at $x = 0$, and because $\log(1 + x) \le x - \frac{1}{6}x^2$ and $-\log(1 - x) \ge x + \frac{1}{2}x^2$ for $0 < x < 1$, it follows that

$$\inf_{\overline{N}_\delta \cap \Theta_1} U(H) \ge \min(2\delta - \log(1 + 2\delta), \qquad -2\delta - \log(1 - 2\delta)) > \frac{1}{2}\delta^2. \qquad (7.8)$$

On the other hand, from the inequality $|\log(1 + x)| \le 2|x|$ for $|x| \le \frac{1}{2}$ we deduce that, for any nonnegative random variable Y,

$$P(|\log Y| \le \varepsilon) \le 2P(|Y - 1| \le 2\varepsilon) \quad \text{when } \varepsilon \le 1, \qquad (7.9)$$

and thus that $\sup_{\Theta_1} |T(H)| \to_p 0$ if

$$\sup_{\Theta_1}\left|\frac{\hat{G}(H) - G(H)}{G(H)}\right| \qquad (7.10)$$

is $o_p(1)$, while

$$\sup_{\Theta_1}\left|\frac{2(H - H_0) + 1}{m}\sum_1^m \left(\frac{j}{m}\right)^{2(H-H_0)} - 1\right| \qquad (7.11)$$

and

$$\left|\frac{1}{m}\sum_1^m \log j - (\log m - 1)\right| \qquad (7.12)$$

are both $o(1)$. Now (7.11) is $O(m^{-2(\Delta - H_0)-1}) \to 0$ as $m \to \infty$ from Lemma 1 (see Section 7.5), while (7.12) is $O(\log m/m)$ from Lemma 2 (see Section 7.5). Write

$$\frac{\hat{G}(H) - G(H)}{G(H)} = \frac{A(H)}{B(H)}, \qquad (7.13)$$

where

$$A(H) = \frac{2(H - H_0) + 1}{m} \sum_1^m \left(\frac{j}{m}\right)^{2(H-H_0)} \left(\frac{I_j}{g_j} - 1\right),$$

$$B(H) = \frac{2(H - H_0) + 1}{m} \sum_1^m \left(\frac{j}{m}\right)^{2(H-H_0)},$$

(7.14)

for $g_j = G_0 \lambda_j^{1-2H_0}$. Now

$$\inf_{\Theta_1} B(H) \geq 1 - \sup_{\Theta_1} \left| \frac{2(H - H_0) + 1}{m} \sum_1^m \left(\frac{j}{m}\right)^{2(H-H_0)} - 1 \right| \geq \frac{1}{2}, \quad (7.15)$$

for all sufficiently large m, by Lemma 1. By summation by parts

$$|A(H)| \leq \frac{3}{m} \left| \sum_{r=1}^{m-1} \left\{ \left(\frac{r}{m}\right)^{2(H-H_0)} - \left(\frac{r+1}{m}\right)^{2(H-H_0)} \right\} \sum_{j=1}^r \left(\frac{I_j}{g_j} - 1\right) \right|$$

$$+ \frac{3}{m} \left| \sum_1^m \left(\frac{I_j}{g_j} - 1\right) \right|.$$

(7.16)

Because $|(1 + 1/r)^{2(H-H_0)} - 1| \leq 2/r$ on Θ_1 when $r > 0$, the first term on the right of (7.16) has supremum on Θ_1 bounded by

$$6 \sup_{\Theta_1} \sum_1^{m-1} \left(\frac{r}{m}\right)^{2(H-H_0)+1} \frac{1}{r^2} \left| \sum_1^r \left(\frac{I_j}{g_j} - 1\right) \right|$$

$$\leq 6 \sum_1^{m-1} \left(\frac{r}{m}\right)^{2(\Delta - H_0)+1} \frac{1}{r^2} \left| \sum_1^r \left(\frac{I_j}{g_j} - 1\right) \right|, \quad (7.17)$$

the inequality being due to $0 < 2(\Delta - H_0) + 1 \leq 2(H - H_0) + 1$ on Θ_1. Now

$$\frac{I_j}{g_j} - 1 = \left(1 - \frac{g_j}{f_j}\right) \frac{I_j}{g_j} + \frac{1}{f_j}(I_j - |\alpha_j|^2 I_{\varepsilon j}) + (2\pi I_{\varepsilon j} - 1), \quad (7.18)$$

where $I_{\varepsilon j} = I_\varepsilon(\lambda_j) = |\omega_\varepsilon(\lambda_j)|^2$, $\omega_\varepsilon(\lambda) = (2\pi n)^{-1/2} \sum_{t=1}^n \varepsilon_t e^{it\lambda}$, $f_j = f(\lambda_j)$ and $\alpha_j = \alpha(\lambda_j) = \sum_{l=0}^\infty \alpha_l e^{il\lambda_j}$. For any $\eta > 0$, Assumptions A1 and A4 imply that n can be chosen such that

$$\left| 1 - \frac{g_j}{f_j} \right| \leq \eta, \quad j = 1, \ldots, m. \quad (7.19)$$

Let C be a generic finite positive constant. Assumptions A1, A2, and A4 imply that for n sufficiently large,

$$E \left| \frac{I_j}{g_j} \right| \leq C, \quad j = 1, \ldots, m, \quad (7.20)$$

in view of the proof of theorem 2 of Robinson (1995). Thus

$$E\left\{\sum_1^{m-1}\left(\frac{r}{m}\right)^{2(\Delta-H_0)+1}\frac{1}{r^2}\left|\sum_1^r\left(1-\frac{g_j}{f_j}\right)\frac{I_j}{g_j}\right|\right\} \le \frac{C\eta}{m}\sum_1^m\left(\frac{r}{m}\right)^{2(\Delta-H_0)}$$

$$\le \frac{C\eta}{2(\Delta-H_0)+1},$$

because $\sum_1^m r^a \le m^{a+1}/(a+1)$ for $a > -1$. Next,

$$E|I_j - |\alpha_j|^2 I_{\varepsilon j}| \le E\{|w_j - \alpha_j w_{\varepsilon j}||w_j + \alpha_j w_{\varepsilon j}|\}$$

$$\le (EI_j - \alpha_j E w_{\varepsilon j}\overline{w}_j - \overline{\alpha}_j E\overline{w}_{\varepsilon j}w_j + |\alpha_j|^2 EI_{\varepsilon j})^{1/2}$$

$$\times (EI_j + \alpha_j E w_{\varepsilon j}\overline{w}_j + \overline{\alpha}_j E\overline{w}_{\varepsilon j}w_j + |\alpha_j|^2 EI_{\varepsilon j})^{1/2},$$

$$(7.21)$$

where the first inequality is due to $||a|^2 - |b|^2| = |\text{Re}\{(a - b)(\overline{a} + \overline{b})\}| \le |(a - b)(\overline{a} + \overline{b})| \le |a - b||a + b|$ and the second to the Schwarz inequality. It follows from the proof of theorem 2 of Robinson (1995) that, as $n \to \infty$,

$$EI_j = f_j\left(1 + O\left(\frac{\log(j+1)}{j}\right)\right),$$

$$E w_j \overline{w}_{\varepsilon j} = \frac{\alpha_j}{2\pi} + O\left(\frac{\log(j+1)}{j}\lambda_j^{1/2-H_0}\right),$$

$$EI_{\varepsilon j} = \frac{1}{2\pi} + O\left(\frac{\log(j+1)}{j}\right)$$

uniformly in $j = 1, \ldots, m$. Thus (7.21) is $O(f_j(\log(j+1)/j)^{1/2})$ as $n \to \infty$ and

$$E\left\{\sum_1^{m-1}\left(\frac{r}{m}\right)^{2(\Delta-H_0)+1}\frac{1}{r^2}\left|\sum_1^r\frac{1}{f_j}(I_j - |\alpha_j|^2 I_{\varepsilon j})\right|\right\}$$

$$\le C\sum_1^m\left(\frac{r}{m}\right)^{2(\Delta-H_0)+1}\frac{1}{r^2}\sum_1^r\left(\frac{\log j}{j}\right)^{1/2}$$

$$\le \frac{C}{m^{2(\Delta-H_0)+1}}\sum_1^m r^{2(\Delta-H_0)-1/2}(\log r)^{1/2}$$

$$= O\left(m^{2(H_0-\Delta)-1} + (\log m)^{3/2}m^{-1/2}\right)$$

$$= o(1) \quad \text{as } n \to \infty,$$

where the penultimate equality results from separate consideration of the cases $2(\Delta - H_0) - \frac{1}{2} < -1$ and $2(\Delta - H_0) - \frac{1}{2} \ge -1$. To deal with the final contribution to (7.18), write

$$2\pi I_{\varepsilon j} - 1 = \frac{1}{n} \sum_{1}^{n} (\varepsilon_t^2 - 1) + \frac{1}{n} \sum_{s \neq t} \sum \cos\{(s - t)\lambda_j\} \varepsilon_s \varepsilon_t,$$

so that

$$\sum_{1}^{m-1} \left(\frac{r}{m}\right)^{2(\Delta-H_0)+1} \frac{1}{r^2} \left| \sum_{1}^{r} (2\pi I_{\varepsilon j} - 1) \right|$$

$$\leq \left| \frac{1}{n} \sum_{1}^{n} (\varepsilon_t^2 - 1) \right| \sum_{1}^{m} \left(\frac{r}{m}\right)^{2(\Delta-H_0)+1} \frac{1}{r}$$

$$+ \sum_{1}^{m} \left(\frac{r}{m}\right)^{2(\Delta-H_0)+1} \frac{1}{r^2 n} \left| \sum_{s \neq t} \sum \varepsilon_s \varepsilon_t \sum_{1}^{r} \cos\{(s - t)\lambda_j\} \right|. \qquad (7.22)$$

Under Assumption A3,

$$\frac{1}{n} \sum_{1}^{n} (\varepsilon_t^2 - 1) \to_p 0 \qquad (7.23)$$

from theorem 1 of Heyde and Seneta (1972), so the first term on the right of (7.22) is $o_p(1)$. Assumption A3 also implies that

$$E \left(\sum_{s \neq t} \sum \varepsilon_s \varepsilon_t \sum_{1}^{r} \cos\{(s - t)\lambda_j\} \right)^2$$

$$= 2 \sum_{s \neq t} \sum \left(\sum_{1}^{r} \cos\{(s - t)\lambda_j\} \right)^2$$

$$= 2 \sum_{1}^{r} \sum_{1}^{r} \left[\sum_{1}^{n} \sum_{1}^{n} \cos\{(s - t)\lambda_j\} \cos\{(s - t)\lambda_k\} - n \right]$$

$$= rn^2 - 2r^2 n \qquad (7.24)$$

for $1 \leq r < \frac{1}{2}n$, so that the second term on the right of (7.22) is

$$o \left(\sum_{1}^{m} \left(\frac{r}{m}\right)^{2(\Delta-H_0)+1} r^{-3/2} \right) = O(m^{2(H_0-\Delta)-1} + (\log m)m^{-1/2}) = o(1)$$

as $n \to \infty$. Because η is arbitrary, we have proved that (7.17) is $o_p(1)$. Using the same techniques, as $n \to \infty$,

$$\left| \frac{1}{m} \sum_{1}^{m} \left(\frac{I_j}{g_j} - 1\right) \right| = O_p \left(\eta + \frac{1}{m} \sum_{1}^{m} \left(\frac{\log j}{j}\right)^{1/2} \right) + o_p(1) = o_p(1),$$

so the second term on the right of (7.16) is $o_p(1)$. Thus as $n \to \infty$, $\sup_{\Theta_1} |A(H)| \to_p 0$ and, with (7.13) and (7.15), $\sup_{\Theta_1} |\hat{G}(H)/G(H) - 1| \to_p 0$. In view of (7.8) it follows that (7.7) $\to 0$ as $n \to \infty$.

In case $H_0 < \frac{1}{2} + \Delta_1$, the proof of the theorem is completed. In case $H_0 \geq \frac{1}{2} + \Delta_1$, the second probability on the right of (7.6) can be nonzero. Put $p = p_m = \exp(m^{-1} \sum_1^m \log j)$ and $S(H) = \log\{\hat{D}(H)/\hat{D}(H_0)\}$, where

$$\hat{D}(H) = \frac{1}{m} \sum_1^m \left(\frac{j}{p}\right)^{2(H-H_0)} j^{2H_0-1} I_j.$$

Because $1 \leq p \leq m$ and $\inf_{\Theta_2} (j/p)^{2(H-H_0)} \geq (j/p)^{2(\Delta-H_0)}$ for $1 \leq j \leq p$, while $\inf_{\Theta_2} (j/p)^{2(H-H_0)} \geq (j/p)^{2(\Delta_1-H_0)}$ for $p < j \leq m$, it follows that

$$\inf_{\Theta_2} \hat{D}(H) \geq \frac{1}{m} \sum_1^m a_j j^{2H_0-1} I_j,$$

where

$$a_j = \begin{cases} \left(\frac{j}{p}\right)^{2(\Delta-H_0)}, & 1 \leq j \leq p, \\ \left(\frac{j}{p}\right)^{2(\Delta_1-H_0)}, & p < j \leq m. \end{cases}$$

Thus,

$$P\left(\inf_{\Theta_2} S(H) \leq 0\right) \leq P\left(\frac{1}{m} \sum_1^m (a_j - 1) j^{2H_0-1} I_j \leq 0\right). \tag{7.25}$$

As $m \to \infty$, $p \sim \exp(\log m - 1) = m/e$ and

$$\sum_{1 \leq j \leq p} a_j \sim p^{2(H_0-\Delta)} \int_0^p x^{2(\Delta-H_0)} \, dx$$

$$= \frac{p}{2(\Delta - H_0) + 1} \sim \frac{m/e}{2(\Delta - H_0) + 1}. \tag{7.26}$$

It follows that

$$\frac{1}{m} \sum_1^m (a_j - 1) \geq \frac{1}{m} \sum_{1 \leq j \leq p} a_j - 1 \sim \frac{1}{e(2(\Delta - H_0) + 1)} - 1 \quad \text{as } m \to \infty.$$

Choose $\Delta < H_0 - \frac{1}{2} + \frac{1}{4}e$, which we may do with no loss of generality. Then for all sufficiently large m, $m^{-1} \sum_1^m (a_j - 1) \geq 1$ and thus (7.25) is bounded by

$$P\left(\left|\frac{1}{m} \sum_1^m (a_j - 1)\left(\frac{I_j}{g_j} - 1\right)\right| \geq 1\right).$$

Now apply (7.18) again and first note from (7.19) and (7.20) that

$$\left| \frac{1}{m} \sum_1^m (a_j - 1)\left(1 - \frac{g_j}{f_j}\right)\frac{I_j}{g_j} \right| = O_p\left(\frac{\eta}{m}\sum_1^m (a_j + 1)\right) = O_p(\eta),$$

using also (7.21) and

$$\sum_{p < j \leq m} a_j \sim p^{2(H_0 - \Delta_1)} \int_p^m x^{2(\Delta_1 - H_0)}\, dx = O(m). \qquad (7.27)$$

Next we have from (7.26) and theorem 2 of Robinson (1995) that

$$\left| \frac{1}{m} \sum_1^m \frac{(a_j - 1)}{f_j}(I_j - |\alpha_j|^2 I_{\varepsilon j}) \right| = O_p\left(\frac{1}{m}\sum_1^m (a_j + 1)\left(\frac{\log(j+1)}{j}\right)^{1/2} \right)$$

$$= O_p\left(\frac{\log m}{m}\left(\sum_1^m a_j^2 + m\right)^{1/2} \right) \qquad (7.28)$$

by the Cauchy inequality. Because

$$\sum_1^m a_j^2 \leq p^{4(H_0 - \Delta)}\sum_1^p j^{4(\Delta - H_0)} + p^{4(H_0 - \Delta_1)}\sum_p^m j^{4(\Delta_1 - H_0)}$$

$$= O(m^{4(H_0 - \Delta)} + m \log m),$$

it follows that (7.28) is $O_p((\log m)^{3/2}(m^{2(H_0 - \Delta)-1} + m^{-1/2})) = o_p(1)$. Finally,

$$\frac{1}{m}\sum_1^m (a_j - 1)(2\pi I_{\varepsilon j} - 1) = \frac{1}{n}\sum_1^n (\varepsilon_t^2 - 1)\frac{1}{m}\sum_1^m (a_j - 1)$$

$$+ \frac{1}{n}\sum_{s \neq t}\sum \varepsilon_s \varepsilon_t \frac{1}{m}\sum_1^m (a_j - 1)\cos\{(s - t)\lambda_j\}.$$

From (7.23), (7.24), (7.26), and (7.27) the first term on the right is $o_p(1)$, while the second has variance

$$\frac{2}{n^2}\sum_{s \neq t}\sum \left(\frac{1}{m}\sum_1^m (a_j - 1)\cos\{(s - t)\lambda_j\} \right)^2$$

$$= \frac{1}{m^2}\sum_1^m (a_j - 1)^2 - \frac{2}{m^2 n}\left(\sum_1^m (a_j - 1)\right)^2 \to 0$$

as $n \to \infty$. The proof is complete. ∎

7.4 Asymptotic Normality of Estimates

We now show that under conditions somewhat stronger than those of the previous section,

$$m^{1/2}(\hat{H} - H_0) \to_d N(0, \tfrac{1}{4}). \tag{7.29}$$

The variance in the limiting distribution is thus constant over H_0 and indeed completely free of unknown parameters, so (7.29) is simple to use in approximate rules of inference. (7.29) indicates that $\hat{H}$ is asymptotically more efficient (for the same m sequence) than the estimate of Geweke and Porter-Hudak (1983) as modified by Robinson (1992). Robinson (1992) developed a class of estimates whose limiting variance after $m^{1/2}$ norming has upper bound $\pi^2/24$ and lower bound $\tfrac{1}{4}$, though the lower bound is not precisely attainable by members of this class. Moreover, the limiting distribution theory of Robinson (1992) employed Gaussianity of x_t, and it seems unlikely that this assumption could be replaced by assumptions which are as mild or comprehensible as those under which we shall establish (7.29). The analogy with parametric problems suggests that $\tfrac{1}{4}$ can be identified with an efficiency bound for semiparametric estimation of H_0 in the Gaussian case with a given m sequence, but the verification of this conjecture remains an open question. The question of optimal choice of m as a function of n is also important and left for future research; some formulae for optimally choosing m in another statistic of interest in the study of long-range dependence were derived by Robinson (1994c).

We have found it necessary to strengthen each of the assumptions A1–A4 in order to obtain (7.29). The new assumptions are described as follows.

Assumption A1$'$ For some $\beta \in (0, 2]$,

$$f(\lambda) \sim G_0 \lambda^{1-2H_0}(1 + O(\lambda^\beta)) \quad \text{as } \lambda \to 0+,$$

where $G_0 \in (0, \infty)$ and $H_0 \in [\Delta_1, \Delta_2]$.

Assumption A2$'$ In a neighbourhood $(0, \delta)$ of the origin, $\alpha(\lambda)$ is differentiable and

$$\frac{d}{d\lambda}\alpha(\lambda) = O\left(\frac{|\alpha(\lambda)|}{\lambda}\right) \quad \text{as } \lambda \to 0+.$$

Assumption A3$'$ Assumption A3 holds and also

$$E\left(\varepsilon_t^3 | F_{t-1}\right) = \mu_3, \quad \text{a.s.}, \quad E\left(\varepsilon_t^4 | F_{t-1}\right) = \mu_4, \quad t = 0, \pm 1, \ldots,$$

for finite constants μ_3 and μ_4.

Assumption A4$'$ As $n \to \infty$,

$$\frac{1}{m} + \frac{m^{1+2\beta}(\log m)^2}{n^{2\beta}} \to 0.$$

Assumption A1$'$ is equivalent to one employed by Robinson (1995), imposing a rate of convergence on (7.1) analogous to the smoothness conditions used in the asymptotic theory of power spectral density estimates. Assumption A2$'$ implies A2 because $f(\lambda) = |\alpha(\lambda)|^2/2\pi$. Assumption A3′ implies x_t is fourth-order stationary, and holds in case of independent and identically distributed ε_t with finite fourth moment. Assumption A4′, like A4, allows m to increase arbitrarily slowly with m, but it also imposes an upper bound on the rate of increase of m with n, the weakest version arising when $\beta = 2$.

Theorem 2 Let Assumptions A1′–A4′ hold. Then

$$m^{1/2}(\hat{H} - H_0) \rightarrow_d N(0, \tfrac{1}{4}) \quad \text{as } n \rightarrow \infty.$$

Proof Theorem 1 holds under the current conditions and implies that with probability approaching 1, as $n \rightarrow \infty$, $\hat{H}$ satisfies

$$0 = \frac{\mathrm{d}R(\hat{H})}{\mathrm{d}H} = \frac{\mathrm{d}R(H_0)}{\mathrm{d}H} + \frac{\mathrm{d}^2 R(\tilde{H})}{\mathrm{d}H^2}(\hat{H} - H_0), \tag{7.30}$$

where $|\tilde{H} - H_0| \leq |\hat{H} - H_0|$. Now

$$\frac{\mathrm{d}R(H)}{\mathrm{d}H} = 2\frac{\hat{G}_1(H)}{\hat{G}(H)} - \frac{2}{m}\sum_1^m \log \lambda_j,$$

$$\frac{\mathrm{d}^2 R(H)}{\mathrm{d}H^2} = \frac{4\{\hat{G}_2(H)\hat{G}(H) - \hat{G}_1^2(H)\}}{\hat{G}^2(H)},$$

where

$$\hat{G}_k(H) = \frac{1}{m}\sum_1^m (\log \lambda_j)^k \lambda_j^{2H-1} I_j.$$

Defining also

$$\hat{F}_k(H) = \frac{1}{m}\sum_1^m (\log j)^k \lambda_j^{2H-1} I_j, \qquad \hat{E}_k(H) = \frac{1}{m}\sum_1^m (\log j)^k j^{2H-1} I_j,$$

elementary calculation gives

$$\frac{\mathrm{d}^2 R(H)}{\mathrm{d}H^2} = \frac{4\{\hat{F}_2(H)\hat{F}_0(H) - \hat{F}_1^2(H)\}}{\hat{F}_0^2(H)}$$

$$= \frac{4\{\hat{E}_2(H)\hat{E}_0(H) - \hat{E}_1^2(H)\}}{\hat{E}_0^2(H)}. \tag{7.31}$$

Fix $\varepsilon > 0$ and choose n so that $2\varepsilon < (\log m)^2$. On the set $M = \{H\colon (\log m)^3 \times |H - H_0| \leq \varepsilon\}$,

$$|\hat{E}_k(H) - \hat{E}_k(H_0)| \leq \frac{1}{m}\sum_1^m |j^{2(H-H_0)} - 1|(\log j)^k j^{2H_0-1} I_j$$

$$\leq 2e|H - H_0|\hat{E}_{k+1}(H_0)$$

$$\leq 2e\varepsilon(\log m)^{k-2}\hat{E}_0(H_0),$$

where the second inequality uses $\frac{1}{2}||j^{2(H-H_0)}-1|/|H-H_0| \leq (\log j)m^{2|H-H_0|} \leq (\log j)m^{1/\log m} = e \log j$ on M. Thus for $\eta > 0$,

$$P\left(|\hat{E}_k(\tilde{H}) - \hat{E}_k(H_0)| > \eta \left(\frac{2\pi}{n}\right)^{1-2H_0}\right)$$

$$\leq P\left(\hat{G}(H_0) > \frac{\eta}{2e\varepsilon}(\log m)^{2-k}\right) + P((\log m)^3|\tilde{H} - H_0| > \varepsilon). \quad (7.32)$$

For any $\eta > 0$ and $k = 0, 1, 2$, the first probability on the right-hand side tends to 0 for ε sufficiently small because $\hat{G}(H_0) \to_p G_0 \in (0, \infty)$ is implied by the proof that $\sup_{\Theta_1} |A(H)| \to_p 0$ in the proof of Theorem 1. The second probability is bounded by

$$P\left(\inf_{\Theta_1 \cap \bar{N}_\delta \cap \overline{M}} S(H) \leq 0\right) + P\left(\inf_{\Theta_1 \cap N_\delta} S(H) \leq 0\right) + P\left(\inf_{\Theta_2} S(H) \leq 0\right),$$

where $\overline{M} = (-\infty, \infty) - M$. We have already shown in the proof of Theorem 1 that the last two probabilities tend to 0. The first probability is bounded by

$$P\left(\sup_{\Theta_1 \cap N_\delta} |T(H)| \geq \inf_{\Theta_1 \cap N_\delta \cap \overline{M}} U(H)\right). \quad (7.33)$$

By applying (7.8) and then (7.9) and the orders of magnitude for (7.11) and (7.12) noted in the proof of Theorem 1, it follows from (7.13) that (7.33) tends to 0 if

$$\sup_{\Theta_1 \cap N_\delta} \left|\frac{\hat{G}(H) - G(H)}{G(H)}\right| = o_p((\log m)^{-6}). \quad (7.34)$$

Making further use of the notation of (7.14) we have $\inf_{\Theta_1 \cap N_\delta} B(H) \geq \inf_{\Theta_1} B(H) \geq \frac{1}{2}$ for all large enough m from (7.15), and (cf. (7.16) and (7.17))

$$\sup_{\Theta_1 \cap N_\delta} |A(H)| \leq 6 \sum_1^m \left(\frac{r}{m}\right)^{1-2\delta} \frac{1}{r^2} \left|\sum_1^r \left(\frac{I_j}{g_j} - 1\right)\right|$$

$$+ \frac{3}{m} \left|\sum_1^m \left(\frac{I_j}{g_j} - 1\right)\right|. \quad (7.35)$$

We now state the following properties, to be established later:

$$\sum_1^r \left(\frac{I_j}{g_j} - 2\pi I_{\varepsilon j}\right) = O_p(r^{1/3}(\log r)^{2/3} + r^{\beta+1}n^{-\beta} + r^{1/2}n^{-1/4}) \quad \text{as } n \to \infty,$$

$$(7.36)$$

and

$$\sum_1^r (2\pi I_{\varepsilon j} - 1) = O_p(r^{1/2}) \quad \text{as } n \to \infty, \quad (7.37)$$

for $1 \leq r \leq m$. Using A4', we then deduce that (7.35) is $O_p(m^{-1/2})$, so it follows from (7.13) and (7.33) that $P(\inf_{\Theta_1 \cap \overline{N}_\delta \cap \overline{M}} S(H) \leq 0) \to 0$. The detailed nature of the bounds in (7.36) and (7.37) will be of further use below. We have established that (7.32) tends to 0 as $n \to \infty$. Thus

$$\frac{d^2 R(\tilde{H})}{dH^2}$$

$$= \frac{4[\{\hat{E}_2(H_0) + o_p(n^{2H_0-1})\}\{\hat{E}_0(H_0) + o_p(n^{2H_0-1})\} - \{\hat{E}_1(H_0) + o_p(n^{2H_0-1})\}^2]}{\{\hat{E}_0(H_0) + o_p(n^{2H_0-1})\}^2}$$

$$= \frac{4\{\hat{F}_2(H_0)\hat{F}_0(H_0) - \hat{F}_1^2(H_0)\}}{\hat{F}_0^2(H_0)} + o_p(1) \quad \text{as } n \to \infty.$$

For $k \geq 0$,

$$\left| \hat{F}_k(H_0) - G_0 \frac{1}{m} \sum_1^m (\log j)^k \right|$$

$$\leq \frac{G_0}{m} \sum_1^{m-1} \left| (\log r)^k - (\log(r+1))^k \right| \left| \sum_1^r \left\{ \frac{I_j}{g_j} - 1 \right\} \right|$$

$$+ \frac{G_0}{m} (\log m)^k \left| \sum_1^m \left\{ \frac{I_j}{g_j} - 1 \right\} \right|$$

$$= O_p \left(\frac{(\log m)^2}{m^{1/2}} \right) \quad \text{as } n \to \infty,$$

using summation by parts, (7.36), (7.37) and $|(\log r)^k - (\log(r+1))^k| \leq (\log(r+1))^{k-1}/r$ for $k = 1, 2$. Thus from (7.31), as $n \to \infty$,

$$\frac{d^2 R(\tilde{H})}{dH^2} = 4 \left\{ \frac{1}{m} \sum_1^m (\log j)^2 - \left(\frac{1}{m} \sum_1^m \log j \right)^2 \right\} (1 + o_p(1)) + o_p(1) \to_p 4. \tag{7.38}$$

Now because $\hat{G}(H_0) \to_p G_o$,

$$m^{1/2} \frac{dR(H_0)}{dH} = 2m^{-1/2} \sum_1^m v_j \frac{\lambda_j^{2H_0-1} I_j}{G_0 + o_p(1)}$$

$$= 2m^{-1/2} \left(1 - \frac{o_p(1)}{G_0 + o_p(1)} \right) \sum_1^m v_j \frac{I_j}{g_j}$$

$$= 2m^{-1/2} \sum_1^m v_j \left(\frac{I_j}{g_j} - 1 \right) (1 + o_p(1)), \tag{7.39}$$

where $v_j = \log j - m^{-1} \sum_1^m \log j$ satisfies $\sum_1^m v_j = 0$. From summation by parts, (7.36) and $\sum_1^m \log j = O(m \log m)$, (7.39) is

$$\left\{ 2m^{-1/2} \sum_1^m v_j (2\pi I_{\varepsilon j} - 1) \right.$$

$$\left. + O_p \left(\left((\log m)^{2/3} m^{-1/6} + m^{\beta+1/2} n^{-\beta} + n^{-1/4} \right) \log m \right) \right\} (1 + o_p(1)),$$

which from Assumption A4′ and $\sum_1^m v_j = 0$ is $\left(2 \sum_1^n z_t + o_p(1) \right)(1 + o_p(1))$, where $z_1 = 0$ and, for $t \geq 2$,

$$z_t = \varepsilon_t \sum_{s=1}^{t-1} \varepsilon_s c_{t-s},$$

$$c_s = 2n^{-1} m^{-1/2} \sum_1^m v_j \cos(s\lambda_j),$$

supressing reference to n in z_t and c_s. Now the z_t form a zero-mean martingale difference array, and from a standard martingale clt we can deduce that $\sum_1^n z_t$ tends in distribution to a $N(0, 1)$ random variable if

$$\sum_1^n E(z_t^2 | F_{t-1}) - 1 \to_p 0, \tag{7.40}$$

$$\sum_1^n E(z_t^2 I(|z_t| > \delta)) \to 0 \quad \text{for all } \delta > 0. \tag{7.41}$$

The left-hand side of (7.40) is

$$\left\{ \sum_{t=2}^n \sum_{s=1}^{t-1} \varepsilon_s^2 c_{t-s}^2 - 1 \right\} + \sum_{t=2}^n \sum_{r \neq s} \varepsilon_r \varepsilon_s c_{t-r} c_{t-s}. \tag{7.42}$$

The term in braces is

$$\left\{ \sum_1^{n-1} (\varepsilon_t^2 - 1) \sum_{s=1}^{n-t} c_s^2 \right\} + \left\{ \sum_1^{n-1} \sum_1^{n-t} c_s^2 - 1 \right\}. \tag{7.43}$$

Now

$$\sum_1^{n-1} \sum_1^{n-t} c_s^2 = \frac{4}{mn^2} \sum_{j,k=1}^m v_j v_k \sum_1^{n-1} \sum_1^{n-t} \cos(s\lambda_j) \cos(s\lambda_k) \tag{7.44}$$

$$= \frac{4}{mn^2} \sum_1^m v_j^2 \sum_1^{n-1} \sum_1^{n-t} \cos^2(s\lambda_j) + \frac{2}{mn^2} \sum_{j \neq k} v_j v_k$$

$$\times \sum_1^{n-1} \sum_1^{n-t} [\cos\{s(\lambda_j + \lambda_k)\} + \cos\{s(\lambda_j - \lambda_k)\}]. \tag{7.45}$$

From the trigonometric identities (e.g. see Zygmund 1977: p. 49),

$$\sum_{t=1}^{r} \cos \theta t = \frac{\sin(r + \frac{1}{2})\theta}{2 \sin \frac{1}{2}\theta} - \frac{1}{2},$$

$$\sum_{t=1}^{r} \sin \theta t = \frac{\cos \frac{1}{2}\theta - \cos(r + \frac{1}{2})\theta}{2 \sin \frac{1}{2}\theta}, \quad \theta \neq 0, \ \mathrm{mod}(2\pi),$$

and trigonometric addition formulae, we deduce

$$\sum_{r=1}^{q-1} \sum_{t=1}^{q-r} \cos \theta t = \frac{(\cos \theta - \cos q\theta)}{4 \sin^2 \frac{1}{2}\theta} - \frac{q-1}{2}. \tag{7.46}$$

Thus, for $j = 1, \ldots, m < \frac{1}{2}n$,

$$\sum_{1}^{n-1} \sum_{1}^{n-t} \cos^2(s\lambda_j) = \frac{1}{2} \sum_{1}^{n-1} \sum_{1}^{n-t} \{1 + \cos(2s\lambda_j)\}$$

$$= \frac{n(n-1)}{4} - \frac{n-1}{4} = \frac{(n-1)^2}{4}.$$

Because

$$\frac{1}{m} \sum_{1}^{m} v_j^2 = 1 + O\left(\frac{(\log m)^2}{m}\right),$$

it follows that (7.44) tends to 1 as $n \to \infty$. Using (7.46) again, for $j, k = 1, \ldots, m < \frac{1}{2}n, j \neq k$,

$$\sum_{1}^{n-1} \sum_{1}^{n-t} \left[\cos\{s(\lambda_j + \lambda_k)\} + \cos\{s(\lambda_j - \lambda_k)\}\right] = -n,$$

so that (7.45) is

$$\frac{-2}{mn} \sum \sum_{j \neq k} v_j v_k = O\left(\frac{\sum_{1}^{m} v_j^2}{n}\right) = O\left(\frac{m}{n}\right) \to 0$$

as $n \to \infty$. Thus the second component of (7.43) tends to 0 as $n \to \infty$. The first component of (7.43) has zero mean and variance

$$O\left(\sum_{1}^{n-1} \left(\sum_{s=1}^{n-t} c_s^2\right)^2\right). \tag{7.47}$$

Now

$$|c_s| \leq 2m^{-1/2}n^{-1} \sum_{1}^{m} |v_j| = O\left(\frac{m^{1/2} \log m}{n}\right), \tag{7.48}$$

whereas $c_s = c_{n-s}$ and, by summation by parts,

$$
|c_s| = \left| 2m^{-1/2}n^{-1} \sum_1^{m-1}(v_j - v_{j+1}) \sum_1^j \cos(s\lambda_l) + 2m^{-1/2}n^{-1}v_m \sum_1^m \cos(s\lambda_j) \right|
$$

$$
\leq Cm^{-1/2}s^{-1}\left(\left| \sum_1^{m-1}\log\left(1 + \frac{1}{j}\right) \right| + \log m \right)
$$

$$
= O(m^{-1/2}s^{-1}\log m), \tag{7.49}
$$

for $1 \leq s \leq n/2$, because $|\sum_1^j \cos(s\lambda_l)| = O(n/s)$ for such s (Zygmund 1977: p. 2), and because $|\log(1 + 1/j)| \leq 1/j$ for $j \geq 1$. The bound in (7.49) is at least as good as that in (7.48) for $n/m < s \leq n/2$. From (7.48) and (7.49),

$$
\sum_{s=1}^n c_s^2 = O\left(\frac{n}{m}\left(\frac{m^{1/2}\log m}{n} \right)^2 + \frac{(\log m)^2}{m}\sum_{s>n/m} s^{-2} \right)
$$

$$
= O\left(\frac{(\log m)^2}{n} \right) \tag{7.50}
$$

and so (7.47) $= O((\log m)^4/n)$. We have shown that (7.43) is $o_p(1)$. The second component of (7.42) has mean zero and variance

$$
2\sum_{t,u=2}^n \sum_{r \neq s}^{\min(t-1,u-1)} (c_{t-r}c_{t-s}c_{u-r}c_{u-s})
$$

$$
= 2\sum_{t=2}^n \sum_{r \neq s} c_{t-r}^2 c_{t-s}^2 + 4\sum_{t=3}^n \sum_{u=2}^{t-1} \sum_{r \neq s}^{u-1} c_{t-r}c_{t-s}c_{u-r}c_{u-s}.
$$

From (7.50) the first term on the right is $O((\log m)^4/n)$. The second term has absolute value bounded by

$$
4\sum_3^n \sum_2^{t-1}\left(\sum_1^{u-1}c_{t-r}^2 \sum_1^{u-1}c_{u-r}^2 \right) \leq 4\left(\sum_1^n c_t^2 \right)\left(\sum_3^n \sum_2^{t-1}\sum_{t-u+1}^{t-1} c_r^2 \right), \tag{7.51}
$$

and the final bracketed factor is

$$\sum_{j=1}^{n-2} j(n-j-1)c_j^2 \le 2n \sum_{1}^{[n/2]} j c_j^2$$

$$\le 2n \sum_{2}^{[n/m^{2/3}]} j c_j^2 + 2n \sum_{[n/m^{2/3}]+1}^{[n/2]} j c_j^2$$

$$= O\left(\frac{n^3}{m^{4/3}} \left(\frac{m^{1/2} \log m}{n} \right)^2 + n^2 \left(\frac{\log m}{m^{1/2}} \right)^2 \sum_{[n/m^{2/3}]}^{\infty} s^{-2} \right)$$

$$= O\left(\frac{n(\log m)^2}{m^{1/3}} \right)$$

as $n \to \infty$, so (7.51) is $O((\log m)^4/m^{1/3})$ in view of (7.50). We have thus verified (7.40). We shall prove (7.41) by checking the sufficient condition

$$\sum_{1}^{n} E\left(z_t^4 \right) \to 0 \quad \text{as } n \to \infty.$$

The left-hand side of this equals

$$\mu_4 \sum_{2}^{n} E\left(\sum_{1}^{t-1} \varepsilon_s c_{t-s} \right)^4$$

$$\le C \sum_{2}^{n} E\left(\sum\sum\sum\sum_{1}^{t-1} \varepsilon_s \varepsilon_r \varepsilon_q \varepsilon_p c_{t-s} c_{t-r} c_{t-q} c_{t-p} \right)$$

$$\le C \sum_{1}^{n} \left(\sum_{1}^{n} c_{t-s}^4 \right) + C \sum_{1}^{n} \sum\sum_{1}^{t-1} c_{t-s}^2 c_{t-r}^2$$

$$\le Cn \left(\sum_{1}^{n} c_t^2 \right)^2$$

$$= O\left(\frac{(\log m)^4}{n} \right)$$

from (7.50). We have shown that $2\sum_{1}^{n} z_t$, and thus (7.39), is asymptotically $N(0, 4)$, so that in view of (7.30) and (7.38) the proof of the theorem will be completed on verifying (7.36) and (7.37). To prove (7.37), we have

$$\sum_{1}^{r} (2\pi I_{\varepsilon j} - 1) = \frac{r}{n} \sum_{1}^{n} (\varepsilon_t^2 - 1) + \sum_{2}^{n} \varepsilon_t \sum_{s=1}^{t-1} \varepsilon_s d_{t-s}, \qquad (7.52)$$

where $d_s = (2/n) \sum_1^r \cos(s\lambda_j)$. We have $|d_s| \leq 2r/n$, on the one hand, and $d_s = d_{n-s}$, $|d_s| \leq 2/\pi s$ for $1 \leq s \leq n/2$. Both terms on the right of (7.52) have zero mean, and they have variances, respectively, $O(r^2/n)$ and

$$O\left(n \sum_1^n d_s^2\right) = O\left(n \frac{n}{r} \left(\frac{r}{n}\right)^2 + n \sum_{[n/r]}^{\infty} s^{-2}\right) = O(r).$$

To prove (7.36), first choose an integer $l < r$. From (7.20) and $E(2\pi I_{\varepsilon j}) = 1$,

$$E\left|\sum_1^l \left(\frac{I_j}{g_j} - 2\pi I_{\varepsilon j}\right)\right| = O(l) \quad \text{as } n \to \infty. \tag{7.53}$$

From (7.20) and A1′,

$$E\left|\sum_{l+1}^r \left(\frac{I_j}{g_j} - \frac{I_j}{f_j}\right)\right| \leq C \sum_{l+1}^r \left|1 - \frac{g_j}{f_j}\right|$$

$$= O\left(\frac{r^{\beta+1}}{n^\beta}\right) \quad \text{as } n \to \infty.$$

Write $u_j = w_j/|\alpha_j|$, $v_j = w_{\varepsilon j}$. Then we are left with consideration of

$$E\left\{\sum_{l+1}^r \left(\frac{I_j}{f_j} - 2\pi I_{\varepsilon j}\right)\right\}^2 = (2\pi)^2 (a + b),$$

where

$$a = \sum_{l+1}^r \left(E|u_j|^4 - 2E|u_j v_j|^2 + E|v_j|^4\right),$$

$$b = 2\sum_{\substack{j<k \\ l+1}}^r \left(E|u_j u_k|^2 - E|u_j v_k|^2 - E|u_k v_j|^2 + E|v_j v_k|^2\right).$$

We then have $a = a_1 + a_2$ and $b = b_1 + b_2$, where

$$a_1 = \sum_{l+1}^{r} \left\{ 2\big(E|u_j|^2\big)^2 + |E\big(u_j^2\big)|^2 - 2|E(u_j v_j)|^2 - 2|E(u_j \bar{v}_j)|^2 \right.$$
$$\left. - 2E|u_j|^2 E|v_j|^2 + 2\big(E|v_j|^2\big)^2 + |E(v_j^2)|^2 \right\},$$

$$a_2 = \sum_{l+1}^{r} \left\{ \text{cum}\big(u_j, u_j, \bar{u}_j, \bar{u}_j\big) - 2\,\text{cum}\big(u_j, v_j, \bar{u}_j, \bar{v}_j\big) + \text{cum}\big(v_j, v_j, \bar{v}_j, \bar{v}_j\big) \right\},$$

$$b_1 = 2\sum_{\substack{j<k \\ l+1}}^{r}\sum \left\{ E|u_j|^2 E|u_k|^2 + |E(u_j u_k)|^2 + |E(u_j \bar{u}_k)|^2 - E|u_j|^2 E|v_k|^2 \right.$$
$$- |E(u_j v_k)|^2 - |E(u_j \bar{v}_k)|^2 - E|u_k|^2 E|v_j|^2 - |E(u_k v_j)|^2$$
$$\left. - |E(u_k \bar{v}_j)|^2 + E|v_j|^2 E|v_k|^2 + |E(v_j v_k)|^2 + |E(v_j \bar{v}_k)|^2 \right\},$$

$$b_2 = 2\sum_{\substack{j<k \\ l+1}}^{r}\sum \left\{ \text{cum}(u_j, u_k, \bar{u}_j, \bar{u}_k) - \text{cum}(u_j, v_k, \bar{u}_j, \bar{v}_k) \right.$$
$$\left. - \text{cum}(v_j, u_k, \bar{v}_j, \bar{u}_k) + \text{cum}(v_j, v_k, \bar{v}_j, \bar{v}_k) \right\},$$

where $\text{cum}(\cdot, \cdot, \cdot, \cdot)$ is the joint cumulant of the argument random variables. Because $E|v_j|^2 \equiv 1$,

$$a_1 = \sum_{l+1}^{r} \left\{ 2\big(E|u_j|^2 - 1\big)^2 + 2\big(E|u_j|^2 - 1\big) + |E\big(u_j^2\big)|^2 - 2|E(u_j v_j)|^2 \right.$$
$$\left. - 2\big|E\big(u_j \bar{v}_j\big) - 1\big|^2 - 2\big(E u_j \bar{v}_j - 1\big) - 2\big(E\bar{u}_j v_j - 1\big) + |E(v_j^2)|^2 \right\}$$
$$= O\left(\sum_{l+1}^{r} \left(\frac{\log j}{j} \right) \right) = O\big((\log r)^2\big),$$

$$b_1 = \sum_{\substack{j<k \\ j=l+1}}^{r}\sum \left\{ \big(E|u_j|^2 - 1\big)\big(E|u_k|^2 - 1\big) + |E(u_j u_k)|^2 + |E\big(u_j \bar{u}_k\big)|^2 \right.$$
$$- |E(u_j v_k)|^2 - |E(u_j \bar{v}_k)|^2 - |E(u_k v_j)|^2 - |E(u_k \bar{v}_j)|^2$$
$$\left. + |E(v_j v_k)|^2 + |E\big(v_j \bar{v}_k\big)|^2 \right\}$$

$$= O\left(\sum_{\substack{j<k \\ j=l+1}}^{r}\sum\left\{\left(\frac{\log j}{j}\right)\left(\frac{\log k}{k}\right) + \left(\frac{\log k}{j}\right)^2\right\}\right)$$

$$= O\left((\log r)^2 \sum_{k=l+2}^{r}\sum_{j=l+1}^{k-1} j^{-2}\right) = O\left(\frac{r(\log r)^2}{l}\right) \tag{7.54}$$

as $n \to \infty$, using theorem 2 of Robinson (1995). Choosing $l \sim r^{1/3}(\log r)^{2/3}$ in (7.53) and (7.54) gives the $O_p(r^{1/3}(\log r)^{2/3})$ component in (7.36). Now consider a_2 and b_2. Applying formulae of Brillinger (1975: eqn (2.6.3), p. 26; eqn (2.10.3), p. 39), we deduce after straightforward calculation that the summand of $(2\pi)^5 b_2$ is

$$\frac{\kappa}{n^2}\iiint_{-\pi}^{\pi}\left\{\frac{\alpha(\lambda+\mu+\zeta)\alpha(-\mu)}{|\alpha_j|^2} - 1\right\}\left\{\frac{\alpha(-\lambda)\alpha(-\zeta)}{|\alpha_k|^2} - 1\right\} \tag{7.55}$$

$$\times E_{jk}(\lambda,\mu,\zeta)\,d\lambda\,d\mu\,d\zeta,$$

where $\kappa = \mu_4 - 3$ and

$$E_{jk}(\lambda,\mu,\zeta) = D(\lambda_j - \lambda - \mu - \zeta)D(\lambda_k + \lambda)D(\mu - \lambda_j)D(\zeta - \lambda_k),$$

$$D(\lambda) = \sum_{t=1}^{n}e^{it\lambda}.$$

Now use the identity

$$(c_1 c_2 - 1)(c_3 c_4 - 1) = \prod_{1}^{4}(c_j - 1) + \sum_{i=1}^{4}\prod_{\substack{j=1 \\ j\neq i}}^{4}(c_j - 1)$$

$$+ \sum_{i,j=1}^{2}(c_i - 1)(c_{j+2} - 1)$$

to observe that (7.55) has components of three types. The first is

$$\frac{\kappa}{n^2}\iiint_{-\pi}^{\pi}\left\{\frac{\alpha(\lambda+\mu+\zeta)}{\alpha_j} - 1\right\}\left\{\frac{\alpha(-\mu)}{\overline{\alpha}_j} - 1\right\}\left\{\frac{\alpha(-\lambda)}{\alpha_k} - 1\right\}\left\{\frac{\alpha(-\zeta)}{\overline{\alpha}_k} - 1\right\}$$

$$\times E_{jk}(\lambda,\mu,\zeta)\,d\lambda\,d\mu\,d\zeta.$$

By the Schwarz inequality and periodicity, this is bounded in absolute value by

$$(2\pi)^3\kappa P_j P_k, \tag{7.56}$$

where

$$P_j = \int_{-\pi}^{\pi}\left|\frac{\alpha(\lambda)}{\alpha_j} - 1\right|^2 K(\lambda - \lambda_j)\,d\lambda$$

and

$$K(\lambda) = \frac{|D(\lambda)|^2}{2\pi n}$$

is Fejèr's kernel. The second sort of component of (7.55) is typified by

$$\frac{k}{(2\pi n)^2} \int\!\!\int\!\!\int_{-\pi}^{\pi} \left\{ \frac{\alpha(\lambda + \mu + \zeta)}{\alpha_j} - 1 \right\} \left\{ \frac{\alpha(-\mu)}{\overline{\alpha}_j} - 1 \right\} \left\{ \frac{\alpha(-\zeta)}{\overline{\alpha}_k} - 1 \right\}$$

$$\times E_{jk}(\lambda, \mu, \zeta) \, d\lambda \, d\mu \, d\zeta.$$

The modulus of this is bounded by

$$(2\pi)^3 \kappa P_j P_k^{1/2} \tag{7.57}$$

because

$$\int_{-\pi}^{\pi} K(\lambda) \, d\lambda = 1. \tag{7.58}$$

An example of the third type of component of (7.55) is

$$\frac{\kappa}{(2\pi n)^2} \int\!\!\int\!\!\int_{-\pi}^{\pi} \left\{ \frac{\alpha(\lambda + \mu + \zeta)}{\alpha_j} - 1 \right\} \left\{ \frac{\alpha(-\lambda)}{\alpha_k} - 1 \right\} \times E_{jk}(\lambda, \mu, \zeta) \, d\lambda \, d\mu \, d\zeta$$

$$= \frac{\kappa}{(2\pi n)^2} \int\!\!\int\!\!\int_{-\pi}^{\pi} \left\{ \frac{\alpha(\theta)}{\alpha_j} - 1 \right\} \left\{ \frac{\alpha(-\lambda)}{\alpha_k} - 1 \right\}$$

$$\times E_{jk}(\lambda, \theta - \lambda - \zeta, \zeta) \, d\lambda \, d\mu \, d\zeta$$

$$= \frac{\kappa}{n^2} \int\!\!\int_{-\pi}^{\pi} \left\{ \frac{\alpha(\theta)}{\alpha_j} - 1 \right\} \left\{ \frac{\alpha(-\lambda)}{\alpha_k} - 1 \right\}$$

$$\times D(\lambda_j - \theta) D(\lambda_k + \lambda) D(\theta - \lambda - \lambda_j - \lambda_k) \, d\theta \, d\lambda, \tag{7.59}$$

because

$$\int_{-\pi}^{\pi} D(u + \lambda) D(v - \lambda) \, d\lambda = 2\pi D(u + v).$$

The absolute value of (7.59) is bounded by

$$\frac{(2\pi)^2 k}{n^{1/2}} P_j^{1/2} P_k^{1/2}. \tag{7.60}$$

The summand of a_2 is (7.55) with $j = k$ and has components bounded by (7.56), (7.57) or (7.60), with $j = k$. Applying Lemma 3, we deduce that

$$a_2 = O\left(\sum_1^r (j^{-2} + j^{-3/2} + n^{-1/2}j^{-1})\right) = O(1),$$

$$b_2 = O\left(\sum_{\substack{j<k \\ l+1}}^r (j^{-1}k^{-1} + j^{-1}k^{-1/2} + j^{-1/2}k^{-1} + n^{-1/2}j^{-1/2}k^{-1/2})\right)$$

$$= O((\log r)^2 + r^{1/2}\log r + rn^{-1/2}),$$

to complete the proof of the theorem. ∎

7.5 Technical Lemmas

Lemma 1 For $\varepsilon \in (0, 1]$ and $C \in (\varepsilon, \infty)$, as $m \to \infty$,

$$\sup_{C \geq \gamma \geq \varepsilon} \left| \frac{\gamma}{m} \sum_1^m \left(\frac{j}{m}\right)^{\gamma-1} - 1 \right| = O\left(\frac{1}{m^\varepsilon}\right).$$

Proof Because $\int_0^a x^{\gamma-1}\,dx = a^\gamma/\gamma$ for $\gamma > 0$,

$$\left| \frac{\gamma}{m} \sum_1^m \left(\frac{j}{m}\right)^{\gamma-1} - 1 \right| \leq \gamma \int_0^{1/m} \left\{ \left(\frac{1}{m}\right)^{\gamma-1} + x^{\gamma-1} \right\} dx$$

$$+ \gamma \sum_2^m \left| \int_{(j-1)/m}^{j/m} \left\{ \left(\frac{j}{m}\right)^{\gamma-1} - x^{\gamma-1} \right\} dx \right|$$

$$\leq \frac{\gamma}{m^\gamma} + \frac{1}{m^\gamma} + \frac{\gamma|\gamma-1|}{m^2} \sum_1^m \left(\frac{j}{m}\right)^{\gamma-2}$$

by the mean-value theorem. The last term is $O(\gamma^2 m^{-1})$ for $\gamma > 1$, zero for $\gamma = 1$ and $O(m^{-\gamma})$ for $0 < \gamma < 1$, whence the conclusion follows straightforwardly. ∎

Lemma 2 For all $m \geq 2$,

$$\left| \frac{1}{m} \sum_1^m \log j - \log m + 1 \right| \leq \frac{2 + \log(m-1)}{m}. \tag{7.61}$$

Proof Because $\int_0^m \log x \, dx = m(\log m - 1)$, the left-hand side of (7.61) is

$$\left| \frac{1}{m} \int_0^1 \log x \, dx - \frac{1}{m} \sum_2^m \int_{j-1}^j \log\left(\frac{j}{x}\right) dx \right|$$

$$\leq \frac{1}{m} + \frac{1}{m} \sum_1^{m-1} \frac{1}{j} \leq \frac{2 + \log(m-1)}{m}. \quad \blacksquare$$

Lemma 3 Under A1 and A2′, uniformly in integer j such that $j/n \to 0$ as $n \to \infty$,

$$\int_{-\pi}^{\pi} \left| \frac{\alpha(\lambda)}{\alpha(\lambda_j)} - 1 \right|^2 K(\lambda - \lambda_j) \, d\lambda = O\left(\frac{1}{j}\right) \quad \text{as } n \to \infty.$$

Proof We split the integral up as follows:

$$\int_{-\pi}^{-\delta} + \int_{-\delta}^{-\lambda_j/2} + \int_{-\lambda_j/2}^{-\lambda_j/2} + \int_{\lambda_j/2}^{2\lambda_j} + \int_{2\lambda_j}^{\delta} + \int_{\delta}^{\pi},$$

for $\delta \in (2\lambda_j, \pi)$. For n sufficiently large, A1 and A2′ imply that we can choose δ such that, for some $C < \infty$,

$$|\alpha(\lambda)| \leq C|\lambda|^{1/2-H_0}, \quad |\alpha'(\lambda)| \leq C|\lambda|^{-1/2-H_0}, \quad 0 < |\lambda| < \delta.$$

Now

$$\left| \int_{-\pi}^{-\delta} \right| \leq \frac{2}{f_j} \left\{ \int_{-\pi}^{-\delta} f(\lambda) K(\lambda - \lambda_j) \, d\lambda + f_j \int_{-\pi}^{-\delta} K(\lambda - \lambda_j) \, d\lambda \right\}$$

$$\leq \frac{1}{\pi f_j} \left\{ \frac{1}{n\delta^2} \int_{-\pi}^{\pi} f(\lambda) \, d\lambda + f_j \frac{2\pi}{n\delta^2} \right\}$$

$$= O\left(\frac{1}{n} \left(\frac{1}{f_j} + 1 \right) \right) = O\left(\frac{\lambda_j^{2H}}{j} + \frac{1}{n} \right) = O\left(\frac{1}{j} \right),$$

where the second inequality uses the property (Zygmund 1977: pp. 49–51)

$$|D(\lambda)| < 2/|\lambda|, \quad 0 < \lambda < \pi; \tag{7.62}$$

also $|\int_{\delta}^{\pi}|$ has the same bound by the same proof. Using (7.62) again,

$$\left| \int_{-\delta}^{-\lambda_j/2} \right| \leq \frac{1}{f_j} \left\{ \int_{-\delta}^{-\lambda_j/2} f(\lambda) K(\lambda - \lambda_j) \, d\lambda + f_j \int_{-\delta}^{-\lambda_j/2} K(\lambda - \lambda_j) \, d\lambda \right\}$$

$$\leq \frac{1}{2\pi n f_j} \left\{ \int_{\lambda_j/2}^{\pi} \lambda^{-1-2H} \, d\lambda + f_j \int_{\lambda_j/2}^{\pi} \lambda^{-2} \, d\lambda \right\}$$

$$= O\left(\frac{\lambda_j^{-2H}}{n f_j} \right) = O\left(\frac{1}{j} \right);$$

$|\int_{2\lambda_j}^{\delta}|$ has the same bound by the same proof. Next,

$$\left|\int_{-\lambda_j/2}^{\lambda_j/2}\right| \leq f_j^{-1} \max_{|\lambda|\leq\lambda_j/2} K(\lambda-\lambda_j) \left\{ \int_{-\lambda_j/2}^{\lambda_j/2} f(\lambda)\,d\lambda + \lambda_j f_j \right\}$$

$$= O\left(f_j^{-1}\frac{1}{n\lambda_j^2}\lambda_j f_j\right) = O\left(\frac{1}{j}\right).$$

Finally, by the mean-value theorem,

$$\left|\int_{\lambda_j/2}^{2\lambda_j}\right| \leq \frac{1}{|\alpha_j|^2} \max_{\lambda_j/2\leq\lambda\leq 2\lambda_j} \left|\frac{d\alpha(\lambda)}{d\lambda}\right|^2 \int_{\lambda_j/2}^{2\lambda_j} |\lambda-\lambda_j|^2 K(\lambda-\lambda_j)\,d\lambda$$

$$= O\left(\frac{1}{n\lambda_j}\right) = O\left(\frac{1}{j}\right). \blacksquare$$

7.6 Numerical Work

The finite-sample behaviour of $\hat{H}$ was investigated in a small Monte Carlo study. $\hat{H}$ was also compared with a simple closed-form estimate:

$$\tilde{H} = \frac{1}{2}\left\{1 - \frac{\sum_{j=1}^m \log I_j (\log j - (1/m)\sum_{l=1}^m \log l)}{\sum_{j=1}^m \log j (\log j - (1/m)\sum_{l=1}^m \log l)}\right\}. \tag{7.63}$$

This estimate is a slightly simplified version of the one proposed by Geweke and Porter-Hudak (1983). Robinson (1995) derived asymptotic theory for a modified form of $\tilde{H}$, in which the contribution from a slowly increasing (with m) number l of the lowest frequencies $\lambda_1, \ldots, \lambda_l$ are deleted from (7.63). (Künsch (1986) earlier suggested such a trimming.) It is not known whether the trimming is necessary in order to achieve the desirable asymptotic property

$$m^{1/2}(\tilde{H} - H) \to_d N\left(0, \frac{\pi^2}{24}\right) \tag{7.64}$$

obtained by Robinson (1995); the bulk of the many empirical applications of the Geweke–Porter-Hudak approach have used all of the m lowest frequencies. (We omit the zero subscript on H throughout this section and Tables 7.1–7.8.) In view of this, and for simplicity and ease of comparison with respect to degrees of freedom, we employ the untrimmed estimate (7.63) here.

Using an algorithm of Davies and Harte (1987) and random number generator G05DDF from the NAG library, Gaussian time series were generated with mean zero, variance unity and lag-j autocovariance

$$\gamma_j = \tfrac{1}{2}(|j+1|^{2H} - 2|j|^{2H} + |j-1|^{2H}). \tag{7.65}$$

We call this Model A. The corresponding spectral density satisfies (7.1); indeed it satisfies Assumption A1$'$ with $\beta = 2$. Five values of H were employed: $H = 0.1$,

Table 7.1. Bias, Model A

H	n = 64			n = 128			n = 256		
	m = n/16	m = n/8	m = n/4	m = n/16	m = n/8	m = n/4	m = n/16	m = n/8	m = n/4
0.1 log	−0.008	−0.064	−0.123	−0.015	−0.066	−0.125	−0.016	−0.070	−0.124
eff	0.132	0.018	−0.049	0.048	−0.022	−0.071	0.006	−0.044	−0.081
0.3 log	−0.009	−0.008	−0.038	−0.004	−0.013	−0.032	−0.001	−0.014	−0.033
eff	0.041	−0.011	−0.050	−0.006	−0.028	−0.044	−0.016	−0.026	−0.037
0.5 log	−0.001	−0.005	−0.003	0.007	0.001	0.002	−0.001	−0.004	0.001
eff	−0.031	−0.026	−0.020	−0.022	−0.020	−0.011	−0.022	−0.013	−0.005
0.7 log	0.012	0.010	0.018	0.012	0.003	0.015	0.002	0.012	0.014
eff	−0.099	−0.044	−0.004	−0.041	−0.017	−0.004	−0.020	−0.001	0.009
0.9 log	0.008	0.029	0.035	0.025	0.021	0.031	0.016	0.015	0.027
eff	−0.181	−0.084	−0.019	−0.088	−0.030	0.005	−0.034	−0.007	0.016

Table 7.2. Standard deviation, Model A

H	$n = 64$			$n = 128$			$n = 256$		
	$m = n/16$	$m = n/8$	$m = n/4$	$m = n/16$	$m = n/8$	$m = n/4$	$m = n/16$	$m = n/8$	$m = n/4$
0.1 log	0.610	0.349	0.218	0.350	0.215	0.139	0.214	0.143	0.093
eff	0.307	0.173	0.084	0.190	0.104	0.051	0.121	0.071	0.034
0.3 log	0.613	0.344	0.213	0.347	0.209	0.136	0.215	0.136	0.091
eff	0.345	0.239	0.159	0.244	0.161	0.109	0.166	0.109	0.073
0.5 log	0.606	0.344	0.210	0.348	0.209	0.135	0.209	0.134	0.089
eff	0.366	0.270	0.175	0.268	0.175	0.111	0.173	0.110	0.071
0.7 log	0.606	0.347	0.211	0.347	0.216	0.136	0.209	0.133	0.090
eff	0.362	0.266	0.173	0.264	0.175	0.111	0.172	0.109	0.073
0.9 log	0.597	0.345	0.210	0.340	0.212	0.135	0.211	0.136	0.089
eff	0.333	0.222	0.134	0.221	0.138	0.091	0.142	0.094	0.065

Table 7.3. MSE, Model A

H	$n = 64$			$n = 128$			$n = 256$		
	$m = n/16$	$m = n/8$	$m = n/4$	$m = n/16$	$m = n/8$	$m = n/4$	$m = n/16$	$m = n/8$	$m = n/4$
0.1 log	0.372	0.126	0.063	0.123	0.050	0.035	0.046	0.025	0.024
eff	0.112	0.030	0.010	0.039	0.011	0.008	0.015	0.007	0.008
0.3 log	0.376	0.118	0.047	0.120	0.044	0.019	0.046	0.019	0.009
eff	0.121	0.057	0.028	0.060	0.027	0.014	0.028	0.013	0.007
0.5 log	0.367	0.118	0.044	0.121	0.044	0.018	0.044	0.018	0.008
eff	0.135	0.073	0.031	0.073	0.031	0.012	0.031	0.012	0.005
0.7 log	0.367	0.121	0.045	0.121	0.047	0.019	0.044	0.018	0.008
eff	0.141	0.073	0.030	0.072	0.031	0.012	0.030	0.012	0.005
0.9 log	0.356	0.120	0.045	0.116	0.045	0.019	0.045	0.019	0.009
eff	0.144	0.056	0.018	0.057	0.020	0.008	0.021	0.009	0.004

Table 7.4. Relative efficiency, Model A

H	$n = 64$			$n = 128$			$n = 256$		
	$m = n/16$	$m = n/8$	$m = n/4$	$m = n/16$	$m = n/8$	$m = n/4$	$m = n/16$	$m = n/8$	$m = n/4$
0.1	0.300	0.240	0.151	0.314	0.226	0.220	0.317	0.276	0.322
0.3	0.321	0.487	0.594	0.494	0.611	0.706	0.601	0.677	0.725
0.5	0.368	0.620	0.705	0.600	0.714	0.679	0.701	0.694	0.645
0.7	0.385	0.604	0.669	0.593	0.661	0.662	0.687	0.669	0.647
0.9	0.403	0.470	0.406	0.488	0.443	0.433	0.474	0.475	0.517

Table 7.5. 95% coverage probabilities, Model A

H	$n = 64$			$n = 128$			$n = 256$		
	$m = n/16$	$m = n/8$	$m = n/4$	$m = n/16$	$m = n/8$	$m = n/4$	$m = n/16$	$m = n/8$	$m = n/4$
0.1 log	0.725	0.816	0.800	0.806	0.851	0.760	0.869	0.846	0.648
eff	0.841	0.938	0.991	0.907	0.975	0.997	0.949	0.988	1.000
0.3 log	0.722	0.825	0.865	0.806	0.874	0.888	0.862	0.895	0.900
eff	0.847	0.915	0.823	0.912	0.841	0.859	0.831	0.873	0.864
0.5 log	0.733	0.817	0.870	0.816	0.870	0.900	0.879	0.907	0.924
eff	0.637	0.766	0.839	0.762	0.840	0.885	0.843	0.883	0.914
0.7 log	0.728	0.813	0.868	0.813	0.868	0.900	0.871	0.902	0.922
eff	0.797	0.852	0.840	0.860	0.837	0.883	0.838	0.889	0.910
0.9 log	0.738	0.814	0.860	0.814	0.869	0.890	0.869	0.895	0.909
eff	0.798	0.865	0.924	0.861	0.915	0.954	0.909	0.940	0.974

Table 7.6. 99% coverage probabilities, Model A

H	n = 64			n = 128			n = 256		
	m = n/16	m = n/8	m = n/4	m = n/16	m = n/8	m = n/4	m = n/16	m = n/8	m = n/4
0.1 log	0.832	0.903	0.903	0.908	0.932	0.885	0.946	0.932	0.817
eff	0.900	0.972	0.997	0.956	0.990	1.000	0.984	0.997	1.000
0.3 log	0.837	0.917	0.942	0.908	0.953	0.957	0.946	0.966	0.966
eff	0.905	0.960	0.988	0.956	0.985	0.934	0.981	0.951	0.948
0.5 log	0.847	0.910	0.948	0.913	0.953	0.966	0.950	0.969	0.981
eff	1.000	0.868	0.926	0.873	0.928	0.954	0.931	0.958	0.977
0.7 log	0.849	0.911	0.947	0.905	0.948	0.967	0.946	0.972	0.976
eff	0.857	0.909	0.955	0.923	0.948	0.956	0.952	0.964	0.973
0.9 log	0.843	0.913	0.947	0.917	0.950	0.967	0.952	0.969	0.977
eff	0.862	0.924	0.965	0.921	0.962	0.983	0.956	0.978	0.991

Table 7.7. Bias, Model B

H	$n = 64$			$n = 128$			$n = 256$		
	$m = n/16$	$m = n/8$	$m = n/4$	$m = n/16$	$m = n/8$	$m = n/4$	$m = n/16$	$m = n/8$	$m = n/4$
0.1 log	0.193	−0.175	0.048	0.230	−0.050	0.055	0.253	0.013	0.058
eff	0.218	−0.042	−0.034	0.212	−0.025	−0.042	0.231	0.003	−0.036
0.3 log	0.031	−0.260	−0.044	0.045	−0.160	−0.044	0.054	−0.113	−0.047
eff	0.040	−0.196	−0.166	0.024	−0.170	−0.169	0.033	0.136	−0.161
0.5 log	−0.028	−0.279	−0.083	−0.026	−0.191	−0.084	−0.021	−0.151	−0.088
eff	−0.074	−0.289	−0.228	−0.062	−0.229	−0.213	−0.045	−0.179	−0.197
0.7 log	−0.039	−0.285	−0.104	−0.137	−0.198	−0.102	−0.030	−0.158	−0.104
eff	−0.154	−0.337	−0.259	−0.091	−0.244	−0.232	−0.057	−0.186	−0.214
0.9 log	−0.061	−0.381	−0.169	−0.049	−0.270	−0.161	−0.038	−0.218	−0.160
eff	−0.244	−0.450	−0.375	−0.137	−0.324	−0.334	−0.080	−0.252	−0.306

Table 7.8. Bias, Model C

H	$n = 64$			$n = 128$			$n = 256$		
	$m = n/16$	$m = n/8$	$m = n/4$	$m = n/16$	$m = n/8$	$m = n/4$	$m = n/16$	$m = n/8$	$m = n/4$
0.1 log	0.193	0.050	−0.254	0.208	0.092	−0.205	0.222	0.115	−0.177
eff	0.218	0.070	−0.092	0.192	0.081	−0.095	0.201	0.104	−0.096
0.3 log	0.045	−0.027	−0.275	0.038	−0.007	−0.237	0.041	0.002	−0.219
eff	0.048	−0.042	−0.256	0.019	−0.030	−0.258	0.020	−0.012	−0.254
0.5 log	0.002	−0.045	−0.278	−0.008	−0.031	−0.242	−0.008	−0.025	−0.225
eff	−0.055	−0.089	−0.347	−0.046	−0.061	−0.318	−0.032	−0.039	−0.290
0.7 log	−0.004	−0.048	−0.299	−0.012	−0.033	−0.259	−0.011	−0.027	−0.241
eff	−0.131	−0.110	−0.398	−0.069	−0.065	−0.344	−0.038	−0.041	−0.310
0.9 log	−0.010	−0.080	−0.426	−0.012	−0.054	−0.371	−0.009	−0.044	−0.344
eff	−0.214	−0.167	−0.539	−0.111	−0.097	−0.464	−0.056	−0.062	−0.415

0.3, 0.5, 0.7, and 0.9; $H = 0.5$ corresponds to white noise in this model. The sample sizes chosen were $n = 64$, 128 and 256, and for each of these, three values of m were tried: $n/16$, $n/8$ and $n/4$. For each (H, n, m) combination, 5000 replications were generated. From each of these, $\tilde{H}$ and $\hat{H}$ were computed, in the latter case using a simple golden section search applied to the first derivative of the objective function. No difficulties were encountered in computing $\hat{H}$, and for selected replications $R(H)$ was plotted and always found either to have a single relative minimum or (in some cases when $H = 0.1$ or 0.9) to be minimized at 0.001 or 0.999, these being our chosen values of Δ_1 and Δ_2. Tables 7.1–7.6 give the Monte Carlo biases, standard deviations, mean squared errors (MSE), relative efficiencies (ratios of the $\hat{H}$ and $\tilde{H}$ MSEs) and 95 per cent and 99 per cent coverage frequencies based on the limit distributions in (7.29) and (7.64). In Tables 7.1–7.3 and 7.5–7.6, 'log' refers to the log-periodogram estimate $\tilde{H}$ and 'eff' refers to the more efficient estimate $\hat{H}$.

For the most part, $\hat{H}$ seems more biased than $\tilde{H}$, though this effect tends to be reversed when $m = n/8$ and $n/4$ for $n = 256$, and when $m = n/4$ for the other values of n. $\tilde{H}$ is apt to be negatively biased for small H and positively biased for large H, but a negative bias in $\hat{H}$ is more pervasive. Unsurprisingly, bias tends to increase with m. The standard deviations in Table 7.2 all diminish as both m and n increase, and the $\hat{H}$ standard deviations are always decisively the smaller. Table 7.3 presents a similar picture, and in only one case does MSE show an increase with m. The entries in Table 7.4 are to be compared with 0.608, which is the asymptotic relative efficiency (from (7.29) and (7.64)). Though the finite-sample superiority of $\hat{H}$ tends not to be as great for the central values of H, for $H = 0.9$ and especially $H = 0.1$, it is much better. Much the same can be said of the relative accuracy of the 95 per cent and 99 per cent interval estimates, but whereas $\hat{H}$ is best in 24 of the 45 cases in Table 7.5, in Table 7.6 it is best on 29 occasions, with one tie.

Model A, (7.65), is favourably disposed toward both $\hat{H}$ and $\tilde{H}$ for any value of m, because the approximation

$$f(\lambda) \overset{.}{\sim} G\lambda^{1-2H} \tag{7.66}$$

is good for all $\lambda \in (0, \pi]$. However, both $\hat{H}$ and $\tilde{H}$ are motivated by the far wider range of circumstances (7.1). In many of these there is the possibility that (7.66) is not good over $(0, 2\pi m/n]$, and this is a potential source of bias.

Consider, in particular, the process

$$x_t = y_t + z_t, \tag{7.67}$$

where $\{y_t\}$ is Gaussian with autocovariances given by (7.65), and $\{z_t\}$ is a second-order autoregressive (AR) process

$$z_t + a_1 z_{t-1} + a_2 z_{t-2} = \eta_t, \tag{7.68}$$

such that η_t is an i.i.d. $N(0, 1)$ sequence. When the quadratic $x^2 + a_1 x + a_2$ has complex zeroes, the spectral density of z_t has a peak at a frequency in $(0, \pi]$, in particular, at $\tilde{\lambda} = \arccos\{-a_1(1 + a_2)/4a_2\}$. For example,

$$a_1 = -1.34, \quad a_2 = 0.9 \quad \Rightarrow \quad \tilde{\lambda} = \pi/4, \qquad (7.69)$$

$$a_1 = -0.725, \quad a_2 = 0.9 \quad \Rightarrow \quad \tilde{\lambda} = 3\pi/8 \qquad (7.70)$$

and the amplitude of the peak is very sharp (the zeroes of $x^2 + a_1 x + a_2$ have moduli 0.948 in both cases). Because it is just the sum of the spectra of y_t and z_t, $f(\lambda)$ will have a peak at around $\tilde{\lambda}$, while at $\lambda = 0$ it will be infinite when $H > 0.5$ and small but positive when $H < 0.5$. The values of m and n chosen in the Monte Carlo study reported above entail the frequency bands $(0, \pi/8]$, $(0, \pi/4]$ and $(0, \pi/2]$, respectively, over which $f(\lambda)$ corresponding to (7.67), (7.68) and (7.69) or (7.70) would be influenced in various ways by the peak at $\tilde{\lambda}$.

The Monte Carlo experiment was repeated for (7.67)–(7.70), using the same values of H, m and n and the same number of replications as before. Only biases are reported. Tables 7.7 and 7.8, respectively contain results for the parameter values in (7.69) and (7.70), referring to these as, respectively, Models B and C. For Model B, (7.69), the AR peak happens to occur around the $m = n/8$ cutoff point, and the serious negative biases when $H > 0.5$ are not unexpected. For $m = n/4$, the whole peak is included and the biases are less, owing to some cancellation effect, but evidently meaningful estimates of H are not achieved. For $m = n/16$, the AR peak has some negative impact on bias when $H > 0.5$, as comparison with Table 7.1 indicates, but it is not nearly as significant as in the other cases. When $H < 0.5$ in (7.65) for Models B and C, (7.1) is actually misspecified (or rather, it has $H = 0.5$), and this appears to be dominant in producing the positive biases in Table 7.7 when $H = 0.1$ and $m = n/16$, while the AR peak is responsible for the large negative biases when $H = 0.3$ and $m = n/8$. In Table 7.8, the more distant AR peak at $3\pi/8$ produces a larger bias for $m = n/4$, but a smaller one for $m = n/8$. In both Tables 7.7 and 7.8, the overwhelming tendency when $H \geq 0.5$ is for $\tilde{H}$ to be less biased than $\hat{H}$ and for bias to decrease in n, though the reverse phenomenon predominates when $H = 0.1$. To place these results in perspective, it should be observed that if m had been chosen according to some optimal bandwidth scheme (see, e.g. Robinson 1994c), it would increase more slowly with n than the proportionate increase employed in the Monte Carlo, while a preliminary plot of the logged periodogram can help to avoid pitfalls.

In the long-range dependence literature, various parametric and semiparametric estimates of H have been reported for the time series of 663 annual minimum water levels of the River Nile measured at the Roda Gorge near Cairo during the years 622 through 1284. (The data are in Toussoun 1925; for subsequent years there are missing observations.) For $m = 41$, we obtained $\tilde{H} = 1.033$ and $\hat{H} = 0.941$, so that $\tilde{H}$ is outside the stationary region. For $m = 82$, $\tilde{H} = 0.920$ and $\hat{H} = 0.905$. For $m = 164$, the estimates are again smaller but the order is reversed: $\tilde{H} = 0.855$ and $\hat{H} = 0.866$.

References

Brillinger, D. R. (1975), *Time Series, Data Analysis and Theory*. Holden-Day, San Francisco.

Cheung, Y.-W. and Diebold, F. X. (1994), 'On maximum-likelihood estimation of the differencing parameter of fractionally-integrated noise with unknown mean', *Journal of Econometrics*, 62, 317–50.

Dahlhaus, R. (1989), 'Efficient parameter estimation of self-similar processes', *Annals of Statistics*, 17, 1749–66.

Davies, R. B. and Harte, D. S. (1987), 'Tests for Hurst effect', *Biometrika*, 74, 95–101.

Fox, R. and Taqqu, M. S. (1986), 'Large-sample properties of parameter estimates for strongly dependent stationary Gaussian time series', *Annals of Statistics*, 14, 517–32.

Geweke, J. and Porter-Hudak, S. (1983), 'The estimation and application of long memory time series models', *Journal of Time Series Analysis*, 4, 221–38.

Giraitis, L. and Surgailis, D. (1990), 'A central limit theorem for quadratic forms in strongly dependent linear variables and its application to asymptotic normality of Whittle's estimate', *Probability Theory and Related Fields*, 86, 87–104.

Heyde, C. C. and Seneta, E. (1972), 'Estimation theory for growth and immigration rates in a multiplicative process', *Journal of Applied Probability*, 9, 235–56.

Künsch, H. R. (1986), 'Discrimination between monotonic trends and long-range dependence', *Journal of Applied Probability*, 23, 1025–30.

—— (1987), 'Statistical aspects of self-similar processes', in *Proceedings of the First World Congress of the Bernoulli Society* (Yu. Prohorov and V. V. Sazanov, eds) 1, 67–74. VNU Science Press, Utrecht.

Lobato, I. and Robinson, P. M. (1996), 'Averaged periodogram estimate of long memory', *Journal of Econometrics*, 73, 303–24.

Robinson, P. M. (1994*a*), 'Time series with strong dependence', In *Advances in Econometrics. Sixth World Congress* (C. A. Sims, ed.) 1, 47–95. Cambridge University Press.

—— (1994*b*), 'Semiparametric analysis of long-memory time series', *Annals of Statistics*, 22, 515–39.

—— (1994*c*), 'Rates of convergence and optimal bandwidth in spectral analysis of processes with long-range dependence', *Probability Theory and Related Fields* 99, 443–73.

—— (1995), 'Log-periodogram regression of time series with long-range dependence', *Annals of Statistics*, 23, 1048–72.

Taqqu, M. S. (1975), 'Weak convergence to fractional Brownian motion and to the Rosenblatt process', *Zeitschrift für Wahrscheinlichkeitstheorie und Verwandte Gebiete*, 31, 287–302.

Toussoun, O. (1925), Mémoire sur l'histoire du Nil. *Mémoires de l'Institut d'Egypte (Cairo)* 9.

Yong, C. H. (1974), *Asymptotic Behaviour of Trigonometric Series*. Chinese University Hong Kong.

Zygmund, A. (1977), *Trigonometric Series*. Cambridge University Press.

8

Testing for Strong Serial Correlation and Dynamic Conditional Heteroskedasticity in Multiple Regression

P. M. ROBINSON

8.1 Introduction

Most of the many diagnostic tests for serial correlation (SC) and dynamic conditional heteroskedasticity (DCH) of regression disturbances have good asymptotic local power against finite autoregressive (AR) or moving average (MA) alternatives. In testing for SC, statistics typically depend on finitely many, p, least squares (LS) residual sample autocorrelations. One test, that rejects when sample size N times the sum of squares of the first p LS residual sample autocorrelations lies in the upper tail of the χ_p^2 distribution (e.g. Box and Pierce 1970), has a Lagrange multiplier (LM) interpretation relative to Gaussian AR(p) and MA(p) alternatives, and can be justified as asymptotically locally most powerful unbiased (ALMPU) against these. Tests that are optimal against one-sided AR(1) alternatives (e.g. Durbin and Watson 1950) have long been popular, and other alternatives, such as constrained AR(4) and AR(5) models, have also featured in empirical work.

A popular LM test for DCH is ALMPU against both an AR(p) and MA(p) alternative (with intercept) for the squared disturbances (i.e. an ARCH(p) or GARCH(p) model), and is based on the first p sample autocorrelations of the squared LS residuals (e.g. Engle 1982; Weiss 1986); this is often called an ARCH test.

In some applications of tests for SC it has seemed sufficient to take $p = 1$ and in some others a natural finite AR parameterization may be indicated, for example when seasonal effects seem likely. Otherwise no alternative may dominate on prior grounds, and applied workers calculate statistics for several, possibly many, values of p. It is not uncommon for some of these statistics to be significant, and some insignificant. Drawing conclusions from such a sequence can be a delicate business, and unless the statistics take account of sample autocorrelations at long lags there is always the possibility that relevant information is being neglected, leading perhaps to unwarranted confidence in the usual standard errors of LS estimates of the

This article is based on research funded by the Economic and Social Research Council (ESRC), reference numbers B00232156 and R000231441.

regression coefficients, while in models containing lagged dependent variables there will be a failure to detect a source of inconsistency of the LS estimates.

'Long-memory' covariance stationary time series have been characterized by the property that autocorrelations decay, but so slowly as not to be summable, or similarly that a spectral density exists but is unbounded. They represent an alternative direction of approach to unit root behaviour compared with the usual AR one. Many series are too short, or span too brief a period, to enable reliable distinction of long-memory autocorrelation from AR autocorrelation with a near-unit root. However, some evidence of such long-memory behaviour in economic and other time series has been reported. Unlike tests against AR(p) or MA(p) SC for specific p, tests against long-memory behaviour use all the sample autocorrelations, and might be expected to have good power against AR or MA alternatives of long, but unknown order. Tests against AR or MA SC will generally be consistent against long-memory behaviour, but inefficient.

In ARCH tests, dependence on p causes similar difficulty. To avoid this we can resort to concepts of long-memory DCH, in which the Wold decomposition innovations have conditional variances whose dependence on remote innovations decays very slowly. Standard ARCH tests are often consistent against such alternatives, but more efficient tests are available.

This chapter considers classes of statistics that allow account to be taken of possible long-memory behaviour in moments of order two, three, and four in deciding whether SC or DCH are present. We find it convenient to index members of the classes by conditionally Gaussian parametric alternatives, deriving statistics via the LM principle. The classes thus include some currently used statistics, with good power against the 'short-memory' AR, MA, ARCH, and GARCH models, as well as new ones with good power against long-memory models, not to mention many in which autocorrelations are summable but decay slower than exponentially (see, e.g. Robinson 1978). For the usual asymptotic distribution theory to hold, our conditions on the dependence structure of the alternative models seem relatively weak. Realistically, an applied worker may have no grounds for believing in a particular SC or DCH alternative, and nor will he or she necessarily wish to construct an array of statistics, each designed to detect a different type of departure. Thus a particular practical goal of this chapter is to suggest specific diagnostics for SC and for DCH that are of relatively simple form and appear to complement the usual ones. These emerge in relation to one of two popular long-memory time-series models. Generally the use of the LM principle leads to computationally convenient statistics, with use of the fast Fourier transform (FFT) recommended when N is large.

Throughout we focus on the static linear regression model with stochastic or nonstochastic regressors,

$$y_t = \beta' x_t + u_t, \quad t = 1, 2, \ldots, \tag{8.1}$$

where y_t and x_t are, respectively, a scalar and a $k \times 1$ vector of observables, β is a $k \times 1$ unknown vector, and u_t is an unobservable disturbance, assumed throughout to be strictly stationary and ergodic at the relevant null model, with zero mean and variance σ^2. Stationarity and ergodicity could be traded for other, somewhat more complicated but not strictly weaker, conditions. The x_t are either stochastic or nonstochastic, and in the former case x_t is independent of u_s for all t, s. We assume throughout that the matrix $(x_1, \ldots, x_n)$ has full rank almost surely (a.s.). Analogous statistics can be, and with respect to some alternatives already have been, constructed in other settings, such as nonlinear regressions, linear models with lagged dependent variables, and higher-order alternatives to null SC or DCH models, and it would doubtless be possible to present some sort of treatment that includes all or many of these cases in addition to (8.1). However, this would likely be lengthy and involve tedious formulae, while a unified treatment of asymptotic statistical properties would be unlikely to yield conditions that are both mild and primitive with respect to specific cases. Linear regression affords a relatively succinct account and includes the simple case of testing for SC or DCH in stationary time series, as well as allowing for possibly trending explanatory variables (which can be difficult in more general models), indeed our asymptotic theory puts no restriction at all on the regressors as $N \to \infty$, beyond that implied by the usual assumption introduced above, necessary in order to compute LS estimates, that the matrix of regressors for the N observations has full rank.

We emphasize first-order asymptotic approximations to the distributions of the statistics because of their practical convenience and approximate validity in the presence of wide distributional assumptions on the disturbances.

Although the LM-type tests are derived from an objective function that is the likelihood when the disturbances are conditionally Gaussian, of course no such precise distributional assumptions are employed in asymptotically justifying the tests, indeed the moment conditions we require seem minimal and null models are described in terms of martingale difference, not independence, assumptions. We aim for conditions that are not only mild but primitive, separating different features of the problem in as comprehensible way as possible. Future workers might wish to investigate the use of our statistics within a conditionally Gaussian environment, in which computationally more complicated rules with stronger finite-sample justification are available.

The following section introduces two statistics for testing against disturbance SC, and derives their null asymptotic distributions. One test assumes there is no DCH, the other is robust to DCH of unknown form. Section 8.3 introduces statistics for testing against disturbance DCH of two types, and derives null asymptotic distributions. Versions which do and do not assume constancy of conditional dynamic skewness and kurtosis are considered. Corresponding simultaneous tests against both SC and DCH are presented in Section 8.4. In Section 8.5 we discuss the precise specification of test statistics when good power against long-memory alternatives is desired.

8.2 Testing for Serial Correlation

Denote by F_t the σ-field of events generated by $u_s, s < t$. In this section we test the hypothesis

$$\mathrm{E}(u_t | F_t) = 0 \quad \text{a.s.} \tag{8.2}$$

against the alternative

$$E(u_t|F_t) = \sum_{j=1}^{\infty} \varphi_j(\theta) u_{t-j} \quad \text{a.s.} \tag{8.3}$$

Concerning the $\varphi_j(\theta)$ sequence we introduce the following assumptions.

A.1 The $\varphi_j(\theta)$ are uniquely-defined functions of a $p \times 1$ vector θ such that $\varphi_j(\theta) = 0$, for all $j \geq 1$, if and only if $\theta = 0$. The partial derivatives at $\theta = 0$, $\varphi_j = (\partial/\partial\theta)\varphi_j(0)$, are square-summable.

A.2 The matrix $\Phi = \sum_{j=1}^{\infty} \varphi_j \varphi_j'$ is nonsingular.

In view of A.1 we may recast (8.2) as a parametric hypothesis,

$$H_0: \quad \theta = 0. \tag{8.4}$$

The assumptions on the $\varphi_j(\theta)$ seem necessary and sufficient both for the construction of our test statistics and for the properties derived. A related specification was used by Robinson *et al.* (1985) in testing for disturbance SC in the Tobit model, though the actual alternatives discussed there were the AR(p) and MA(p), familiar special cases of our specification. Condition A.2 is a kind of identifiability condition at the null model, ruling out, for example, standard mixed autoregressive moving-average alternatives. Versions of the $\varphi_j(\theta)$ suitable for long-memory alternatives are discussed in Section 8.5.

The first version of our test is based also on the following assumption, which implies no DCH, as was standard in the earlier literature on testing for SC:

A.3
$$E(u_t^2|F_t) = \sigma^2 \quad \text{a.s.} \tag{8.5}$$

Define, for $0 \leq j < N$, the lag-j residual sample autocovariance based on observations at $t = 1, \ldots, N$,

$$c_j = N^{-1} \sum \hat{u}_t \hat{u}_{t+j}, \qquad \hat{u}_t = y_t - b'x_t,$$

where b is the LS estimate of β, and unless indicated otherwise $\sum$ is a sum over t from 1 through $N - j$. Our LM statistics will all be of form

$$\Lambda_i = \lambda_i' \lambda_i, \tag{8.6}$$

and we shall establish under regularity conditions that

$$\Lambda_i \to_d \chi_p^2. \tag{8.7}$$

It is easy to see that a LM statistic for testing (8.2) against (8.3) based on a Gaussian likelihood is of form Λ_1 with

$$\lambda_1 = N^{1/2} \Phi^{-1/2} \sum_{j=1}^{N-1} \varphi_j c_j / c_0, \tag{8.8}$$

where Φ could be replaced, if desired, by $\sum_1^{N-1} \varphi_j \varphi_j'$, for example. We have introduced the factor λ_1 explicitly because when $p = 1$ a one-sided test of (8.4) against the alternatives $\theta > 0$, say, may be preferred, and would be based on λ_1 or $-\lambda_1$, depending on the parameterization; the proof of Theorem 1 below implicitly justifies a limiting null $N(0, I)$ distribution for λ_1, where I is the p-rowed identity matrix. The usual type of invariance property holds, (8.8) arising also from a LM argument based on an infinite MA alternative for u_t, with MA coefficients $\varphi_j(\theta)$. Moreover, Λ_1 is invariant to linear transformation of the form $\varphi_j \to H\varphi_j$ where H is nonsingular.

Theorem 1 Under A.1–A.3 and the null hypothesis (8.2), it follows that (8.7) holds for $i = 1$.

Proof Since the limiting distribution is independent of the x_t, it will suffice to show that (8.7) holds conditionally on the x_t; in all proofs, expectations will be conditional on the x_t. For $j \geq 0$,

$$c_j = \bar{c}_j + N^{-1} \sum_{t=1}^{N-j} (v_t u_{t+j} + u_t v_{t+j} + v_t v_{t+j}),$$

where $v_t = \hat{u}_t - u_t, \bar{c}_j = N^{-1} \sum u_t u_{t+j}$. Now

$$\mathrm{E} \left| \sum u_t v_{t+j} \right| \leq \mathrm{E} \left| \sum u_t^2 b_{t,t+j} \right| + \left\{ \mathrm{E} \left(\sum u_t \sum_{s \neq t} u_s b_{s,t+j} \right)^2 \right\}^{1/2}$$

$$\leq 3\sigma^2 \left(\sum b_{tt} \sum b_{t+j,t+j} \right)^{1/2} \leq 3k\sigma^2, \tag{8.9}$$

$$\mathrm{E} \left| \sum v_t v_{t+j} \right| \leq \left(\sum \mathrm{E} v_t^2 \sum \mathrm{E} v_{t+j}^2 \right)^{1/2}$$

$$\leq \sigma^2 \left(\sum b_{tt} \right)^{1/2} \left(\sum b_{t+j,t+j} \right)^{1/2} \leq k\sigma^2, \tag{8.10}$$

where $b_{uv} = x_u'(\sum_{t=1}^N x_t x_t')^{-1} x_v$, and we use $b_{tt} \geq 0, \sum_{t=1}^N b_{tt} = k$, A.3, and (8.2). Thus $\mathrm{E}|c_j - \bar{c}_j| = \mathrm{O}(N^{-1})$ uniformly in j. Next, using A.1,

$$\sum_1^{N-1} |\varphi_j| \leq \left(\sum_1^M |\varphi_j|^2 M \right)^{1/2} + \left(N \sum_M^\infty |\varphi_j|^2 \right)^{1/2} = \mathrm{o}(N^{1/2}),$$

choosing $M \to \infty, M/N \to 0$. It follows that $N^{1/2} \sum_1^{N-1} \varphi_j(c_j - \bar{c}_j) = \mathrm{o}_p(1)$. From A.2 and the martingale central limit theorem of Brown (1971), $(\sum_1^m \varphi_j \varphi_j')^{-1/2} N^{1/2} \sum_1^m \varphi_j \bar{c}_j / \sigma^2 \to_d N(0, I)$ for fixed m. As $m \to \infty$, $\sum_m^\infty \varphi_j \varphi_j' \to 0$. Because $N\mathrm{E}\bar{c}_j^2 \leq \sigma^4$ and $N\mathrm{E}\bar{c}_j \bar{c}_k = 0, 0 < j < k$, it follows that $N\mathrm{E}| \sum_m^{N-1} \varphi_j \bar{c}_j |^2 \leq \sigma^4 N \sum_m^\infty |\varphi_j|^2 \to 0$ as $m \to \infty$. Finally because the $u_t^2 - \sigma^2$ are stationary integrable martingale differences, $\bar{c}_0 \to_p \sigma^2$. ∎

When u_t is Gaussian, standard methods can be employed to derive the exact finite-sample distribution of λ_1 and Λ_1, because λ_1 is a ratio of quadratic forms in the u_t. Using Pitman's (1937) result, exact moments of Λ_1 and λ_1 can be obtained under a Gaussian null model, and used to improve the asymptotic approximations.

Unless the φ_j are eventually all 0 (as in case of AR(p) or MA(p) alternatives) computation of λ_1 or Λ_1 entails O(N^2) operations. This is unlikely to pose a serious computational drawback nowadays unless N is very large, in which case a double use of the FFT enables all c_j to be computed in something like the order of $N \log N$ operations. Similar savings are available if λ_1 is approximated in the frequency domain. Let $h(\omega) = \sum_{j=1}^{\infty} \varphi_j \cos j\omega$, $w(\omega) = N^{-1/2} \sum_{t=1}^{N} \hat{u}_t e^{it\omega}$ and, with no effect on the asymptotic distributional properties we discuss, replace λ_1 by

$$\bar{\lambda}_1 = \Phi^{-1/2} N^{-1/2} \sum_{s}{}' h(\omega_s) |w(\omega_s)|^2 / c_0,$$

where $\omega_s = 2\pi s/N$ and the primed sum omits from $s = 1, \ldots, N$ any terms for which h is unbounded. The $w(\omega_s)$ can be computed via the FFT. By Parseval's equality $\Phi = \frac{1}{4}\pi^{-1} \int_{-\pi}^{\pi} h(\omega) h'(-\omega) \, d\omega$, which can be approximated by $\frac{1}{2} N^{-1} \sum_s{}' h(\omega_s) h'(-\omega_s)$.

The behaviour of Λ_1 when (8.2) is false is also of interest. For fixed $\theta \neq 0$, long-memory parameterizations of (8.3) may result in the $N^{1/2}(\bar{c}_j - \gamma_j)$ not being asymptotically normal, where $\gamma_j = E(u_t u_{t+j})$, and thus a nonnormal limiting distribution for the correspondingly centred λ_1 is a possibility (see e.g. Fox and Taqqu 1985). A condition for consistency of a test based on rejecting (8.4) when Λ_1, is large, against a general fixed alternative autocovariance sequence $\{\gamma_j\}$, is $\sum_{j=1}^{\infty} \varphi_j \gamma_j \neq 0$. For a sequence of Pitman-type alternatives, $\zeta = z/N^{1/2}$, where $E(u_t | F_t) = \sum_{j=1}^{\infty} \tau_j(\zeta) u_{t-j}$ and the $s \times 1$ vector ζ is 0 if and only if $\tau_j(\zeta) = 0$, for all $j \geq 1$, the asymptotic relative efficiency of our test is $(\sum_j^{\infty} \rho_j \varphi_j' \Phi^{-1} \sum_j^{\infty} \varphi_j \rho_j) / \sum_j^{\infty} \rho_j^2$, where $\rho_j = z'(\partial/\partial\zeta)\tau_j(0)$.

Theorem 1 will not be robust to departures from the no-DCH Assumption A.3. To justify a robust test, replace A.3 by

A.4

$$E(u_1 u_{1+j} u_{1+k}^2) = 0, \quad k > j \geq 1, \tag{8.11}$$

$$E(u_1^4) < \infty, \tag{8.12}$$

$$\inf_{j \geq 1} E(u_1^2 u_{1+j}^2) > 0. \tag{8.13}$$

We can check A.4 for versions on ARCH models, for example, such as $u_t | F_t \sim N(0, \alpha_0 + \alpha_1 u_{t-1}^2)$ when $\alpha_0 > 0$ and $0 < \alpha_1 < \sqrt{3}$. Of course (8.11) and (8.13) would be implied by (8.2) and (8.5), but because of (8.12), A.4 is not strictly weaker than A.3, which required existence of only second moments.

Our robustified LM statistic (cf. Weiss 1986) is of form Λ_2 in (8.6), with

$$\lambda_2 = N^{1/2} \left(\sum_{j=1}^{N-1} \varphi_j \varphi_j' d_j \right)^{-1/2} \sum_{j=1}^{N-1} \varphi_j c_j,$$

where

$$d_j = N^{-1} \sum \hat{u}_t^2 \hat{u}_{t+1}^2.$$

Theorem 2 Under A.1, A.2, A.4, and the null hypothesis (8.2), it follows that (8.7) holds for $i = 2$.

Proof The proof that $E|c_j - \bar{c}_j| = O(N^{-1})$ uniformly, differs from that for Theorem 1 only in that $E| \sum u_t v_{t+j}| \leq 3k\{E(u_1^4)\}^{1/2}$ replaces (8.9). For fixed m, using (8.11) $\{\sum_1^m \varphi_j \varphi_j' \tilde{d}_j\}^{-1/2} N^{1/2} \sum_1^m \varphi_j \bar{c}_j \to_d N(0, I)$, where $\tilde{d}_j = E(u_1^2 u_{1+j}^2)$. We can bound $| \sum (\hat{u}_t^2 \hat{u}_{t+j}^2 - u_t^2 u_{t+j}^2)|$ by a sum of terms, each of form

$$\sum |u_t|^a |u_{t+j}|^b |v_t|^c |v_{t+j}|^d, \quad a, b, c, d \geq 0, \quad a + b + c + d = 4,$$

$$c + d > 0. \tag{8.14}$$

The fourth power of the sum in (8.14) is bounded by

$$\left(\sum u_t^4 \right)^a \left(\sum u_{t+j}^4 \right)^b \left(\sum v_t^4 \right)^c \left(\sum v_{t+j}^4 \right)^d$$

$$= O_p(N^{a+b}) \left(\sum v_t^2 \right)^{2c} \left(\sum v_{t+j}^2 \right)^{2d}$$

$$= O_p(N^4), \tag{8.15}$$

from (8.10) and A.4. Thus $\sum_1^m \varphi_j \varphi_j' (d_j - \tilde{d}_j) \to_p 0$ for fixed m, where $\tilde{d}_j = N^{-1} \sum u_t^2 u_{t+j}^2$. By ergodicity $\sum_1^m \varphi_j \varphi_j' \{\tilde{d}_j - \bar{d}_j\} \to_p 0$ for fixed m. We have easily $NE\bar{c}_j^2 \leq E(u_1^4)$, $NE(\bar{c}_j \bar{c}_k) = 0, 0 < j < k$, so that $NE| \sum_m^{N-1} \varphi_j \bar{c}_j|^2 \to 0$ as $m \to \infty$, as in Theorem 1's proof. Because of (8.12) and thus $E|\bar{d}_j| = O(1)$, $E|d_j| = O(1)$ uniformly, it follows that $\sum_m^{N-1} \varphi_j \varphi_j' (E|d_j| + E|\bar{d}_j| + |\tilde{d}_j|) \to 0$ as $m \to \infty$. Finally note that A.2 and (8.13) imply $\sum_1^\infty \varphi_j \varphi_j' d_j$ is nonsingular. ∎

8.3 Testing for Dynamic Conditional Heteroskedasticity

The setting (8.1) is unchanged except that no SC, (8.2), is now a maintained assumption whereas no DCH, (8.5), is to be tested. We assume:

A.5 The $\psi_j(\theta)$ are uniquely-defined functions of a $p \times 1$ vector θ such that $\psi_j(\theta) = 0$, for all $j \geq 1$, if and only if $\theta = 0$. The partial derivatives at $\theta = 0$, $\psi_j = (\partial/\partial \theta)\psi_j(0)$, are square-summable.

A.6 The matrix $\Psi = \sum_{j=1}^{\infty} \psi_j \psi_j'$ is nonsingular.

Two distinct classes of alternatives to (8.5) are considered,

$$V(u_t | F_t) = \left\{ 1 + \sum_{j=1}^{\infty} \psi_j^2(\theta) \right\}^{-1} \left\{ \sigma + \sum_{j=1}^{\infty} \psi_j(\theta) u_{t-j} \right\}^2 \quad \text{a.s.} \quad (8.16)$$

and

$$V(u_t | F_t) = \sigma^2 + \sum_{j=1}^{\infty} \psi_j(\theta)(u_{t-j}^2 - \sigma^2) \quad \text{a.s.} \quad (8.17)$$

We assume first that u_t behaves like an independent sequence so far as third and fourth moments are concerned.

A.7 $E(u_t^3 | F_t) = \mu_3$, $E(u_t^4 | F_t) = \mu_4$, where μ_3 and μ_4 are finite constants.

The form (8.17) can be viewed as generalizing the original ARCH model and some of its existing generalizations: in the ARCH(p), $\psi_j(\theta) = 0$, $j > p$; in stationary GARCH(p) models, the $\psi_j(\theta)$ decay exponentially. We allow for tests against more general, possibly more slowly-decaying, $\psi_j(\theta)$. The form (8.16) violates Engle's (1982) 'symmetry' property. However, in general there is no reason in principle why u_t should depend only on the magnitude of past shocks u_{t-j}. To detect a difference in the consequences of (8.16) and (8.17) one need not look beyond third moments: when $\mu_3 = 0$, $E(u_{1+j}^2 u_1) = 0$ for all j under (8.17), but only for $j = 0$, in general, under (8.16). As is well known, (8.16) and (8.17) do not cover all possibilities, because $V(u_t | F_t)$ need not be quadratic, but nor, in Section 8.2, need $E(u_t | F_t)$ have been linear. The alternatives (8.16) and (8.17) are based on the hope that departures from (8.5) can be detected from, respectively, third and fourth moments.

A LM-type statistic for alternative (8.16), based on a likelihood with conditionally Gaussian u_t, is readily seen to be of form Λ_3 given by (8.6) with

$$\lambda_3 = N^{1/2} \Phi^{-1/2} \sum_{j=1}^{N-1} \psi_j e_j / e, \quad (8.18)$$

where

$$e = c_0^{1/2} \left\{ N^{-1} \sum_{t=1}^{N} (\hat{u}_t^2 - c_0)^2 \right\}^{1/2}, \qquad e_j = N^{-1} \sum \hat{u}_t (\hat{u}_{t+j}^2 - c_0). \quad (8.19)$$

Our test for DCH, like those of Weiss (1986), is robust to departures from Gaussianity, though in case of Gaussian kurtosis of u_t, $\mu_4 = 3\sigma^4$, e can be replaced by $(2\hat{\sigma}^3)^{1/2}$, and at a given non-Gaussian model, a more efficient test may be available. A LM argument also gives (8.18) if (8.16) is replaced by the model $u_t = \{1 + \sum_{j=1}^{\infty} \varphi_j^2(\theta)\}^{-1/2} \{1 + \sum_1^{\infty} \varphi_j(\theta)\varepsilon_{t-j}\}\varepsilon_j$, for a standard normal white-noise sequence ε_t.

Theorem 3 Under (8.2), A.5–A.7, and the null hypothesis (8.5), it follows that (8.7) holds for $i = 3$.

Proof Write, for $j \geq 1$.

$$e_j = \bar{e}_j + N^{-1} \sum \{(u_{t+j}^2 - \sigma^2)v_t + 2u_{t+j}v_{t+j}u_t$$
$$+ 2u_{t+j}v_{t+j}v_t + v_{t+j}^2 u_t + v_{t+j}^2 v_t - (c_0 - \sigma^2)\hat{u}_t\},$$

where $\bar{e}_j = N^{-1} \sum u_t(u_{t+j}^2 - \sigma^2)$. The proofs that $\sum(u_{t+j}^2 - \sigma^2)v_t$ and $\sum u_{t+j}v_{t+j}u_t$ are uniformly $O_p(1)$ follow (8.9), using A.7. We have

$$\left| \sum u_{t+j}v_{t+j}v_t \right| \leq \left(\sum u_{t+j}^2 v_t^2 \sum v_{t+j}^2 \right)^{1/2},$$

where $\sum v_t^2 = O_p(1)$ was established in (8.10) and

$$\sum u_{t+j}^2 v_t^2 = \sum u_{t+j}^4 b_{t+j,t}^2 + 2 \sum u_{t+j}^3 b_{t+j,t} \sum_{s \neq t+j} u_s b_{st}$$
$$+ \sum u_{t+j}^2 \sum \sum_{r \neq s \neq t+1} u_r u_s b_{rt} b_{st}$$

has bounded expectation. We can handle $\sum v_{t+j}^2 u_t$ in the same way, while $|\sum v_{t+j}^2 v_t| \leq \sum v_{t+j}^2 (\sum v_t^2)^{1/2} = O_p(1)$ using (8.10). The proof that $(c_0 - \sigma^2) \sum \hat{u}_t = O_p(1)$ uses previous arguments and the fact that now $\bar{c}_0 - \sigma^2 = O_p(N^{-1/2})$, by A.7. It follows that $e_j - \bar{e}_j = O_p(N^{-1})$ uniformly. Because the $u_t(u_{t+j}^2 - \sigma^2)$ are stationary square-integrable martingale differences the remainder of the proof follows that in Section 8.2, and we only need demonstrate that $e \to_p (\mu_4 - \sigma^4)^{1/2}\sigma$. We have

$$e^2/c_0 = N^{-1} \sum (u_t^4 + 4u_t^3 v_t + 6u_t^2 v_t^2 + 4u_t v_t^3 + v_t^4 - c_0^2).$$

Clearly $N^{-1} \sum u_t^4 \to_p \mu_4$, and since $c_0 \to_p \sigma^2$ is already known and the remaining terms are $o_p(1)$ in view of (8.15), the proof is completed. ∎

A test for DCH against (8.16) that is robust to departures from A.7 will be justified under:

A.8

$$E(u_1 u_{1+j} u_{1+k}^4) = 0, \quad k > j \geq 1, \tag{8.20}$$

$$E(u_1^6) < \infty, \tag{8.21}$$

$$\inf_{j \geq 1} \{E(u_1^2(u_{1+j}^2 - \sigma^2)^2\} > 0. \tag{8.22}$$

We note that (8.20) and (8.22) would be implied by A.7 if A.7 were extended to include (8.21). Let

$$\lambda_4 = N^{1/2} \left(\sum_1^{N-1} \varphi_j \varphi_j' f_j \right)^{-1/2} \sum_1^{N-1} \varphi_j e_j,$$

where

$$f_j = N^{-1} \sum \hat{u}_t^2 \left(\hat{u}_{t+j}^2 - c_0 \right)^2.$$

Theorem 4 Under (8.2), A.4, A.5, A.8, and the null hypothesis (8.5), it follows that (8.7) holds for $i = 4$.

Proof The proof that $e_j - \bar{e}_j = O_p(N^{-1})$ only differs in minor ways from that in the proof of Theorem 3, swapping the moment condition (8.21) for the martingale difference assumptions in A.7. The $N^{1/2} \bar{e}_j$ are still asymptotically normal with zero mean, but the limiting covariance between $N^{1/2} \bar{e}_j$ and $N^{1/2} \bar{e}_k$ is $E(u_1^2(u_{1+j}^2 - \sigma^2)^2)$ when $k = j$, and zero otherwise, using (8.20). Since (8.22) uniformly bounds the latter expectation away from 0 it remains to prove that $f_j \to_p E\{u_1^2(u_{1+j}^2 - \sigma^2)^2\}$ for fixed j. In view of earlier proofs and ergodicity it suffices to prove that $N^{-1} \sum(\hat{u}_t^2 \hat{u}_{t+j}^4 - u_t^2 u_{t+j}^4) \to_p 0$. But this follows in the same way as $N^{-1} \sum(\hat{u}_t^2 \hat{u}_{t+j}^2 - u_t^2 u_{t+j}^2) \to_p 0$ in the proof of Theorem 2, replacing 2 by 4 in proof. The rest of the proof also follows Theorem 2's. ∎

We now turn to the class of DCH alternatives (8.17). A LM statisic is Λ_5 given by (8.6) and

$$\lambda_5 = N^{1/2} \Phi^{-1.2} \sum_1^{N_1} \varphi_j g_j / g_0,$$

where

$$g_j = N^{-1} \sum (\hat{u}_t^2 - c_0)(\hat{u}_{t+j}^2 - c_0).$$

The statistic Λ_5 can also be regarded as the LM statistic when the alternatives are given by the model $u_t = \sigma\{1 + \sum_1^\infty \varphi_j(\theta)(\varepsilon_{t-j}^2 - 1)\}^{1/2}\varepsilon_t$, for a standard normal white noise sequence ε_t.

Theorem 5 Under (8.2), A.5–A.7, and the null hypothesis (8.5), it follows that (8.7) holds for $i = 5$.

Proof This is very similar to previous ones. We can expand g_j much as before and demonstrate that, for $j \geq 1, g_j = \bar{g}_j + O_p(N^{-1})$ uniformly, where $\bar{g}_j = N^{-1} \sum(u_t^2 - \sigma^2)(u_{t+j}^2 - \sigma^2)$. This property holds under the fourth-moment condition of A.4 for much the same reason that second moments sufficed in Section 8.2. By way of a specific illustration, $\sum(u_t^2 - \sigma^2)v_{t+j}^2$ depends on the u_t^4, but we have already seen that this is $O_p(1)$. We omit remaining details. ∎

Finally, we robustify Λ_5, replacing it by λ_6 given by (8.6) with

$$\lambda_6 = N^{1/2} \left(\sum_1^{N-1} \varphi_j \varphi_j' h_j \right)^{-1/2} \sum_1^{N-1} \varphi_j g_j,$$

where

$$h_j = N^{-1} \sum (\hat{u}_t^2 - c_0)^2 (\hat{u}_{t+j}^2 - c_0)^2.$$

We introduce condition:

A.9

$$E((u_1^2 - \sigma^2)(u_{1+j} - \sigma^2)u_{1+k}^4) = 0, \quad k > j \geq 1, \qquad (8.23)$$

$$E(u_1^8) < \infty, \qquad (8.24)$$

$$\inf_{j \geq 1} E\{(u_1^2 - \sigma^2)^2 (u_{1+j}^2 - \sigma^2)^2\} > 0.$$

Theorem 6 Under (8.2), A.4, A.5, A.9, and the null hypothesis (8.5), it follows that (8.7) holds for $i = 6$,

Proof Omitted. ■

A frequency-domain representation of our tests against DCH is also possible, along similar lines to those mentioned in Section 8.2.

8.4 Simultaneous Testing for Serial Correlation and Dynamic Conditional Heteroskedasticity

The possibility of both SC and DCH may be entertained. One option is to apply the robust test for SC given by λ_2 followed by one of the tests for DCH. Another is to apply one of the tests for DCH; if it rejects the robust test for SC is used, if it fails to reject the test for SC given by λ_1 might be used. Alternatively one might test simultaneously for SC and DCH. The two alternatives to the null hypothesis that both (8.2) and (8.5) are true are

$$(8.3) \quad \text{and} \quad (8.16), \qquad (8.25)$$

$$(8.3) \quad \text{and} \quad (8.17). \qquad (8.26)$$

These models extend the notion of AR models with ARCH errors. Considering (8.25) first, we initially impose the independence-up-to-fourth-moments condition A.7. We assume also:

A.10 $\Phi + \Psi$ is nonsingular.

Condition A.10 is implied by either A.2 or A.6, but if the $\varphi_j(\theta)$ and $\psi_j(\theta)$ are disjointly parameterized, as is likely, A.2 or A.6 would not hold. A LM statistic for alternative (8.25), based on a likelihood with conditionally Gaussian u_t, is of form Λ_7 given by (8.6) with

$$\lambda_7 = N^{1/2}\hat{\Delta}^{-1/2} \sum_{j=1}^{N-1} \{\varphi_j c_j/c_0 + \psi_j e_j/c_0^{3/2}\},$$

where

$$\hat{\Delta} = \Phi + (e_0/c_0^{3/2})(\Gamma + \Gamma') + e^2/c_0^3 \Psi,$$

and e and the e_j (including $j = 0$) are given in (8.19). Note that $\hat{\Delta}$ is guaranteed nonnegative definite. The form of λ_7 is simpler than might appear when the $\varphi_j(\theta)$ and $\psi_j(\theta)$ are separately parameterized, in particular $\Gamma = 0$.

Theorem 7 Under A.1, A.5, A.7, A.10, and the null hypothesis (8.2) plus (8.5), it follows that (8.7) holds for $i = 7$.

Proof Omitted. ∎

A robust version of our test against (8.25) that is robust to departures from A.7 will be justified under:

A.11 The conditions (8.20), (8.21) hold, along with

$$E(u_1 u_{1+j} u_{1+k}^3) = 0 \quad k > j \geq 1,$$

and

$$\inf_{j \geq 1} \min_{\|a\|=1} a' E\left\{ u_1^2 \begin{bmatrix} \sigma u_{1+j} \\ u_{1+j}^2 - \sigma^2 \end{bmatrix} \begin{bmatrix} \sigma u_{1+j} \\ u_{1+j}^2 - \sigma^2 \end{bmatrix}' \right\} a > 0.$$

The last inequality extends (8.22). Let

$$\lambda_8 = N^{1/2}\tilde{\Delta}^{-1/2} \sum_{j=1}^{N-1} \{\varphi_j c_j/c_0 + \psi_j e_j/c_0^{3/2}\},$$

where

$$\tilde{\Delta} = \sum_{j=1}^{N-1} \frac{1}{N} \sum \left\{ \varphi_j \frac{\hat{u}_t \hat{u}_{t+j}}{c_0} + \psi_j \frac{\hat{u}_t(\hat{u}_{t+j}^2 - c_0)}{c_0^{3/2}} \right\}$$

$$\times \left\{ \varphi_j \frac{\hat{u}_t \hat{u}_{t+j}}{c_0} + \psi_j \frac{\hat{u}_t(\hat{u}_{t+j}^2 - c_0)}{c_0^{3/2}} \right\}'.$$

Clearly $\tilde{\Delta}$ is always nonnegative definite.

Theorem 8 Under A.1, A.5, A.10, A.11, and the null hypothesis (8.2) plus (8.5), it follows that (8.7) holds for $i = 8$.

Proof Omitted. ∎

Now we turn to the alternatives (8.26). For a statistic based on A.7 let

$$\lambda_9 = N^{1/2}\hat{\Pi}^{-1/2}\sum_{j=1}^{N-1}\{2\varphi_j c_j/c_0 - \psi_j g_j/c_0^2\},$$

where

$$\hat{\Pi} = 4\Phi + 2(e_0^2/c_0^3)(\Gamma + \Gamma') + (e_4/c_0^6)\Psi.$$

Theorem 9 Under A.1, A.5, A.7, A.10, and the null hypothesis (8.2) plus (8.5), it follows that (8.7) holds for $i = 9$.

Proof Omitted. ■

A robustified version is justified under:

A.12 The conditions (8.23), (8.24) hold, along with

$$E(u_1(u_{1+j}^2 - \sigma^2)u_{1+k}^3) = 0, \quad 1 \le j \ne k \ge 1,$$

and

$$\inf_{j \ge 1}\min_{\|a\|=1} a'E\left\{\left[\frac{2\sigma^2 u_1 u_{1+j}}{(u_1^2 - \sigma^2)(u_{1+j}^2 - \sigma^2)}\right] \times \left[\frac{2\sigma^2 u_1 u_{1+j}}{(u_1^2 - \sigma^2)(u_{1+j}^2 - \sigma^2)}\right]\right\} a > 0.$$

Let

$$\lambda_{10} = N^{1/2}\tilde{\Pi}^{-1/2}\sum_{j=1}^{N-1}\{2\varphi_j c_j/c_0 + \psi_j g_j/c_0^2\},$$

where

$$\tilde{\Pi} = \sum_{j=1}^{N-1}\frac{1}{N}\sum\left\{2\varphi_j\frac{\hat{u}_t\hat{u}_{t+j}}{c_0} + \psi_j\frac{(\hat{u}_t^2 - c_0)(\hat{u}_{t+j}^2 - c_0)}{c_0^2}\right\}$$
$$\times \left\{2\varphi_t\frac{\hat{u}_t\hat{u}_{t+j}}{c_0} + \psi_j\frac{(\hat{u}_t^2 - c_0)(\hat{u}_{t+j}^2 - c_0)}{c_0^2}\right\}'.$$

Theorem 10 Under A.1, A.5, A.10, A.12, and the null hypothesis (8.2) plus (8.5), it follows that (8.7) holds for $i = 10$.

Proof Omitted. ■

8.5 Long-memory Examples

Two parameterizations of long-memory process that have proved successful in modelling a variety of time-series data, including economic series, will be shown to generate alternatives which satisfy the requirement that Φ or Ψ be finite. For a real even sequence indexed by θ, $\gamma_j(\theta)$, $j = 0, \pm 1, \ldots$, with $\gamma_0(\theta) = 1$, define $f(\omega; \theta) = (2\pi)^{-1}\sum_{j=-\infty}^{\infty}\gamma_j(\theta)e^{-ijw}$, which need not be bounded. We can interpret $\gamma_j(\theta)$ and $f(\omega; \theta)$ as the autocovariance and spectral density of a covariance-stationary process with zero mean and unit variance.

The first model is most simply described by

$$\text{Model 1:} \quad f(\omega; \theta) = (2\pi)^{-1}|1 - e^{i\omega}|^{-2\theta}, \qquad (8.27)$$

where the scalar $\theta < \frac{1}{2}$. Model 1 is a fractional ARIMA $(0, \theta, 0)$; Adenstedt (1974) gave an expression for $\gamma_j(\theta)$. The second model is most simply described by

$$\text{Model 2:} \quad \gamma_j(\theta) = \frac{1}{2}\{|j - 1|^{2\theta+1} - 2|j|^{2\theta+1} + |j + 1|^{2\theta+1}\}, \qquad (8.28)$$

where $\theta < \frac{1}{2}$. Model 2 describes the increments of a stationary self-similar process; Sinai (1976) gave an expression for $f(\omega; \theta)$. In both models, when $\theta \neq 0, \theta \geq -\frac{1}{2}$, the $\gamma_j(\theta)$ are not absolutely summable, and $f(0; \theta)$ is unbounded when $\theta > 0$ and zero for $\theta < 0$. Estimation of θ has been studied, a recent treatment being Fox and Taqqu (1986); efficient estimates entail nonlinear optimization, so there is some advantage in a test of (8.4) based on the LM principle. Some statistical properties of b in the presence of disturbances with the second-order properties of Model 1 were studied by, for example, Adenstedt (1974), Samarov and Taqqu (1988), Yajima (1988). For $\theta > 0$ and regressors that are polynomials in t, it seems that b is not asymptotically efficient [as it is against weakly dependent disturbances, see Grenander (1954)] but its asymptotic relative efficiency is good. For $\theta < 0$, which corresponds to fractional overdifferencing, the evidence when x_t consists solely of an intercept is that b's relative efficiency can be very poor indeed.

For either (8.27) or (8.28) there are $\xi_j(\theta)$ satisfying

$$f(\omega; \theta) = \frac{1}{2\pi}\left|1 - \sum_{j=1}^{\infty}\xi_j(\theta)e^{ijw}\right|^{-2}.$$

We can use these $\xi_j(\theta)$ as $\varphi_j(\theta)$ or $\psi_j(\theta)$ in the alternatives (8.3), (8.16), (8.17), (8.25), and (8.26). Expressions for the $\xi_j(\theta)$ are available in case of Model 1, and the $\dot{\xi}_j = (\partial/\partial\theta)\xi_j(0)$ can be easily computed for either model by means of the readily obtainable identity

$$\dot{\xi}_j = \frac{\partial\gamma_j(0)}{\partial\theta} = \int_{-\pi}^{\pi}\frac{\partial f(\omega; 0)}{\partial\theta}e^{ijw}d\omega.$$

Routine calculation of the $\dot{\xi}_j$ and $\Xi = \sum_{j=1}^{\infty}\dot{\xi}_j^2$ gives

$$\text{Model 1:} \quad \dot{\xi}_j = \frac{1}{j}, \quad \Xi = \frac{\pi^2}{6},$$

$$\text{Model 2:} \quad \dot{\xi}_j = \sum_{m=1}^{\infty}\frac{j^{1-2m}}{m(2m+1)} = \frac{1}{j} + o\left(\frac{1}{j}\right) \quad \text{as } j \to \infty.$$

We have no closed-form expression for Ξ under Model 2, though this can be approximated to any desired degree of accuracy by direct computation (see also Davies and Harte 1987). Both models satisfy the conditions A.1, A.2, A.5, A.6, on Φ and Ψ, and by only a small margin, justifying the care we took to avoid a stronger assumption than square-summability on the φ_j and ψ_j. It is also clear that weights ξ_i for Models 1 and 2 are similar, and so it would suffice to use Model 1 for diagnostic testing purposes with its simple hyperbolically decaying weights and closed form Ξ. Whistler (1990) has applied our test for DCH with Model 1 to exchange rate data. In case of testing for SC, because θ is scalar, a one-sided test based on comparing λ_1 or λ_2 with the upper tail of the standard normal is likely to be preferred to a two-sided test, if strong positive dependence is the likely alternative. On the other hand, if overdifferencing is feared, a one-sided test on the lower tail is appropriate. A frequency domain version of the test based on Model 1 takes $h(\omega) = -\log|1 - e^{i\omega}| = -\frac{1}{2}\log\{2\sin^2(\omega/2)\}$. In view of the discussion of relative efficiency in Section 8.2, an AR(1)-based diagnostic, such as that of Durbin and Watson (1950), has efficiency $6/\pi^2$ if Model 1 describes the true alternative, and more generally an AR(p)-based diagnostic (e.g. Box and Pierce 1970) has efficiency $1 - 6\psi'(p + 1)/\pi^2$, where ψ' is the derivative of Euler's psi function.

References

Adenstedt, R. K. (1974), 'On large-sample estimation for the mean of a stationary random sequence', *Annals of Statistics*, 2, 1095–107.

Box, G. E. P. and Pierce, D. A. (1970), 'Distribution of residual autocorrelations in autoregressive-integrated moving average time series models', *Journal of the American Statistical Association*, 65, 1509–26.

Brown, B. M. (1971), 'Martingale central limit theorems', *Annals of Mathematical Statistics* 42, 59–66.

Davies, R. B. and Harte, D. S. (1987), 'Tests for Hurst effect', *Biometrika*, 74, 95–101.

Durbin, J. and Watson, G. S. (1950), 'Testing for serial correlation in least squares regression: I', *Biometrika*, 37, 409–28.

Engle, R. F. (1982), 'Autoregressive conditional heteroscedasticity with estimates of the variance of United Kingdom inflation', *Econometrica*, 50, 987–1007.

Fox, R. and Taqqu, M. S. (1985), 'Noncentral limit theorems for quadratic forms in random variables having long-range dependence', *Annals of Probability*, 13, 428–46.

—— and Taqqu, M. S. (1986), 'Large-sample properties of parameter estimates for strongly dependent stationary Gaussian time series', *Annals of Statistics*, 14, 517–32.

Grenander, U. (1954), 'On the estimation of regression coefficients in the case of an autocorrelated disturbance', *Annals of Mathematical Statistics*, 25, 252–72.

Pitman, E. J. G. (1937), 'The "closest" estimates of statistical parameters', *Proceedings of the Cambridge Philosophical Society*, 33, 212–22.

Robinson, P. M. (1978), 'Alternative models for stationary stochastic processes', *Stochastic Processes and Their Applications*, 8, 141–52.

——, A. K. Bera, and Jarque, C. M. (1985), 'Tests for serial dependence in limited dependent variable models', *International Economic Review*, 26, 629–38.

Samarov, A. and Taqqu, M. S. (1988), 'On the efficiency of the sample mean in long memory noise', *Journal of Time Series Analysis*, 9, 191–200.

Sinai, Y. G. (1976), 'Self-similar probability distributions', *Theory of Probability and Its Applications*, 21, 64–80.

Weiss, A. A. (1986), 'Asymptotic theory for ARCH models estimation and testing', *Econometric Theory*, 2, 107–31.

Whistler, D. E. N. (1990), Semiparametric models of daily and intra-daily exchange rate volatility, Ph.D. thesis submitted to the University of London.

Yajima, Y. (1988), 'On estimation of a regression model with long-memory stationary errors', *Annals of Statistics*, 16, 791–807.

9

The Detection and Estimation of Long Memory in Stochastic Volatility

F. JAY BREIDT, NUNO CRATO, AND PEDRO DE LIMA

9.1 Introduction

A large body of research suggests that the conditional volatility of asset prices displays long-memory or long-range persistence.[1] Furthermore, as we demonstrate below, this type of persistence cannot be appropriately modelled by autoregressive conditional heteroskedastic (ARCH), generalized ARCH (GARCH), exponential GARCH (EGARCH) or standard (short-memory) stochastic volatility models.[2]

In light of these recent findings and the limitations of short-memory models of stochastic volatility in this chapter we propose a new time series representation of persistence in conditional volatility that we call a long-memory stochastic volatility model (LMSV).[3] The LMSV model is constructed by incorporating an ARFIMA process in a standard stochastic volatility scheme. We show that the parameters of the LMSV models can be estimated by applying a frequency domain likelihood estimator.[4] The finite sample properties of the spectral-likelihood estimator are evaluated by means of a Monte Carlo study.

The LMSV model has several advantages. First, because it is well-defined in the mean square sense many of its stochastic features are easy to establish. Second, because it has well-known counterparts in models for level series it inherits most of the statistical properties of those models.

The rest of the paper proceeds as follows. In Section 9.2 we review models of persistence in volatility (i.e. fractional GARCH and EGARCH models) and introduce the long-memory stochastic volatility model. In Section 9.3 we present further empirical evidence on the relevance of long memory by testing the null of short memory for proxies of the conditional variances in an extensive set of US stock return indexes. In Section 9.4 we discuss a Whittle-type estimator for the LMSV model parameters obtained by maximizing the spectral approximation to the Gaussian likelihood. We present finite-sample simulation evidence about the

This chapter was previously published as Breidt, F. J., Crato, N, and de Lima, P. (1998), 'The detection and estimation of long memory in stochastic volatility', *Journal of Econometrics*, 73, 325–48. Copyright © 1998 Elsevier Science S.A., reproduced by kind permission.

This work was presented as a paper at a conference in honour of Carl Christ held at the Johns Hopkins University in April 1995. We are thankful to our discussant Patrick Asea, seminar participants at the University of Pennsylvania, Francis Diebold, Thomas Epps, and two anonymous referees at Elsevier for helpful comments. Any remaining errors are our own.

properties of the estimators and, as an example, we study the daily returns for the value-weighted CRSP market index. In Section 9.5 we conclude. Proofs are found in the Appendices A and B.

9.2 Models of Persistence in Volatility

Following Brockwell and Davis (1991), we state that a weakly stationary process has *short memory* when its autocorrelation function (ACF), say $\rho(h)$, is geometrically bounded

$$|\rho(h)| \leqslant Cr^{|h|} \quad \text{for some } C > 0, \; 0 < r < 1.$$

In contrast to a short-memory process with a geometrically decaying ACF, a weakly stationary process has *long memory* if its ACF $\rho(\cdot)$ has a hyperbolic decay,

$$p(h) \sim Ch^{2d-1} \quad \text{as } h \to \infty,$$

where $C \neq 0$ and $d < 0.5$ (e.g. Brockwell and Davis 1991, section 13.2). Alternatively, we can say the process has long memory if its spectrum $f(\lambda)$ has the asymptotic decay

$$f(\lambda) \sim C|\lambda|^{-2d} \quad \text{as } \lambda \to 0 \text{ with } d \neq 0. \tag{9.1}$$

If, in addition, $d > 0$, then the autocorrelations are not absolutely summable, $\sum |\rho(h)| = \infty$, and the spectrum diverges at zero, $f(\lambda) \uparrow \infty$ as $\lambda \to 0$. In this case we conclude that the process is *persistent*. For a discussion of alternative long-memory characterizations see sections 2.2 and 2.3 of Baillie (1996).

9.2.1 GARCH and EGARCH models
Following Engle (1982), Bollerslev (1986) and Nelson (1991), let the prediction error y_t satisfy

$$y_t = \sigma_t \xi_t,$$

where $\{\xi_t\}$ is independent and identically distributed (i.i.d.) with mean zero and variance one, and σ_t^2 is the variance of y_t given information at time $t - 1$. Among the most successful specifications for the conditional variance σ_t^2 are the GARCH and EGARCH models. A GARCH specification is given by

$$\sigma_t^2 = \omega + \sum_{j=1}^{q} b_j \sigma_{t-j}^2 + \sum_{j=1}^{p} a_j y_{t-j}^2, \tag{9.2}$$

where $\omega > 0$. Constraints on $\{b_j\}$ and $\{a_j\}$ are discussed below. More compactly, we can write eqn (9.2) as

$$b(B)\sigma_t^2 = \omega + a(B)y_t^2,$$

where B is the backshift operator ($B^j v_t = v_{t-j}, \, j = 0, \pm 1, \pm 2, \ldots$), $b(z) = 1 - b_1 z - \cdots - b_q z^q$ and $a(z) = a_1 z + \cdots + a_p z^p$.

As an alternative to the GARCH specification, Nelson (1991) proposes the Exponential GARCH (EGARCH) model

$$\log \sigma_t^2 = \mu_t + \sum_{j=0}^{\infty} \psi_j g(\xi_{t-j-1}), \quad \psi_0 \equiv 1, \tag{9.3}$$

where no restriction is needed for the signs of the coefficients. The function $g(\cdot)$ may be chosen to allow for asymmetric changes, depending on the sign of ξ_t.

It is known that the GARCH model (Bollerslev 1986) can also be written in an ARMA($\max\{p, q\}, q$) form, with the process $\{y_t^2\}$ being driven by the noise $v_t = y_t^2 - \sigma_t^2$. From this representation it is clear that the autocorrelation function for $\{y_t^2\}$ has a short-memory geometric decay.

EGARCH models have the general representation as in (9.3), but they are also usually parameterized with weights $\{\psi_j\}$ corresponding to an ARMA(p, q). Thus, the usual EGARCH specification can be written as:

$$\phi(B)(\log \sigma_t^2 - \mu_t) = \theta(B)g(\xi_{t-j-1}),$$

where $\phi(z) = 1 - \phi_1 z - \cdots - \phi_p z^p \neq 0$ for $|z| \leqslant 1$ is an autoregressive polynomial, $\theta(z) = 1 + \theta_1 z + \cdots + \theta_q z^q$ is a moving average polynomial and $\theta(z)$ has no roots in common with $\phi(z)$.

The empirical evidence previously discussed points in the direction of long memory, both in the squared process $\{y_t^2\}$ and in the process of log squares $\{\log y_t^2\}$. This contrasts with the usual short-memory formulations of GARCH and EGARCH models. We will look for the formulation of models with persistent properties.

9.2.2 Long-memory GARCH models

In order to accommodate the findings of long memory, a sensible approach is to generalize GARCH models by using fractional differences, along lines earlier suggested by Robinson (1991: p. 82). The fractional differencing operator is defined through the expansion:

$$(1 - B)^d = \sum_{j=0}^{\infty} \frac{\Gamma(j - d)}{\Gamma(j + 1)\Gamma(-d)} B^j,$$

from which Baillie *et al.* (1996) formulated the fractionally integrated GARCH (FIGARCH) model

$$(1 - B)^d b(B)(\sigma_t^2 - \mu) = a(B)(y_t^2 - \mu).$$

Baillie *et al.* (1996) suggested quasi-maximum-likelihood estimation methods for this model.

In order to have a well-defined process, the parameters $\{a_j\}$, $\{b_j\}$, and d are constrained so that the coefficients ψ_j in the representation

$$\sigma_t^2 = \mu + (1 - B)^{-d} a(B) b^{-1}(B) y_t^2 = \mu + \sum_{j=0}^{\infty} \psi_j (y_{t-1-j}^2 - \mu), \qquad (9.4)$$

are all nonnegative. Otherwise we have $\sigma_t^2 < 0$ with a positive probability. This implies that the parameters $\{a_j\}$ and $\{b_j\}$ are constrained as in the standard GARCH models. This also implies that the parameter d is constrained to be positive, and so $\sum_{j=0}^{\infty} \psi_j = \infty$. However, this means that the sum of all coefficients is greater than one. It follows from a now standard result in Bollerslev (1986) that $\{y_t\}$ is not covariance stationary. Consequently, the autocovariance function (ACVF) of the process $\{y_t^2\}$ is not defined,[5] the series $\sum_{j=0}^{\infty} \psi_j (y_{t-1-j}^2 - \mu)$ is not defined in L^2, and the use of spectral and time-domain autocorrelation methods is not justifiable in a standard way. In addition, initializing the quasi-likelihood, which is usually done with unconditional moments of out-of-sample σ_t^2, can be problematical, although Baillie *et al.* (1996) reported good results for the quasi-maximum-likelihood estimation method.

As an alternative, Nelson (1991: p. 352) notes that persistence can be modelled in the log squares with a long-memory specification of an EGARCH model. Bollerslev and Mikkelsen (1996) have explicitly formulated a fractionally integrated EGARCH model of the form

$$\log \sigma_t^2 = \mu_t + \theta(B) \phi(B)^{-1} (1 - B)^{-d} g(\xi_{t-1}), \qquad (9.5)$$

where $\phi(z)$ and $\theta(z)$ are defined above. This generalization of EGARCH with fractional noise gives a strictly stationary and ergodic process. The condition for the covariance stationarity of $\{\log \sigma_t^2 - \mu_t\}$ is $\sum_{j=0}^{\infty} \psi_j^2 < 1$ (Nelson 1991: th. 2.1), which is satisfied for a parameter value $d < \frac{1}{2}$.

EGARCH models have the convenient feature that the coefficients in the moving average expansion (9.5) are not restricted to be positive. However, asymptotic results about the estimators have proven to be extremely hard to obtain, even when $d = 0$.

9.2.3 A long-memory stochastic volatility model

In this section we introduce a different approach, based on stochastic volatility (SV) models similar to those discussed by Melino and Turnbull (1990) and Harvey *et al.* (1994).

The stochastic volatility model is defined by

$$y_t = \sigma_t \xi_t, \qquad \sigma_t = \sigma \exp(v_t/2), \qquad (9.6)$$

where $\{v_t\}$ is independent of $\{\xi_t\}$, $\{\xi_t\}$ is independent and identically distributed (i.i.d.) with mean zero and variance one, and $\{v_t\}$ is an ARMA model.

The long-memory stochastic volatility (LMSV) model we now introduce is defined by (9.6) with $\{v_t\}$ being a stationary long-memory process.

Restricting our attention to a Gaussian $\{v_t\}$, it follows that $\{y_t\}$ is both covariance and strictly stationary. Denote by $\gamma(\cdot)$ the ACVF of $\{v_t\}$. The covariance structure of y_t is obtained from properties of the lognormal distribution:

$$E[y_t] = 0 \quad \mathrm{Var}(y_t) = \exp\{\gamma(0)/2\}\sigma^2 \quad \text{and}$$

$$\mathrm{Cov}(y_t, y_{t+h}) = 0 \quad \text{for } h \neq 0,$$

so that $\{y_t\}$ is a white noise sequence. In fact, $\{y_t\}$ is a martingale difference, a property inherited from $\{\xi_t\}$. An appealing property of this model in terms of its empirical relevance is the excess kurtosis displayed by y_t, which is

$$\frac{E[y_t^4]}{E[y_t^2]^2} - 3 = 3(\exp\{\gamma(0)\} - 1)$$

when the driving noise $\{\xi_t\}$ is Gaussian.

The process $\{y_t^2\}$ is also both covariance and strictly stationary. Moments of y_t^2 are again obtained from properties of the lognormal distribution:

$$E[y_t^2] = \exp\{\gamma(0)/2\}\sigma^2,$$

$$\mathrm{Var}(y_t^2) = \sigma^4[\{1 + \mathrm{Var}(\xi_t^2)\} \exp\{2\gamma(0)\} - \exp\{\gamma(0)\}],$$

$$\mathrm{Cov}(y_t^2, y_{t+h}^2) = \sigma^4[\exp\{\gamma(0) + \gamma(h)\} - \exp\{\gamma(0)\}] \quad \text{for } h \neq 0.$$

The series is simple to analyse after it is transformed to the stationary process

$$x_t = \log y_t^2 = \log \sigma^2 + E[\log \xi_t^2] + v_t + (\log \xi_t^2 - E[\log \xi_t^2])$$
$$= \mu + v_t + \varepsilon_t,$$

where $\{\varepsilon_t\}$ is i.i.d. with mean zero and variance σ_ε^2. For example, if ξ_t is standard normal, then $\log \xi_t^2$ is distributed as the log of a χ_1^2 random variable, $E[\log \xi_t^2] = -1.27$ and $\sigma_\varepsilon^2 = \pi^2/2$ (Wishart 1947).

The process $\{x_t\}$ is thus a long-memory Gaussian signal plus an i.i.d. non-Gaussian noise, with $E[x_t] = \mu$ and

$$\gamma_x(h) = \mathrm{Cov}(x_t, x_{t+h}) = \gamma(h) + \sigma_\varepsilon^2 I_{\{h=0\}}, \tag{9.7}$$

where $I_{\{h=0\}}$ is one if $h = 0$ and zero otherwise. It turns out that the ACVF of the process $\{\log^2 y_t^2\}$ is the same as that of a fractionally integrated EGARCH model whenever $\delta_2 = 0$ (see Appendix 9.2).

A simple long-memory model for $\{v_t\}$ is the fractionally integrated Gaussian noise defined as the unique stationary solution of the difference equations

$$(1 - B)^d v_t = \eta_t, \qquad \{\eta_t\} \text{ i.i.d. N}(0, \sigma_\eta^2), \tag{9.8}$$

where $d \in (-0.5, 0.5)$. The spectral density, ACVF, and ACF of $\{v_t\}$, denoted by $f(\cdot)$, $\gamma(\cdot)$, and $\rho(\cdot)$, respectively, are given by

$$f(\lambda) = \frac{\sigma_\eta^2}{2\pi} |1 - e^{-i\lambda}|^{-2d}, \quad -\pi \leqslant \lambda \leqslant \pi,$$

$$\gamma(0) = \sigma_\eta^2 \Gamma(1 - 2d)/\Gamma^2(l - d),$$

$$\rho(h) = \frac{\Gamma(h+d)\Gamma(1-d)}{\Gamma(h-d+1)\Gamma(d)}, \quad h = 1, 2, \ldots$$

(e.g. Brockwell and Davis, 1991: p. 522).

More generally, $\{v_t\}$ can be modelled as an ARFIMA(p, d, q), defined as the unique stationary solution of the difference equations

$$(1 - B)^d \phi(B) v_t = \theta(B) \eta_t, \quad \{\eta_t\} \text{ i.i.d. N}(0, \sigma_\eta^2). \quad (9.9)$$

The spectral density of $\{x_t\}$, denoted by $f_\beta(\cdot)$, is then given by

$$f_\beta(\lambda) = \frac{\sigma_\eta^2 |\theta(e^{-i\lambda})|^2}{2\pi |1 - e^{-i\lambda}|^{2d} |\phi(e^{-i\lambda})|^2} + \frac{\sigma_\varepsilon^2}{2\pi}, \quad -\pi \leqslant \lambda \leqslant \pi, \quad (9.10)$$

where $\beta = (d, \sigma_\eta^2, \sigma_\varepsilon^2, \phi_1, \ldots, \phi_p, \theta_1, \ldots, \theta_q)'$.

9.3 Evidence of Long Memory in Volatility

In this section we formally test for the existence of long memory in the volatilities of stock markets' series. This is achieved by analysing two traditional volatility proxies, namely the squared series and the logarithm of the squared series.

9.3.1 Testing for long memory in volatility

There are several methods to test for long memory, ranging from fully parametric to nonparametric approaches. The present paper uses both a semiparametric and a nonparametric test.

The first test is implemented by regressing the logarithm of the periodogram at low frequencies on a function of the frequencies; the expected slope is dependent on the long-memory parameter d, as can be seen from eqn (9.1). This method was introduced by Geweke and Porter-Hudak (1983) and developed by Robinson (1993).

Geweke and Porter-Hudak suggest the use of only the first ordinates of the periodogram, up to m_U, say, and argue that the resulting regression estimator for d could capture the long-memory behaviour without being 'contaminated' by the short-memory behaviour of the process. Further, Robinson suggests an additional truncation of the very first ordinates, up to m_L, say, in order to avoid biases. However, no clear rule exists about the choice of either m_U or m_L. Therefore, we adhere to the common practice of experimenting with a few different values. To test the null hypothesis of short memory against long-memory alternatives, we perform the usual t-test for the hypothesis that $d = 0$ against $d \neq 0$. The standard deviation is obtained from the output of the regression.

It should be emphasized that we will apply this regression as a test of short memory without assuming any particular form of long-memory alternatives. The asymptotics in eqn (9.1) define long-memory processes.

The second test used in this paper is the *normalized rescaled range*, the R/S statistic (e.g. see Beran 1994). The *adjusted range R* is defined as

$$R(n) = \max_{1 \leqslant k \leqslant n} \left\{ \sum_{i=1}^{k} X_i - k\bar{X} \right\} - \min_{1 \leqslant k \leqslant n} \left\{ \sum_{i=1}^{k} X_i - k\bar{X} \right\},$$

where $\bar{X}$ represents the sample mean. The *normalization factor S* can be defined as the square root of a consistent estimator for the variance, given by

$$S^2(n, q) = \sum_{j=-q}^{q} w_q(j)\hat{\gamma}(j),$$

with $\hat{\gamma}(j)$ representing the usual estimators for the autocovariances. The weights $w_q(j)$ we used are those from the Bartlett window. The R/S statistic is

$$Q(n, q) = \frac{R(n)}{S(n, q)},$$

and when $q = 0$ we have the classical R/S statistic of Hurst. The so-called Hurst exponent J is estimated as

$$\hat{J}(n, q) = \frac{\log Q(n, q)}{\log n}.$$

If only short memory is present, then $\hat{J}(n, q)$ converges to $\frac{1}{2}$. If persistent long memory is present, then $\hat{J}(n, q)$ converges to a value larger than $\frac{1}{2}$ (e.g. see Mandelbrot and Taqqu 1979).

If a process satisfies a set of regularity conditions, including the existence of moments of order $4 + \delta$, with $\delta > 0$, Lo (1991) shows that under the short memory null the statistic $V = n^{-1/2} Q(n, q)$ converges weakly to the range of the Brownian bridge on the unit interval. The distribution function for this range, say F_V, is

$$F_V(v) = \sum_{k=-\infty}^{\infty} (1 - 4v^2 k^2) e^{-2v^2 k^2}.$$

If a short-memory process does not have finite second-order moments then the classical Hurst estimate $\hat{J}(n, 0)$ still converges to $\frac{1}{2}$, as discussed in Mandelbrot and Taqqu (1979). Therefore, the estimate $\hat{J}(n, 0)$ continues to provide an indication of long memory. However, no distribution theory is available in this case.

In the absence of clear rules for the choice of q, we experimented with a few values. First, we used $q = 0$, corresponding to the classical estimate. Second, we used $q = q^*$, chosen by Andrews' (1991) data-dependent formula as in Lo (1991: p. 1302). Finally, we tried $q = 200$ in an attempt to yield statistics which are more robust against short-memory effects.

9.3.2 Finite sample performance of the long-memory tests

In this section we consider the finite sample performance of the spectral regression test and the R/S analysis under both short and long memory. The generated processes were long-memory ($d \neq 0$) and short-memory ($d = 0$) stochastic volatility models, defined in eqns (9.6) and (9.9) above. Here, we focus on detecting long memory in the log-squared observations; results for the squared observations are qualitatively similar. An analogous Monte Carlo study for long memory in the levels series is reported by Cheung (1993).

In designing this simulation experiment, we chose as a short-memory benchmark the first-order autoregressive stochastic volatility (ARSV) model given by eqns (9.6) and (9.9) with $p = 1$, $d = 0$ and $q = 0$. This model has been studied extensively; see Jacquier *et al.* (1994) and the references therein. We chose four ARSV parameter settings from table 4 of Jacquier *et al.* (1994). To make the LMSV results comparable, we chose for each ARSV model an ARFIMA$(0, d, 0)$ LMSV model which matched the ARSV parameterizations in two ways: first, the ratio

$$\mathrm{Var}(\sigma_t^2)/\mathrm{E}^2[\sigma_t^2]$$

is the same for both models (implying that the excess kurtosis of y_t is the same for both models), and second, the lag-one autocorrelation of v_t is the same for both models. Under these parameterizations, the job of distinguishing long and short memory is quite challenging. Finally, we consider an ARFIMA$(1, d, 0)$ LMSV model similar to the one fitted to the value-weighted CRSP data in Section 9.4.3. All processes are simulated with ξ_t and η_t Gaussian.

Simulation means and standard deviations over 1000 simulated realizations of each model are given for the spectral regression test in Table 9.1. The table also presents the proportion of rejections of the short-memory null hypothesis $d = 0$, using the standard t-test with nominal significance level 0.05. Some conclusions from the results reported in Table 9.1 are as follows:

- Under the short-memory null, the size of the test is not far from nominal if the upper truncation is taken to be less than $[n^{0.5}]$; $[n^{0.45}]$ seems to be an appropriate choice. For this sample size, larger upper truncations have little value: they distort the size under short memory and bias the point estimates under all models considered.

- The spectral regression test has high power against all the long-memory models considered. Power is lower for the third and fourth LMSV models, since these models have a weaker long-memory signal (i.e. a smaller value of $\mathrm{Var}(\sigma_t^2)/\mathrm{E}^2[\sigma_t^2]$) than the first and second LMSV models.

Table 9.1. Finite sample performance of the spectral regression tests and the R/S analysis under stochastic volatility

Model	$\hat{d}_{u=0.45}$	$\hat{d}_{u=0.50}$	$\hat{d}_{u=0.55}$	$\hat{J}(n,0)$	$\hat{J}(n,q^*)$	$\hat{J}(n,200)$
ARSV: $\phi = 0.9$, $\sigma_\eta^2 = 0.45$	0.031	0.061	0.112	0.627	0.549	0.522
	(0.118)	(0.089)	(0.070)	(0.025)	(0.025)	(0.022)
Rejection proportion	0.057	0.110	0.388		0.190	0.024
LMSV: $d = 0.47$, $\sigma_\eta^2 = 0.37$	0.423	0.392	0.356	0.707	0.669	0.562
	(0.120)	(0.087)	(0.067)	(0.039)	(0.032)	(0.023)
Rejection proportion	0.923	0.992	0.999		0.997	0.428
ARSV: $\phi = 0.95$, $\sigma_\eta^2 = 0.23$	0.111	0.189	0.278	0.661	0.571	0.522
	(0.115)	(0.088)	(0.069)	(0.027)	(0.026)	(0.023)
Rejection proportion	0.145	0.562	0.978		0.507	0.030
LMSV: $d = 0.49$, $\sigma_\eta^2 = 0.19$	0.384	0.348	0.307	0.688	0.667	0.564
	(0.121)	(0.089)	(0.070)	(0.039)	(0.034)	(0.023)
Rejection proportion	0.885	0.965	0.988		0.995	0.438
ARSV: $\phi = 0.9$, $\sigma_\eta^2 = 0.13$	0.026	0.049	0.085	0.585	0.556	0.523
	(0.119)	(0.085)	(0.067)	(0.026)	(0.025)	(0.021)
Rejection proportion	0.054	0.083	0.238		0.285	0.025
LMSV: $d = 0.47$, $\sigma_\eta^2 = 0.11$	0.302	0.263	0.224	0.651	0.643	0.562
	(0.121)	(0.089)	(0.068)	(0.039)	(0.036)	(0.024)
Rejection proportion	0.704	0.832	0.907		0.967	0.399
ARSV: $\phi = 0.95$, $\sigma_\eta^2 = 0.07$	0.092	0.157	0.221	0.614	0.581	0.523
	(0.117)	(0.086)	(0.067)	(0.027)	(0.026)	(0.022)
Rejection proportion	0.133	0.425	0.906		0.651	0.026
LMSV: $d = 0.49$, $\sigma_\eta^2 = 0.05$	0.255	0.212	0.176	0.629	0.626	0.560
	(0.120)	(0.091)	(0.069)	(0.038)	(0.037)	(0.023)
Rejection proportion	0.587	0.665	0.746		0.929	0.362
LMSV: $d = 0.44$, $\sigma_\eta^2 = 0.003$, $\phi = 0.93$	0.459	0.455	0.442	0.717	0.677	0.560
	(0.121)	(0.092)	(0.068)	(0.038)	(0.030)	(0.024)
Rejection proportion	0.957	0.998	1.000		0.998	0.366

Simulation means and standard deviations (in parentheses) over 1000 replications of estimated d parameters and of Hurst exponents $\hat{J}(n,q)$. Also reported is the proportion of rejections of the short-memory null hypotheses $d = 0$ and $J = \frac{1}{2}$ using two-sided tests with nominal significance level 0.05. The d parameters were estimated by spectral regression using the periodogram of the log squares. Indices of the Fourier frequencies used in the regression have a lower truncation at $m_L = [n^{0.1}]$ and different upper truncations $m_U = [n^u]$ with $u = 0.45$, 0.50, and 0.55. Hurst exponents $\hat{J}(n,q)$ were estimated from log squares with $q = 0$, $q = q^*$, which is the value chosen by Andrew's data-dependent formula, and $q = 200$. Sample size is $n = 6144$.

- The point estimates of d under long-memory have large negative biases, which increase with m_U, reflecting contamination by short-memory effects. Even with this downward bias, estimates of d under long memory are clearly different from those under short memory.

Simulation means and standard deviations for the R/S analysis over 1000 simulated realizations of each model are also provided in Table 9.1. Some general conclusions that follow from the results in Table 9.1 are as follows:

- Long and short memory are distinguishable.
- The classical Hurst exponent is substantially larger than $\frac{1}{2}$ under the short-memory models we have considered.
- Andrews' data-dependent formula for choice of q goes a long way toward reducing the bias of the classical Hurst exponent, though $\hat{J}(n, q^*)$ is still above $\frac{1}{2}$ on average.
- The Hurst exponents estimated with values of $q = 200$ provide some robustness against even highly correlated short memory.

The overall conclusions from these tables are that the spectral regression tests and the R/S analyses can be useful indicators of long memory in stochastic volatility, but as with any asymptotic tests, they should be interpreted with caution. We recommend that additional diagnostics, in particular the shape of the estimated autocorrelation function, be used to help assess the usefulness of LMSV in any particular application.

9.3.3 Empirical evidence

The tests for long memory were performed over several market indexes' daily returns. The data and the designations used in the tables are as follows.

From the Center for Research in Security Prices (CRSP) tapes we used series starting on the first trading day of July 1962 and ending on the last trading day of July 1989. We computed returns for both the equally weighted and the value-weighted data, here denoted ECRSP and VCRSP, respectively.

Using the same raw data, we also constructed the excess returns series based on the monthly Treasury bill returns. We followed the usual simplification of assuming the riskless returns were constant within each month and subtracted these latter returns from the ones in the stock market indexes.

We have also used the long series constructed by Schwert (1990), complemented with the more recent CRSP value-weighted index. This series, here denoted SCHWERT, spans from the first trading day of February 1885 to the last trading day of 1990.

In each case, in order to whiten the series of returns, we followed the usual practice of first removing any apparent correlation in the data, namely the day-of-the-week and the month-of-the-year effects, by applying standard filters.

For each of the series we applied the long-memory tests over the squared returns and the logarithms of the squared returns.

In the first three columns of Table 9.2 we show the results of the spectral regression tests. We immediately note that in almost all cases and all series the tests are highly significant, even when the high-frequency cut-off is severe ($u = 0.40$). Interestingly, the memory of the volatilities is reduced when the excess returns are computed. In the case of the equally weighted index the tests are less significant.

Table 9.2. Results of the spectral tests and the R/S analysis*

Series	$\hat{d}_{u=0.45}$	$\hat{d}_{u=0.50}$	$\hat{d}_{u=0.55}$	$\hat{J}(n,0)$	$\hat{J}(n,q^*)$	$\hat{J}(n,200)$
VCRSPU	0.295	0.314	0.435	0.740	0.671	0.567
(Jul62–Jul89)	(0.003)	(0.000)	(0.000)		(0.000)	(0.036)
lnVCRSP	0.382	0.342	0.365	0.732	0.696	0.575
(Jul62–Jul89)	(0.002)	(0.000)	(0.000)		(0.000)	(0.015)
ECRSP	0.333	0.218	0.295	0.696	0.619	0.538
(Jul62–Jul89)	(0.024)	(0.029)	(0.000)		(0.000)	(0.232)
lnECRSP	0.263	0.186	0.248	0.703	0.667	0.565
(Jul62–Jul89)	(0.020)	(0.018)	(0.000)		(0.000)	(0.044)
ExRt-VCRSP	0.075	0.011	0.153	0.618	0.566	0.517
(Jul62–Jul89)	(0.140)	(0.006)	(0.000)		(0.039)	(0.591)
lnExRt-VCRSP	0.446	0.391	0.347	0.746	0.703	0.583
(Jul62–Jul89)	(0.001)	(0.000)	(0.000)		(0.000)	(0.005)
ExRt-ECRSP	0.032	0.063	0.101	0.588	0.519	0.490
(Jul62–Jul89)	(0.316)	(0.099)	(0.003)		(0.562)	(0.908)
lnExRt-ECRSP	0.346	0.322	0.303	0.660	0.650	0.557
(Jul62–Jul89)	(0.005)	(0.001)	(0.000)		(0.000)	(0.061)
SCHWERT	0.781	0.482	0.407	0.742	0.667	0.560
(Feb1885–Dec1990)	(0.000)	(0.000)	(0.000)		(0.000)	(0.000)
lnSCHWERT	0.582	0.540	0.501	0.736	0.684	0.593
(Feb1885–Dec1990)	(0.000)	(0.000)	(0.000)		(0.000)	(0.000)
VCRSP	0.399	0.432	0.409	0.667	0.652	0.546
(Jan78–Sep87)	(0.005)	(0.000)	(0.000)		(0.000)	(0.240)
lnVCRSP	0.437	0.386	0.384	0.653	0.653	0.550
(Jan78–Sep87)	(0.003)	(0.000)	(0.000)		(0.000)	(0.197)
ECRSP	−0.019	−0.072	−0.017	0.654	0.603	0.561
(Jan78–Sep87)	(0.460)	(0.287)	(0.427)		(0.002)	(0.106)
lnECRSP	0.165	0.168	0.142	0.660	0.650	0.557
(Jan78–Sep87)	(0.164)	(0.107)	(0.073)		(0.000)	(0.131)
ExRt-VCRSP	0.399	0.435	0.410	0.667	0.652	0.546
(Jan78–Sep87)	(0.005)	(0.000)	(0.000)		(0.000)	(0.238)
lnExRt-VCRSP	0.408	0.351	0.335	0.654	0.654	0.551
(Jan78–Sep87)	(0.011)	(0.002)	(0.000)		(0.000)	(0.189)
ExRt-ECRSP	−0.017	−0.071	−0.016	0.654	0.603	0.561
(Jan78–Sep87)	(0.464)	(0.228)	(0.121)		(0.002)	(0.106)
lnExRt-ECRSP	0.244	0.240	0.190	0.661	0.651	0.555
(Jan78–Sep87)	(0.082)	(0.047)	(0.032)		(0.000)	(0.149)

*The integration parameters d are estimated with a lower truncation at $m_L = [n^{0.1}]$ and different upper truncations $m_U = [n^u]$ with $u = 0.45, 0.50$, and 0.55. Hurst exponents $\hat{J}(n,q)$ are estimated with $q = 0, q = q^*$, which is the value chosen by Andrew's data-dependent formula, and $q = 200$. Unilateral test p-values for d and for $V = n^{-1/2}Q(n,q)$ are displayed within parentheses.
VCRSP: squared returns from the filtered value-weighted CRSP series; ECRSP: squared returns from the filtered equally weighted CRSP series; ln: logarithms of the squared return series; ExRt: series of excess returns.

In some cases, they do not reject the null of sole existence of short memory in the volatilities. We note, however, that the equally weighted indexes are economically much less sensible as representatives of the overall financial markets' activity than the value-weighted ones. Finally, it is interesting to note that the log squared series reveal the existence of a considerably more significant long-memory component.

In the last three columns of Table 9.2 we show the estimates $\hat{J}(n, q)$ and the p-values for the statistic $V = n^{-1/2} Q(n, q)$ for all the series described. All estimates point in the direction of persistent long memory. The J estimates computed with Andrews' (1991) data-dependent formula are highly significant for all series but the squared excess returns of the equally-weighted index. When the number of lags q increases, the significance of all statistics is reduced, as it is natural to expect in persistent processes. Even so, most of the computations show J estimates significantly larger than $\frac{1}{2}$.

These tests can be questioned on the grounds that long data sets may display nonstationarity in the variances and that we may be detecting nonstationarity instead of long memory. In particular, the evidence indicates this for the Schwert long indexes, since some d estimates are larger than $\frac{1}{2}$. Diebold (1986), among others, interprets the findings of persistence in volatility as the outcome of shifts in the unconditional variances.

We complement the results with tests for shorter series, beginning with January 1978 and ending in September 1987. These shorter series avoid the crashes of 1976 and 1987 and display a period known for its relatively stable volatility.

The same tests, reproduced in the second part of Table 9.2, continue to reveal long memory in the conditional variances, with the exception of the already noted equally weighted index. This fact is significant and suggests that long-memory models provide an alternative to nonstationarity for volatility modelling.

9.4 Estimating an LMSV Model

The exact likelihood of the parameter vector β given $(y_1, \ldots, y_n)$ involves an n-dimensional integral and as a consequence is extremely difficult to evaluate. Jacquier *et al.* (1994) have developed a Markov chain simulation methodology for likelihood-based inference in an autoregressive stochastic volatility model (ARSV). Their algorithm, a cyclic independence Metropolis chain, requires specification of prior distributions on all parameters and relies heavily on the special Markovian structure of pure autoregressive processes. Other simulation-based estimation methods for the first-order ARSV exist, see, for example, Danielsson (1994), but it is not clear whether they apply to more general stochastic volatility models. However, these methods are computationally intensive. Here we consider simpler estimation strategies since the LMSV model is far more complicated than the ARSV model.

Other methods for estimation from SV models have been proposed. A method of moments (MM) estimator, which avoids the problem of evaluating the likelihood function, was suggested by Taylor (1986) and Melino and Turnbull (1990). While easy to implement, MM estimators for parameters in the ARSV model have a number of disadvantages. The MM method seems relatively inefficient when some kind of persistence in the autocorrelation is present, as it is in the case of nearly nonstationary AR models (see Jacquier *et al.* 1994; Anderson and Sorenson 1994, for a discussion). Moreover, the choice of appropriate moments can be problematic.

Though the process $\{x_t\}$ is non-Gaussian, a reasonable estimation procedure is to maximize the quasi-likelihood, or likelihood computed as if $\{x_t\}$ was Gaussian with ACVF $\gamma_x(h)$ (see Nelson (1988) and Harvey *et al.* (1994) for discussion of QML in the context of short-memory stochastic volatility models). In the context of ARFIMA models, exact computation of the quasi-likelihood is possible (e.g. Sowell 1992). However, it presents convergence problems and is extremely slow especially for long time series. An alternative version of this method can be conceived for LMSV models. However, the computational problems are likely to be amplified.

We suggest a spectral-domain estimator. This is a computationally simple method for which we provide an asymptotic characterization.

9.4.1 The spectral-likelihood estimator

A simple alternative to maximizing the time-domain Gaussian likelihood is to maximize its frequency-domain representation, as discussed in a long-memory context by Fox and Taqqu (1986), Dahlhaus (1989), and Giraitis and Surgailis (1990). The simulation results of Cheung and Diebold (1994) suggest that spectral-likelihood estimators have efficiency comparable to exact QML estimators when the process has an unknown mean. The following result gives the strong consistency of estimators obtained by minimizing the negative of the logarithm of the spectral-likelihood function,

$$\mathcal{L}_n(\beta) = 2\pi n^{-1} \sum_{k=1}^{[n/2]} \left\{ \log f_\beta(\omega_k) + \frac{I_n(\omega_k)}{f_\beta(\omega_k)} \right\}, \tag{9.11}$$

where $[\cdot]$ denotes the integer part, $\omega_k = 2\pi k n^{-1}$ is the kth Fourier frequency, and

$$I_n(\omega_k) = \frac{1}{2\pi n} \left(\sum_{t=1}^n x_t \cos \omega_k t \right)^2 + \frac{1}{2\pi n} \left(\sum_{t=1}^n x_t \sin \omega_k t \right)^2$$

is the kth normalized periodogram ordinate. For a general justification of the method see, for example, Beran (1994: ch. 6).

Theorem 1 Assume that the parameter vector

$$\beta = (d, \sigma_\eta^2, \sigma_\eta^2, \phi_1, \ldots, \phi_p, \theta_1, \ldots, \theta_q)'$$

is an element of the compact parameter space Θ and assume that $f_{\beta_1}(\omega) \equiv f_{\beta_2}(\omega)$ for all ω in $[0, \pi]$ implies that $\beta_1 = \beta_2$, where $f_\beta(\cdot)$ is defined in (9.10). Let $\hat{\beta}_n$ minimize (9.11) over Θ and let β_0 denote the vector of true parameter values. Then $\hat{\beta}_n \to \beta_0$ almost surely.

The proof is provided in Appendix 9.1.

Remarks

1. The proof follows Dahlhaus (1989) in avoiding the special parameterization of Fox and Taqqu (1986). Dahlhaus' (1989) result is not directly applicable to our case because his explicit assumptions include Gaussianity and his objective function is an integral version of (9.11). For the non-Gaussian case, we verify Dahlhaus' remark (p. 1753) that his results extend to the function (9.11).

2. The component $|1 - e^{-i\lambda}|^{-2d} = (\sqrt{2 - 2\cos\lambda})^{-2d}$ of $f_\beta(\lambda)$ introduces in the likelihood a term proportional to

$$d \sum_{k=1}^{[n/2]} \log(2 - 2\cos\omega_k) 2\pi n^{-1}; \qquad (9.12)$$

   the corresponding integral is improper, but converges to zero (see Appendix 9.1). In the course of the proof, we show that the effect on the estimators of dropping the term (9.12) is negligible.

3. The identifiability condition in Theorem 1 is met if σ_ε^2 is known from an assumed distribution for ξ_t; for example, $\xi_t \sim N(0, 1)$ implies $\sigma_\varepsilon^2 = \pi^2/2$. If σ_ε^2 is not known, the model is identifiable only if the ARFIMA component is not white noise; that is, if $\phi_p \neq 0$ for some p, $\theta_q \neq 0$ for some q, or $d \neq 0$.

9.4.2 Finite sample properties of the spectral-likelihood estimator

This subsection presents a simulation study of the finite sample properties of the maximum-likelihood spectral estimator previously proposed. In this experiment we consider two different sample sizes ($n = 1024$ and $n = 4096$) and three classes of LMSV models given, respectively, by ARFIMA$(0, d, 0)$, ARFIMA$(1, d, 0)$, and ARFIMA$(1, d, 1)$. Within each class of models, several combinations for the parameters of these models are considered—see Table 9.3. The variance of the i.i.d. innovations in the ARFIMA component are set to one as well as the variance of the noise component.

All the results reported in this section are obtained from 1000 realizations of each model. Table 9.3 presents simulation means and standard deviations for the parameter estimates. Fig. 9.1 presents box plots for some of the models considered in the simulation. The cases considered in this figure are representative of the overall results.

Some general conclusions from the table and the box plots are:

- Maximum-likelihood estimation in the spectral domain perform well for relatively large samples, such as those found in the high frequency financial markets' data.

- The biases are relatively small and decrease uniformly from $n = 1024$ to $n = 4096$. The increase in the sample size also reduced significantly the dispersion of the results.

Table 9.3. Finite-sample results for the spectral-likelihood estimator*

Parameters ϕ, d, θ	ϕ		d		θ	
	$n = 1024$	$n = 4096$	$n = 1024$	$n = 4096$	$n = 1024$	$n = 4096$
$(0, -0.4, 0)$			−0.550	−0.419		
			(0.550)	(0.189)		
$(0, -0.2, 0)$			−0.337	−0.223		
			(0.521)	(0.152)		
$(0, 0.0, 0)$			−0.0678	−0.0169		
			(0.296)	(0.107)		
$(0, 0.2, 0)$			0.189	0.196		
			(0.187)	(0.042)		
$(0, 0.4, 0)$			0.407	0.401		
			(0.086)	(0.036)		
$(0.8, -0.4, 0)$	0.756	0.781	−0.369	−0.389		
	(0.169)	(0.114)	(0.234)	(0.172)		
$(0.8, -0.2, 0)$	0.771	0.795	−0.211	−0.215		
	(0.157)	(0.081)	(0.207)	(0.129)		
$(0.8, 0.0, 0)$	0.773	0.797	−0.0213	−0.0142		
	(0.147)	(0.063)	(0.188)	(0.101)		
$(0.8, 0.2, 0)$	0.774	0.798	0.180	0.187		
	(0.158)	(0.057)	(0.191)	(0.096)		
$(0.8, 0.4, 0)$	0.774	0.797	0.381	0.394		
	(0.147)	(0.052)	(0.196)	(0.085)		
$(0.4, -0.4, 0)$	0.366	0.391	−0.420	−0.400		
	(0.300)	(0.228)	(0.389)	(0.247)		
$(0.4, -0.2, 0)$	0.398	0.435	−0.242	−0.255		
	(0.293)	(0.212)	(0.271)	(0.213)		
$(0.4, 0.0, 0)$	0.434	0.427	−0.0828	−0.0423		
	(0.279)	(0.169)	(0.233)	(0.145)		
$(0.4, 0.2, 0)$	0.425	0.403	0.142	0.191		
	(0.250)	(0.121)	(0.172)	(0.059)		
$(0.4, 0.4, 0)$	0.373	0.390	0.382	0.399		
	(0.240)	(0.112)	(0.129)	(0.046)		
$(0.8, 0.2, 0.3)$	0.788	0.800	0.161	0.186	0.319	0.134
	(0.128)	(0.050)	(0.175)	(0.083)	(0.521)	(0.378)
$(0.8, 0.2, -0.3)$	0.908	0.839	0.167	0.183	0.308	0.144
	(0.251)	(0.134)	(0.172)	(0.093)	(0.530)	(0.368)
$(0.4, 0.2, 0.3)$	0.345	0.371	0.128	0.188	0.303	0.114
	(0.410)	(0.243)	(0.204)	(0.067)	(0.525)	(0.373)
$(0.4, 0.2, -0.3)$	0.399	0.394	0.123	0.187	0.277	0.0864
	(0.374)	(0.229)	(0.210)	(0.060)	(0.555)	(0.400)
$(0.8, 0.4, 0.3)$	0.790	0.802	0.367	0.390	0.320	0.147
	(0.113)	(0.050)	(0.177)	(0.088)	(0.548)	(0.403)
$(0.8, 0.4, -0.3)$	0.978	0.893	0.369	0.388	0.349	0.188
	(0.316)	(0.190)	(0.172)	(0.089)	(0.557)	(0.422)
$(0.4, 0.4, 0.3)$	0.302	0.365	0.362	0.393	0.371	0.177
	(0.419)	(0.220)	(0.151)	(0.050)	(0.551)	(0.426)
$(0.4, 0.4, -0.3)$	0.395	0.374	0.348	0.394	0.297	0.147
	(0.352)	(0.221)	(0.172)	(0.051)	(0.587)	(0.427)

*For each model and set of parameters, 1000 replications were performed with length $n = 1024$ and $n = 4096$. The LMSV model parameters are given within parentheses. The values in the table represent the simulation means and standard deviations (in parentheses) for the estimated parameters.

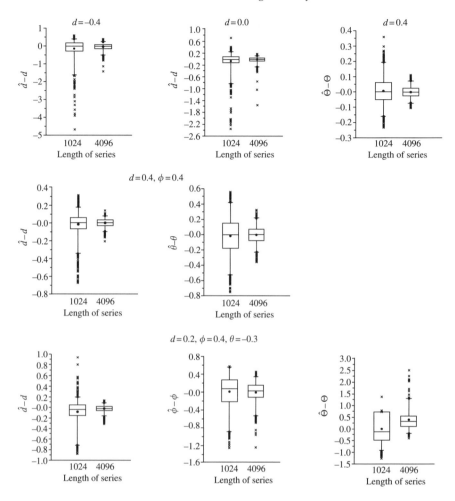

Fig. 9.1. Box plots represent the deviations of the estimated parameters from the true values.

- The box plots in Fig. 9.1 also show that some less positive aspects of the results obtained for $n = 1024$ tend to be smoothed out for $n = 4096$, namely, the asymmetry of the distribution of the estimates. An extreme case was the LMSV model with an ARFIMA(0, -0.4, 0) component.

- The maximum-likelihood spectral estimator provides less biased and more precise parameter estimates in processes in which the fractional parameter d was positive. This includes both the estimate of d and the estimates of the other parameters in the model.

- The performance of the maximum-likelihood spectral estimator in small samples might be less than ideal as is illustrated by the box plots of the smaller sample size. Moreover, some very large outliers occurred when $n = 1024$.

- The procedure encountered some difficulties in estimating the moving average term, even when the number of observations was 4096 (although the magnitude of the problem decreased for the larger sample size).

Given the overall good performance of the estimator when $n = 4096$, these sampling experiments indicate that maximum-likelihood spectral estimation of LMSV models may be a very effective method for the type of financial applications that have led to this line of research.

Moreover, this maximum-likelihood estimator is easy to implement. Convergence for a LMSV model with an ARFIMA(0, d, 0) component and $n = 4096$ was typically attained in less than 20 iterations and less than 4 s of CPU time on a Pentium 100 MHz.[6]

9.4.3 Modelling volatility of stock returns

Nelson (1991) introduces the EGARCH mode using as an example the daily returns for the value-weighted market index from the CRSP tapes for July 1962–December 1987. He selects an ARMA(2, 1) model for log σ_t^2 and finds the largest estimated AR root to be 0.99962, suggesting substantial persistence.

For comparison, we fitted a model with long-memory stochastic volatility to the log squares of the VCRSP series described in Section 9.3.3 above. This log squared series, denoted $\{x_t\}$, consists of $n = 6801$ mean-corrected observations modelled as

$$x_t = v_t + \varepsilon_t,$$

where $\{\varepsilon_t\}$ is i.i.d. $(0, \sigma_\varepsilon^2)$ independent of $\{v_t\}$ and $\{v_t\}$ is the ARFIMA(1, d, 0),

$$(1 - B)^d (1 - \phi B) v_t = \eta_t,$$

with $\{\eta_t\}$ i.i.d. $(0, \sigma_\eta^2)$.

The spectral likelihood for the x_t's was formed as in eqn (9.11) replacing (9.12) with zero, which we have found useful though this needs further investigation. The resulting likelihood was maximized with respect to the parameters σ_η^2, d, ϕ and σ_ε^2 yielding the estimates $\hat{\sigma}_\eta^2 = 0.00318$, $\hat{d} = 0.444$, $\hat{\phi} = 0.932$ and $\hat{\sigma}_\varepsilon^2 = 5.238$.

Fig. 9.2 shows the empirical and fitted autocorrelations for the series $\{x_t\}$. The empirical autocorrelations show a slow decay, remaining non-negligible for hundreds of lags. The ACF of the fitted LMSV model was derived from the ARFIMA(1, d, 0) formulae in Hosking (1981) and adjusted for the bias due to the existence of long memory as in theorem 5 of Hosking (1995). In Fig. 9.2, the bias-adjusted ACF for LMSV accurately reflects the slow decay of the empirical ACF.

We also fitted short-memory GARCH and EGARCH models as well as an IGARCH model to the same VCRSP series. In order to compare the properties of fitted GARCH, IGARCH, EGARCH, and LMSV models with the observations, we computed the autocorrelations of the fitted models and plotted them against the sample autocorrelations of the series. The order of the models was selected by SIC.

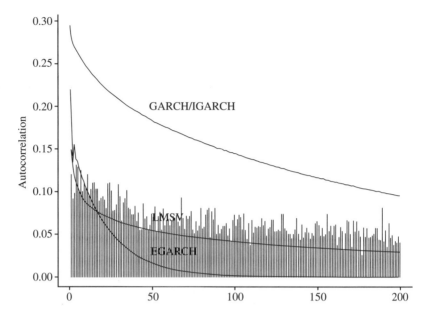

Fig. 9.2. Empirical and fitted autocorrelation functions for the log squares
VCRSP series.

As it is often observed in practice, the fitted GARCH and IGARCH are very similar: SIC selected a GARCH(1, 2) and an IGARCH(1, 2). The fitted GARCH had parameter estimates $\hat{a}_1 = 0.923, \hat{b}_1 = 0.143$, and $\hat{b}_2 = -0.067$. Their sum is 0.999. The fitted IGARCH has parameter estimates $\hat{a} = 0.923, \hat{b}_1 = 0.145$, and $\hat{b}_2 = -0.068$. Using the GARCH parameter estimates, we simulated 1000 GARCH realizations, each of length $n = 6801$, and computed the sample ACF of the log squares for each realization. The same simulations were done for IGARCH. The average of the GARCH sample ACF's, plotted in Fig. 9.2 and labeled 'GARCH/IGARCH', is almost indistinguishable from the average of the IGARCH sample ACF's (not plotted). The GARCH models, nearly integrated or integrated, seem 'too persistent' to model these data.

The SIC criterion selected an EGARCH(2, 0). The fitted EGARCH had parameter estimates $\hat{\delta}_1 = 0.0185, \hat{\delta}_2 = 0.200, \hat{\phi}_1 = 0.577$ and $\hat{\phi}_2 = 0.359$. The ACF for the log squares corresponding to the fitted EGARCH model was obtained theoretically through the formulae derived in Appendix 9.2. The short-memory EGARCH model clearly fails to reflect the slow decay of the empirical ACF.

9.5 Conclusions

Empirical evidence suggests that the recent interest in long-memory conditional variance models for stock market indices is well-founded. We find evidence of long memory in variance proxies using both a nonparametric and semiparametric test for many series. A simulation exercise shows that these tests are able to distinguish long from short memory in the volatilities.

The long-memory stochastic volatility (LMSV) model is an analytically tractable model of this persistence in the conditional variances. The LMSV is easily fitted and analysed using standard tools for weakly stationary processes. In particular, the LMSV model is built from the widely used ARFIMA class of long-memory time series models, so that many of its properties are well-understood. The spectral-likelihood estimator proposed for this model is strongly consistent and finite-sample simulation results show it has reasonable properties for series of the length usually found in financial data.

An example with a long series of stock prices shows that short-memory models are unable to reproduce more than the short-term structure of the autocorrelations. In contrast, a parsimonious LMSV model fit to the data is able to reproduce closely the empirical autocorrelation structure of the conditional volatilities.

These results are encouraging and suggest some avenues for future research. We believe it would be interesting to investigate further the empirical relevance of the LMSV model, namely, its relevance for estimating and forecasting the volatilities and pricing derivatives. We also believe that it would be useful to compare properties of the LMSV with other models of persistence in the volatilities.

Appendix 9.1 Proof of Strong Consistency for Spectral-likelihood Estimators

Let $\hat{\beta}_n$ minimize (9.11) and let

$$\mathcal{L}(\beta) = 2 \int_0^\pi \left\{ \log f_\beta(\omega) + \frac{f_{\beta_0}(\omega)}{f_\beta(\omega)} \right\} d\omega,$$

where β_0 denotes the vector of true parameter values. Then

$$|\mathcal{L}_n(\beta) - \mathcal{L}(\beta)| \leqslant \left| 2\pi n^{-1} \sum_{k=1}^{[n/2]} g_\beta(\omega_k) - 2 \int_0^\pi g_\beta(\omega) \, d\omega \right|$$

$$+ \left| 2\pi n^{-1} \sum_{k=1}^{[n/2]} d \log(2 - 2\cos\omega_k) - 2 \int_0^\pi d \log(2 - 2\cos\omega) \, d\omega \right|$$

$$+ \left| 2\pi n^{-1} \sum_{k=1}^{[n/2]} \frac{I_n(\omega_k)}{f_\beta(\omega_k)} - 2 \int_0^\pi \frac{f_{\beta_0}(\omega)}{f_\beta(\omega)} \, d\omega \right|$$

$$= M_{1n}(\beta) + M_{2n}(\beta) + M_{3n}(\beta),$$

where

$$g_\beta(\lambda) = \log \left\{ \frac{\sigma_\eta^2 |\theta(e^{-i\lambda})|^2 + \sigma_\varepsilon^2 |\phi(e^{-i\lambda})|^2 |1 - e^{-i\lambda}|^{2d}}{2\pi |\phi(e^{-i\lambda})|^2} \right\}.$$

Now $M_{1n}(\beta)$ converges to zero uniformly in β by Riemann integrability of $g_{\hat{\beta}}(\omega)$, continuity in β of the integral and compactness of Θ. Given $\delta > 0$, $M_{1n}(\beta)$ can be bounded above and below by the upper Riemann sum plus δ and the lower Riemann sum minus δ for a partition $\mathcal{P}_n$ of $[0, \pi]$, where $\mathcal{P}_n$ contains the nth-order Fourier frequencies and $\mathcal{P}_n \subset \mathcal{P}_{n+1} \subset \cdots$. For each β, these bounds converge to zero $\pm \delta$ monotonically, and so uniform convergence in β follows by Dini's theorem.

Next, $M_{2n}(\beta)$ can be bounded uniformly in β by

$$0.5 \left| 2\pi n^{-1} \sum_{k=1}^{[n/2]} \log(2 - 2\cos\omega_k) - 2 \int_0^\pi \log(2 - 2\cos\omega) \, d\omega \right|,$$

which converges to zero since

$$\int_0^\pi \log(2 - 2\cos\omega) \, d\omega = 0.$$

Finally, $M_{3n}(\beta)$ can be shown to converge almost surely (a.s.) to zero uniformly in β by modifying lemma 1 of Hannan (1973) (see also lemma 1 of Fox and Taqqu 1986; Dahlhaus 1989). First, $1/f_\beta(\omega)$ satisfies the continuity condition of Hannan (1973) and so the Césaro sum of its Fourier series converges uniformly in (ω, β) for $\beta \in \Theta$. Second, the process $\{x_t\}$ is ergodic since $\{v_t\}$ is a linear process with i.i.d. innovations and square-summable coefficients (e.g. Hannan: 1970: p. 204) and $\{\varepsilon_t\}$ is i.i.d., independent of $\{v_t\}$. From these two facts, lemma 1 of Hannan (1973) follows.

Hence,

$$\sup_{\beta \in \Theta} |\mathcal{L}_n(\beta) - \mathcal{L}(\beta)| \to 0 \text{ a.s.}$$

Since $-\log x \geqslant 1 - x$, with equality holding if and only if $x = 1$,

$$\mathcal{L}(\beta) = 2 \int_0^\pi \left\{ -\log \frac{f_{\beta_0}(\omega)}{f_\beta(\omega)} + \log f_{\beta_0}(\omega) + \frac{f_{\beta_0}(\omega)}{f_\beta(\omega)} \right\} d\omega$$

$$\geqslant 2 \int_0^\pi \left\{ 1 - \frac{f_{\beta_0}(\omega)}{f_\beta(\omega)} + \log f_{\beta_0}(\omega) + \frac{f_{\beta_0}(\omega)}{f_\beta(\omega)} \right\} d\omega$$

$$= 2 \int_0^\pi \left\{ \log f_{\beta_0}(\omega) + \frac{f_{\beta_0}(\omega)}{f_{\beta_0}(\omega)} \right\} d\omega$$

$$= \mathcal{L}(\beta_0),$$

and so (using the identifiability condition) β_0 uniquely minimizes $\mathcal{L}(\beta)$. Thus

$$\mathcal{L}_n(\hat{\beta}_n) \leqslant \mathcal{L}_n(\beta_0) \quad \text{and} \quad \mathcal{L}(\beta_0) \leqslant \mathcal{L}(\hat{\beta}_n),$$

which implies that $\mathcal{L}(\hat{\beta}_n) \to \mathcal{L}(\beta_0)$ a.s. and therefore also $\hat{\beta}_n \to \beta_0$ a.s. by compactness of Θ. ∎

Appendix 9.2 Autocovariance Function of Log Squares under EGARCH

Under an EGARCH model for $\{y_t\}$, the ACVF for the series $\{x_t\} = \{\log y_t^2 - \mu_t\}$, where $\{\mu_t\}$ are the deterministic volatility changes in eqn (9.3), can be computed as follows:

$$\text{Cov}(x_t, x_{t+h})$$

$$= \text{Cov}\left(\sum_{j=0}^{\infty} \psi_j g(\xi_{t-j-1}) + \log \xi_t^2, \sum_{j=0}^{\infty} \psi_j g(\xi_{t+h-j-1}) + \log \xi_{t+h}^2\right)$$

$$= \text{Var}\{g(\xi_t)\}\gamma(h) + \psi_{h-1}\text{E}[g(\xi_t) \log \xi_t^2] + \text{Var}(\log \xi_t^2)I_{\{h=0\}},$$

where $\gamma(h)$ is the autocovariance function

$$\gamma(h) = \sum_{k=0}^{\infty} \psi_j \psi_{j+h}$$

and $\psi_{-1} := 0$. If, as it was originally suggested by Nelson (1991), the function $g(\cdot)$ is chosen to be

$$g(\xi_t) = \delta_1 \xi_t + \delta_2(|\xi_t| - \text{E}|\xi_t|),$$

then we have

$$\text{Var}\{g(\xi_t)\} = \delta_1^2 + \delta_2^2(1 - \text{E}^2|\xi_t|).$$

For Gaussian ξ_t, $\text{E}|\xi_t| = \sqrt{2/\pi}$, $\text{Var}(\log \xi_t^2) = \pi^2/2$ and

$$\text{E}[g(\xi_t) \log \xi_t^2] = \frac{2\delta_2}{\sqrt{2\pi}}(\log 2 - \kappa + 1.27),$$

where $\kappa \simeq 0.577216$ is Euler's constant. Thus, if $\delta_2 = 0$, the Gaussian EGARCH ACVF has the same form as the SV ACVF in (9.7).

Notes

1. See Ding *et al.* (1993), de Lima and Crato (1993) and Bollerslev and Mikkelsen (1996) for evidence that persistence in stock markets' volatility can be characterized as a long-memory process.
2. See Bollerslev *et al.* (1992) for a review of ARCH and GARCH-type models, and Taylor (1994) for a recent review of stochastic volatility models.
3. Harvey (1993) independently proposed a stochastic volatility model driven by fractional noise and applied it to exchange rate series, obtaining smoothed estimates of the underlying volatilities.
4. This work was directly motivated by the empirical results of de Lima and Crato (1993).
5. However, this model displays the important property of having a bounded cumulative impulse–response function for any $d < 1$ as Baillie *et al.* (1996) have shown.
6. In these simulations we used the set of routines for maximum likelihood estimation provided by the GAUSS programming language. The algorithm used is the derivative-based procedure of Broyden, Fletcher, Goldfarb, and Shanno, as described in the GAUSS manual. Analytical derivatives were provided. The code is available from the authors upon request.

References

Andrews, D. W. Q. (1991), 'Heteroskedasticity and autocorrelation consistent covariance matrix estimation', *Econometrica*, 59, 817–58.

Anderson, T. G. and Sørenson, B. E. (1994), 'GMM and QML asymptotic standard deviations in stochastic volatility models: a response to Ruiz', Working Paper 189, Department of Finance, Kellog School, Northwestern University.

Baillie, R. T. (1996), 'Long-memory processes and fractional integration in econometrics', *Journal of Econometrics*, 73, 5–59.

——, Bollerslev, T., and Mikkelsen, H. O. (1996), 'Fractionally integrated generalized autoregressive conditional heteroskedasticity', *Journal of Econometrics*, 74, 3–30.

Beran, J. (1994), *Statistics for Long-memory Processes*. Chapman & Hall, New York.

Bollerslev, T. (1986), 'Generalized autoregressive conditional heteroskedasticity', *Journal of Econometrics*, 31, 307–27.

——, Chou, R. Y., and Kroner, K. F. (1992), 'ARCH modeling in finance', *Journal of Econometrics*, 52, 5–59.

—— and Mikkelsen, H. O. A. (1996), 'Modeling and pricing long-memory in stock market volatility', *Journal of Econometrics*, 73, 151–84.

Breidt, F. J., Crato, N., and de Lima, P. (1994), 'Modeling long memory stochastic volatility'. Working Papers in Economics No. 323, Johns Hopkins University.

Brockwell, P. J. and Davis, R. A. (1991), *Time Series: Theory and Methods*, 2nd edn. Springer, New York.

Cheung, Y. W. (1993), 'Tests for fractional integration: a Monte Carlo investigation', *Journal of Time Series Analysis*, 14, 331–45.

—— and Diebold, F. X. (1994), 'On maximum likelihood estimation of the differencing parameter of fractionally-integrated noise with unknown mean', *Journal of Econometrics*, 62, 301–36.

Dahlhaus, R. (1989), 'Efficient parameter estimation for self-similar processes', *Annals of Statistics*, 17, 1749–66.

Danielsson, J. (1994), 'Stochastic volatility in asset prices: estimation with simulated maximum likelihood. *Journal of Econometrics*, 64, 375–400.

de Lima, P. J. F. and Crato, N. (1993), 'Long-range dependence in the conditional variance of stock returns'. August 1993 Joint Statistical Meetings, San Francisco. Proceedings of the Business and Economic Statistics Section.

Diebold, F. X. (1986), 'On modeling the persistence of conditional variances', *Econometric Reviews*, 5, 51–6.

Ding, Z., Granger, C., and Engle, R. F. (1993), 'A long memory property of stock market returns and a new model'. *Journal Empirical Finance*, 1, 83–106.

Engle, R. F. (1982), 'Autoregressive conditional heteroskedasticity with estimates of the variance of United Kingdom inflation', *Econometrica*, 50, 987–1007.

Fox, R. and Taqqu, M. S. (1986), 'Large-sample properties of parameter estimates for strongly dependent stationary Gaussian time series', *Annals of Statistics*, 14, 517–32.

Gcwckc, J. and Porter-Hudak, S. (1983), 'The estimation and application of long memory time series models', *Journal of Time Series Analysis*, 4, 221–38.

Giraitis, L. and Surgailis, D. (1990), 'A central limit theorem for quadratic forms in strongly dependent linear variables and its application to asymptotical normality of Whittle's estimate', *Probability Theory and Related Fields*, 86, 87–104.

Granger, C. W. and Joyeux, R. (1980), 'An introduction to long-memory time series models and fractional differencing', *Journal of Time Series Analysis*, 1, 15–29.

Hannan, E. J. (1970), *Multiple Time Series*. Wiley, New York.

—— (1973), 'The asymptotic theory of linear time-series models', *Journal of Applied Probability*, 10, 130–45.

Harvey, A. C. (1993), 'Long memory in stochastic volatility'. Mimeo, London School of Economics.

——, Ruiz, E., and Shephard, N. (1994), 'Multivariate stochastic variance models', *Review of Economic Studies*, 61, 247–64.

Hosking, J. R. M. (1981), 'Fractional differencing', *Biometrika*, 68, 165–76.

—— (1995), 'Asymptotic distributions of the sample mean, autocovariances and autocorrelations of long-memory time series', *Journal of Econometrics*, forthcoming.

Jacquier, E., Polson, N. G., and Rossi, P. E. (1994), 'Bayesian analysis of stochastic volatility models (with discussion)', *Journal of Business and Economic Statistics*, 12, 371–417.

Lo, A. W. (1991), 'Long-term memory in stock market prices', *Econometrica*, 59, 1279–313.

Mandelbrot, B. B. and Taqqu, M. (1979), 'Robust R/S analysis of long-run serial correlation'. Proceedings of the 42nd Session of the International Statistical Institute, International Statistical Institute.

Melino, A. and Turnbull, S. M. (1990), 'Pricing foreign currency options with stochastic volatility', *Journal of Econometrics*, 45, 239–65.

Nelson, D. B. (1988), Time series behavior of stock market volatility and returns. Ph.D. dissertation, Economics Department, Massachusetts Institute of Technology.

—— (1991), 'Conditional heteroskedasticity in asset returns: a new approach', *Econometrica*, 59, 347–70.

Robinson, P. M. (1991), 'Testing for strong serial correlation and dynamic conditional heteroskedasticity in multiple regression', *Journal of Econometrics*, 47, 67–84.

—— (1995), 'Log-periodogram regression of time series with long range dependence', *Annals of Statistics*, 23, 1048–72.

Schwert, G. W. (1990), 'Indexes of US stock prices from 1802 to 1987', *Journal of Business*, 63, 399–426.

Sowell, F. (1992), 'Maximum likelihood estimation of stationary univariate fractionally integrated time series models', *Journal of Econometrics*, 53, 165–88.

Taylor, S. (1986), *Modelling Financial Time Series*, Wiley, New York.

—— (1994), 'Modeling stochastic volatility: a review and comparative study', *Mathematical Finance*, 4, 183–204.

Wishart, J. (1947), 'The cumulants of the z and of the logarithmic χ^2 and t distributions', *Biometrika*, 34, 170–78.

10

Efficient Tests of Nonstationary Hypotheses

P. M. ROBINSON

10.1 Introduction

The score, likelihood-ratio, and Wald principles are widely applied in deriving tests of nested parametric hypotheses. In many problems the resulting test statistic has a simple, easy-to-use null limit distribution, namely the χ_p^2 distribution, where p is the number of restrictions tested, and the test is efficient within a large class of tests, in the sense of having a limiting noncentral χ_p^2 distribution against Pitman drift, with maximal noncentrality parameter. Such problems will be called 'standard'. Versions of the score, likelihood-ratio, and Wald principles have been much used in testing for a unit root in a time series against autoregressive (AR) alternatives that are stationary or explosive. This is not a 'standard' problem: the test statistics often have nonstandard null and local asymptotic distributions. Sufficiency considerations suggest that some unit root tests developed against AR alternatives may have good power against such alternatives. In specialized situations they can have optimal finite-sample properties, assuming Gaussianity and certain nuisance parameterizations (see, e.g. Bhargava 1986; Dufour and King 1991; and Sargan and Bhargava 1983). Typically, critical values must be calculated numerically on a case-by-case basis, and there can be interest in a more convenient approach, whereas in some problems greater generality is required and only a large-sample theory may be feasible. Though there is scope for improving existing tests for the unit root hypothesis (see Elliott *et al.* 1996), there can be difficulties in identifying a test in a specific model that is asymptotically efficient within a given class, let alone in building a comprehensive theory that covers many models and will automatically deliver an efficient test.

The AR model is only one of many models that nest a unit root. Consider a scalar real-valued sequence $X_t, t = 1, 2, \ldots$, hypothesized to satisfy

$$\varphi(B)X_t = U_t, \quad t = 1, 2, \ldots, \tag{10.1}$$

$$X_t = 0, \quad t \leq 0, \tag{10.2}$$

This chapter was previously published as Robinson, P. M. (1994), 'Efficient tests of nonstationary hypotheses', *Journal of the American Statistical Association*, 89, 1420–37. Copyright © 1994 American Statistical Association, reproduced by kind permission.

The author thanks Joris Pinkse, and especially Ignacio Lobato, for carrying out the computations. This work is based on research funded by the Economic and Social Research Council, reference number R000233609.

where φ is a prescribed function of the backshift operator B and $\{U_t\}$ is a covariance stationary sequence with zero mean and weak parametric autocorrelation. Consider a given function $\varphi(z; \theta)$ of the complex variate z and the p-dimensional vector θ of real-valued parameters, where $\varphi(z; \theta) = \varphi(z)$ for all z such that $|z| = 1$ if and only if

$$H_0 : \theta = 0 \tag{10.3}$$

holds, where there is no loss of generality in using the vector of zeros instead of an arbitrary given vector. We have cast (10.1) in terms of a nested composite parametric null hypothesis, within the class of alternatives

$$\varphi(B; \theta)X_t = U_t, \quad t = 1, 2, \ldots. \tag{10.4}$$

In the simplest unit root problem, $\varphi(z) = 1 - z$. An AR alternative is described by $\varphi(z; \theta) = 1 - (\theta + 1)z$, with $p = 1$, but there exist any number of functions $\varphi(z; \theta)$ with our prescribed properties. This chapter stresses that the 'nonstandard' asymptotic behaviour of commonly used unit root tests is a consequence of the AR alternative and provides a different, and unusually unified, treatment of testing unit roots (and many other hypotheses) as a 'standard' problem.

The distinguishing features of our work are as follows:

1. We adopt the normalization $\varphi(0; \theta) = 1$ for all θ and assume that $\varphi(z; \theta)$ is differentiable in θ on a neighbourhood of $\theta = 0$ for all $|z| = 1$ and that

$$0 < \det(\Psi) < \infty, \tag{10.5}$$

where

$$\Psi = (2\pi)^{-1} \int_{-\pi}^{\pi} \psi(\lambda)\psi(\lambda)' \, d\lambda,$$

$$\psi(\lambda) = \mathrm{Re}\left\{ \frac{\partial}{\partial\theta} \log \varphi(e^{i\lambda}; 0) \right\} \tag{10.6}$$

for real λ. Then we define a test statistic that has an asymptotic null χ_p^2 distribution and is efficient when X_t is Gaussian and, more generally, more efficient than other statistics that, like ours, are based on sample second moments of X_t. Thus our problem is 'standard'. The right-side inequality in (10.5) is not satisfied by the AR alternative $\varphi(z; \theta) = 1 - (1 + \theta)z$ but is satisfied by the 'fractional' alternative $\varphi(z; \theta) = (1 - z)^{1+\theta}$, for example. [The left-side inequality in (10.5) is essentially an assumption of identifiability under H_0.] Our tests will typically be consistent against AR and other alternatives not satisfying (10.5).

2. Our $\varphi(z)$ need not have a zero at $z = 1$, but can have a finite number of zeros in any specified locations on the unit circle, to describe seasonal and cyclic nonstationarity. The orders of these null unit roots need not be unity, but can include other positive integers, such as 2, and negative integers, as in case of overdifferencing. They generally can be any real numbers, to include fractional

nonstationary and stationary null hypotheses. Our framework is sufficiently general to also cover weakly autocorrelated null hypotheses, when unit roots can be said to have zero orders; there is a vast literature on testing such hypotheses (e.g. see Hosking 1980), but our work affords some slightly increased generality even here. But we are motivated largely by the nonstationary hypotheses described previously. For only a few specialized cases of these have tests been previously derived, and AR alternatives have been stressed. This limited experience suggests that no comparably general theory is available for AR alternatives; the nonstandard, null, and local limiting distributional behaviour seeming to vary with the form of $\varphi(z)$ and to require case-by-case evaluation. Compare, for example, Dickey and Fuller's (1979) test for a non-seasonal unit root with Dickey, Hasza, and Fuller's (1984) test for a seasonal one.

3. X_t need not be observable but can be the errors in a multiple regression model,

$$Y_t = \beta'Z_t + X_t, \quad t = 1, 2, \ldots, \tag{10.7}$$

where Y_t and the $k \times 1$ vector Z_t, are observable and β is an unknown $k \times 1$ vector. The elements of Z_t are assumed to be nonstochastic, such as polynomials in t, to include the null hypothesis of a unit root with drift, for example. The limiting null and local distributions of our test statistic are unaffected by the presence of such regressors. In contrast, asymptotic distributions of test statistics for a unit root null for X_t in (10.7) against AR alternatives seem liable, on the basis of the few special cases that have been analysed, to be dependent on characteristics of the Z_t sequence (e.g. see Schmidt and Phillips 1992). For simplicity, we treat only linear regression, but undoubtedly a nonlinear regression will also leave our limit distributions unchanged, under standard regularity conditions.

4. The initial discussion assumes that the U_t in (10.1) are white noise, so the only nuisance parameters are β and the variance of U_t, and we are concerned with such null hypotheses as a random walk. The test statistic is relatively simple to compute in these circumstances. Unlike tests based on AR alternatives, ours cannot be robustified to allow for weak nonparametric autocorrelation in U_t, but we do include an extension to the case of weak parametric autocorrelation in U_t, of quite general form to cover stationary and invertible autoregressive moving average (ARMA) behaviour and the exponential-spectrum model of Bloomfield (1973) (where our test statistic takes on a rather convenient form), as well as autocorrelations that decay fairly slowly.

5. Though undoubtedly the same asymptotic behaviour can be expected of Wald and likelihood-ratio tests, we use only score tests in this article, with the usual computational motivation that they entail estimation only under the null hypothesis (10.3). Efficient estimates of fractional models have been studied (e.g. see Fox and Taqqu 1986), but they require numerical optimization, have not been very widely used, and are not featured in the most widely used time series software packages. One definition of the score statistic for H_0 (10.3) is as follows. Consider a column vector of parameters $\eta = (\theta', \nu')'$, and let $L(\eta)$ be an objective function, such as the negative of the log-likelihood, which has a minimum under H_0 (10.3) at

$\tilde{\eta} = (0', \tilde{v}')'$. A score statistic (see Rao 1973: p. 418) is

$$\frac{\partial L(\eta)}{\partial \eta'} \left[E_0 \left(\frac{\partial L(\eta)}{\partial \eta} \frac{\partial L(\eta)}{\partial \eta'} \right) \right]^{-1} \frac{\partial L(\eta)}{\partial \eta} \Bigg|_{\theta=0, \eta=\tilde{v}}, \tag{10.8}$$

where the expectation is taken under H_0 (10.3) prior to substitution of $\tilde{v}$. In 'standard' problems such as ours, the null and local limit distributions and efficiency properties of our test are unaffected if the inverted matrix is replaced by certain alternatives, such as by a sample average, or by the Hessian. In contrast, for unit root hypotheses nested in AR alternatives, such modifications generally can lead to different null and local limit distributions, as illustrated in Section 10.8.

6. The ease with which approximate critical values for our tests can be obtained is offset by substantially greater computational complexity of the test statistics relative to many unit root tests against AR alternatives, due to the alternatives against which our tests are directed. But the computations are hardly onerous by the standards of many statistical calculations carried out nowadays. Time domain versions of our tests are available and might be preferred by workers unused to the frequency domain, but they are not necessarily computationally more desirable. We stress a frequency domain approach because of its comparative elegance; its use of $\psi(\lambda)$, which is in general more immediately written down than its time domain Fourier inverse; the ease with which it accommodates autocorrelation corrections for U_t, especially the Bloomfield spectral model; and the natural way in which it exploits the fast Fourier transform in case of long time series.

The following section discusses the null and alternative hypotheses. Sections 10.3 and 10.4 introduce the test statistics and present null and local limit distribution theory, in case of white noise U_t, and Sections 10.5 and 10.6 do the same for autocorrelated U_t. Mathematical proofs are relegated to four Appendixes. Section 10.7 reports applications to empirical data. Section 10.8 uses Monte Carlo simulation to study finite-sample behaviour of versions of our tests and compare them with some existing unit root tests. Finally, Section 10.9 presents some brief concluding remarks.

10.2 Null and Alternative Hypotheses

The null hypotheses of principal interest entail φ of form

$$\varphi(z) = (1 - z)^{\gamma_1} (1 + z)^{\gamma_2} \prod_{j=3}^{h} (1 - 2\cos\omega_j z + z^2)^{\gamma_j} \tag{10.9}$$

for given h, for given distinct real numbers $\omega_j, j = 3, \ldots, h$ on the interval $(0, \pi)$, and for given real numbers $\gamma_j, j = 1, \ldots, h$. If U_t is an ARMA process, then (10.1) and (10.9) combine to include the class of processes studied by Box and Jenkins (1970) and many other authors. We briefly indicate some special cases of interest:

a. '$I(1)$': $\varphi(z) = 1 - z$. Then X_t is a random walk when U_t is a white noise sequence.

b. '$I(2)$': $\varphi(z) = (1-z)^2$.

c. 'Cyclic $I(1)$': $\varphi(z) = 1 - 2\cos\omega z + z^2$, for $0 < \omega < \pi$.

d. 'Quarterly $I(1)$': $\varphi(z) = 1 - z^4 = (1-z)(1+z)(1+z^2)$.

e. '$1/f$ noise': $\varphi(z) = (1-z)^{1/2}$. Then X_t given by (10.1) is a fractionally differenced process that is 'just non-stationary'. X_t being stationary when $\varphi(z) = (1-z)^d$, $d < 1/2$. Theoretical and empirical evidence in physics, hydrology, traffic flow, and other fields suggests the usefulness of $d = 1/2$ (e.g. see Marinari *et al.* 1983).

The last two hypotheses are stationary ones of particular interest:

f. '$1/f^{1/2}$ noise': $\varphi(z) = (1-z)^{1/4}$. The limiting distribution of sample auto-covariances of $X_t = (1-B)^{-d} U_t$ tends to be normal for $d < 1/4$ but nonnormal for $d \geq 1/4$ (when the spectrum of X_t is not square integrable) (e.g. see Fox and Taqqu 1985).

g. 'Overdifferenced': $\varphi(z) = (1-z)^{-1}$.

In place of AR alternatives, research on long-memory processes suggests the fractional alternatives

$$\varphi(z; \theta) = (1-z)^{\gamma_1 + \theta_{i_1}} (1+z)^{\gamma_2 + \theta_{i_2}} \prod_{j=3}^{h} (1 - 2\cos\omega_j z + z^2)^{\gamma_j + \theta_{i_j}}, \quad (10.10)$$

where for each j, $\theta_{i_j} = \theta_l$ for some l, and for each l there is at least one j such that $\theta_{i_j} = \theta_l$; thus $h \geq p$. In the foregoing examples, $p = h = 1$ in each case except for (d), where $h = 3$ and $p = 1, 2$, or 3; for example,

$$\varphi(z; \theta) = (1 - z^4)^{1+\theta} \qquad (10.11)$$

or

$$\varphi(z; \theta) = (1-z)^{1+\theta_1} (1+z)^{1+\theta_2} (1+z^2)^{1+\theta_3}. \qquad (10.12)$$

The function $\psi(\lambda)$ in (10.6) is immediately calculated from (10.10) to have lth element

$$\delta_{1l} \log\left|2\sin\frac{1}{2}\lambda\right| + \delta_{2l} \log\left(2\cos\frac{1}{2}\lambda\right) + \sum_{j=3}^{n} \delta_{jl} \log|2(\cos\lambda - \cos\omega_j)|, \quad (10.13)$$

for $l = 1, \ldots, p$ and $|\lambda| < \pi$, where $\delta_{jl} = 1$ if $\theta_{1_j} = \theta_l$ and 0 otherwise. Thus $\psi(\lambda)$ is $\log\left|2\sin\frac{1}{2}\lambda\right|$ in examples (a), (b), (e), (f), and (g), it is $\log\left(2|\cos\lambda - \cos\omega|\right)$ in example (c), and in example (d) it is $\log\left|2\sin\frac{1}{2}\lambda\right| + \log\left(2\cos\frac{1}{2}\lambda\right) + \log\left|2\cos\lambda\right|$ corresponding to (10.11) and $\left(\log|2\sin\frac{1}{2}\lambda|, \quad \log\left(2\cos\frac{1}{2}\lambda\right), \log|2\cos\lambda|\right)'$ corresponding to (10.12).

Notice that (10.13) is unbounded: $\log|2\sin\frac{1}{2}\lambda| \to -\infty$, as $\lambda \to 0\pm$; $\log(2\cos\frac{1}{2}\lambda) \to -\infty$, as $\lambda \to \pm\pi$; $\log|2(\cos\lambda - \cos\omega_j)| \to -\infty$, as $\lambda \to \pm\omega_j$. This property adds significantly to the difficulty of our proofs of asymptotic properties, but we are able to use the fact that these functions are square integrable on $(-\pi, \pi]$ (e.g. $|\log|2\sin\frac{1}{2}\lambda|| = O(|\lambda|^{-\varepsilon})$ for any $\varepsilon > 0$) as required by (10.4). Related tests for white noise against stationary fractional alternatives were developed by Davies and Harte (1987) and by Robinson (1991). Empirical applications of seasonal long-memory models were reported by, for example, Porter-Hudak (1990). Fractional cyclic models were discussed by Gray, Zhang, and Woodward (1989). It must be stressed, however, that we are not concerned with estimation of long-memory models, nor is the distribution theory for such estimates (Fox and Taqqu 1985; Geweke and Porter-Hudak 1983; Robinson 1995) of direct relevance. Our test statistics are functions of the hypothesized differenced series, which have short memory under the null.

10.3 Score Test Under White Noise

In this section we develop a score test of H_0 (10.3) in the model given by (10.2), (10.4), and (10.7) under the presumption that U_t in (10.4) is a sequence of zero-mean uncorrelated random variables with unknown variance σ^2.

We take L in (10.8) with $\eta = (\theta', \beta', \sigma^2)'$ to be the negative of the log-likelihood based on Gaussian U_t, though the null and local limit distribution theory for our test will not assume Gaussianity. In Appendix 10.1 it is shown that (10.8) takes the form $\bar{R} = (n/\tilde{\sigma}^4)\bar{a}'\bar{A}^{-1}\bar{a}$, where $\bar{a} = -\sum_{l=1}^{n-1}\psi_l C_{\tilde{U}}(l)$, $\bar{A} = \sum_{l=1}^{n-1}(1 - l/n)\psi_l\psi_l'$, $C_{\tilde{U}}(l) = (n - l)^{-1}\sum_{t=1}^{n-l}\tilde{U}_t\tilde{U}_{t+l}$, $\tilde{U}_t = \varphi(B)(Y_t - \tilde{\beta}'Z_t)$, $\tilde{\beta} = (\sum_{t=1}^{n}W_t W_t')^{-1}\sum_{t=1}^{n}W_t\varphi(B)Y_t$, $\tilde{\sigma}^2 = n^{-1}\sum_{t=1}^{n}\tilde{U}_t^2$, $W_t = \varphi(B)Z_t$, and ψ_l is given by expanding $\psi(\lambda)$ in (10.6) as $\psi(\lambda) = \sum_{l=1}^{\infty}\psi_l\cos l\lambda$.

Concise formulas for ψ_l are available in some simple cases; for example, $\psi_l = -l^{-1}$ when $\varphi(z; \theta) = (1 - z)^{\gamma+\theta}$. But for reasons given in Section 10.1, we prefer a frequency domain version of the score statistic. Slight modification to the proofs of our limit theorems shows that the same results hold for the time domain statistic $\bar{R}$ and its extension to autocorrelated U_t; indeed, the proofs for the time domain versions are somewhat simpler.

Routine calculation gives $\bar{a} = -\int_{-\pi}^{\pi}\psi(\lambda)I_{\tilde{U}}(\lambda)\,d\lambda$, where $I_{\tilde{U}}$ is the periodogram of the $\tilde{U}_t$ sequence, $I_{\tilde{U}}(\lambda) = (2\pi n)^{-1}|\sum_{t=1}^{n}\tilde{U}_t e^{it\lambda}|^2$, and $\psi_0 = 0$ is implied by $\varphi(0; \theta) \equiv 1$. We approximate $\bar{R}$ by $\tilde{R} = (n/\tilde{\sigma}^4)\tilde{a}'\tilde{A}^{-1}\tilde{a} = \tilde{r}'\tilde{r}$, where $\tilde{r} = (n^{1/2}/\tilde{\sigma}^2)\tilde{A}^{-1/2}\tilde{a}$, $\tilde{a} = -(2\pi/n)\sum_j'\psi(\lambda_j)I_{\tilde{U}}(\lambda_j)$, and

$$\tilde{A} = \sum_{l=1}^{n-1}\left(1 - \frac{l}{n}\right)\psi_l\psi_l', \quad \text{or} \quad 2\Psi, \quad \text{or} \quad \frac{2}{n}\sum_j'\psi(\lambda_j)\psi(\lambda_j)', \quad (10.14)$$

in which $\lambda_j = 2\pi j/n$ and the primed sums are over $\lambda_j \in M$, where $M = \{\lambda : -\pi < \lambda < \pi, \lambda \notin (\rho_l - \lambda_1, \rho_l + \lambda_1), l = 1, \ldots, s\}$, such that $\rho_l, l = 1, \ldots, s < \infty$ are the distinct poles of $\psi(\lambda)$ on $(-\pi, \pi]$. Thus we use a discrete approximation to the integral $\bar{a}$, omitting the contribution from the finitely many λ_j in an open λ_1-neighbourhood of any of the ρ_l.

To summarize the computation of $\tilde{a}$: $\tilde{U}_t$ and $\tilde{\sigma}^2$ are obtained via least squares estimation of the null differenced regression model $\varphi(B)Y_t = \beta'W_t + U_t$; the periodograms $I_{\tilde{U}}(\lambda_j)$ are computed (which can be rapidly accomplished by means of the fast Fourier transform even when n is large); then the weighted average $\tilde{a}$ is computed, the function $\psi(\lambda)$ having been deduced from $\varphi(z; \theta)$.

Notice that by Parseval's equality,

$$\Psi = \frac{1}{2} \sum_{l=1}^{\infty} \psi_l \psi_l', \qquad (10.15)$$

so that asymptotic equivalence of the first two formulas in (10.14) readily follows. Sometimes a simple closed form is available for Ψ; for example, $\Psi = \pi^2/12$, when $\varphi(z; \theta) = (1 - z)^{\gamma+\theta}$. More generally, for example, when $\varphi(z; \theta)$ has complex zeros, a simple formula may be unavailable, and the first expression in (10.14) may be cumbersome to calculate if the ψ_l are not of simple form, in which case the final option in (10.14) may be preferred.

10.4 Distribution Theory Under White Noise

Because $\tilde{R}$ involves a ratio of quadratic forms, its exact null distribution can be calculated under Gaussianity via Imhof's algorithm. A simple test is approximately valid under much wider distributional assumptions. Theorem 1 is a large-sample justification for rejecting H_0 at the $100\alpha\%$ level when $\tilde{R} > \chi^2_{p,\alpha}$, where $P(\chi^2_p > \chi^2_{p,\alpha}) = \alpha$. It also justifies a one-sided test when $p = 1$: H_0 is rejected in favour of H_1: $\theta > 0$ at the $100\alpha\%$ level, when $\tilde{r} > z_\alpha$, where the probability that a standard normal variate exceeds z_α is α and an arbitrary convention on the parameterization is imposed such that positive values of $\tilde{a}$ correspond to positive θ. A one-sided test is of interest in example (e) of Section 10.2, when $\varphi(z; \theta) = (1 - z)^{(1/2)+\theta}$. Here we can interpret a test of $\theta = 0$ as a discriminator between stationarity and nonstationarity. In this case we form

$$\tilde{U}_t = (1 - B)^{1/2} X_t = \tilde{X}_t - \sum_{j=1}^{t-1} \left(\frac{1}{2}\right)\left(-\frac{1}{2}\right) \cdots \left(\frac{3}{2} - j\right) \tilde{X}_{t-j}/j!. \quad (10.16)$$

A significantly negative $\tilde{r}$ suggests stationarity. On the other hand, a significantly positive $\tilde{r}$ suggests nonstationarity.

Theorem 1 describes the null limit distribution of the test proposed in the previous section. Its proof is contained in Appendix 10.2. The statement of the theorem refers to classes F, G, and H, which are defined in that Appendix to avoid overburdening the body of the article with technical detail. Class F imposes a martingale difference assumption on the white noise U_t, that is substantially weaker than the Gaussianity used in motivating the test statistic and in particular requires a (second) moment condition that is clearly minimal. Class G imposes a mild lack-of-multicollinearity assumption on the W_t that is satisfied by, for example, Z_t with elements that are polynomials in t. Class H includes technical restrictions on ψ that we find necessary to justify approximating integrals by sums but seem practically costless, being readily verified by any ψ that we have been able to think of; ψ will always be known to the applied worker.

Theorem 1 Let $\{U_t : t = 0, \pm 1, \ldots\} \in F$, let $\{Z_t; t = 0, \pm 1, \ldots\} \in G$, let $\varphi(z; \theta) \in H$, and let (10.2) and (10.7) be true. Then under H_0 defined by (10.3) and (10.4), condition (10.5) is sufficient for $\tilde{r} \to_d N(0, I_p)$, as $n \to \infty$, where I_p is the p-rowed identity matrix.

Under Gaussianity, an approximate finite-sample optimality theory might be constructed along the lines of that of Bhargava (1986), when $p = 1$. For general p, the approximate score interpretation suggests that $\tilde{R}$ will be optimal in the sense of providing an asymptotically most powerful test against local alternatives of form

$$H_1 : \theta = \theta_n \stackrel{\text{def}}{=} \delta n^{-(1/2)}, \tag{10.17}$$

where δ is any non-null $p \times 1$ vector. Because our setting is a long way from the classical ones in which Pitman sequences (10.17) have been used, and we know of no reasonably close reference, and because we have stressed this optimality property in comparing our tests to rival ones, we also prove (see Appendix 10.3) the following theorem.

Theorem 2 Let $\{U_t, t = 0, \pm 1, \ldots\} \in F$, let $\{Z_t, t = 0, \pm 1, \ldots\} \in G$ and let

$$Y_{tn} = \beta' Z_t + X_{tn}, \tag{10.18}$$

where

$$\varphi(B; \theta_n) X_{tn} = U_t, \quad t \geq 1, \qquad X_{tn} = 0, \quad t \leq 0, \tag{10.19}$$

where θ_n satisfies (10.17) and $\varphi(z; \theta) \in J$ (defined in Appendix 10.3). Let $\tilde{a}$, $\tilde{A}$, and $\tilde{\sigma}^2$ now be defined in terms of X_{tn} rather than X_t. Then condition (10.5) is sufficient for $\tilde{r} \to_d N(-\Psi^{(1/2)}\delta, I_p)$, as $n \to \infty$.

Class J entails a strengthening of the restrictions on φ, but it is readily checked in case of (10.10). Theorem 2 implies that under local alternatives, $\tilde{R} \to_d \chi_p^{2'}(\delta' \Psi \delta)$, indicating a noncentral χ_p^2 distribution with noncentrality parameter $\delta' \Psi \delta$. Optimality under Gaussianity of U_t then follows from the information matrix formula (A.1) in Appendix 10.1. In non-Gaussian environments the test is of course no longer fully efficient, but it is still the most efficient test based on quadratic functions of the data.

10.5 Score Test Under Weak Autocorrelation

The test developed in the preceding section will tend to be consistent against stationary autocorrelated U_t, as well as against alternatives (10.10) to the null hypothesis (10.9). Thus it may be prudent to carry out a test that corrects for possible autocorrelation in U_t. Let U_t be covariance stationary with spectral density of form $f(\lambda; \tau, \sigma^2) = (\sigma^2/2\pi)g(\lambda; \tau)$, $-\pi < \lambda \leq \pi$, where g is a known function of λ and the unknown $q \times 1$ vector τ, such that τ and σ^2 are not a priori related. Note that σ^2 is generally no longer the variance of U_t, but rather the variance of the innovation sequence in a normalized Wold representation for U_t. Technical assumptions are made on g in the statements of Theorems 3 and 4; their principal practical implications being that though U_t is capable of exhibiting a much stronger degree of autocorrelation than stationary and invertible ARMA processes, its spectrum is bounded and bounded away from zero, so that it cannot be a fractional process with positive or negative differencing parameter. The latter '$I(0)$' property is typically taken for granted in unit root tests with a nonparametric autocorrelation correction. Though q can be arbitrarily large, we assume it is finite, thus treating only parametric alternatives. A unit root test against fractional alternatives with nonparametric autocorrelation under the null would have negligible efficiency relative to parametric autocorrelation. Intuitively, this can be explained as follows. If there were a score test with parametric rates of convergence, we would expect a Wald test with the same rates to exist. But this Wald test would be based on estimates of differencing parameters in a model that is not fully parametric; these estimates would thus use an asymptotically negligible fraction of the data and lack parametric convergence rates, even under the null hypothesis (cf. Geweke and Porter-Hudak 1983; Robinson 1995). Unit root tests against AR alternatives that allow for weak nonparametric autocorrelation are sensitive to choice of a particular form of spectral estimate and of a smoothing parameter, which in practice is not very different from choice of q and a parametric functional form, and a parametric approach might in any case be preferred when n is not very large.

By extending the argument in Section 10.3 and Appendix 10.3, under Gaussianity of U_t an approximate score statistic for testing (10.3) in (10.2), (10.4), and (10.6) is $\hat{R} = (n/\hat{\sigma}^4)\hat{a}'\hat{A}^{-1}\hat{a} = \hat{r}'\hat{r}$, where $\hat{r} = (n^{1/2}/\hat{\sigma}^2)\hat{A}^{-1/2}\hat{a}$, $\hat{a} = -(2\pi/n)\sum_j' \psi(\lambda_j)g(\lambda_j; \hat{\tau})^{-1}I_{\hat{U}}(\lambda_j)$ and $\hat{A}$ is either $2(\Psi - \Phi\Xi^{-1}\Phi')$ or

$$\frac{2}{n}\sum_j' \psi(\lambda_j)\psi(\lambda_j)' - \frac{2}{n}\sum_j' \psi(\lambda_j)\hat{\xi}(\lambda_j)' \left\{\frac{1}{n}\sum_j' \hat{\xi}(\lambda_j)\hat{\xi}(\lambda_j)'\right\}^{-1}$$

$$\times \frac{1}{n}\sum_j' \hat{\xi}(\lambda_j)\psi(\lambda_j)'. \tag{10.20}$$

Here, $\Phi = (2\pi)^{-1}\int_{-\pi}^{\pi} \psi(\lambda)\xi(\lambda)'\,d\lambda$, $\Xi = (2\pi)^{-1}\int_{-\pi}^{\pi} \xi(\lambda)\xi(\lambda)'\,d\lambda$. $\hat{\xi}(\lambda) = (\partial/\partial_\tau)\log g(\lambda; \tau)$, $\hat{\xi}(\lambda) = (\partial/\partial_\tau)\log g(\lambda; \hat{\tau})$, $\hat{\tau} = \arg\min\sigma^2(\cdot) + o_p(n^{-(1/2)})$, $\hat{\sigma}^2 = \sigma^2(\hat{\tau}) + o_p(1)$, and $\sigma^2(\cdot) = (2\pi/n)\sum_{j-1}^{n-1} g(\lambda_j; \cdot)^{-1}I_{\hat{U}}(\lambda_j)$, and the minimization is carried out over a compact subset of q-dimensional Euclidean space.

Notice that $\hat{R}$ is based on least squares residuals. In general $\tilde{\beta}$ does not have the same limiting distribution as the Gaussian maximum likelihood estimate when U_t is autocorrelated (though it does if W_t consists of polynomials in t). But this makes no difference to the limiting null and local asymptotic distributions of $\hat{R}$ because of orthogonality (cf. (A.1) in Appendix 10.1), so we stress the computationally simpler $\tilde{\beta}$. The estimates $\hat{\tau} = \arg\min \sigma^2(\cdot)$ and $\hat{\sigma}^2 = \sigma^2(\hat{\tau})$ may be the most natural ones in view of the form of the test statistic, but other Whittle-type estimates of τ having the same limiting distribution (such as the time domain Gaussian maximum likelihood estimate), and other consistent estimates of σ^2 could be used, as indicated by the $o_p(n^{-(1/2)})$ and $o_p(1)$ error terms allowed for in the definitions of $\hat{\tau}$ and σ.

10.6 Distribution Theory Under Weak Autocorrelation

The following theorem strictly relaxes the conditions of Theorem 1, so that, for example, we continue to require finiteness of only second moments of U_t.

Theorem 3 Let $\{U_t, t = 0, \pm 1, \ldots\} \in K$, let $\{Z_t, t = 0, \pm 1, \ldots\} \in G$, let $\varphi \in H$, let $g \in L$, and let (10.2) and (10.7) be true. Then, under H_o defined by (10.3) and (10.4), the condition

$$0 < \det(\Psi - \Phi \Xi^{-1} \Phi') < \infty \qquad (10.21)$$

is sufficient for $\hat{r} \to_d N(0, I_p)$, as $n \to \infty$.

The proof and the definitions of classes K and L are contained in Appendix 10.4.

Theorem 2 can likewise be extended. The proof of the following theorem is omitted, because it straightforwardly combines arguments in the proofs of Theorems 2 and 3.

Theorem 4 Let $\{U_t, t = 0, \pm 1, \ldots\} \in K$, let $\{Z_t, t = 0, \pm 1, \ldots\} \in G$, and let (10.18) and (10.19) hold where θ_n satisfies (10.17) and $\varphi(z; \theta) \in J$. Let $\hat{a}$, $\hat{A}$, and $\hat{\sigma}^2$ now be defined in terms of X_{tn} rather than X_t. Then condition (10.21) is sufficient for $\hat{r} \to_d N(-(\Psi - \Phi \Xi^{-1} \Phi')^{1/2} \delta, I_p)$, as $n \to \infty$.

Perhaps the most obvious choice of a time series model for U_t satisfying our conditions is a stationary and invertible ARMA, where relatively simple formulas for g and ξ are available; for example, in the pure AR case, $g(\lambda; \tau) = |1 - \sum_{j=1}^q \tau_j e^{ij\lambda}|^{-2}$, $\xi(\lambda) = (2\{\cos l\lambda - \sum_{j=1}^q \tau_j \cos(l - j)\lambda\} g(\lambda; \tau))$, where the lth element of $\xi(\lambda)$ is indicated. There are grounds for preferring the exponential spectrum model of Bloomfield (1973):

$$g(\lambda; \tau) = \exp\left\{2 \sum_{j=1}^q \tau_j \cos j\lambda\right\}, \qquad -\pi < \lambda \leq \pi. \qquad (10.22)$$

Bloomfield (1973) found that (10.22) has a somewhat cumbersome time domain representation but it leads to an especially neat version of our frequency domain test statistic. We have $\xi(\lambda) = (2 \cos l\lambda)$, $\Xi = 2I_q$, $\Phi = (\psi_1, \ldots, \psi_q)$, and $\Psi - \Phi \Xi^{-1} \Phi' = \frac{1}{2} \sum_{j=q+1}^\infty \psi_j \psi_j'$. Bloomfield (1973) exploited the simplicity of Ξ to give a particularly convenient Newton-type iteration for $\hat{\tau}$. The approximate independence of the elements of $\hat{\tau}$ is also helpful in determining a parsimonious parameterization in (10.22). Unlike in the AR model, $\xi(\lambda)$, and thus $\Psi - \Phi \Xi^{-1} \Phi'$, are free of the nuisance parameter vector τ, and (10.20) correspondingly simplifies. The loss of efficiency due to overparameterization in (10.22) is also free of τ, unlike in the AR model. If $\tau_j = 0$ for $j > q_0$, where $0 \le q_0 < q$, then the asymptotic relative efficiency under (10.17) is $1 - \sum_{j=q_0+1}^{q} (\delta' \psi_j)^2 / \sum_{j=q_0+1}^{\infty} (\delta' \psi_j)^2$. For example, consider testing for a simple unit root (i.e. example (a) of Section 10.2), where $\varphi(z; \theta) = (1 - z)^{1+\theta}$; if $q_0 = 0$, so that U_t is actually white noise, then the efficiency is $1 - (6/\pi^2)(1 + 2^{-2} + \cdots + q^{-2})$, which is 0.392 for $q = 1, 0.240$ for $q = 2$, and 0.173 for $q = 3$.

10.7 Empirical Illustration

To illustrate application of the methods to real data, another look was taken at time series analysed by Box and Jenkins (1970), who used their techniques of identifying and estimating seasonal and nonseasonal nonfractionally integrated ARMA models. We test the integer degrees of integration that Box and Jenkins (1970) suggested as relevant for these series, as well as some other null hypotheses. On the whole, the series seem long enough to justify use of large-sample inference rules. Instead of describing the stationary autocorrelation in the series by an ARMA, we use (10.22). The time series have been analysed by a number of subsequent authors including Said and Dickey (1985), who applied unit root tests against ARMA alternatives, and Bloomfield (1973), who fitted versions of (10.22) after carrying out the differencing proposed by Box and Jenkins (1970) but did not test the suitability of this differencing. Bloomfield (1973) suggested values of q that produce a parsimonious fit; we carried out tests for $q = 1, 2$, and 3, as well as the test of Section 10.3 that takes U_t to be white noise under the null, which we refer to here as the case $q = 0$.

Summary details of the series and computations are contained in Table 10.1. The computations were carried out using the programming language C on LSE's

Table 10.1. One-sided test statistics for $q = 0, 1, 2, 3$

Series	n	$\varphi(z; \theta)$	$q = 0$	$q = 1$	$q = 2$	$q = 3$
A	107	$(1-z)^{1+\theta}$	-5.19	-3.45	-3.32	-2.65
		$(1-z)^{1/2+\theta}$	-1.53	-6.85	-9.33	-10.60
B	369	$(1-z)^{1+\theta}$	1.64	0.33	0.44	1.45
		$(1-z)^{1/2+\theta}$	42.91	8.28	2.71	0.04
B'	255	$(1-z)^{1+\theta}$	-0.97	0.14	0.53	0.21
		$(1-z)^{1/2+\theta}$	25.76	-4.62	-7.71	-9.27
C	226	$(1-z)^{2+\theta}$	-4.76	-2.93	-2.74	-2.42
		$(1-z)^{3/2+\theta}$	0.75	0.41	-0.49	-0.58
D	310	$(1-z)^{1+\theta}$	-2.27	-2.87	-3.09	-3.10
		$(1-z)^{1/2+\theta}$	6.14	-4.92	-7.52	-8.96
G	144	$(1-z)^{1+\theta}(1-z^{12})^{1+\theta}$	-6.21	-6.18	-6.22	-5.93
		$(1-z)^{1/2+\theta}(1-z^{12})^{1+\theta}$	-5.92	-5.95	-5.92	-5.71

VAX computer. The series A, B, B', C, D, E, and G are labelled as in Box and Jenkins (1970) (like them, we took logs of series G). The top $\varphi(z; \theta)$ for each series indicates Box and Jenkins's suggested differencing when $\theta = 0$ ($p = 1$ throughout). But for some series they also reported fits of stationary models, and some of their results could be interpreted to suggest a degree of overdifferencing, providing some motivation for the bottom $\varphi(z; \theta)$ for each series, based on example (e) of Section 10.2. Because the algorithm proposed by Bloomfield (1973) was found to not always converge, a modified algorithm in fitting (10.22) was used, entailing discrete averaging over frequency rather than integration, and an approximate Newton–Raphson, rather than Gauss–Newton, iteration. A satisfactory degree of convergence was always achieved within seven iterations. The statistics tabulated are $\tilde{r}$ ($q = 0$ column) and $\hat{r}$ for $q = 1, 2, 3$. Increasing q sometimes leads to a more significant test statistic due to the reduction in $\hat{\sigma}^2$. Box and Jenkins's (1970) models are not rejected in series B and B', but they are rejected in the other series, for each q, and generally the results provide some empirical illustration of the good theoretical power properties of our tests. The tests based on 1/2 fractional differencing also nearly always reject, even for series B and B', and these models seem inferior to those of Box and Jenkins (1970) in several cases—although for series C, the differencing $(1 - B)^{3/2}$ is not rejected and a decisive improvement over $(1 - B)^2$ is indicated.

10.8 Finite-Sample Performance and Comparison

This section examines the finite-sample behaviour of versions of the new test by means of Monte Carlo simulations and compares them with a number of leading unit root tests. The computations were carried out using Fortran and the NAG library's random number generator, on LSE's VAX computer.

Many unit root tests use a random walk null, or a null that is a random walk about an intercept or a linear trend. Using the same data, seven existing tests that have such properties, and were developed with AR alternatives in mind, are compared with versions of the new test. The null model consists of (10.1), (10.2), and (10.7), with $\varphi(z) = 1 - z$ and the U_t in (10.1) correctly assumed to be white noise. The test denoted S1 in Tables 10.2–10.9 is a test of Section 10.3 where $\beta = 0$ is correctly assumed (i.e. $Y_t = X_t$) and $\varphi(z; \theta) = (1 - z)^{1+\theta}$. Test S2 is the corresponding test with unknown β, where $Z_t = (1, t)'$ in (10.7); it is invariant with respect to β. The $\hat{\rho}$ and $\hat{\tau}$ tests are due to Fuller (1976) and to Dickey and Fuller (1979); they assume that $\beta = 0$ and are designed to be particularly sensitive to AR alternatives, $\varphi(z; \theta) = 1 - (1 + \theta)z$. Likewise, the $\hat{\rho}_\tau$ and $\hat{\tau}_\tau$ tests of Fuller (1976) and Dickey and Fuller (1979) take $Z_t = (1, t)'$ in (10.7) but assume that the second element of β is zero and are invariant only with respect to the first element. The $\tilde{\rho}$ and $\tilde{\tau}$ tests are due to Schmidt and Phillips (1992); these result from application of a version of the score principle to (10.2), (10.4), and (10.7), with $\varphi(z) = 1 - (1 + \theta)z$, $Z_t = (1, t)'$, and they are invariant with respect to β. The F test, from Robinson (1993), is an exact test under Gaussianity when $\beta = 0$ in (10.7) and was shown to be consistent against fractional, as well as

Table 10.2. Rejection frequencies for upper-tailed 5% test and fractional X_t

n	θ	S1	S2	$\hat{\rho}$	$\hat{\tau}$	$\hat{\rho}_\tau$	$\hat{\tau}_\tau$	$\tilde{\rho}$	$\tilde{\tau}$
25	0	0.016	0.016	0.053	0.049	0.046	0.044	0.048	0.063
	0.05	0.028	0.026	0.076	0.082	0.067	0.064	0.072	0.090
	0.1	0.047	0.047	0.107	0.130	0.093	0.084	0.109	0.126
	0.2	0.120	0.117	0.181	0.241	0.149	0.132	0.193	0.220
	0.3	0.230	0.225	0.265	0.362	0.227	0.192	0.306	0.335
	0.5	0.516	0.511	0.424	0.574	0.379	0.306	0.544	0.578
	0.7	0.772	0.775	0.563	0.703	0.524	0.393	0.735	0.763
	0.9	0.909	0.916	0.662	0.785	0.633	0.465	0.859	0.873
50	0	0.023	0.023	0.054	0.050	0.053	0.052	0.056	0.061
	0.05	0.063	0.063	0.080	0.092	0.088	0.080	0.096	0.102
	0.1	0.125	0.124	0.113	0.148	0.126	0.116	0.152	0.164
	0.2	0.323	0.324	0.197	0.288	0.237	0.195	0.321	0.332
	0.3	0.583	0.579	0.290	0.435	0.360	0.285	0.495	0.508
	0.5	0.906	0.902	0.455	0.651	0.565	0.423	0.769	0.778
	0.7	0.991	0.992	0.585	0.771	0.694	0.508	0.914	0.919
	0.9	0.999	1.000	0.681	0.836	0.746	0.525	0.973	0.975
100	0	0.030	0.030	0.049	0.050	0.046	0.046	0.052	0.057
	0.05	0.100	0.101	0.086	0.111	0.092	0.087	0.109	0.116
	0.1	0.233	0.232	0.140	0.187	0.156	0.138	0.199	0.209
	0.2	0.631	0.628	0.240	0.358	0.309	0.244	0.442	0.454
	0.3	0.897	0.896	0.338	0.516	0.483	0.361	0.670	0.679
	0.5	0.998	0.997	0.498	0.715	0.718	0.513	0.915	0.920
	0.7	1.000	1.000	0.626	0.823	0.779	0.559	0.986	0.988
	0.9	1.000	1.000	0.705	0.872	0.796	0.553	0.998	0.998
200	0	0.030	0.030	0.055	0.052	0.051	0.050	0.054	0.054
	0.05	0.168	0.169	0.096	0.114	0.112	0.100	0.134	0.135
	0.1	0.447	0.449	0.152	0.214	0.193	0.160	0.254	0.256
	0.2	0.910	0.911	0.272	0.416	0.400	0.307	0.572	0.574
	0.3	0.995	0.995	0.375	0.588	0.611	0.430	0.818	0.818
	0.5	1.000	1.000	0.529	0.780	0.814	0.574	0.982	0.983
	0.7	1.000	1.000	0.639	0.866	0.838	0.580	0.999	0.999
	0.9	1.000	1.000	0.719	0.906	0.826	0.565	1.000	1.000

AR, alternatives. The foregoing descriptions imply that all tests have asymptotic validity with respect to the same null hypothesis $Y_t = X_t$, $(1 - B)X_t = U_t$, U_t white noise.

It turns out that the $\hat{\rho}$, $\hat{\tau}$, and F tests have score interpretations when $\beta = 0$ is correctly assumed, just as the S1 test is known to, and just as the S2, $\tilde{\rho}$ and $\tilde{\tau}$ tests are known to when β is treated as unknown. To see this, in (10.8) take $L(\eta)$ to be the negative of the log-likelihood based on

Table 10.3. Rejection frequencies for upper-tailed 1% test and fractional X_t

n	θ	S1	S2	$\hat{\rho}$	$\hat{\tau}$	$\hat{\rho}_\tau$	$\hat{\tau}_\tau$	$\tilde{\rho}$	$\tilde{\tau}$
25	0	0.004	0.003	0.011	0.009	0.008	0.008	0.010	0.012
	0.05	0.009	0.010	0.017	0.019	0.013	0.013	0.017	0.021
	0.1	0.015	0.016	0.024	0.039	0.020	0.019	0.029	0.039
	0.2	0.051	0.050	0.048	0.108	0.042	0.041	0.069	0.089
	0.3	0.124	0.120	0.077	0.213	0.076	0.074	0.143	0.168
	0.5	0.374	0.362	0.153	0.441	0.171	0.164	0.350	0.385
	0.7	0.652	0.644	0.239	0.610	0.270	0.254	0.573	0.604
	0.9	0.840	0.847	0.317	0.717	0.366	0.336	0.743	0.771
50	0	0.007	0.006	0.013	0.010	0.009	0.008	0.011	0.013
	0.05	0.020	0.020	0.020	0.028	0.019	0.019	0.028	0.032
	0.1	0.057	0.057	0.032	0.059	0.033	0.034	0.061	0.068
	0.2	0.195	0.193	0.062	0.159	0.082	0.076	0.157	0.168
	0.3	0.428	0.426	0.099	0.296	0.149	0.137	0.318	0.330
	0.5	0.832	0.831	0.180	0.554	0.296	0.270	0.626	0.639
	0.7	0.975	0.975	0.268	0.705	0.407	0.365	0.836	0.841
	0.9	0.997	0.998	0.358	0.790	0.471	0.419	0.930	0.934
100	0	0.009	0.009	0.011	0.010	0.008	0.008	0.012	0.013
	0.05	0.039	0.039	0.021	0.032	0.022	0.020	0.033	0.034
	0.1	0.123	0.125	0.035	0.080	0.045	0.041	0.074	0.076
	0.2	0.468	0.467	0.078	0.234	0.123	0.110	0.245	0.251
	0.3	0.805	0.805	0.121	0.397	0.210	0.186	0.475	0.482
	0.5	0.993	0.993	0.209	0.641	0.398	0.346	0.822	0.826
	0.7	1.000	1.000	0.292	0.773	0.498	0.429	0.954	0.955
	0.9	1.000	1.000	0.387	0.840	0.547	0.467	0.989	0.989
200	0	0.009	0.009	0.011	0.013	0.010	0.010	0.011	0.011
	0.05	0.070	0.070	0.024	0.044	0.029	0.027	0.044	0.047
	0.1	0.270	0.270	0.044	0.098	0.064	0.059	0.114	0.117
	0.2	0.822	0.821	0.093	0.292	0.166	0.147	0.372	0.378
	0.3	0.988	0.988	0.144	0.482	0.287	0.249	0.656	0.663
	0.5	1.000	1.000	0.234	0.725	0.504	0.409	0.945	0.946
	0.7	1.000	1.000	0.311	0.832	0.582	0.475	0.994	0.994
	0.9	1.000	1.000	0.404	0.878	0.596	0.498	1.000	1.000

the model $Y_t = X_t$, $(1 - (1 + \theta)B)X_t = U_t$, for $t > 0$, $X_0 = 0$, $U_t \sim$ NID$(0, \sigma^2)$, so $\eta = (\theta, \sigma^2)'$. A score statistic for testing $\theta = 0$ of forms $s_1 = \{\sum_{t=1}^n Y_{t-1}(Y_t - Y_{t-1})\}^2 / a\tilde{\sigma}^4$ is deduced, where $\tilde{\sigma}^2 = n^{-1} \sum_{t=1}^n (Y_t - Y_{t-1})^2$ and $a = E_0(\partial L(\eta)/\partial\theta)^2$ is found to be $E_0(\sum_{t=1}^n Y_{t-1}^2/\sigma^2) = (1/2)n(n-1)$. Routine algebra indicates that $s_1 = 1/2n(n-1)\{1 + n/(F-1)\}^2$, where $F = (n-1)Y_n^2/\{n\tilde{\sigma}^2 - Y_n^2\}$ is the F test statistic referred to previously. Were this a 'standard' problem, replacing a by $-\partial^2 L(\tilde{\eta})/\partial\theta^2$, or indeed

Table 10.4. Rejection frequencies for upper-tailed 5% test and AR X_t

n	θ	S1	S2	$\hat{\rho}$	$\hat{\tau}$	$\hat{\rho}_\tau$	$\hat{\tau}_\tau$	$\tilde{\rho}$	$\tilde{\tau}$
25	0	0.016	0.016	0.053	0.049	0.046	0.044	0.048	0.063
	0.05	0.014	0.015	0.105	0.123	0.047	0.044	0.046	0.060
	0.1	0.014	0.014	0.288	0.325	0.049	0.049	0.046	0.059
	0.2	0.135	0.173	0.709	0.691	0.391	0.378	0.240	0.268
	0.3	0.879	0.890	0.963	0.956	0.939	0.941	0.901	0.907
	0.5	0.982	0.985	0.993	0.991	0.991	0.992	0.985	0.987
	0.7	0.999	0.999	1.000	0.999	1.000	1.000	0.999	0.999
	0.9	1.000	1.000	1.000	1.000	1.000	1.000	1.000	1.000
50	0	0.023	0.023	0.054	0.050	0.053	0.052	0.056	0.061
	0.05	0.022	0.022	0.202	0.251	0.049	0.048	0.047	0.052
	0.1	0.232	0.247	0.730	0.711	0.417	0.409	0.309	0.319
	0.2	0.924	0.926	0.970	0.963	0.954	0.954	0.933	0.934
	0.3	0.999	0.999	1.000	0.999	0.999	0.999	0.999	0.999
	0.5	1.000	1.000	1.000	1.000	1.000	1.000	1.000	1.000
	0.7	1.000	1.000	1.000	1.000	1.000	1.000	1.000	1.000
	0.9	1.000	1.000	1.000	1.000	1.000	1.000	1.000	1.000
100	0	0.030	0.030	0.049	0.050	0.046	0.046	0.052	0.057
	0.05	0.087	0.091	0.602	0.594	0.211	0.208	0.144	0.153
	0.1	0.937	0.937	0.972	0.968	0.961	0.963	0.945	0.946
	0.2	0.999	0.999	1.000	1.000	1.000	1.000	1.000	1.000
	0.3	1.000	1.000	1.000	1.000	1.000	1.000	1.000	1.000
	0.5	1.000	1.000	1.000	1.000	1.000	1.000	1.000	1.000
	0.7	1.000	1.000	1.000	1.000	1.000	1.000	1.000	1.000
	0.9	1.000	1.000	1.000	1.000	1.000	1.000	1.000	1.000
200	0	0.030	0.030	0.055	0.052	0.051	0.050	0.054	0.054
	0.05	0.833	0.833	0.938	0.926	0.902	0.902	0.864	0.865
	0.1	0.999	0.999	1.000	1.000	0.999	0.999	0.999	0.999
	0.2	1.000	1.000	1.000	1.000	1.000	1.000	1.000	1.000
	0.3	1.000	1.000	1.000	1.000	1.000	1.000	1.000	1.000
	0.5	1.000	1.000	1.000	1.000	1.000	1.000	1.000	1.000
	0.7	1.000	1.000	1.000	1.000	1.000	1.000	1.000	1.000
	0.9	1.000	1.000	1.000	1.000	1.000	1.000	1.000	1.000

by $\{\partial^2 L(\tilde{\eta})/\partial\theta^2\}^2/E_0(\partial L(\eta)/\partial\theta)^2$, would not change the null limit distribution. But because the alternatives are AR(1), this is not a 'standard' problem. In fact, $-\partial^2 L(\tilde{\eta})/\partial\theta^2 = \sum_{t=1}^{n} Y_{t-1}^2/\tilde{\sigma}^2$, which leads to a score statistic $s_2 = 1/2(1 - n^{-1})\{n(\hat{\rho} - 1)\}^2$, where $\hat{\rho} = \sum_{t=1}^{n} Y_t Y_{t-1}/\sum_{t=1}^{n} Y_{t-1}^2$; s_2 has the same null limit distribution as the square of the $\hat{\tau}$ test statistic. Also, $\{\partial^2 L(\tilde{\eta})/\partial\theta^2\}^2/E_0(\partial L(\eta)\partial\theta)^2 = (\sum_{t=1}^{n} Y_{t-1}^2)^2/\{1/2n(n-1)\tilde{\sigma}^4\}$, which leads to a score statistic $s_3 = 1/2(1 - n^{-1})\{n(\hat{\rho} - 1)\}^2$, where $n(\hat{\rho} - 1)$ is the $\hat{\rho}$ test

Table 10.5. Rejection frequencies for upper-tailed 1% test and AR X_t

n	θ	S1	S2	$\hat{\rho}$	$\hat{\tau}$	$\hat{\rho}_\tau$	$\hat{\tau}_\tau$	$\tilde{\rho}$	$\tilde{\tau}$
25	0	0.004	0.003	0.011	0.009	0.008	0.008	0.010	0.014
	0.05	0.003	0.003	0.020	0.036	0.008	0.009	0.010	0.013
	0.1	0.004	0.003	0.058	0.178	0.009	0.010	0.009	0.012
	0.2	0.069	0.105	0.481	0.595	0.238	0.232	0.108	0.126
	0.3	0.839	0.860	0.956	0.940	0.918	0.918	0.843	0.855
	0.5	0.979	0.980	0.992	0.989	0.989	0.989	0.978	0.980
	0.7	0.999	0.999	1.000	0.999	1.000	1.000	0.999	0.999
	0.9	1.000	1.000	1.000	1.000	1.000	1.000	1.000	1.000
50	0	0.007	0.006	0.013	0.010	0.009	0.008	0.011	0.013
	0.05	0.005	0.006	0.044	0.129	0.011	0.010	0.009	0.011
	0.1	0.163	0.177	0.541	0.627	0.280	0.274	0.184	0.196
	0.2	0.908	0.913	0.965	0.956	0.941	0.941	0.910	0.912
	0.3	0.999	0.999	1.000	0.999	0.999	0.999	0.999	0.999
	0.5	1.000	1.000	1.000	1.000	1.000	1.000	1.000	1.000
	0.7	1.000	1.000	1.000	1.000	1.000	1.000	1.000	1.000
	0.9	1.000	1.000	1.000	1.000	1.000	1.000	1.000	1.000
100	0	0.009	0.009	0.011	0.010	0.008	0.008	0.012	0.013
	0.05	0.039	0.042	0.298	0.495	0.098	0.093	0.055	0.056
	0.1	0.928	0.929	0.969	0.962	0.952	0.953	0.930	0.931
	0.2	0.999	0.999	1.000	1.000	1.000	1.000	0.999	0.999
	0.3	1.000	1.000	1.000	1.000	1.000	1.000	1.000	1.000
	0.5	1.000	1.000	1.000	1.000	1.000	1.000	1.000	1.000
	0.7	1.000	1.000	1.000	1.000	1.000	1.000	1.000	1.000
	0.9	1.000	1.000	1.000	1.000	1.000	1.000	1.000	1.000
200	0	0.009	0.009	0.011	0.013	0.010	0.010	0.011	0.011
	0.05	0.807	0.808	0.922	0.910	0.869	0.869	0.821	0.822
	0.1	0.999	0.999	1.000	0.999	0.999	0.999	0.999	0.999
	0.2	1.000	1.000	1.000	1.000	1.000	1.000	1.000	1.000
	0.3	1.000	1.000	1.000	1.000	1.000	1.000	1.000	1.000
	0.5	1.000	1.000	1.000	1.000	1.000	1.000	1.000	1.000
	0.7	1.000	1.000	1.000	1.000	1.000	1.000	1.000	1.000
	0.9	1.000	1.000	1.000	1.000	1.000	1.000	1.000	1.000

statistic. Here s_1, s_2, and s_3 have different null and local limit distributions, to illustrate point 5 of Section 10.1.

For the seven tests directed against AR alternatives, finite-sample critical values derived from the (Monte Carlo-based) tables of Fuller (1976) and Schmidt and Phillips (1992) for the $\hat{\rho}$, $\hat{\tau}$, $\hat{\rho}_\tau$, $\hat{\tau}_\tau$, $\tilde{\rho}$, and $\tilde{\tau}$ tests, and from the standard F tables for the F test, are used. For S1 and S2, the asymptotic critical values given by the normal tables are used. Of course, the finite-sample sizes of S1 and S2 will

Table 10.6. Rejection frequencies for lower-tailed 5% test and fractional X_t

n	θ	S1	S2	$\hat{\rho}$	$\hat{\tau}$	$\hat{\rho}_\tau$	$\hat{\tau}_\tau$	$\tilde{\rho}$	$\tilde{\tau}$	F
25	−0.9	0.983	0.981	1.000	0.998	0.933	0.879	0.905	0.886	0.310
	−0.7	0.938	0.934	0.964	0.955	0.773	0.675	0.740	0.709	0.260
	−0.5	0.819	0.816	0.710	0.682	0.507	0.416	0.484	0.446	0.179
	−0.3	0.566	0.565	0.327	0.306	0.244	0.203	0.232	0.203	0.116
	−0.2	0.418	0.417	0.198	0.181	0.158	0.132	0.147	0.128	0.088
	−0.1	0.279	0.275	0.109	0.095	0.095	0.086	0.084	0.072	0.068
	−0.05	0.224	0.217	0.079	0.071	0.069	0.067	0.066	0.057	0.058
	0	0.175	0.173	0.056	0.049	0.051	0.052	0.049	0.041	0.047
50	−0.9	1.000	1.000	1.000	1.000	1.000	1.000	1.000	1.000	0.443
	−0.7	0.998	0.998	1.000	1 000	0.997	0.991	0.989	0.986	0.376
	−0.5	0.976	0.975	0.932	0.925	0.888	0.840	0.866	0.853	0.254
	−0.3	0.763	0.761	0.496	0.484	0.479	0.420	0.461	0.439	0.147
	−0.2	0.542	0.539	0.278	0.268	0.260	0.221	0.252	0.237	0.108
	−0.1	0.297	0.295	0.130	0.125	0.120	0.114	0.114	0.108	0.071
	−0.05	0.196	0.195	0.085	0.081	0.081	0.083	0.077	0.070	0.059
	0	0.117	0.117	0.053	0.051	0.054	0.057	0.048	0.043	0.051
100	−0.9	1.000	1.000	1.000	1.000	1.000	1.000	1.000	1.000	0.594
	−0.7	1.000	1.000	1.000	1.000	1.000	1.000	1.000	1.000	0.502
	−0.5	1.000	1.000	0.996	0.996	0.998	0.994	0.995	0.994	0.339
	−0.3	0.960	0.960	0.679	0.667	0.768	0.712	0.754	0.741	0.174
	−0.2	0.772	0.768	0.380	0.370	0.439	0.388	0.435	0.420	0.115
	−0.1	0.387	0.387	0.158	0.154	0.178	0.161	0.175	0.167	0.072
	−0.05	0.209	0.213	0.096	0.091	0.100	0.095	0.106	0.101	0.058
	0	0.097	0.097	0.049	0.047	0.056	0.057	0.057	0.053	0.047
200	−0.9	1.000	1.000	1.000	1.000	1.000	1.000	1.000	1.000	0.768
	−0.7	1.000	1.000	1.000	1.000	1.000	1.000	1.000	1.000	0.672
	−0.5	1.000	1.000	1.000	1.000	1.000	1.000	1.000	1.000	0.459
	−0.3	1.000	0.999	0.820	0.815	0.951	0.927	0.933	0.931	0.213
	−0.2	0.959	0.958	0.496	0.487	0.666	0.608	0.631	0.626	0.135
	−0.1	0.561	0.562	0.199	0.198	0.256	0.224	0.249	0.245	0.088
	−0.05	0.265	0.267	0.108	0.105	0.124	0.112	0.122	0.119	0.067
	0	0.085	0.085	0.048	0.048	0.053	0.052	0.050	0.049	0.045

differ from the asymptotic ones, so inevitably these tests will appear worse than the others on the basis of size. This must also be borne in mind when comparing rejection frequencies under alternatives. Finite-sample critical values for S1 and S2, for a given U_t, marginal distribution, could of course be obtained by Monte Carlo, and because S1 and S2 involve ratios of quadratic forms, it also seems possible in principle, assuming Gaussianity, to derive them via Imhof's algorithm or to improve the normal or χ^2 approximation, as was done long ago with classical

Table 10.7. Rejection frequencies for lower-tailed 1% test and fractional X_t

n	θ	S1	S2	$\hat{\rho}$	$\hat{\tau}$	$\hat{\rho}_\tau$	$\hat{\tau}_\tau$	$\tilde{\rho}$	$\tilde{\tau}$	F
25	−0.9	0.793	0.788	0.981	0.976	0.724	0.600	0.677	0.639	0.055
	−0.7	0.607	0.595	0.799	0.775	0.464	0.340	0.425	0.385	0.055
	−0.5	0.364	0.356	0.391	0.368	0.216	0.153	0.197	0.166	0.036
	−0.3	0.156	0.152	0.120	0.108	0.078	0.053	0.069	0.057	0.022
	−0.2	0.092	0.089	0.057	0.050	0.041	0.032	0.038	0.031	0.018
	−0.1	0.049	0.047	0.025	0.022	0.022	0.018	0.017	0.013	0.013
	−0.05	0.035	0.034	0.016	0.014	0.016	0.015	0.011	0.008	0.010
	0	0.024	0.024	0.011	0.010	0.011	0.011	0.007	0.006	0.010
50	−0.9	0.994	0.994	1.000	1.000	0.999	0.998	0.992	0.991	0.096
	−0.7	0.957	0.955	0.995	0 995	0.961	0.929	0.933	0.924	0.080
	−0.5	0.761	0.756	0.745	0.733	0.657	0.576	0.619	0.598	0.050
	−0.3	0.348	0.348	0.239	0.230	0.202	0.167	0.193	0.178	0.029
	−0.2	0.164	0.166	0.098	0.095	0.090	0.076	0.081	0.074	0.022
	−0.1	0.059	0.057	0.037	0.035	0.035	0.029	0.028	0.025	0.018
	−0.05	0.033	0.033	0.019	0.020	0.020	0.018	0.017	0.015	0.013
	0	0.018	0.017	0.013	0.013	0.013	0.012	0.010	0.009	0.013
100	−0.9	1.000	1.000	1.000	1.000	1.000	1.000	1.000	1.000	0.137
	−0.7	1.000	1.000	1.000	1.000	1.000	1.000	1.000	1.000	0.110
	−0.5	0.990	0.990	0.955	0.953	0.978	0.963	0.961	0.959	0.075
	−0.3	0.749	0.746	0.426	0.415	0.499	0.432	0.482	0.465	0.037
	−0.2	0.384	0.381	0.172	0.164	0.200	0.166	0.194	0.184	0.020
	−0.1	0.101	0.101	0.049	0.046	0.059	0.051	0.057	0.054	0.013
	−0.05	0.041	0.041	0.024	0.024	0.029	0.027	0.028	0.027	0.010
	0	0.015	0.015	0.012	0.010	0.012	0.013	0.012	0.012	0.008
200	−0.9	1.000	1.000	1.000	1.000	1.000	1.000	1.000	1.000	0.201
	−0.7	1.000	1.000	1.000	1.000	1.000	1.000	1.000	1.000	0.157
	−0.5	1.000	1.000	0.999	0.998	1.000	1.000	0.999	0.999	0.099
	−0.3	0.984	0.985	0.630	0.623	0.825	0.768	0.773	0.771	0.046
	−0.2	0.776	0.777	0.272	0.266	0.405	0.335	0.360	0.355	0.029
	−0.1	0.220	0.221	0.063	0.065	0.092	0.073	0.082	0.079	0.019
	−0.05	0.067	0.067	0.027	0.027	0.033	0.027	0.031	0.030	0.011
	0	0.013	0.013	0.010	0.010	0.010	0.010	0.010	0.009	0.009

tests for serial correlation. But these options would involve significant additional computational effort, and the class of tests proposed here covers a much wider class of null hypotheses and alternatives than is feasible to address in a Monte Carlo study, whereas the test statistics of Section 10.5 are not ratios of quadratic forms. To provide some comparison with existing unit root tests, a focus on specific, simple, models for which AR based tests have been derived is necessary, and here it seems reasonable to assume that the user has access to the published finite-sample

Table 10.8. Rejection frequencies for lower-tailed 5% test and AR X_t

n	θ	S1	S2	$\hat{\rho}$	$\hat{\tau}$	$\hat{\rho}_\tau$	$\hat{\tau}_\tau$	$\tilde{\rho}$	$\tilde{\tau}$	F
25	−0.9	0.978	0.974	1.000	0.999	0.926	0.863	0.901	0.881	0.309
	−0.7	0.703	0.694	0.928	0.913	0.403	0.303	0.412	0.370	0.231
	−0.5	0.422	0.418	0.599	0.567	0.170	0.128	0.173	0.145	0.181
	−0.3	0.299	0.293	0.350	0.325	0.109	0.085	0.099	0.084	0.155
	−0.2	0.210	0.211	0.161	0.141	0.066	0.061	0.064	0.054	0.108
	−0.1	0.183	0.183	0.097	0.087	0.055	0.054	0.055	0.045	0.083
	−0.05	0.175	0.175	0.071	0.062	0.052	0.053	0.050	0.041	0.066
	0	0.175	0.173	0.056	0.049	0.051	0.052	0.049	0.041	0.047
50	−0.9	1.000	1.000	1.000	1.000	1.000	1.000	1.000	1.000	0.449
	−0.7	0.922	0.920	1.000	1.000	0.933	0.883	0.931	0.924	0.350
	−0.5	0.613	0.607	0.975	0.969	0.505	0.410	0.539	0.516	0.265
	−0.3	0.393	0.390	0.782	0.772	0.247	0.194	0.270	0.247	0.212
	−0.2	0.217	0.213	0.332	0.326	0.102	0.092	0.101	0.092	0.152
	−0.1	0.150	0.150	0.147	0.145	0.067	0.065	0.063	0.057	0.114
	−0.05	0.124	0.125	0.083	0.081	0.057	0.058	0.050	0.046	0.077
	0	0.117	0.117	0.053	0.051	0.054	0.057	0.048	0.043	0.051
100	−0.9	1.000	1.000	1.000	1.000	1.000	1.000	1.000	1.000	0.595
	−0.7	0.997	0.997	1.000	1.000	1.000	1.000	1.000	1.000	0.462
	−0.5	0.891	0.888	1.000	1.000	0.982	0.959	0.983	0.980	0.366
	−0.3	0.630	0.630	0.998	0.998	0.743	0.642	0.763	0.770	0.309
	−0.2	0.285	0.284	0.772	0.770	0.243	0.192	0.277	0.264	0.220
	−0.1	0.162	0.159	0.317	0.316	0.106	0.092	0.115	0.106	0.164
	−0.05	0.109	0.107	0.124	0.124	0.067	0.065	0.068	0.063	0.099
	0	0.097	0.097	0.049	0.047	0.056	0.057	0.057	0.053	0.047
200	−0.9	1.000	1.000	1.000	1.000	1.000	1.000	1.000	1.000	0.769
	−0.7	1.000	1.000	1.000	1.000	1.000	1.000	1.000	1.000	0.629
	−0.5	0.995	0.995	1.000	1.000	1.000	1.000	1.000	1.000	0.511
	−0.3	0.899	0.895	1.000	1.000	1.000	0.999	0.997	0.997	0.421
	−0.2	0.470	0.470	0.999	0.998	0.728	0.625	0.765	0.876	0.313
	−0.1	0.221	0.221	0.759	0.760	0.245	0.191	0.271	0.266	0.218
	−0.05	0.118	0.119	0.239	0.240	0.091	0.073	0.090	0.089	0.133
	0	0.085	0.085	0.048	0.048	0.053	0.052	0.050	0.049	0.045

tables for these. On the other hand, this chapter has stressed large-sample theory and suggested only large-sample approximate critical values for the tests, and it also seems more realistic to use these rather than attempt a size correction of S1 and S2, especially bearing in mind the many other situations covered by the tests of Sections 10.3–10.6.

Because each of the nine tests is motivated by either fractionally differenced or AR alternatives, the performances of all tests is evaluated against

Table 10.9. Rejection frequencies for lower-tailed 1% test and AR X_t

n	θ	S1	S2	$\hat{\rho}$	$\hat{\tau}$	$\hat{\rho}_\tau$	$\hat{\tau}_\tau$	$\tilde{\rho}$	$\tilde{\tau}$	F
25	−0.9	0.756	0.746	0.989	0.984	0.693	0.563	0.659	0.611	0.056
	−0.7	0.224	0.215	0.601	0.568	0.143	0.087	0.137	0.116	0.045
	−0.5	0.087	0.085	0.207	0.186	0.044	0.027	0.039	0.031	0.040
	−0.3	0.051	0.048	0.090	0.078	0.022	0.015	0.020	0.016	0.032
	−0.2	0.030	0.031	0.029	0.027	0.013	0.011	0.009	0.007	0.022
	−0.1	0.026	0.026	0.018	0.015	0.012	0.011	0.007	0.006	0.016
	−0.05	0.024	0.024	0.014	0.011	0.010	0.011	0.007	0.005	0.013
	0	0.024	0.024	0.011	0.010	0.011	0.011	0.007	0.006	0.010
50	−0.9	0.991	0.991	1.000	1.000	0.999	0.998	0.993	0.992	0.099
	−0.7	0.566	0.562	0.994	0.992	0.684	0.577	0.679	0.658	0.075
	−0.5	0.192	0.192	0.716	0.710	0.188	0.133	0.208	0.191	0.055
	−0.3	0.083	0.084	0.334	0.332	0.071	0.050	0.076	0.066	0.042
	−0.2	0.033	0.034	0.087	0.083	0.023	0.017	0.023	0.021	0.029
	−0.1	0.022	0.021	0.034	0.034	0.014	0.011	0.013	0.011	0.025
	−0.05	0.018	0.017	0.019	0.018	0.011	0.010	0.011	0.010	0.015
	0	0.018	0.017	0.013	0.013	0.013	0.012	0.010	0.009	0.013
100	−0.9	1.000	1.000	1.000	1.000	1.000	1.000	1.000	1.000	0.135
	−0.7	0.955	0.952	1.000	1.000	1.000	1.000	0.997	0.997	0.093
	−0.5	0.514	0.511	0.999	0.999	0.839	0.746	0.851	0.840	0.074
	−0.3	0.220	0.218	0.925	0.923	0.371	0.285	0.419	0.401	0.066
	−0.2	0.061	0.059	0.319	0.314	0.070	0.050	0.080	0.075	0.046
	−0.1	0.027	0.027	0.084	0.082	0.026	0.019	0.027	0.025	0.035
	−0.05	0.018	0.017	0.027	0.026	0.013	0.013	0.016	0.015	0.020
	0	0.015	0.015	0.012	0.010	0.012	0.013	0.012	0.012	0.008
200	−0.9	1.000	1.000	1.000	1.000	1.000	1.000	1.000	1.000	0.199
	−0.7	1.000	1.000	1.000	1.000	1.000	1.000	1.000	1.000	0.140
	−0.5	0.925	0.924	1.000	1.000	1.000	1.000	1.000	1.000	0.109
	−0.3	0.587	0.587	1.000	1.000	0.986	0.963	0.970	0.968	0.087
	−0.2	0.149	0.151	0.924	0.922	0.361	0.261	0.386	0.379	0.072
	−0.1	0.049	0.049	0.323	0.323	0.068	0.045	0.075	0.071	0.045
	−0.05	0.021	0.021	0.059	0.057	0.020	0.016	0.019	0.018	0.027
	0	0.013	0.013	0.010	0.010	0.010	0.010	0.010	0.009	0.009

data generated by both types of model. For both the fractional alternative $\varphi(z; \theta) = (1 - z)^{1+\theta}$ and the AR alternative $\varphi(z; \theta) = 1 - (1 + \theta)z$, the values $\theta = 0, \pm 0.05, \pm 0.1, \pm 0.2, \pm 0.3, \pm 0.5, \pm 0.7$, and ± 0.9 are used—thus covering the null unit root model as well as stationary, less nonstationary, and more nonstationary fractional alternatives and stationary and explosive AR alternatives. Diebold and Rudebusch (1991) reported a Monte Carlo study of $\hat{\rho}$ and $\hat{\tau}$ only, for both these choices of φ. The present values of θ are somewhat different from theirs, and the focus here is on one-sided tests rather than their two-sided versions of the $\hat{\rho}$ and $\hat{\tau}$ tests, while somewhat different, and on average smaller, values of n, namely $n = 25, 50, 100$, and 200 in place of their 50, 100, and 250, are used. Like Diebold and Rudebusch (1991) we generate Gaussian series, with 5000 replications of each case. The finite-sample critical values of Fuller (1976) and Schmidt and Phillips (1992) are all apparently based on Gaussian series.

Tables 10.2 and 10.3 contain Monte Carlo rejection frequencies for one-sided tests against fractional alternatives $\theta > 0$, with nominal sizes 5 and 1 per cent. Tables 10.4 and 10.5 correspond, with AR alternatives. Tables 10.6 and 10.7 cover fractional, and Tables 10.8 and 10.9 AR, alternatives $\theta < 0$. Tables 10.2–10.5 omit the F test, because this test covers only alternatives $\theta < 0$.

The sizes of the new tests S1 and S2 against $\theta > 0$ and $\theta < 0$ are too small and too large but tend to improve as n increases. The sizes against $\theta > 0$ are poor for the 5 per cent test but satisfactory for the 1 per cent test when $n \geq 100$. There is effectively no difference in size between S1 and S2, though one might expect the large-sample approximation for S1 to be slightly better than that for S2. The discrepancies in sizes for the other tests from the nominal ones can be attributed to a difference in random number generators from those used in the published tables.

Looking again at Table 10.2, the small sizes are associated with some inferior rejection frequencies for S1 and S2 relative to the other tests, despite the fractional data used. This effect is most marked when $n = 25$, though even here (as for the larger n) the rejection frequencies ultimately, for θ large enough, exceed those for the other tests. When $n = 200$ S1 and S2 are best for all $\theta > 0$, and when $n = 100$, they are best except for some cases when $\theta = 0.05$. The rejection frequencies for the other seven tests all increase with both θ and n, but whereas $\tilde{\rho}$ and $\tilde{\tau}$ are competitive with S1 and S2 when both θ and n are large, they are not for smaller θ even when n is large, while $\hat{\tau}_\tau$ performs poorly. The rejection frequencies for S1 and S2 are very close. Table 10.3 presents a similar picture, except that $\hat{\rho}$ now performs worst. S1 and S2 are obviously not efficient with respect to AR alternatives, and as expected they are outperformed in Tables 10.4 and 10.5. This is especially evident for small n and θ. But for $n = 25$, S1 and S2 are almost as good as the other tests for $\theta = 0.5$, and for $n = 200$ they are not much worse, despite their small size, than $\tilde{\rho}$ and $\tilde{\tau}$, when $\theta = 0.05$. Notice that $\hat{\rho}_\tau$ and $\hat{\tau}_\tau$ now tend to perform better than $\tilde{\rho}$ and $\tilde{\tau}$, though less well than $\hat{\rho}$ and $\hat{\tau}$. Overall, bearing in mind that the rejection frequencies for the AR-based tests in Tables 10.2 and 10.3 are frequently less than 1 even when $\theta = 0.9$ (sometimes much less), whereas the rejection frequencies for S1 and S2 in Tables 10.4 and 10.5 are 1 or close to 1 for large θ, S1 and S2 seem to do better against AR alternatives than the other seven tests do against fractional ones.

In assessing the performance of the one-sided tests against $\theta < 0$, it must be recalled that the sizes of S1 and S2 are too large. It is thus all the more striking that for the smallest n and most negative θ, S1 and S2 are beaten on various occasions by other tests, especially $\hat{\rho}$ and $\hat{\tau}$, even for the fractional data. Note in particular Table 10.7 where $\hat{\rho}$ and $\hat{\tau}$ come out decisively best when $n = 25$. Overall, however, the efficiency of S1 and S2 appears to assert itself in Tables 10.6 and 10.7, even taking into account their large sizes. The AR data provide something of an opposite picture in Tables 10.8 and 10.9, because whereas $\hat{\rho}$ and $\hat{\tau}$ again come out best, S1 and S2 frequently have larger rejection frequencies than the other tests for the smaller n and all values of θ, though this may be attributable to their large sizes. In Tables 10.6–10.9 S1 and S2 easily beat the F test, though whereas the latter is very poor for very negative θ, it frequently slightly beats a number of the other AR-based tests (not $\hat{\rho}$ and $\hat{\tau}$) for the less negative θ in Tables 10.8 and 10.9.

The numerical work reported in Section 10.8 and so far in this section, has illustrated the generality of our set-up by covering one sided tests for a seasonal unit root and a fractional one, as well as for the usual unit root hypothesis, and by allowing for nuisance parameters in a regression component and in autocorrelated U_t via Bloomfield's spectral model. In the remaining Monte Carlo work, two-sided tests, corrections for AR autocorrelation in U_t, and departures from Gaussianity in U_t are also illustrated.

Table 10.10 concerns two-sided tests based on S1 for the same (fractional Gaussian) process used in Tables 10.2, 10.3, 10.6 and 10.7, but for $\theta = 0, \pm0.05, \pm0.1, \pm0.2, \pm0.3$, sample sizes $n = 100$ and 200, and 10 and 0.1 per cent as well as the 5 and 1 per cent ones featured in the earlier tables. The sizes now seem substantially closer to the nominal ones. The 10 and 5 per cent tests are clearly biased, especially for $n = 100$. Their rejection frequencies are higher for negative θ than for positive θ, whereas the 1 and 0.1 per cent tests display reverse behaviour. Table 10.11 replaces Gaussianity by a t_3 distribution for the white noise U_t. This distribution is interesting, because it only just satisfies the second-moment condition used in Sections 10.4 and 10.6, its third moments not existing. The results are competitive with the Gaussian ones. The sizes are mostly closer to the nominal ones, and rejection frequencies tend to be lower when $\theta < 0$ but larger when $\theta > 0$.

Finally, we consider some alternative versions of the tests of Section 10.5 to those used in Section 10.7. Tables 10.12 and 10.13 use two-sided tests using

Table 10.10. Rejection frequencies for two-sided S1 test with Gaussian U_t

Size	10%		5%		1%		0.1%	
n	100	200	100	200	100	200	100	200
−0.3	0.960	1.000	0.904	0.997	0.581	0.963	0.116	0.710
−0.2	0.772	0.959	0.609	0.905	0.236	0.640	0.019	0.200
−0.1	0.387	0.561	0.231	0.395	0.048	0.130	0.002	0.013
−0.05	0.215	0.268	0.112	0.154	0.018	0.033	0.001	0.001
0	0.127	0.115	0.063	0.057	0.011	0.012	0.001	0.001
0.05	0.140	0.186	0.084	0.125	0.028	0.050	0.009	0.016
0.1	0.249	0.450	0.117	0.356	0.096	0.219	0.038	0.108
0.2	0.632	0.910	0.555	0.872	0.407	0.783	0.259	0.650
0.3	0.897	0.995	0.859	0.992	0.767	0.983	0.638	0.959

Table 10.11. Rejection frequencies for two-sided S1 test with
t_3-distributed U_t

Size	10%		5%		1%		0.1%	
n	100	200	100	200	100	200	100	200
−0.3	0.960	0.999	0.901	0.998	0.584	0.968	0.097	0.716
−0.2	0.776	0.964	0.606	0.909	0.218	0.643	0.012	0.173
−0.1	0.370	0.555	0.217	0.382	0.037	0.112	0.001	0.009
−0.05	0.202	0.245	0.100	0.133	0.012	0.024	0.000	0.001
0	0.116	0.099	0.053	0.047	0.009	0.009	0.001	0.002
0.05	0.128	0.183	0.075	0.120	0.029	0.050	0.010	0.017
0.1	0.282	0.461	0.180	0.366	0.096	0.219	0.039	0.106
0.2	0.650	0.924	0.577	0.892	0.417	0.803	0.268	0.254
0.3	0.907	0.997	0.874	0.994	0.791	0.985	0.661	0.961

Table 10.12. Rejection frequencies for AR(2)-corrected test for Gaussian
innovations and $n = 200$

Size	10%		5%		1%		0.1%	
τ	(0, 0)	(1, −0.5)	(0, 0)	(1, −0.5)	(0, 0)	(1, −0.5)	(0, 0)	(1, −0.5)
−0.3	0.807	0.928	0.651	0.846	0.281	0.564	0.030	0.173
−0.2	0.563	0.745	0.391	0.604	0.116	0.303	0.007	0.054
−0.1	0.308	0.451	0.178	0.320	0.037	0.099	0.001	0.011
−0.05	0.219	0.315	0.117	0.191	0.018	0.049	0.000	0.004
0	0.167	0.204	0.086	0.118	0.014	0.025	0.001	0.002
0.05	0.148	0.152	0.076	0.084	0.019	0.019	0.004	0.004
0.1	0.156	0.160	0.092	0.098	0.032	0.033	0.011	0.009
0.2	0.223	0.343	0.156	0.262	0.071	0.144	0.035	0.063
0.3	0.263	0.575	0.190	0.495	0.097	0.343	0.041	0.199

Table 10.13. Rejection frequencies for AR(2)-corrected test for t_3 distributed
innovations and $n = 200$

Size	10%		5%		1%		0.1%	
τ	(0, 0)	(1, −0.5)	(0, 0)	(1, −0.5)	(0, 0)	(1, −0.5)	(0, 0)	(1, −0.5)
−0.3	0.818	0.929	0.664	0.849	0.278	0.552	0.029	0.157
−0.2	0.559	0.746	0.383	0.601	0.103	0.273	0.008	0.044
−0.1	0.299	0.424	0.163	0.285	0.032	0.086	0.001	0.010
−0.05	0.200	0.282	0.106	0.176	0.019	0.041	0.000	0.004
0	0.148	0.185	0.077	0.099	0.011	0.073	0.002	0.002
0.05	0.134	0.134	0.067	0.070	0.018	0.017	0.005	0.003
0.1	0.147	0.155	0.087	0.096	0.029	0.032	0.010	0.008
0.2	0.215	0.362	0.153	0.277	0.077	0.151	0.032	0.065
0.3	0.262	0.610	0.187	0.521	0.098	0.367	0.092	0.209

$\varphi(z; \theta) = (1 - z)^{1+\theta}$ and $g(\lambda; \tau) = |1 - \tau_1 e^{i\lambda} - \tau_2 e^{2i\lambda}|^{-2}$; that is, an AR(2) correction to U_t. Both white noise (AR(0)) and AR(2) U_t were generated, where in the latter case the process was $U_t - U_{t-1} + 0.5U_{t-2} = \varepsilon_t$ (implying complex roots $1/2(1 \pm i)$), where the white noise ε_t are distributed as $N(0, 1)$ in Table 10.12 and t_3 in Table 10.13. Comparing Tables 10.10 and 10.12, there is a very noticeable loss of power caused by unnecessarily introducing the two nuisance parameters τ_1 and τ_2, especially for positive fractional alternatives θ; this is particularly worrisome because the sizes tend to be significantly larger. When used with the AR(2) data, the AR(2)-based test seems to be better, though the sizes are even worse and there is still a significant cost relative to the AR(0)-based test of Table 10.10. Although there is no marked difference between Tables 10.12 and 10.13, the sizes are again better when t_3 distributed ε_t (implying similar long-tailedness in U_t) are used.

10.9 Final Comments

This chapter has presented tests for a wide variety of possible hypotheses for time series data, with emphasis on nonstationary null hypotheses, in particular unit root ones. There are already numerous unit root tests, most of which are motivated by AR alternatives, which entail an immediate, drastic shift in the long-run behaviour of the series, to stationarity or explosiveness; there is a notable 'discontinuity' in the asymptotic theory, and classical large-sample testing theory does not apply. The alternatives against which our tests are directed are of fractional type, where small shifts from a unit root do not disturb the nonstationary but nonexplosive character of the series, and classical testing theory applies. In particular, we are able to construct asymptotically locally most powerful tests based on chi-squared critical values. The choice of what sort of alternatives to use can depend on the objectives of the study. The AR-based tests have the attractive potential to suggest on the one hand stationarity (indeed, exponentially decaying autocorrelations) and on the other hand explosivity when the tests reject. But there are many other kinds of stationarity and nonstationarity, and rejection of the null must be interpreted with care. Our alternatives cover different forms of stationarity and nonstationarity. These may or may not be of particular interest in a given application, but in any event our approach provides a flexible diagnostic tool for departures from unit root and many other forms of behaviour. The emphasis in the literature on unit roots may have rather obscured the fact that this is an extremely specialized form of nonstationarity. Our approach allows the testing of various other forms of nonstationarity (e.g. testing for a fractional degree of integration of 1/2), which can help provide an answer to the basic question of whether a series is stationary or nonstationary.

The frequency domain formulation of our tests is not significant; there always exists an asymptotically equivalent time domain formulation. Likewise, the score principle can be replaced by the Wald principle or the likelihood-ratio principle, with no difference in asymptotic properties. Of course, in both cases finite-sample properties will be affected; these remain to be studied. Extensions of our tests to multivariate time series will be of practical interest, along with their implications for studying cointegration (including fractional and seasonal cointegration).

The Monte Carlo simulations that we have reported show that a large-sample approximate distribution theory can leave a great deal to be desired, while also displaying encouraging features. Though the sample sizes used are representative of those encountered in macroeconomics, financial time series (where indeed the forms of long memory stressed by our null and alternative hypotheses have been of significant recent interest) can be very much larger, and our asymptotic theory will be more relevant. On the other hand, it is possible, for the tests in Section 10.3, to construct a finite-sample theory under Gaussianity or obtain Monte Carlo finite-sample critical values, but we have instead justified large-sample tests under mild conditions and for a broad range of situations. This is a prerequisite for a next step, justifying bootstrap versions of our tests.

Appendix 10.1 Derivation of Score Statistic $\bar{R}$

The negative of the log-likelihood under (10.2), (10.4), (10.7), Gaussianity of the white noise U_t, and $Z_t = 0$ for $t \leq 0$, is $L(\theta, b, s^2) = 1/2n \log(2\pi s^2) + (2s^2)^{-1} \sum_{t=1}^{n} \{\varphi(B; \theta) X_t(b)\}^2$, for any admissible b and s^2, where $X_t(b) = Y_t - b' Z_t$. Now

$$\frac{\partial L(\theta, b, s^2)}{\partial \theta} = s^{-2} \sum_{t=1}^{n} \left\{ \frac{\partial \varphi(B; \theta)}{\partial \theta} X_t(b) \right\} \{\varphi(B; \theta) X_t(b)\},$$

so that $(\partial/\partial\theta) L(0, b, s^2) = s^{-2} \sum_{t=1}^{n} \{\zeta(B) U_t(b)\} U_t(b)$, where $\zeta(z) = (\partial/\partial\theta) \log \varphi(z; 0)$ and $U_t(b) = \varphi(B) X_t(b)$. It is easily seen that

$$\frac{\partial L(0, b, s^2)}{\partial b} = -\frac{1}{s^2} \sum_{t=1}^{n} \{W_t U_t(b)\},$$

$$\frac{\partial L(0, b, s^2)}{\partial s^2} = -\frac{1}{2s^4} \sum_{t=1}^{n} \{U_t(b)^2 - s^2\},$$

which vanish for $b = \tilde{\beta}$ and $s^2 = \tilde{\sigma}^2$. Thus $(\partial/\partial\theta) L(0, \tilde{\beta}, \tilde{\sigma}^2) = -n\tilde{a}/\tilde{\sigma}^2$. It also follows that the expectation in (10.8) is

$$n \begin{bmatrix} \bar{A} & 0 & 0 \\ 0 & D\sigma^{-2} & 0 \\ 0 & 0 & \frac{1}{2}\sigma^{-4} \end{bmatrix}, \tag{A.1}$$

where $D = \sum_{t=1}^{n} W_t W_t'$, so that under positive definiteness of D (explicitly assumed in Theorem 1), we deduce $\bar{R}$.

Appendix 10.2 Proof of Theorem 1

We first define the classes F, G, and H.

Definition 1 F is the class of sequences $\{V_t, t = 0, \pm 1, \ldots\}$ of stationary random variables V_t, satisfying $E(V_t|B_{t-1}) = 0$, $E(V_t^2|B_{t-1}) = \sigma^2$ almost surely, where $0 < \sigma^2 < \infty$ and B_t is the σ field of events generated by $V_s, s \leq t$.

F includes the class of independent and identically distributed random variables with zero mean and finite variance.

Definition 2 G is the class of $k \times 1$ vector sequences $\{Z_t, t = 0, \pm 1, \ldots\}$ such that $Z_t = 0, t \leq 0$ and D defined in Appendix 10.1 is positive definite for sufficiently large n.

G imposes no rate of increase on D; different elements can increase at different rates, and indeed D need not tend to infinity as $n \to \infty$ (so that $\tilde{\beta}$ need not be consistent). If D is positive definite for $n = n_0$, then it is positive definite for all $n > n_0$. A leading case in the unit root literature (e.g. see Schmidt and Phillips 1992) is $Z_t = (1, t)'$ for $t > 0$, mentioned in Section 10.1. Here $W_1 = (1, 1)'$, $W_t = (0, 1)'$ for $t \geq 2$, and D is positive definite for $n \geq 2$ because it has elements one except the bottom right one, which is n.

Definition 3 H is the class of functions $\varphi(z; \theta)$ such that $\varphi(0, \theta) = 1$ for all θ, and $\psi(\lambda)$ given in (10.6) has finitely many poles $\rho_l, l = 1, \ldots, r$, on $(-\pi, \pi]$ such that $\|\psi(\lambda)\|$ is monotonically increasing as $\lambda \to \rho_{l-}$ and as $\lambda \to \rho_{l+}$, for $l = 1, \ldots, r$, and there exist disjoint intervals $S_l, l = 1, \ldots, r$ such that $\cup_{l=1}^r S_l \in (-\pi, \pi]$, $\rho_l \in S_l$, $\rho_l \notin S_k$ for $l \neq k$, and

$$\sup_{\lambda \in S_k - (\rho_k - \delta_{\rho_k} + \delta)} \left\{ |\lambda - \rho_k| \left\| \psi(\lambda) - \psi\left(\lambda \pm \frac{1}{2}\delta\right) \right\| \right\} = O(\delta^\eta), \quad \text{as } \delta \to 0,$$

(A.2)

for $k = 1, \ldots, r$ and some $\eta > \frac{1}{2}$, where $\|\cdot\|$ denotes Euclidean norm.

The monotonicity property is innocuous. The $\psi(\lambda)$ implied by H include vector functions with elements of form (10.13), these having derivatives that behave like λ^{-1} around 0, $(\pi - \lambda)^{-1}$ around π, and $|\lambda - \omega_j|^{-1}$ around ω_j, so that $\eta = 1$, whereas for all $\varphi \in H$, ψ satisfies a Lipschitz condition of degree η outside neighbourhoods of the ρ_l.

We first establish three technical lemmas pertaining to class H.

Lemma 1 For all $\varphi \in H$ such that the right-side inequality of (10.5) is satisfied, $\lim_{n \to \infty} n^{-(1/2)} \max_{\lambda_j \in M} \|\psi(\lambda_j)\| = 0$.

Proof For n sufficiently large, monotonicity, (A.2), and (10.5) give

$$\max_{\lambda_j \in M} \|\psi(\lambda_j)\| \leq \sum_{l=1}^{r} \{\|\psi(\rho_l - \lambda_1)\| + \|\psi(\rho_l + \lambda_1)\|\}$$

$$\leq \lambda_1^{-1} \sum_{l=1}^{r} \int_{\rho_l - \lambda_1}^{\rho_l + \lambda_1} \|\psi(\lambda)\| \, d\lambda$$

$$\leq (2/\lambda_1)^{1/2} \sum_{l=1}^{r} \left\{ \int_{\rho_l - \lambda_1}^{\rho_l + \lambda_1} \|\psi(\lambda)\|^2 \, d\lambda \right\}^{1/2}$$

$$= o(n^{1/2}). \quad \blacksquare$$

Lemmas 2 and 3 are results on approximating integrals by sums that are doubtless known but that we have been unable to locate in the literature. Define $\hat{\psi}_k = n^{-1} \sum_j' \psi(\lambda_j) \cos k\lambda_j$.

Lemma 2 For all $\varphi \in H$ such that the right-side inequality of (10.5) is satisfied, $\lim_{n \to \infty} n^{1/2} \max_{0 \leq k < n} \|\hat{\psi}_k - \psi_k\| = 0$.

Proof We have

$$\hat{\psi}_k - \psi_k = 2\frac{1}{\pi} \sum_j' \int_{\lambda_j - (1/2)\lambda_1}^{\lambda_j + (1/2)\lambda_1} \{\psi(\lambda_j) \cos k\lambda_j - \psi(\lambda) \cos k\lambda\} \, d\lambda$$

$$- 2\frac{1}{\pi} \int {}'' \psi(\lambda) \cos k\lambda \, d\lambda,$$

where the double-primed range of integration is $O(n^{-1})$ because M excludes $O(1)$ of the λ_j. Now

$$\psi(\lambda_j) \cos k\lambda_j - \psi(\lambda) \cos k\lambda$$
$$= \psi(\lambda_j)(\cos k\lambda_j - \cos k\lambda) + \{\psi(\lambda_j) - \psi(\lambda)\} \cos k\lambda.$$

We have

$$\max_k \left\| \sum_j' \int_{\lambda_j - (1/2)\lambda_1}^{\lambda_j + (1/2)\lambda_1} \{\psi(\lambda_j) - \psi(\lambda)\} \cos k\lambda \, d\lambda \right\|$$

$$\leq \sum_{l=1}^{r} \sum_j^{(l)} \int_{\lambda_j - (1/2)\lambda_1}^{\lambda_j + (1/2)\lambda_1} \|\psi(\lambda_j) - \psi(\lambda)\| \, d\lambda, \quad \text{(A.3)}$$

where $\sum_j^{(l)}$ is a sum over j such that $\lambda_j \in S_l \cap M$; that is, $\lambda_j \in S_l - (\rho_l - \lambda_1, \rho_l + \lambda_1)$. Thus, as $n \to \infty$, (A.3) is

$$O\left(\sum_{l=1}^{r}\sum_{j}^{(l)}\int_{\lambda_j-(1/2)\lambda_1}^{\lambda_j+(1/2)\lambda_1}\frac{n^{-\alpha}}{|\lambda_j-\rho_l|}\,d\lambda\right)$$

$$= O\left(n^{-\alpha}\sum_{j=1}^{n}j^{-1}\right) = O(n^{-\alpha}\log n) = o(n^{-1/2}). \qquad (A.4)$$

For integers u and v, Abel summation by parts gives

$$\sum_{j=u+1}^{n}\psi(\lambda_j)\int_{\lambda_j-(1/2)\lambda_1}^{\lambda_j-(1/2)\lambda_1}(\cos k\lambda_j - \cos k\lambda)\,d\lambda$$

$$= \sum_{j=u+1}^{n-1}\{\psi(\lambda_j)-\psi(\lambda_{j+1})\}D_k(u,j)+\psi(\lambda_v)D_k(u,v),$$

where, for $j > u$,

$$D_k(u,j) = \frac{2\pi}{n}\sum_{i=u+1}^{j}\cos k\lambda_i - \int_{\lambda_u-(1/2)\lambda_1}^{\lambda_j+(1/2)\lambda_1}\cos k\lambda\,d\lambda$$

$$= \frac{2\pi}{n}\left\{\frac{\sin\left(j+\frac{1}{2}\right)\lambda_k}{2\sin\frac{1}{2}\lambda_k}-\frac{\sin\left(u+\frac{1}{2}\right)\lambda_k}{2\sin\frac{1}{2}\lambda_k}\right\}$$

$$\quad -\left\{\frac{\sin\left(j+\frac{1}{2}\right)\lambda_k}{k}-\frac{\sin\left(u+\frac{1}{2}\right)\lambda_k}{k}\right\},$$

which is $O(kn^{-2})$, uniformly, because $|\sin x - x| = O(|x|^3)$. Making suitable choices of u and v,

$$\sum_j{}'\int_{\lambda_j-(1/2)\lambda_1}^{\lambda_j+(1/2)\lambda_1}\psi(\lambda_j)(\cos k\lambda_j - \cos k\lambda)\,d\lambda$$

$$= O\left(\frac{k}{n^2}\left\{\sum_j{}'\|\psi(\lambda_j)-\psi(\lambda_{j+1})\|+\max_{j\in M}\|\psi(\lambda_j)\|\right\}\right)$$

$$= O\left(\frac{1}{n}\sum_{l=1}^{r}\sum_{j}^{(l)}\frac{n^{-\alpha}}{|\lambda_j-\rho_l|}\right)+o(n^{-(1/2)})-o(n^{-(1/2)})$$

uniformly, using properties of H and Lemma 1, and proceeding as in (A.4). Finally,
$\| \int^v \psi(\lambda) \cos k\lambda \, d\lambda \| = O(n^{-(1/2)} \{ \int^v \| \psi(\lambda) \|^2 \, d\lambda \}^{1/2}) = (n^{-(1/2)})$. ∎

Lemma 3 For all $\psi \in H$ such that the right-side inequality of (10.5) is satisfied.
$n^{-1} \sum_j' \psi(\lambda_j) \psi(\lambda_j)' \to \Psi$, as $n \to \infty$.

Proof Write $n^{-1} \sum_j' \psi(\lambda_j) \psi(\lambda_j)' - \Psi$ as

$$\frac{1}{2\pi} \sum_j' \int_{\lambda_j - (1/2)\lambda_1}^{\lambda_j + (1/2)\lambda_1} \{ \psi(\lambda_j) \psi(\lambda_j)' - \psi(\lambda) \psi(\lambda)' \} \, d\lambda = \frac{1}{2} \int^v \psi(\lambda) \psi(\lambda)' \, d\lambda.$$

The first integrand can be written as

$$\{ \psi(\lambda_j) - \psi(\lambda) \} \{ \psi(\lambda_j) - \psi(\lambda) \}'$$
$$+ \{ \psi(\lambda_j) - \psi(\lambda) \} \psi(\lambda)' + \psi(\lambda) \{ \psi(\lambda_j) - \psi(\lambda) \}' \qquad (A.5)$$

As $n \to \infty$,

$$\sum_j' \int_{\lambda_j - (1/2)\lambda_1}^{\lambda_j + (1/2)\lambda_1} \psi(\lambda) \{ \psi(\lambda_j) - \psi(\lambda) \}' \, d\lambda$$

$$= O\left(\left\{ \int_{-\pi}^{\pi} \| \psi(\lambda) \|^2 \, d\lambda \sum_j' \int_{\lambda_j - (1/2)\lambda_1}^{\lambda_j + (1/2)\lambda_1} \| \psi(\lambda_j) - \psi(\lambda) \|^2 \, d\lambda \right\}^{1/2} \right)$$

$$= O\left(\left\{ \sum_{l=1}^{r} \sum_j^{(l)} \int_{\lambda_j - (1/2)\lambda_1}^{\lambda_j + (1/2)\lambda_1} \frac{n^{-2\alpha}}{(\rho_l - \lambda_j)^2} \, d\lambda \right\}^{1/2} \right)$$

$$= O\left(\left\{ n^{1-2\alpha} \sum_{j=1}^{n} j^{-2} \right\}^{1/2} \right) = O(n^{(1/2)-\alpha}) \to 0.$$

The contribution from the middle term in (A.5) is clearly identical, whereas that for the first term is $O(n^{1-2\alpha})$ by almost the same proof. Finally, $\| \int^v \psi(\lambda) \psi(\lambda)' \, d\lambda \| \leqslant \int^v \| \psi(\lambda) \|^2 d\lambda \to 0$. ∎

Proof of Theorem 1 We have $\tilde{a} = (\tilde{a} - a) + (a - a^*) + a^*$, where $a = -(2\pi/n) \sum_j' \psi(\lambda_j) I_U(\lambda_j)$ and $a^* = -\sum_{k=1}^{n-1} \psi_k C_U(k)$, where I_U and C_U are defined like $I_{\tilde{U}}$ and $C_{\tilde{U}}$ but with $\tilde{U}_t$ replaced by U_t. Now $a - \tilde{a} = -(2\pi/n) \sum_j' \psi(\lambda_j) [(\tilde{\beta} - \beta)' I_W(\lambda_j)(\tilde{\beta} - \beta) - 2 \, \text{Re}\{(\tilde{\beta} - \beta)' I_{WU}(\lambda_j)\}]$, where $I_W(\lambda) = (2\pi n)^{-1} \sum \sum_{s,t=1}^{n} W_s W_t' \bar{e}^{i(s-t)\lambda}$ and $I_{WU}(\lambda) = (2\pi n)^{-1} \sum \sum_{s,t=1}^{n} W_s U_t e^{i(s-t)\lambda}$. It is easily seen that $E \| D^{1/2} (\tilde{\beta} - \beta) \|^2 = O(1)$

as $n \to \infty$. Then

$$
E \left\| n^{-(1/2)} \sum_j{}' \psi(\lambda_j)(\tilde{\beta} - \beta)' I_W(\lambda_j)(\tilde{\beta} - \beta) \right\|
$$

$$
\leqslant n^{-(1/2)} \sum_j{}' \|\psi(\lambda_j)\| \operatorname{tr}\{D^{-(1/2)} I_W(\lambda_j) D^{-(1/2)} \operatorname{Cov}(D^{(1/2)}(\tilde{\beta} - \beta))\}
$$

$$
= O\left(n^{-(1/2)} \max_{\lambda_j \in M} \|\psi(\lambda_j)\| \sum_{j=1}^{n} \operatorname{tr}\{D^{-(1/2)} I_W(\lambda_j) D^{-(1/2)}\} \right) \to 0,
$$

applying Lemma 1 and $\sum_{j=1}^{n} I_W(\lambda_j) = (2\pi)^{-1} D$. Using uncorrelatedness of the U_t.

$$
E \left\| n^{-(1/2)} \sum_j{}' \psi(\lambda_j) D^{-(1/2)} I_{WU}(\lambda_j) \right\|^2
$$

$$
= (\sigma^2/2\pi n) \sum_j{}' \|\psi(\lambda_j)\|^2 \operatorname{Tr}\{D^{-(1/2)} I_W(\lambda_j) D^{-(1/2)}\} \to 0
$$

for the same reasons. Thus $a - \tilde{a} = O_p(n^{-(1/2)})$. From the definition of the periodogram $a = -1/2 \hat{\psi}_0 C_U(0) - \sum_{k=1}^{n-1} \hat{\psi}_k C_U(k)$. Because $\psi_0 = 0$, $E\|\hat{\psi}_0 C_U(0)\| \leqslant \|\hat{\psi}_0 - \psi_0\| \sigma^2 = o(n^{-(1/2)})$ by Lemma 2. On the other hand, $E\|\sum_{k=1}^{n-1}(\hat{\psi}_k - \psi_k) C_U(k)\|^2 = O(n^{-1} \sum_{k=1}^{n-1} \|\hat{\psi}_k - \psi_k\|^2) = o(n^{-1})$, because $E C_U(k)^2 = (\sigma/n)^2(n - k)$, $E\{C_U(k) C_U(l)\} = 0, 0 < k < l$, and Lemma 2 applies. Thus $a - a^* = o_p(n^{-(1/2)})$. For $m < n - 1$, we deduce that $E\|n^{1/2} \sum_{l=m+1}^{n-1} \psi_l C_U(l)\|^2 = O(\sum_{l-m+1}^{\infty} \|\psi_l\|^2)$, which can be made arbitrarily small by choosing m large enough in view of Parseval's equality and (10.5). Straightforward application of a martingale difference central limit theorem (e.g. Brown 1971) shows that for fixed m, $n^{1/2} \sum_{i=1}^{m} \psi_l C_U(l) \to_d N(0, \sigma^4 \sum_{l=1}^{m} \psi_l \psi_l')$, so $a^* \to_d N(0, \sigma^4 \Psi)$ by Bernstein's lemma. The proof that $\tilde{\sigma}^2 \to_p \sigma^2$ is implied by $C_U(0) \to_p \sigma^2$, which follows because the $U_t^2 - \sigma^2$ are stationary martingale differences, and by $C_U(0) - C_U(0) \to_p 0$, which follows from the argument that $a - \tilde{a} = O_p(n^{-(1/2)})$. Finally, $\tilde{A} \to A$ for the first and last terms in (10.14) from (10.15) and Lemma 3. ∎

Appendix 10.3 Proof of Theorem 2

We first introduce the following definition.

Definition 4 J is a subclass of H such that for all $\varphi \in J$, $\zeta(z; \theta) = \{\varphi(z)/\varphi(z; \theta)\}\{(\partial/\partial\theta) \log \varphi(z; \theta)\}$ is continuous in θ at $\theta = 0$ for almost all z

such that $|z| = 1$, and for a neighbourhood N of $\theta = 0$, $\int_{-\pi}^{\pi} \sup_{\in N} \|\zeta(e^{i\lambda}; \theta)\|^2$ $d\lambda < \infty$.

Proof of Theorem 2 Define $U_{tn} = \varphi(L)X_{tn}$ and replace U_t by U_{tn} in the definitions of $\hat{a}$ and $\hat{a}^*$. By the mean value theorem, $U_{tn} = U_t - \theta'_n \zeta(B)U_t - \theta'_n \{\zeta(B; \bar{\theta}_{tn}) - \zeta(B)\}U_t$, where $\zeta(z) = \zeta(z; 0)$ and $0 \leqslant \|\bar{\theta}_{tn}\| \leqslant \|\theta_n\|$. The last term has variance $O(n^{-1} \int_{-\pi}^{\pi} \|\zeta(e^{i\lambda}; \bar{\theta}_{tn}) - \zeta(e^{i\lambda})\|^2 d\lambda) = o(n^{-1})$, by dominated convergence. It follows that $U_{tn} = U_t - \theta'_n \zeta(B)U_t + o_p(n^{-(1/2)})$, uniformly in t. Put $C_{\bar{U}}(k) = n^{-1} \sum_{l=1}^{n-k} U_{tn}U_{t+k,n}$ for $0 \leqslant k \leqslant n - 1$. Then

$$n^{1/2}C_{\bar{U}}(k) = n^{-(1/2)} \sum_{t=1}^{n-k} \{U_t - \theta'_n \zeta(B)U_t + o_p(n^{-(1/2)})\}$$

$$\times \{U_{t+k} - \theta'_n \zeta(B)U_{t+k} + o_p(n^{-(1/2)})\}$$

$$= n^{(1/2)}C_U(k) - \theta'_n n^{-(1/2)} \sum_{t=1}^{n-k} \{\zeta(B)U_{t+k}\}U_t \quad \text{(A.6)}$$

$$- \theta'_n n^{-(1/2)} \sum_{t=1}^{n-k} \{\zeta(B)U_t\}U_{t+k} + \theta'_n n^{-(1/2)}$$

$$\times \sum_{t=1}^{n-k} \{\zeta(B)U_t\}\{\zeta'(B)U_{t+k}\}\theta_n + o_p(n^{-(1/2)}), \quad \text{(A.7)}$$

by (10.5) and uncorrelatedness of U_t. The second term in (A.6) is $\delta'n^{-1}\psi_k \sum_{t=1}^{n-k} U_t^2 + \delta'n^{-1} \sum_{l=1, l \neq k}^{\infty} \psi_l \sum_{t=1}^{n-k} U_{t+k-1}U_t$, whose second term has variance $n^{-1}(1 - k/n)\sigma^4 \sum_{i=1, j \neq k}^{\infty} (\delta'\psi_l)^2 = O(n^{-1})$ and whose first term is, for all fixed k, $(\delta'\psi_k)\sigma^2(1 - k/n) + o_p(\|\psi_k\|)$. By similar means, (A.7) is shown to be $O_p(n^{-(1/2)})$. It follows that

$$n^{1/2}C_{\bar{U}}(k) = n^{1/2}C_U(k) - (\delta'\psi_k)\sigma^2 \left(1 - \frac{k}{n}\right)$$

$$+ O_p(n^{-(1/2)}) + O_p(\|\psi_k\|), \quad \text{(A.8)}$$

where the orders are uniform in k. For fixed m,

$$n^{1/2} \sum_{k=1}^{m} \psi_k C_{\bar{U}}(k) = n^{1/2} \sum_{k=1}^{m} \psi_k C_U(k) - \sigma^2 \sum_{k=1}^{m} \psi_k \psi'_k \delta \left(1 - \frac{k}{n}\right)$$

$$+ O_p \left(n^{-(1/2)} \sum_{k=1}^{m} \|\psi_k\|\right) + o_p \left(\sum_{k=1}^{m} \|\psi_k\|^2\right),$$

which $\to_d N(-\sigma^2 \sum_{k=1}^m \psi_k \psi_k' \delta, \sigma^4 \sum_{k=1}^m \psi_k \psi_k')$. It readily follows from (A.8) that $n^{1/2} \sum_{k-m+1}^{n-k} \psi_k C_{\bar{U}}(k) \to_p 0$, as $m \to \infty$. Thus $a^* \to_d N(-\sigma^2 \Psi \delta, \sigma^4 \Psi)$. The proofs that $a - \tilde{a} = o_p(n^{-(1/2)})$ and $\tilde{\sigma}^2 \to_p \sigma^2$ follows those in Theorem 2's proof plus application of elementary inequalities to the components due to $U_{tn} - U_t$, which has mean zero and variance $O(n^{-1})$ uniformly in t from the foregoing. To prove that $a - a^* = O_p(n^{-(1/2)})$, we use (A.8) and the proof of Theorem 1 to obtain $E\|\hat{\psi}_0 C_{\bar{U}}(0)\| = o(n^{-(1/2)})$ and

$$
\left\| \sum_{k=1}^{n-1} \{\hat{\psi}_k - \psi_k\} C_{\bar{U}}(k) \right\| \leq \left\| \sum_{k=1}^{n-1} \{\hat{\psi}_k - \psi_k\} C_U(k) \right\|
$$

$$
+ O_p \left(n^{-(1/2)} \max_{1 \leqslant k \leqslant n-1} \|\hat{\psi}_k - \psi_k\| \left[\sum_{k=1}^n \|\psi_k\| + n^{-(1/2)} \right] \right)
$$

$$
= O_p(n^{-(1/2)}). \quad \blacksquare
$$

Appendix 10.4 Proof of Theorem 3

We first introduce the following definitions.

Definition 5 K is the class of sequences $U_t = \sum_{j=0}^{\infty} \alpha_j \varepsilon_{t-j}, t = 0, \pm 1, \ldots$, where $\{\varepsilon_t, t = 0, \pm 1, \ldots\} \in F$ and $\sum_{j=1}^{\infty} j^{1/2}|\alpha_j| < \infty$.

Definition 6 L is the class of functions $g(\lambda; \bar{\tau})$ on $(-\pi, \pi] \times T$ such that $g(\lambda; \bar{\tau}) = |\sum_{j=0}^{\infty} \alpha(j; \bar{\tau}) e^{ij\lambda}|^2, \alpha(0; \bar{\tau}) \equiv 1, \alpha(j; \tau) = \alpha_j, j = 0, 1, \ldots$, where T is a compact subset of q-dimensional Euclidean space and τ is the true value of $\bar{\tau}$; τ is an interior point of T; $g(\lambda; \bar{\tau}) \neq g(\lambda; \tau)$ for $\bar{\tau} \in T - \{\tau\}$; for all λ, $g(\lambda; \bar{\tau})$ is bounded away from zero on a neighbourhood N of τ; $g(\lambda; \bar{\tau})$ is continuous in $(\lambda, \bar{\tau})$ for $\bar{\tau} \in N$ and has first and second derivatives with respect to $\bar{\tau}$ that are also continuous in $(\lambda, \bar{\tau})$ for $\bar{\tau} \in N$; and $g(\lambda; \tau)$ and $(\partial/\partial \tau) g(\lambda; \tau)$ satisfy a Lipschitz condition in λ of order $\eta > 1/2$.

It follows from Hannan (1973) that the conditions of Theorem 3 imply that $\hat{\tau}$ is consistent for τ and $n^{1/2}(\hat{\tau} - \tau)$ converges in distribution to a $N(0, 2\Xi^{-1})$ random variable. Implicit in the latter result is the linearization

$$
\hat{\tau} = \tau - n^{-(1/2)} \frac{2\pi}{\sigma^2} \Xi^{-1} \sum_{j=1}^n \xi(\lambda_j) \frac{I_{\bar{U}}(\lambda_j)}{g(\lambda_j; \tau)} + o_p(n^{-(1/2)}), \quad (A.9)
$$

which is directly applied in the following proof. Definitions 5 and 6 are used not only for the purpose of dealing with the presence of $\hat{\tau}$, however, and we assume that $\sum_{j=1}^{\infty} j^{1/2}|\alpha_j| < \infty$. The latter condition is stronger than the corresponding condition $\sum_{j=1}^{\infty} j\alpha_j^2 < \infty$ of Hannan (1973) and is useful in dealing with the unbounded $\psi(\lambda)$ in our problem, which does not arise in Hannan's problem.

Proof of Theorem 3 Writing $v(\lambda) = \psi(\lambda) - \Phi \Xi^{-1} \xi(\lambda)$ and with $I_\varepsilon(\lambda)$ defined like $I_{\bar{U}}(\lambda)$ with $\bar{U}_t$ replaced by ε_t, we write $-n^{1/2} \hat{a} / 2\pi$ as

$$n^{-(1/2)} \sideset{}{'}\sum_j v(\lambda_j) I_\varepsilon(\lambda_j) \tag{A.10}$$

$$+ n^{-(1/2)} \sideset{}{'}\sum_j v(\lambda_j) \left\{ \frac{I_U(\lambda_j)}{g(\lambda_j; \tau)} - I_\varepsilon(\lambda_j) \right\} \tag{A.11}$$

$$+ n^{-(1/2)} \sideset{}{'}\sum_j \frac{v(\lambda_j)}{g(\lambda_j; \tau)} \{ I_{\bar{U}}(\lambda_j) - I_U(\lambda_j) \} \tag{A.12}$$

$$+ \Phi \left\{ \frac{\sigma^2}{2\pi} n^{1/2}(\hat{\tau} - \tau) + \Xi^{-1} \sideset{}{'}\sum_j \xi(\lambda_j) \frac{I_{\bar{U}}(\lambda_j)}{g(\lambda_j; \tau)} \right\} \tag{A.13}$$

$$+ \frac{\sigma^2}{2\pi} \left\{ n^{-1} \sideset{}{'}\sum_j \psi(\lambda_j) \xi(\lambda_j)' - \Phi \right\} n^{1/2}(\hat{\tau} - \tau) \tag{A.14}$$

$$+ \left[n^{-1} \sideset{}{'}\sum_j \psi(\lambda_j) \xi(\lambda_j)' \left\{ I_\varepsilon(\lambda_j) - \frac{\sigma^2}{2\pi} \right\} \right] n^{1/2}(\hat{\tau} - \tau) \tag{A.15}$$

$$+ \left[n^{-1} \sideset{}{'}\sum_j \psi(\lambda_j) \xi(\lambda_j)' \left\{ \frac{I_U(\lambda_j)}{g(\lambda_j; \tau)} - I_\varepsilon(\lambda_j) \right\} \right] n^{1/2}(\hat{\tau} - \tau) \tag{A.16}$$

$$+ \left[n^{-1} \sideset{}{'}\sum_j \frac{\psi(\lambda_j) \xi(\lambda_j)'}{g(\lambda_j; \tau)} \{ I_{\bar{U}}(\lambda_j) - I_U(\lambda_j) \} \right] n^{1/2}(\hat{\tau} - \tau) \tag{A.17}$$

$$+ \left[n^{-(1/2)} \sideset{}{'}\sum_j \psi(\lambda_j) I_{\bar{U}}(\lambda_j) \left\{ \frac{1}{g(\lambda_j; \hat{\tau})} - \frac{1}{g(\lambda_j; \tau)} \right\} \right.$$

$$\left. - n^{-1} \sideset{}{'}\sum_j \frac{\psi(\lambda_j) \xi(\lambda_j)'}{g(\lambda; \tau)} I_{\bar{U}}(\lambda_j) n^{1/2}(\hat{\tau} - \tau) \right]. \tag{A.18}$$

If follows immediately from Theorem 1 that (A.10) $\rightarrow_d N(0, \sigma^4 (\Psi - \Phi \Xi^{-1} \Phi'))$, and it remains to show that the remaining terms are $o_p(1)$. By far the most novel and difficult challenge is posed by (A.11), where the type of argument used by Hannan (1973) in case of bounded, smooth $v(\lambda)$ does not work and a different approach is taken.

Lemma 4 Let $\{U_t, t = 0, \pm 1, \ldots\} \in K$ and let μ satisfy the conditions on Ψ in Definition 3 and be square-integrable on $(-\pi, \pi]$. Then as $n \to \infty$,

$$n^{-(1/2)} \sum_j{}' \mu(\lambda_j)\{I_U(\lambda_j) - g(\lambda_j; \tau)I_\varepsilon(\lambda_j)\} \to_p 0. \qquad (A.19)$$

Proof We may write

$$I_U(\lambda_j) = (2\pi n)^{-1} \sum_{t,s=1}^{n} \sum_{l,m=0}^{\infty} \alpha_l \alpha_m \varepsilon_{t-l} \varepsilon_{s-m} e^{i(t-s)\lambda_j}$$

$$= (2\pi n)^{-1} \sum_{l,m=-\infty}^{n} \varepsilon_l \varepsilon_m \sum_{t=\max(1,j)}^{n} \sum_{s=\max(1,m)}^{n} \alpha_{t-l} \alpha_{s-m} e^{i(t-s)\lambda_j},$$

$$g(\lambda_j; \tau)I_\varepsilon(\lambda_j) = (2\pi n)^{-1} \sum_{l,m=1}^{n} \varepsilon_l \varepsilon_m \sum_{t,s=0}^{\infty} \alpha_t \alpha_s e^{i(t-s+l-m)\lambda_j}$$

$$= (2\pi n)^{-1} \sum_{l,m=1}^{n} \varepsilon_l \varepsilon_m \sum_{t=l}^{\infty} \sum_{s=m}^{\infty} \alpha_{t-l} \alpha_{s-m} e^{i(t-s)\lambda_j}.$$

The left side of (A.19) is thus

$$n^{-(1/2)} \pi^{-1} \sum_{l,m=-\infty}^{n} \varepsilon_l \varepsilon_m G_{lm}, \qquad (A.20)$$

where

$$G_{lm} = \sum_{t=\max(1,l)}^{n} \sum_{s=\max(1,m)}^{n} \alpha_{t-l} \alpha_{s-m} \hat\mu_{t-s},$$

$$l \leq 0, \quad m \geq 1; \quad \text{or } l \geq 1, \quad m \leq 0; \quad \text{or } l, \quad m \leq 0;$$

$$= - \sum_{t,s=n+1}^{\infty} \alpha_{t-l} \alpha_{s-m} \hat\mu_{t-s}, \quad l, m \geq 1,$$

in which $\hat\mu_l = n^{-1} \sum' \mu(\lambda_j) \cos l\lambda_j$. The contribution from squared terms ε_l^2 to (A.20) has norm with expectation $(\sigma^2/n^{1/2}\pi) \sum_{l=-\infty}^{n} \|G_{ll}\|$. For $l \leq 0$ and any m, $\|G_{lm}\| \leq \bar\mu \sum_{t=1}^{n} |\alpha_{t-l}| \sum_{j=0}^{\infty} |\alpha_j| = O(\bar\mu \sum_{t=1}^{n}(1 - l/t)1/2|\alpha_{t-l}|)$, where $\bar\mu = n^{-1} \sum' \|\mu(\lambda_j)\| \leq (n^{-1} \sum' \|\mu(\lambda_j)\|^2)^{1/2} = O(1)$ by Lemma 3 and Parseval's inequality. For $1 \leq l, m \leq n$, $\|G_{lm}\| \leq \bar\mu \sum_{t=n+1}^{\infty} |\alpha_{t-l}| \sum_{j=0}^{\infty} |\alpha_j| =$

$O((n+1-l)^{-(1/2)}e_{n+1-l})$, where $e_l = \sum_{j=l}^{\infty} j^{1/2}|\alpha_j|$. Then $\sum_{l=-\infty}^{n} \|G_{ll}\|$ is

$$
O\left(\bar{\mu}\left\{\sum_{t=1}^{n} t^{-(1/2)} \sum_{l=0}^{x}|t+l|^{1/2}|\alpha_{t+l}| + \sum_{l=1}^{n}(n+1-l)^{-(1/2)}e_{n+1-l}\right\}\right)
$$

$$
= O(E_n),
$$

where $E_n = n^{-(1/2)}\sum_{t=1}^{n} t^{-(1/2)}e_t$. For any $\varepsilon \in (0,1)$, $E_n = O(n^{-(1/2)}\{\sum_{t=1}^{[\varepsilon m]} t^{-(1/2)} + \varepsilon^{-1/2}e_{[\varepsilon m]}\}) \to 0$, letting $n \to \infty$ and then $\varepsilon \to 0$. It follows that $n^{-(1/2)}\sum_{l=-\infty}^{n}\varepsilon_l^2 G_{ll} \to_p 0$. The contribution to (A.20) of terms $\varepsilon_l, \varepsilon_m$ for $l \neq m$ has zero mean and varience $2\sigma^4/(n\pi^2)\sum\sum_{l,m=-\infty}^{n}\|G_{l,m}\|^2$. By symmetry, for $m \leq 0$ $\|G_{lm}\| = 0(\sum_{t=1}^{n}(1-m/t)^{1/2}|\alpha_{t-m}|)$. Thus

$$
\left\{n^{-1}\sum_{l,m=-\infty}^{0}\sum\|G_{lm}\|^2\right\}^{1/2}
$$

$$
= O\left(n^{-(1/2)}\sum_{l=0}^{\infty}\sum_{t=1}^{n}(1+l/t)^{1/2}|\alpha_{t+l}|\right) = O(E_n) \to 0,
$$

as $n \to \infty$. By symmetry, for $1 \leq l,m \leq n$, $\|G_{lm}\| = O((n+1-m)^{-(1/2)}e_{n+1-m})$. Thus $n^{-1}\sum\sum_{l,m=1}^{n}\|G_{lm}\|^2 = O(E_n^2)$. The contribution to (A.20) for $l \leq 0, m \geq 1$ is more difficult to handle. Uniformly in such l, m, $G_{lm} = O(\sum_{t=1}^{n}(1-l/t)^{t/2}|\alpha_{t-l}|)$. Uniformly in such l,

$$
\sum_{m=1}^{n}\|G_{lm}\| \leq \sum_{j=0}^{\infty}|\alpha_j|\sum_{t=1}^{n}|\alpha_{t-l}|\left(\sum_{s=l}^{n}\|\mu_{t-s}\| + n\max_{0\leq j\leq m}\|\hat{\mu}_j - \mu_j\|\right)
$$

$$
\leq \left\{\sum_{j=0}^{\infty}|\alpha_j|\right\}^2\left\{n^{-1/2}\sum_{j=0}^{\infty}\|\mu_j\|^2 + o(n^{1/2})\right\} = O(n^{1/2}),
$$

by Parseval's equality. It follows that

$$
n^{-1}\sum_{l=-\infty}^{0}\sum_{m=1}^{n}\|G_{lm}\|^2 = O\left(n^{-1}\sum_{l=0}^{\infty}\sum_{t=1}^{n}(1+l/t)^{1/2}|\alpha_{t+l}|\sum_{m=1}^{n}\|G_{lm}\|\right)
$$

$$
= O(E_n) \to 0.
$$

Putting $\mu(\lambda) = v(\lambda)/g(\lambda;\tau)$, it follows from Lemma 4 that (A.11) $\to_p 0$. The proof that (A.12) $\to_p 0$ is an extension of that of $a - \tilde{a} \to_p 0$ in Appendix 10.2. It is only necessary to note that $E\|D^{1/2}(\tilde{\beta} - \beta)\|^2 = 2\pi n\int_{-\pi}^{\pi} f(\lambda;\tau,\sigma^2)\text{tr}\{D^{-(1/2)}I_W(\lambda)D^{-(1/2)}\}\,d\lambda < \infty$ and $E\|n^{-(1/2)}\sum_j'\mu(\lambda_j)D^{-(1/2)}I_{WU}(\lambda_j)\|^2 = O(n^{-1}\sum_j'\|\mu(\lambda_j)\|^2\text{tr}\{D^{-(1/2)}I_W(\lambda_j)\{D^{-(1/2)}\})$ by standard arguments, using boundedness of f and orthogonality of $e^{i\lambda}$. Next, $(A.13) \to_p 0$ from (A.9), because only finitely many terms are omitted from the sum over j. The conditions imply that $\hat{\tau} - \tau = O_p(n^{-(1/2)})$ (as follows from arguments in the rest of the proof or from Hannan 1973), so that $(A.14) \to_p 0$ from Lemma 3. Finally $(A.15)$–$(A.18)$ each $\to_p 0$ using similar arguments to those in the proof of Theorem 1, the proof for $(A.11)$, the proof for $(A.12)$, and the mean value theorem and conditions on g. ∎

References

Bhargava, A. (1986), 'On the theory of testing for unit roots in observed time series', *Review of Economic Studies*, 53, 369–84.

Bloomfield, P. J. (1973), 'An exponential model for the spectrum of a scalar time series', *Biometrika*, 60, 217–26.

Box, G. E. P. and Jenkins, G. M. (1970), *Time Series Analysis, Forecasting and Control*. San Francisco: Holden-Day.

Brown, B. M. (1971), 'Martingale central limit theorems', *Annals of Mathematical Statistics*, 42, 59–66.

Cressie, N. (1988), 'A graphical procedure of determining nonstationarity in time series', *Journal of the American Statistical Association*, 83, 1108–16.

Davies, R. B. and Harte, D. S. (1987), 'Tests for Hurst effect', *Biometrika*, 74, 95–101.

Dickey, D. A., Hasza, D. P., and Fuller, W. A. (1984), 'Testing for unit roots in seasonal time series', *Journal of the American Statistical Association*, 79, 355–67.

—— and Fuller, W. A. (1979), 'Distribution of the estimators for autoregressive time series with a unit root', *Journal of the American Statistical Association*, 74, 427–31.

Diebold, F. X. and Rudebusch, C. (1991), 'On the power of Dickey–Fuller tests against fractional alternatives', *Economics Letters*, 35, 155–60.

Dufour, J.-M. and King, M. L. (1991), 'Optimal invariant tests for the autocorrelation coefficient in linear regressions with stationary or nonstationary AR(1) errors', *Journal of Econometrics*, 47, 115–43.

Elliott, G., Rothenberg, T., and Stock, J. H. (1996), 'Efficient tests of an autoregressive unit root', *Econometrica*, 64, 813–36.

Fox, R. and Taqqu, M. S. (1985), 'Noncentral limit theorems for quadratic forms in random variables having long-range dependence', *Annals of Probability*, 13, 428–46.

—— (1986), 'Large-sample properties of parameter estimates for strongly dependent stationary Gaussian time series', *The Annals of Statistics*, 14, 517–32.

Geweke, J. and Porter-Hudak, S. (1983), 'The estimation and application of long memory time series models', *Journal of time series analysis*, 4, 221–38.

Gray, H. L., Zhang, N.-F., and Woodward, W. A. (1989), 'On generalized fractional processes', *Journal of Time Series Analysis*, 10, 233–57.

Hannan, E. J. (1973), 'The asymptotic theory of linear time series models', *Journal of Applied Probability*, 10, 130–45.

Hosking, J. R. M. (1980). 'Lagrange multiplier tests of time-series models', *Journal of the Royal Statistical Society*, Ser. B, 42, 170–81.

Marinari, E., Parisi, G., Ruelle, D. and Windey, P. (1983), 'On the interpretation of $1/f$ noise', *Communications in Mathematical Physics*, 89, 1–12.

Porter-Hudak, S. (1990), 'An application of the seasonal fractionally differenced model to the monetary aggregates', *Journal of the American Statistical Association*, 85, 338–44.

Rao, C. R. (1973), *Linear Statistical Inference and its Applications*. New York: John Wiley.

Robinson, P. M. (1991), 'Testing for strong serial correlation and dynamic conditional heteroskedasticity in multiple regression', *Journal of Econometrics*, 47, 67–84.

—— (1995), 'Log-periodogram regression of time series with long-range dependence', *Annals of Statistics*, 23, 1048–72.

—— (1993), 'Highly insignificant F-ratios', *Econometrica*, 61, 687–96.

Said, S. E. and Dickey, D. A. (1985), 'Hypothesis testing in ARIMA $(p, 1, q)$ models', *Journal of the American Statistical Association*, 80, 369–74.

Sargan, J. D. and Bhargava, A. (1983), 'Testing residuals from least squares regression for being generated by the Gaussian random walk', *Econometrica*, 51, 153–74.

Schmidt, P. and Phillips, P. C. B. (1992), 'LM tests for a unit root in the presence of deterministic trends', *Oxford Bulletin of Economics and Statistics*, 54, 257–87.

Solo, V. (1984), 'The order of differencing in ARMA models', *Journal of the American Statistical Association*, 79, 916–21.

11

Estimation of the Memory Parameter for Nonstationary or Noninvertible Fractionally Integrated Processes

CLIFFORD M. HURVICH AND BONNIE K. RAY

11.1 Introduction

A key problem in time series analysis is the determination or estimation of the degree of integration, d, of the series. In the case of autoregressive moving-average (ARMA) models, for example, a series with a unit root in the AR polynomial (and all MA roots outside the unit circle) has $d = 1$ and is nonstationary, a series with a unit root in the MA polynomial (and all AR roots outside the unit circle) has $d = -1$ and is noninvertible, while a series with all roots of the AR and MA polynomials outside the unit circle has $d = 0$ and is both stationary and invertible. The Dickey–Fuller test (Dickey and Fuller 1979) is a method for deciding, on the basis of the observed data, whether to use $d = 0$ or $d = 1$ in building a model for the series. The test proposed by Saikkonen and Luukkonen (1993) for a unit root in the MA polynomial is a method for deciding between $d = 0$ and $d = -1$.

The notion of fractional integration (Granger and Joyeux 1980; Hosking 1981), in which d is allowed to take nonintegral values, has broadened the issue of deciding on the stationarity or invertibility of a series, while encompassing the special cases $d = -1$, $d = 0$ and $d = 1$. A fractionally integrated ARMA(p, q) process (ARFIMA(p, d, q)) having all autoregressive and moving-average roots outside the unit circle is stationary but not invertible when $d \leqslant -0.5$. When $d \in (-0.5, 0.5)$, it is both stationary and invertible. When $d \geqslant 0.5$, the ARFIMA process is nonstationary, although for $d \in [0.5, 1)$ it is mean-reverting in the sense that there is no long-run impact of an innovation on the value of the process (Cheung and Lai 1993). The mean-reversion property no longer holds when $d \geqslant 1$.

Geweke and Porter-Hudak (1983) developed a method for explicitly estimating d based on the behavior of the spectral density of a stationary and invertible fractionally integrated process at low frequencies. The spectral density of a stationary

This chapter was previously published as Hurvich, C. M. and Ray, B. K. (1995), 'Estimation of the memory parameter for nonstationary or noninvertible fractionally integrated processes', *Journal of Time Series Analysis*, 16, 17–41. Copyright © 1995 Basil Blackwell Ltd. Reproduced by permission.

The authors wish to thank the referee for comments which helped improve the presentation of this work. They also thank N. Crato for his initial work concerning the periodogram of a random walk process, which inspired us to write this paper. The research of B. Ray was supported, in part, by a separately budgeted research grant from the New Jersey Institute of Technology.

and invertible fractionally integrated process $\{X_t\}$ is

$$f(\lambda) = \left| 2 \sin \left(\frac{\lambda}{2} \right) \right|^{-2d} f^*(\lambda),$$

where f^* denotes the spectral density of any short-memory components of the process. (It is not required that f^* be the spectral density of an ARMA process.) The GPH method regresses the logarithm of the periodogram of a data set $x_0, x_1, \ldots, x_{n-1}$ on $-2 \log |2 \sin(\omega_j/2)|$ for a set of initial Fourier frequencies $\omega_j = 2\pi j/n$. The estimated slope obtained from this regression is used as an estimate of d. The GPH estimator is semi-parametric in that it assumes no knowledge of any short-memory components, although the presence of short-memory components may bias the estimate of d in finite samples (see, for example, Agiakloglou *et al.* 1993).

An important reason for interest in the GPH approach is that using standard tests for a unit root in the presence of fractionally integrated errors may result in misleading conclusions. Diebold and Rudebusch (1991) found, for example, that a two-sided Dickey–Fuller test based on the Dickey–Fuller t statistic rejects the hypothesis $d = 1$ approximately 70 per cent of the time for a series of length $n = 250$ when the true process is ARFIMA$(0, 0.7, 0)$. The standard practice when $d = 1$ is rejected is to assume $d = 0$. Thus the practitioner might be tempted to conclude that a series of this type need not be differenced and would attempt to fit a stationary model to a nonstationary series. They also found that the two-sided Dickey–Fuller test rejects the hypothesis $d = 1$ only 17 per cent of the time for a series of length $n = 250$ when the true process is AR(1) with AR parameter $\phi = 0.98$. In this case the practitioner would usually be led to take a first difference, resulting in a noninvertible series. Although the GPH method would give a positively biased estimate of d in this case, the example provides further illustration that it will often be difficult in practice to know whether one is working with a stationary and invertible series.

Theoretical properties of the GPH estimator have only been investigated for processes having $d \in (-0.5, 0.5)$. Robinson (1991) shows that a modified version of the GPH estimator is consistent and asymptotically normal when $d \in (-0.5, 0.5)$. He suggests that for series having d outside this range, the series can be differenced or integrated an integral number of times until a series is obtained having d in the specified range. In practice, however, the GPH estimation method is often applied to nonstationary or noninvertible series before initial differencing or integration (e.g. see Bloomfield 1991; Agiakloglou *et al.* 1993; Hassler 1993). Thus the question arises as to the properties of the periodogram of a noninvertible or nonstationary fractionally integrated process and the effects on the GPH estimator.

It would be desirable to have an estimator of d which is free of the restriction that $d \in (-0.5, 0.5)$. The GPH estimator itself is not free of this restriction. In fact, as we will show in this paper, the GPH estimator can be quite strongly biased for ARFIMA$(0, d, 0)$ processes which are either nonstationary or noninvertible.

An additional consequence of this result, as we also illustrate, is that the GPH estimator is not, in general, invariant to first differences, that is, the estimated d based on the original data is not in general equal to one plus the estimated d based on the differenced data. For instance, Agiakloglou *et al.* (1993) obtained $\hat{d} = -0.26$ for the first difference of a series of US unemployment figures, whereas their estimate of d for the undifferenced series was $\hat{d} = 0.9997$.

Hurvich and Beltrao (1993) investigated the asymptotic properties of the periodogram at a given Fourier frequency ω_j for a general class of long-memory processes having $d \in (-0.5, 0.5)$. They found that the periodogram suffers from an asymptotic relative bias which decreases with j. The implication for the GPH estimator is that it will be positively biased when the number of Fourier frequencies used in the regression is held fixed as $n \to \infty$.

In this paper we extend the results of Hurvich and Beltrao to the case in which d falls outside the range $(-0.5, 0.5)$ and discuss the implications for the GPH estimator. Although Solo (1990) has obtained results concerning the periodogram of a nonstationary fractional noise process at frequencies that are held constant as $n \to \infty$, his focus on fixed frequencies instead of Fourier frequencies obscures biases in the periodogram which will be important in practice. In addition, our results apply to a more general class of processes. Crato (1992) has derived some properties of the periodogram at the Fourier frequencies, but was concerned solely with the case $d = 1$. We shall consider all $d < 1.5$.

In Section 11.2 we derive the asymptotic relative bias of the periodogram at a given Fourier frequency when $d \in (-\infty, 1.5)$, except for the case $d = -0.5$, which remains unresolved. For $d < -0.5$, the relative bias of the periodogram tends to ∞ with n and depends on the spectrum of any nontrivial short-memory component (e.g. an ARMA component). In this case, the GPH estimator, if based on a fixed number of Fourier frequencies, will have an asymptotic expectation of zero for Gaussian processes. When $d \in [0.5, 1)$, the asymptotic relative bias of the periodogram is finite, and decreases with frequency. When $d = 1$, it is constant for all j. When $d \in (1, 1.5)$, it increases with j. We also derive the asymptotic relative bias of the tapered periodogram for $j > 1$ when a cosine bell taper is applied to the series and $d \in (-2.5, 1.5)$. This bias is close to zero when $d \in (-1, -0.5)$, and is almost always less than that of the nontapered periodogram when $d \in (0.5, 1.5)$.

In Section 11.3, we empirically assess the performance of the GPH estimator in finite samples for ARFIMA$(0, d, 0)$ processes having d outside the range $(-0.5, 0.5)$. The results indicate that the bias of the periodogram ordinates for nonstationary or noninvertible series greatly impacts the estimate of d even when the number of ordinates used in the regression is allowed to depend on n. Additionally, our results indicate that the GPH estimator is not invariant to first differences.

The effects of a data taper on the GPH estimator are also investigated. We find that using a cosine bell taper and dropping the first periodogram ordinate from the regression greatly improves the performance of the GPH estimator for noninvertible series. For nonstationary series, the GPH estimator based on the tapered data often has less bias, is more nearly invariant to first-differencing, but has a larger variance than GPH without tapering.

An appendix is given containing proofs of the theoretical results stated in Section 11.2.

11.2 Theoretical Results

We consider a general class of long-memory processes with memory parameter $d < 1.5$. We define the class in two stages, according to the value of d. For $d \in (-\infty, 0.5)$, we say that the process $\{X_t\}_{t=-\infty}^{\infty}$ has memory parameter d if $\{X_t\}$ is weakly stationary with zero mean and spectral density

$$f(\lambda) = |1 - \exp(-i\lambda)|^{-2d} f^*(\lambda), \tag{11.1}$$

where $f^*(\lambda)$ is a positive, even function on $[-\pi, \pi]$ which is bounded above, bounded away from zero and continuous at $\lambda = 0$. This definition includes the stationary, invertible fractional ARIMA models as a special case: take d to be between -0.5 and 0.5 and take f^* to be the spectral density of a stationary, invertible ARMA process.

For $d \in [0.5, 1.5)$, we say that $\{X_t\}$ has memory parameter d if $\{\Delta X_t\}$ has memory parameter $d - 1$, where $\Delta = 1 - B$ and B is the backshift operator. In this case, if we define $\varepsilon_t = \Delta X_t$, then

$$X_t = \sum_{k=-\infty}^{t} \varepsilon_k + C = \sum_{k=-\infty}^{-1} \varepsilon_k + \sum_{k=0}^{t} \varepsilon_k + C$$

for $t \geqslant 0$, where C does not depend on time. Therefore, for $t \geqslant 0$, we have

$$X_t = \sum_{k=0}^{t} \varepsilon_k + R, \tag{11.2}$$

where R is a random variable not depending on time. Note that $\{X_t\}$ is not stationary when $d \geqslant 0.5$. An important special case is the random walk, for which $d = 1$ and f^* is constant on $[-\pi, \pi]$.

Suppose we have a time series data set $x_0, \ldots, x_{n-1}$ generated by a process $\{X_t\}$ having memory parameter d, where $d < 1.5$. For any such d, define $f(\lambda)$ by eqn (11.1). Note that if $d \in [0.5, 1.5)$, then $f(\cdot)$ is not the spectral density of $\{X_t\}$, since the process is then not stationary and therefore does not have a spectral density. Nevertheless, we shall see that in the nonstationary case $f(\cdot)$ plays the role usually played by the spectral density in determining some of the statistical properties of the periodogram.

Define the jth Fourier frequency by $\omega_j = 2\pi j/n$ for $j = 1, 2, \ldots$. Let J_j denote the discrete Fourier transform at frequency ω_j,

$$J_j = \frac{1}{n} \sum_{t=0}^{n-1} x_t \exp(-i\omega_j t).$$

Let I_j denote the periodogram at frequency ω_j.

$$I_j = \frac{n}{2\pi}|J_j|^2 = \frac{1}{2\pi n}\left|\sum_{t=0}^{n-1} x_t \exp(-i\omega_j t)\right|^2.$$

Define $f_j = f(\omega_j)$. The *normalized periodogram* at frequency ω_j is defined by I_j/f_j. Lemma 1 gives an exact, finite-sample expression for the expectation of the normalized periodogram.

Lemma 1 If $d < 1.5$,

$$E\left(\frac{I_j}{f_j}\right) = \frac{1}{f_j}\int_{-\pi}^{\pi} F_n(\omega_j - \lambda)f(\lambda)\,d\lambda,$$

where $F_n(\cdot)$ is the Fejer Kernel, given by

$$F_n(\omega) = \frac{1}{2\pi n}\left\{\frac{\sin(n\omega/2)}{\sin(\omega/2)}\right\}^2.$$

Note that this result is well known in the stationary case, $d < 0.5$, but it is rather surprising that it continues to hold for $d \in [0.5, 1.5)$. Indeed, one might wonder whether the expectation of the periodogram is finite when $d \geqslant 0.5$. In this case, from eqn (11.2), since R does not depend on time, we find that the periodogram of $\{X_t\}_{t=0}^{n-1}$ is equal to the periodogram of $\left\{\sum_{k=0}^{t}\varepsilon_k\right\}_{t=0}^{n-1}$ at all nonzero Fourier frequencies. From this, the finiteness of $E(I_j)$ follows immediately.

We now state our main results, on the limiting expectation of the normalized periodogram, as $n \to \infty$, with j held fixed. Subtracting one from this limiting expectation yields the asymptotic relative bias of the periodogram $I(\omega_j)$ as an estimate of $f(\omega_j)$. The proofs of these theorems use Lemma 1 as the starting point.

Theorem 1 If $d \in (-0.5, 1.5)$ then

$$\lim_{n\to\infty} E\left(\frac{I_j}{f_j}\right) = \frac{2}{\pi}\int_{-\infty}^{\infty} \frac{\sin^2(\lambda/2)}{(2\pi j - \lambda)^2}\left|\frac{\lambda}{2\pi j}\right|^{-2d}\,d\lambda.$$

In general, the integral above will depend on j, suggesting that the GPH estimator will be asymptotically biased if it is based on a fixed number of the initial Fourier frequencies. An exception is the case $d = 1$, for which the integral above can be shown to equal 2 for all j. For Gaussian processes with $d = 1$, it can be shown that I_j/f_j is asymptotically distributed as $\frac{1}{2}Z_1^2 + \frac{3}{2}Z_2^2$, where Z_1 and Z_2 are independent standard normals, so that the GPH estimator will in this case be asymptotically unbiased if it is based on a fixed number of frequencies. Another case where GPH is unbiased is $d = 0$ (i.e. short memory), for which the integral above is equal to 1 for all j. Figs 11.1–11.4 plot numerical evaluations of $\lim_{n\to\infty} E(I_j/f_j)$ for $j = 1, \ldots, 10$ and d ranging from 0.5 to 1.47 in steps of 0.01. It is seen that for $d \in [0.5, 1)$ the asymptotic mean decreases with j, while

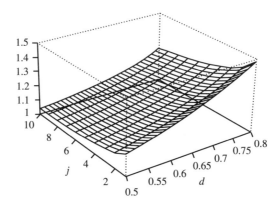

Fig. 11.1. Asymptotic mean of normalized periodogram, $0.5 \leqslant d \leqslant 0.8$.

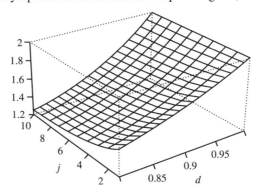

Fig. 11.2. Asymptotic mean of normalized periodogram, $0.8 \leqslant d \leqslant 1$.

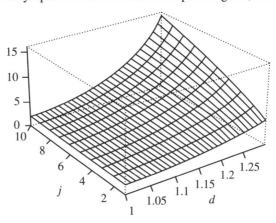

Fig. 11.3. Asymptotic mean of normalized periodogram, $1 \leqslant d \leqslant 1.3$.

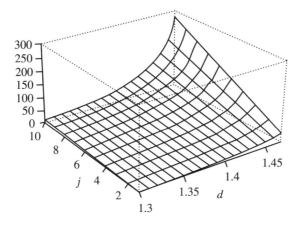

Fig. 11.4. Asymptotic mean of normalized periodogram, $1.3 \leqslant d \leqslant 1.47$.

for $d \in (1, 1.5)$ the asymptotic mean increases with j, especially strongly when d is close to 1.5.

For a heuristic understanding of the connection between the observed patterns in the asymptotic mean of the normalized periodogram and the bias of GPH, note that

$$\log I_j = -2d \log \left| 2 \sin \left(\frac{\omega_j}{2} \right) \right| + \log f^*(\omega_j) + \log \left(\frac{I_j}{f_j} \right)$$

$$\approx -2d \log \left| 2 \sin \left(\frac{\omega_j}{2} \right) \right| + \log f^*(0) + \log \left(\frac{I_j}{f_j} \right)$$

if the largest value of j under consideration is small compared with n. The GPH estimator can be expressed as -0.5 times the slope estimate obtained from a linear least squares regression of $\{\log I_j\}$ on $\{\log |2 \sin(\omega_j/2)|\}$, which is a monotone increasing function of j for $j \leqslant [n/2]$. If $E(I_j/f_j)$ decreases with j, it is plausible that $E\{\log(I_j/f_j)\}$ also decreases with j, thereby producing a positive bias in GPH. Similarly, it is plausible that if $E(I_j/f_j)$ increases with j, GPH will be negatively biased. Therefore, the results of Figs 11.1–11.4 suggest that GPH will be positively biased for $d \in [0.5, 1)$ and negatively biased for $d \in (1, 1.5)$, provided that the sample size is large and the estimator is based on a fixed number of the initial Fourier frequencies. It should be stressed, however, that this last proviso in effect ignores any contribution to the bias in GPH due to the short-memory component f^*, which may be important in small samples, and which has been studied by Agiakloglou *et al.* (1993).

Theorem 2 If $d < -0.5$, then

$$\lim_{n \to \infty} n^{2d+1} E \left(\frac{I_j}{f_j} \right) = \frac{|\pi j|^{2d}}{4\pi f^*(0)} \int_{-\pi}^{\pi} \left| \sin \left(\frac{\lambda}{2} \right) \right|^{-2d-2} f^*(\lambda) \, d\lambda.$$

Thus, if no taper is used and $d < -0.5$, the expected normalized periodogram tends to infinity at a rate $O(n^{-2d-1})$ with a constant of proportionality depending on the short-memory component f^*. Since the integrand above does not depend on j, however, the log of the limiting expectation will be exactly linear in $\log j$ with slope $2d$. Arguing heuristically as above, it is therefore plausible that if n is large, $E\{\log(I_j/f_j)\}$ decreases with j, so that GPH will be positively biased when $d < -0.5$. More rigorously, an argument similar to the proof of Theorem 2 shows that if $d < -0.5$, then $n^{2d+1}E(B_j^2/f_j) \to 0$, where

$$B_j = \frac{1}{(2\pi n)^{1/2}} \sum_{t=0}^{n-1} x_t \sin(\omega_j t).$$

This, combined with Theorem 2, implies that if the process is Gaussian and $d < -0.5$, then $\log(n^{2d+1}I_j/f_j)$ is asymptotically distributed as $2d \log j + C_{d,f^*} + \log \chi_1^2$, where C_{d,f^*} is a constant not depending on j, so that the GPH estimator will in this case have an asymptotic bias of exactly $-d$ if it is based on a fixed number of the initial Fourier frequencies.

Our final theorem concerns the periodogram

$$I_{T,j} = \frac{1}{2\pi \sum w_t^2} \left| \sum_{t=0}^{n-1} w_t x_t \exp(-i\omega_j t) \right|^2$$

of the tapered data $\{w_t x_t\}$, where $\{w_t\}$ is the full cosine bell taper given by $w_t = \frac{1}{2}[1 - \cos\{2\pi(t + 0.5)/n\}]$.

Theorem 3 If $d \in (-2.5, 1.5)$ and a cosine bell taper is used, resulting in a tapered periodogram $I_{T,j}$, then for $j > 1$,

$$\lim_{n\to\infty} E\left(\frac{I_{T,j}}{f_j}\right) = (2\pi j)^{2d} \frac{2}{\pi} \frac{8}{3} \int_{-\infty}^{\infty} \sin^2\left(\frac{\lambda}{2}\right)$$

$$\times \left\{ \frac{0.5}{2\pi j - \lambda} - \frac{0.25}{2\pi(j-1) - \lambda} - \frac{0.25}{2\pi(j+1) - \lambda} \right\}^2 |\lambda|^{-2d} \, d\lambda.$$

(The integral in Theorem 3 diverges when $d \leqslant -2.5$.)

Figures 11.5 and 11.6 present numerical evaluations of the limiting expected normalized tapered periodogram for $j = 2, \ldots, 10$ and d ranging from -1 to 1.47. It is seen that the asymptotic relative bias is close to zero when $d \in (-1, -0.5)$, and seems to be exactly zero when $d = -0.5$. When $d > 0.5$ the asymptotic relative bias is almost always less (often dramatically so) for the tapered periodogram than for the periodogram itself. However, when d is close to 1 the asymptotic relative bias varies more with j for the tapered periodogram than for the periodogram, suggesting that tapering may increase the bias in the GPH estimator in this case if the number of Fourier frequencies used is not large. For all other values of d, the figures suggest that tapering will reduce the bias of GPH.

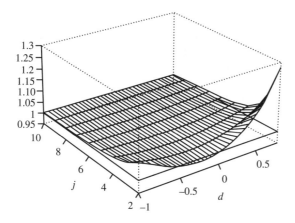

Fig. 11.5. Asymptotic mean of normalized tapered periodogram, $-1 \leqslant d \leqslant 0.8$.

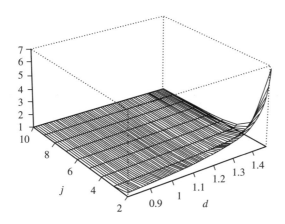

Fig. 11.6. Asymptotic mean of normalized tapered periodogram,
$0.8 \leqslant d \leqslant 1.47$.

In practice, we recommend that the first ordinate $I_{\mathrm{T},1}$ be omitted from the regression in any GPH estimator based on the tapered periodogram. One reason for this, although not the most important one, is that Theorem 3 does not hold when $j = 1$: the above integral diverges when $j = 1$ and $d \geqslant 0.5$. More importantly, the omission of $I_{\mathrm{T},1}$ ensures that the resulting GPH estimator will be invariant to shifts of form $\{x_t\} \rightarrow \{x_t + k\}$, where k is a constant. This property follows from the fact that the tapered discrete Fourier transform at frequency ω_j can be expressed as a linear combination of the nontapered discrete Fourier transform at frequencies ω_{j-1}, ω_j and ω_{j+1} (see Bloomfield 1976: pp. 82–4), together with the fact that the nontapered discrete Fourier transforms J_j (and therefore the nontapered periodograms I_j) of the two series $\{x_t\}_{t=0}^{n-1}$ and $\{x_t + k\}_{t=0}^{n-1}$ agree at all nonzero Fourier frequencies ω_j, $j \geqslant 1$. From this, it is also seen that the ordinary GPH estimator based on the nontapered periodogram at frequencies ω_j, $j \geqslant 1$, is shift-invariant even though I_1 is not omitted from the regression in this case. In addition, it follows from the discussion above that if the data are tapered and $I_{T,1}$ is used in the regression, the resulting GPH estimator will depend on J_0, which is the sample mean $\bar{x}$, and therefore the GPH estimator will not be shift-invariant. We feel that shift-invariance is a very important property for any estimator of d, since it is clear that such a shift does not change the true value of d.

Yet another consequence of the above discussion is that Theorems 1, 2, and 3 continue to hold if the long-memory time series $\{X_t\}$ is replaced by the shifted version $\{X_t + k\}$ for any k.

11.3 Empirical Results

11.3.1 Periodogram regression applied to nonstationary or noninvertible fractionally integrated noise

A Monte Carlo study was performed to analyse the properties of periodogram regression (i.e. the least squares regression of the log periodogram on $\{-2 \log |2 \sin(\omega_j/2)|\}$) for noninvertible or nonstationary series in finite samples. For the noninvertible case, we considered Gaussian ARFIMA$(0, d, 0)$ series having $d = -1.0, -0.9, \ldots, -0.5$. For the nonstationary case, we considered Gaussian ARFIMA$(0, d, 0)$ series having $d = 0.5, 0.6, \ldots, 1.4$. Series having $d < -0.5$ were obtained by generating a stationary and invertible series of length $n' = n + 1$ having memory parameter $d' = d + 1$ and taking first differences. Stationary and invertible series, and series having $d = -0.5$, were generated using the algorithm of Hosking (1984) with the variance of the series set equal to 1. Series having $d \geqslant 0.5$ were obtained by generating a stationary and invertible series of length n having memory parameter $d' = d - 1$ and taking partial sums.

We investigated periodogram regression for series having lengths $n = 100, 400, 1000$. Five hundred replications were performed for each case and four different estimators were computed. The estimates were obtained by applying periodogram regression to the raw data, the tapered data, the differenced data and the differenced tapered data, respectively, using periodogram ordinates at frequencies ω_j for $j = l+1, \ldots, m$. For the raw data, we used $l = 0$. For the tapered data, we used $l = 1$. Based on simulation results, Geweke and Porter-Hudak suggest using $m = [n^{1/2}]$ to mitigate the effect of any short-memory components on the estimate of d. Thus we took $m = [n^{1/2}]$ in our Monte Carlo study and refer to the estimates obtained using the raw data as GPH estimates and the estimates obtained using the tapered data as GPHT estimates. Tables 11.1–11.3 give the average values, $\bar{d}$, standard deviations, $s_{\hat{d}}$, and root mean squared errors (RMSEs) of the GPH and GPHT estimates.

Table 11.1 summarizes our results for noninvertible series. We see that the GPH estimates are very positively biased. This is consistent with the empirical findings of Hassler (1993: table 3). The size of the bias decreases, however, as d gets closer to -0.5, the invertibility boundary. For a given d, the bias does not appear to

Table 11.1. Average estimated values of
$d: d = -1.0, \ldots, -0.5$

| | n | $m = [n^{1/2}]$ | | |
		$\bar{d}$	$s_{\hat{d}}$	RMSE
$d = -1.0$				
GPH	100	−0.7185	0.3616	0.4580
	400	−0.6133	0.2913	0.4839
	1000	−0.5897	0.2689	0.4904
GPHT	100	−1.0194	0.5192	0.5191
	400	−1.0152	0.2849	0.2850
	1000	−0.9924	0.2057	0.2056
$d = -0.9$				
GPH	100	−0.6899	0.3414	0.4005
	400	−0.6183	0.2579	0.3817
	1000	−0.6129	0.2339	0.3701
GPHT	100	−0.9266	0.5223	0.5225
	400	−0.9184	0.2845	0.2848
	1000	−0.8948	0.2062	0.2060
$d = -0.8$				
GPH	100	−0.6503	0.3300	0.3621
	400	−0.6078	0.2339	0.3021
	1000	−0.6137	0.1998	0.2730
GPHT	100	−0.8314	0.5201	0.5205
	400	−0.8209	0.2841	0.2846
	1000	−0.7970	0.2068	0.2066
$d = -0.7$				
GPH	100	−0.5957	0.3187	0.3350
	400	−0.5786	0.2128	0.2448
	1000	−0.5903	0.1725	0.2043
GPHT	100	−0.7360	0.5194	0.5201
	400	−0.7232	0.2841	0.2848
	1000	−0.6991	0.2074	0.2072
$d = -0.6$				
GPH	100	−0.5341	0.3051	0.3119
	400	−0.5337	0.1976	0.2083
	1000	−0.5395	0.1485	0.1602
GPHT	100	−0.6399	0.5192	0.5202
	400	−0.6253	0.2845	0.2854
	1000	−0.6008	0.2081	0.2079
$d = -0.5$				
GPH	100	−0.4280	0.3016	0.3097
	400	−0.4589	0.1760	0.1806
	1000	−0.4577	0.1401	0.1462
GPHT	100	−0.4807	0.5191	0.5189
	400	−0.5311	0.2948	0.2962
	1000	−0.5053	0.1975	0.1974

Table 11.2. Average estimated values of
$d: d = 0.5, \ldots, 0.9$

	n	$m = [n^{1/2}]$		
		$\bar{d}$	$s_{\hat{d}}$	RMSE
$d = 0.5$				
GPH	100	0.5349	0.3176	0.3192
	400	0.5232	0.1687	0.1701
	1000	0.5159	0.1372	0.1380
GPHT	100	0.5488	0.5354	0.5371
	400	0.4821	0.2979	0.2981
	1000	0.5045	0.1969	0.1968
$d = 0.6$				
GPH	100	0.6141	0.2987	0.2988
	400	0.6378	0.1817	0.1854
	1000	0.6225	0.1399	0.1416
GPHT	100	0.6055	0.5123	0.5118
	400	0.6218	0.2829	0.2835
	1000	0.6191	0.1989	0.1996
$d = 0.7$				
GPH	100	0.7492	0.3086	0.3121
	400	0.7363	0.1716	0.1753
	1000	0.7296	0.1376	0.1407
GPHT	100	0.7707	0.5397	0.5437
	400	0.6943	0.2972	0.2970
	1000	0.7132	0.1959	0.1961
$d = 0.8$				
GPH	100	0.8289	0.2820	0.2832
	400	0.8459	0.1757	0.1814
	1000	0.8312	0.1371	0.1404
GPHT	100	0.8338	0.4999	0.5005
	400	0.8351	0.2830	0.2849
	1000	0.8284	0.1972	0.1991
$d = 0.9$				
GPH	100	0.9355	0.2942	0.2961
	400	0.9286	0.1655	0.1678
	1000	0.9253	0.1300	0.1323
GPHT	100	1.0109	0.5219	0.5331
	400	0.9158	0.2885	0.2887
	1000	0.9254	0.1944	0.1959

decrease as the sample size increases. In fact, for very negative values of d, the bias increases with sample size. When $d = -1$, the RMSE also increases with n.

The presence of positive bias in the GPH estimator when $d < -0.5$ is consistent with the remark after Theorem 2 that GPH has expectation zero asymptotically if $d < -0.5$, no taper is used and m is held fixed. The bias results from the flattening

Table 11.3. Average estimated values of
$d : d = 1.0, \ldots, 1.4$

	n	$m = [n^{1/2}]$		
		$\bar{d}$	$s_{\hat{d}}$	RMSE
$d = 1.0$				
GPH	100	1.0063	0.2813	0.2811
	400	1.0047	0.1592	0.1591
	1000	0.9988	0.1200	0.1198
GPHT	100	1.1350	0.5159	0.5328
	400	1.0283	0.2864	0.2875
	1000	1.0326	0.1942	0.1967
$d = 1.1$				
GPH	100	1.0594	0.2604	0.2633
	400	1.0545	0.1554	0.1618
	1000	1.0466	0.1172	0.1287
GPHT	100	1.2618	0.5151	0.5394
	400	1.1426	0.2855	0.2883
	1000	1.1408	0.1945	0.1985
$d = 1.2$				
GPH	100	1.0745	0.2400	0.2706
	400	1.0734	0.1492	0.1956
	1000	1.0697	0.1214	0.1780
GPHT	100	1.3252	0.4886	0.5040
	400	1.2912	0.2762	0.2906
	1000	1.2584	0.1948	0.2031
$d = 1.3$				
GPH	100	1.0988	0.2308	0.3060
	400	1.0747	0.1482	0.2696
	1000	1.0681	0.1315	0.2665
GPHT	100	1.5417	0.4995	0.5545
	400	1.3861	0.2803	0.2929
	1000	1.3677	0.1962	0.2074
$d = 1.4$				
GPH	100	1.0871	0.2912	0.3819
	400	1.0676	0.1479	0.3638
	1000	1.0470	0.1061	0.3686
GPHT	100	1.6657	0.4822	0.5501
	400	1.5797	0.2592	0.3152
	1000	1.5047	0.2005	0.2260

effect of the log periodogram at the first few Fourier frequencies, explained theoretically in the discussion following Theorem 2, and illustrated in Fig. 11.7 for a simulated Gaussian ARFIMA $(0, -1, 0)$ series with $n = 1000$. We see that the plot of the log periodogram versus $\log |2 \sin(\omega_j/2)|$ is essentially horizontal at the first few Fourier frequencies. Thus the estimated coefficient, $-2\hat{d}$, of $\log |2 \sin(\omega_j/2)|$

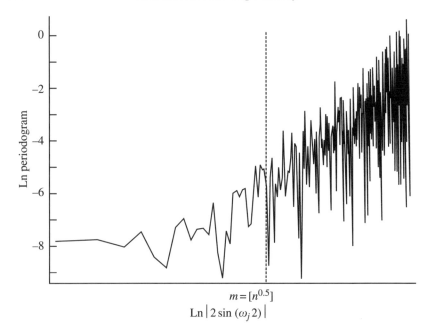

Fig. 11.7. Ln periodogram of simulated ARFIMA $(0, -1, 0)$ process, $n = 1000$.

based on $m = [n^{1/2}]$ Fourier frequencies is smaller than would be obtained using all the Fourier frequencies, resulting in a larger value for $\hat{d}$. For this example, taking $m = [n^{1/2}]$ gives $\hat{d} = -0.430$ and taking $m = [(n - 1)/2]$ gives $\hat{d} = -0.805$.

Table 11.1 also reveals that tapering the series and dropping the first periodogram ordinate eliminates the bias in $\hat{d}$ almost entirely when $d < -0.5$. This is consistent with numerical evaluation of the integrals appearing in Theorem 3 (see Fig. 11.5). For a stationary and invertible short-memory series, it is known that applying a cosine bell taper to a series increases the variance of the smoothed periodogram (Brillinger 1981: p. 151). In light of this fact, it is somewhat surprising that the standard error of GPHT is sometimes smaller than that of GPH when $d \leqslant -0.5$. This phenomenon could be a result of smaller correlations between normalized periodogram ordinates for the tapered data when d is very negative. The standard errors of the GPHT estimates increase relative to GPH as d gets closer to the stationary and invertible range, however. Based on RMSE, the GPHT estimates outperform the GPH estimates when $d = -1.0, -0.9, -0.8$ and $n = 400$ or 1000. The GPH estimates have a smaller RMSE when $d = -0.7, -0.6, -0.5$, perhaps due to the inflated variance of the tapered periodogram. The box-plots shown in Fig. 11.8 illustrate these results for $d = -1.0$ and $d = -0.7$. The broken line indicates the true value of d.

Table 11.2 summarizes our results for series having $d \in [0.5, 1)$. When d is in this region, the GPH estimates have a small positive bias which remains relatively constant for different values of d. Tapering reduces the bias slightly in

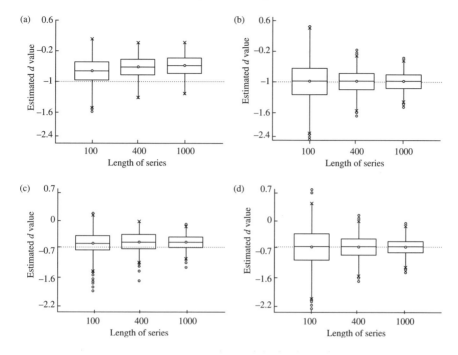

Fig. 11.8. Simulated GPH and GPHT estimates: (a) GPH, $d = -1.0$; (b) GPHT, $d = -1.0$; (c) GPH, $d = -0.7$; (d) GPHT, $d = -0.7$.

large samples, but increases the variance substantially. The box-plots shown in Figs 11.9(a) and (b) illustrate these results for $d = 0.7$. The RMSE of the GPH estimate is smaller than that of the GPHT estimate for each $d = 0.5, \ldots, 0.9$ at all sample sizes.

When $d = 1.0$, the estimates shown in Table 11.3 illustrate an interesting result, which is suggested by the asymptotic result of Theorem 1 and the remarks thereafter. For the random walk process, the GPH estimator is essentially unbiased! The unbiasedness of GPH with $d = 1$ has also been observed experimentally by Hassler (1993: table 3). Tapering in this case introduces a positive bias into the estimate, as well as increasing its variance. Figs 11.9(c) and (d) illustrate these results for $d = 1.0$.

When $d > 1$, the results given in Table 11.3 indicate that the GPH estimate of d has an average value of approximately 1, no matter the value of d, that is $\hat{d}$ appears to be negatively biased, with bias approximately equal to $1 - d$. From Fig. 11.9(e), it appears that $\hat{d}$ is reasonably close to 1 at least 50 per cent of the time when $d = 1.4$. For $n = 1000$, the first quartile of the estimates is 0.991, while the third quartile is 1.060. The conclusion that $E[\hat{d}] \approx 1$ for the GPH estimator when $d > 1$ is consistent with our theoretical results: numerical evaluation of the integrals in Theorem 1 when $d > 1$ reveals that the log of the limiting expected normalized periodogram is very nearly linear in log j with a slope of approximately

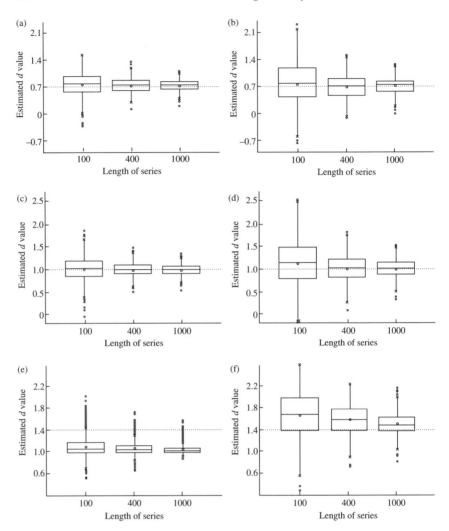

Fig. 11.9. Simulated GPH and GPHT estimates: (a) GPH, $d = 0.7$; (b) GPHT, $d = 0.7$; (c) GPH, $d = 1.0$; (d) GPHT, $d = 1.0$; (e) GPH, $d = 1.4$; (f) GPHT, $d = 1.4$.

$-2(1-d)$. For example, when $d = 1.4$, a least squares regression of the log of the limiting expected normalized periodogram on $\log j$ for $j = 1, \ldots, 100$ yielded $R^2 = 0.9998$ and a slope of 0.7836. The result implies that, in practice, obtaining an estimate of d close to 1 (as was the case in the example of Agiakloglou *et al.* mentioned in Section 11.1) does not necessarily indicate that a series follows a random walk. Tapering gives estimates which are somewhat *positively* biased. When $d = 1.4$, the RMSE of the GPHT estimates is smaller than that of the GPH estimates when $n = 400$ or 1000 due to the reduction in bias obtained by tapering.

Table 11.4. Average estimated values of d for
ARFIMA$(0, -0.7, 1)$ process

n	$m = [n^{1/2}]$		
	$\bar{d}$	$s_{\hat{d}}$	RMSE
$\theta = 0.5$			
GPH 100	-0.5527	0.3180	0.3502
400	-0.4919	0.2489	0.3242
1000	-0.4807	0.2216	0.3116
GPHT 100	-0.8162	0.5224	0.5346
400	-0.7328	0.2763	0.2780
1000	-0.7093	0.2015	0.2015

11.3.2 Periodogram regression applied to noninvertible fractionally integrated processes with short-memory components

Table 11.4 summarizes the performance of GPH and GPHT for an ARFIMA-$(0, -0.7, 1)$ process of the form $(1-B)^{-0.7}X_t = \varepsilon_t - \theta\varepsilon_{t-1}$ having MA parameter $\theta = 0.5$, where $\{\varepsilon_t\}$ is zero-mean Gaussian white noise. The bias of the GPH estimate is strongly positive, appears to be increasing with n and is larger than the bias in the corresponding case with no MA component. These results are not contradicted by our theory: although the discussion following Theorem 2 implies that a plot of the log periodogram versus $\log j$ should be essentially horizontal at the first few Fourier frequencies, the extent and vertical level of this flat stretch will depend on n and f^*. Thus, although the theory for $d < -0.5$ predicts that $E[\hat{d}] \approx 0.0$ if n is large and $\hat{d}$ is based on a fixed number of Fourier frequencies, it is not surprising that in practice, when $\hat{d}$ is based on $m = [n^{1/2}]$ Fourier frequencies, the bias of $\hat{d}$ depends on both n and f^*. The empirical results of Table 11.4 indicate that tapering eliminates the bias when n is large.

Our empirically observed biases for GPH when $d = -0.7$ differ greatly from the biases of -0.12 and -0.03 for $n = 100, 400$, (respectively) obtained from a theoretical calculation of the bias for the MA(1) process $X_t = \varepsilon_t - 0.5\varepsilon_{t-1}$ by Agiakloglou *et al.* (1993: table VII). However, their calculation applies only to the case $d = 0$.

11.3.3 Periodogram regression applied to nonstationary fractionally integrated processes with short-memory components

Table 11.5 summarizes the performance of GPH and GPHT for an ARFIMA-$(0, 0.7, 1)$ process of the form $(1 - B)^{0.7}X_t = \varepsilon_t - \theta\varepsilon_{t-1}$ having MA parameter $\theta = 0.5$ and an ARFIMA$(1, 0.7, 0)$ process of the form $(1-B)^{0.7}(1-\phi B)X_t = \varepsilon_t$ having AR parameter $\phi = 0.5$. For both processes, the bias in GPH is roughly equal to the sum of two contributions: (1) the bias attributable to the short-memory component, obtained theoretically by Agiakloglou *et al.* (1993: tables VII and III) for the MA(1) process $X_t = \varepsilon_t - 0.5\varepsilon_{t-1}$ and the AR(1) process $X_t = 0.5X_{t-1} + \varepsilon_t$;

Table 11.5. Average estimated values of d for
ARFIMA(0, 0.7, 1) and ARFIMA(1, 0.7, 0) processes

	n	$m = [n^{1/2}]$		
		$\bar{d}$	$s_{\hat{d}}$	RMSE
$\theta = 0.5$				
GPH	100	0.6270	0.2895	0.2982
	400	0.7007	0.1838	0.1836
	1000	0.7161	0.1462	0.1469
GPHT	100	0.5572	0.4989	0.5184
	400	0.6644	0.2916	0.2935
	1000	0.7031	0.2079	0.2077
$\phi = 0.5$				
GPH	100	0.8132	0.2979	0.3184
	400	0.7606	0.1815	0.1912
	1000	0.7316	0.1381	0.1415
GPHT	100	0.8901	0.5066	0.5406
	400	0.7677	0.2894	0.2970
	1000	0.7310	0.2063	0.2084

(2) the bias attributable to the long-memory component, obtained empirically from our simulations of the ARFIMA(0, 0.7, 0) model (Table 11.2). This claim can be verified from Tables 11.2 and 11.5 together with the short-memory biases, which are $-0.12, -0.03, -0.01$ for the MA(1) and $0.12, 0.03, 0.01$ for the AR(1) with $n = 100, 400$, and 900, respectively.

For the two processes studied here, compared with GPH, tapering increases the bias when $n = 100$ and $n = 400$. This phenomenon can be explained by the nonconstancy of the short-memory spectral density f^*, together with the fact that GPHT omits the first Fourier frequency while GPH does not. For both processes, the standard errors of $\hat{d}$ for both GPH and GPHT are essentially the same as found for the ARFIMA(0, 0.7, 0) in Table 11.2, in particular, GPHT is seen here to have a larger standard error than GPH.

11.3.4 Invariance to first-differencing of periodogram regression applied to nonstationary fractionally integrated noise

We also investigated the effect of first-differencing the data before computing a periodogram regression estimator. The estimate of d for the original series is then one plus the estimated value of d for the differenced series. As discussed previously, one might be tempted to apply this technique to a series which appears nonstationary. A desirable property of an estimator of d is that it be invariant to differencing. To investigate this property, we looked at the difference between the estimate of d for the nondifferenced data and one plus the estimate of d for the differenced data. The plots in Fig. 11.10 illustrate our results for nonstationary ARFIMA(0, d, 0) series having $d = 0.7, 1.0, 1.4$. In each case, the differences between the estimates

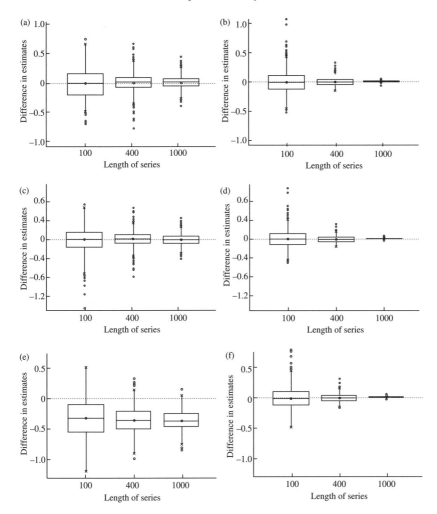

Fig. 11.10. Difference in estimates: (a) GPH, $d = 0.7$; (b) GPHT, $d = 0.7$; (c) GPH, $d = 1.0$; (d) GPHT, $d = 1.0$; (e) GPH, $d = 1.4$; (f) GPHT, $d = 1.4$.

tend to be much closer to zero for GPHT, indicating that tapering the data produces estimates which are more nearly invariant to first-differencing than GPH. Similar results were obtained for nonstationary processes having $d = 0.7$ and an AR or MA component.

Appendix 11.1

Proof of Lemma 1 We shall give a proof for all $d < 1.5$, although in the stationary case the result is already well known, and is easily proved using the spectral representation for $\{X_t\}$. Define $\{\varepsilon_t\} = \{\Delta X_t\}$. Then ε_t has memory parameter $d - 1$. Since $d < 1.5$ the memory parameter of $\{\varepsilon_t\}$ is less than 0.5, so that ε_t is weakly stationary with spectral density

$$f_\varepsilon(\lambda) = |1 - \exp(-i\lambda)|^{-2(d-1)} f^*(\lambda).$$

Furthermore, the process $\{\varepsilon_t\}$ has the spectral representation

$$\varepsilon_t = \int_{-\pi}^{\pi} \exp(i\lambda t)\, dZ_\varepsilon(\lambda).$$

Since the discrete Fourier transform of $\{x_t\}_{t=0}^{n-1}$ is equal to the discrete Fourier transform of $\{\sum_{k=0}^{t} \varepsilon_k\}_{t=0}^{n-1}$ for all integers $j \geq 1$, we obtain

$$J_j = \frac{1}{n} \sum_{t=0}^{n-1} x_t \exp(-i\omega_j t)$$

$$= \frac{1}{n} \sum_{t=0}^{n-1} \sum_{k=0}^{t} \varepsilon_k \exp(-i\omega_j t)$$

$$= \int_{-\pi}^{\pi} \frac{1}{n} \sum_{t=0}^{n-1} \sum_{k=0}^{t} \exp(i\lambda k) \exp(-i\omega_j t)\, dZ_\varepsilon(\lambda)$$

$$= \int_{-\pi}^{\pi} \frac{1}{n} \sum_{t=0}^{n-1} \exp\left(\frac{i\lambda t}{2}\right) \frac{\sin\{(t+1)\lambda/2\}}{\sin(\lambda/2)} \exp(-i\omega_j t)\, dZ_\varepsilon(\lambda)$$

$$= \int_{-\pi}^{\pi} \frac{1}{n \sin(\lambda/2)} \left[\sum_{t=0}^{n-1} \exp\left(\frac{i\lambda t}{2} - i\omega_j t\right) \sin\left\{\frac{(t+1)\lambda}{2}\right\}\right] dZ_\varepsilon(\lambda). \quad \text{(A.1)}$$

The bracketed term in eqn (A.1) is equal to

$$\sum_{t=0}^{n-1} \exp\left\{-it\left(-\frac{\lambda}{2} + \omega_j\right)\right\} \frac{1}{2i} \left[\exp\left\{\frac{i(t+1)\lambda}{2}\right\} - \exp\left\{\frac{-i(t+1)\lambda}{2}\right\}\right]$$

$$= \frac{1}{2i} \sum_{t=0}^{n-1} \left\{\exp\left(-i\omega_j t + \frac{i\lambda}{2} + i\lambda t\right) - \exp\left(-\frac{i\lambda}{2} - i\omega_j t\right)\right\}$$

$$= \frac{1}{2i} \sum_{t=0}^{n-1} \exp\{-i(\omega_j - \lambda)t\} \exp\left(\frac{i\lambda}{2}\right)$$

$$= \frac{1}{2i} \exp\left\{\frac{-i(n-1)(\omega_j - \lambda)}{2}\right\} \frac{\sin\{n(\omega_j - \lambda)/2\}}{\sin\{(\omega_j - \lambda)/2\}} \exp\left(\frac{i\lambda}{2}\right). \quad \text{(A.2)}$$

Therefore, we obtain

$$J_j = \int_{-\pi}^{\pi} \frac{1}{2i} \exp\left\{\frac{-i(n-1)(\omega_j - \lambda)}{2}\right\} \frac{\sin\{n(\omega_j - \lambda)/2\}}{n\sin(\lambda/2)\sin\{(\omega_j - \lambda)/2\}}$$
$$\times \exp\left(\frac{i\lambda}{2}\right) dZ_\varepsilon(\lambda).$$

From the properties of the spectral representation and the definition of the periodogram, we have

$$E(I_j) = E\left(\frac{n}{2\pi}|J_j|^2\right) = \frac{1}{8\pi n}\int_{-\pi}^{\pi} \frac{\sin^2\{n(\omega_j - \lambda)/2\}}{\sin^2(\lambda/2)\sin^2\{(\omega_j - \lambda)/2\}} f_\varepsilon(\lambda)\, d\lambda.$$

Since

$$f_\varepsilon(\lambda) = |1 - \exp(-i\lambda)|^{-2(d-1)} f^*(\lambda) = |1 - \exp(-i\lambda)|^2 f(\lambda) = 4\sin^2\left(\frac{\lambda}{2}\right) f(\lambda)$$

we find that

$$E\left(\frac{I_j}{f_j}\right) = \frac{1}{8\pi n f_j}\int_{-\pi}^{\pi} \frac{\sin^2\{n(\omega_j - \lambda)/2\}}{\sin^2\{(\omega_j - \lambda)/2\}} 4 f(\lambda)\, d\lambda$$
$$= \frac{1}{f_j}\int_{-\pi}^{\pi} F_n(\omega_j - \lambda) f(\lambda)\, d\lambda$$

as was required to show. ■

Proof of Theorem 1 The result was already proved for $d \in (-0.5, 0.5)$ in Hurvich and Beltrao (1993, th. 1). The same proof as given there carries through for the more general case $d \in (-0.5, 1.5)$, as long as the formula

$$E\left(\frac{I_j}{f_j}\right) = \frac{1}{f_j}\int_{-\pi}^{\pi} F_n(\omega_j - \lambda) f(\lambda)\, d\lambda \qquad (A.3)$$

continues to hold in this case. Since (A.3) for $d \in (-0.5, 1.5)$ follows from Lemma 1, the proof is complete. ■

Proof of Theorem 2 We have

$$
n^{2d+1} E\left(\frac{I_j}{f_j}\right)
$$

$$
= \frac{n^{2d+1}}{f_j} \int_{-\pi}^{\pi} F_n(\omega_j - \lambda) f(\lambda)\, d\lambda
$$

$$
= \frac{n^{2d+1}}{|1 - \exp(-i\omega_j)|^{-2d} f^*(\omega_j)} \int_{-\pi}^{\pi} \frac{\sin^2\{n(\omega_j - \lambda)/2\}}{2\pi n \sin^2\{(\omega_j - \lambda)/2\}}
$$

$$
\times |1 - \exp(-i\lambda)|^{-2d} f^*(\lambda)\, d\lambda
$$

$$
= \frac{|n \sin(\omega_j/2)|^{2d}}{2\pi f^*(\omega_j)} \int_{-\pi}^{\pi} \frac{\sin^2(n\lambda/2)}{\sin^2\{(\omega_j - \lambda)/2\}} \left|\sin\left(\frac{\lambda}{2}\right)\right|^{-2d} f^*(\lambda)\, d\lambda
$$

$$
= \frac{|n \sin(\omega_j/2)|^{2d}}{2\pi f^*(\omega_j)} \int_{-\pi}^{\pi} \sin^2\left(\frac{n\lambda}{2}\right)
$$

$$
\times \left[\frac{1}{\sin^2(\lambda/2)} + \frac{1}{\sin^2\{(\omega_j - \lambda)/2\}} - \frac{1}{\sin^2(\lambda/2)}\right] \left|\sin\left(\frac{\lambda}{2}\right)\right|^{-2d} f^*(\lambda)\, d\lambda
$$

so that

$$
\lim_{n\to\infty} n^{2d+1} E\left(\frac{I_j}{f_j}\right)
$$

$$
= \frac{|\pi j|^{2d}}{2\pi f^*(0)} \lim_{n\to\infty} \int_{-\pi}^{\pi} \frac{\sin^2(n\lambda/2)}{\sin^2(\lambda/2)} \left|\sin\left(\frac{\lambda}{2}\right)\right|^{-2d} f^*(\lambda)\, d\lambda
$$

$$
+ \frac{|\pi j|^{2d}}{2\pi f^*(0)} \lim_{n\to\infty} \int_{-\pi}^{\pi} \sin^2\left(\frac{n\lambda}{2}\right) \left[\frac{1}{\sin^2\{(\omega_j - \lambda)/2\}} - \frac{1}{\sin^2(\lambda/2)}\right]
$$

$$
\times \left|\sin\left(\frac{\lambda}{2}\right)\right|^{-2d} f^*(\lambda)\, d\lambda. \tag{A.4}
$$

Next, we use Lebesgue's Dominated Convergence Theorem to show that the second limit on the right-hand side of eqn (A.4) is zero. Since $\sin^2(n\lambda/2) = \sin^2\{n(\omega_j - \lambda)/2\}$, the integrand in the second term on the right-hand side of eqn (A.4) is

$$
g_n(\lambda)
$$

$$
= \sin^2\left\{\frac{n(\omega_j - \lambda)}{2}\right\} \frac{\sin^2(\lambda/2) - \sin^2\{(\omega_j - \lambda)/2\}}{\sin^2\{(\omega_j - \lambda)/2\} \sin^2(\lambda/2)} \left|\sin\left(\frac{\lambda}{2}\right)\right|^{-2d} f^*(\lambda)
$$

$$
= \frac{\sin^2\{n(\omega_j - \lambda)/2\}}{\sin^2\{(\omega_j - \lambda)/2\}} \left\{\sin^2\left(\frac{\lambda}{2}\right) - \sin^2\left(\frac{\omega_j - \lambda}{2}\right)\right\} \left|\sin\left(\frac{\lambda}{2}\right)\right|^{-2d-2} f^*(\lambda). \tag{A.5}
$$

Since

$$\sin^2\left(\frac{\lambda}{2}\right) - \sin^2\left(\frac{\omega_j - \lambda}{2}\right)$$

$$= \sin^2\left(\frac{\lambda}{2}\right) - \left\{\sin\left(\frac{\omega_j}{2}\right)\cos\left(\frac{\lambda}{2}\right) - \cos\left(\frac{\omega_j}{2}\right)\sin\left(\frac{\lambda}{2}\right)\right\}^2$$

$$= \sin^2\left(\frac{\lambda}{2}\right) - \sin^2\left(\frac{\omega_j}{2}\right)\cos^2\left(\frac{\lambda}{2}\right) - \cos^2\left(\frac{\omega_j}{2}\right)\sin^2\left(\frac{\lambda}{2}\right)$$

$$+ 2\sin\left(\frac{\omega_j}{2}\right)\cos\left(\frac{\omega_j}{2}\right)\sin\left(\frac{\lambda}{2}\right)\cos\left(\frac{\lambda}{2}\right)$$

$$= \sin^2\left(\frac{\lambda}{2}\right)\sin^2\left(\frac{\omega_j}{2}\right) - \sin^2\left(\frac{\omega_j}{2}\right)\cos^2\left(\frac{\lambda}{2}\right) + \frac{1}{2}\sin\omega_j\sin\lambda$$

$$= \sin^2\left(\frac{\omega_j}{2}\right)\left\{\sin^2\left(\frac{\lambda}{2}\right) - \cos^2\left(\frac{\lambda}{2}\right)\right\} + \frac{1}{2}\sin\omega_j\sin\lambda,$$

there exist constants k_1, k_2 and k_3 such that for all sufficiently large n,

$$\left|\sin^2\left(\frac{\lambda}{2}\right) - \sin^2\left(\frac{\omega_j - \lambda}{2}\right)\right|$$

$$\leqslant k_1 n^{-2} + k_2 n^{-1}(k_3 n^{-1}\chi_{(\omega_{j-1},\omega_{j+1})} + |\lambda|\chi_{(\omega_{j-1},\omega_{j+1})}c),$$

where χ_A is the indicator function of the set A. From the elementary properties of the Fejér kernel, and since $|\sin\{(\omega_j - \lambda)/2\}| \geqslant \frac{1}{4}|(\omega_j - \lambda)/2|$ for $\lambda \in [-\pi, \pi]$ and for all sufficiently large n, there exist constants k_4 and k_5 such that for all sufficiently large n,

$$\frac{\sin^2\{n(\omega_j - \lambda)/2\}}{\sin^2\{(\omega_j - \lambda)/2\}} \leqslant k_4 n^2 \chi_{(\omega_{j-1},\omega_{j+1})} + \frac{k_5}{(\omega_j - \lambda)^2}\chi_{(\omega_{j-1},\omega_{j+1})}c.$$

Since $\sup_{\lambda\in[-\pi,\pi]}|f^*(\lambda)| \leqslant M$ for some constant M we conclude from eqn (A.5) and the preceding discussion that for all sufficiently large n,

$$|g_n(\lambda)| \leqslant M\left|\sin\left(\frac{\lambda}{2}\right)\right|^{-2d-2}\{k_4 n^2 \chi_{(\omega_{j-1},\omega_{j+1})}$$

$$+ k_5(\omega_j - \lambda)^{-2}\chi_{(\omega_{j-1},\omega_{j+1})}c\}\{k_1 n^{-2} + k_2 n^{-1}(k_3 n^{-1}\chi_{(\omega_{j-1},\omega_{j+1})}$$

$$+ |\lambda|\chi_{(\omega_{j-1},\omega_{j+1})}c)\}$$

$$\leqslant [(k_4 k_1 + k_4 k_2 k_3)\chi_{(\omega_{j-1},\omega_{j+1})} + \{k_5 k_1(\omega_j - \lambda)^{-2}n^{-2}$$

$$+ k_5 k_2|\lambda|(\omega_j - \lambda)^{-2}n^{-1}\}\chi_{(\omega_{j-1},\omega_{j+1})}c]M\left|\sin\left(\frac{\lambda}{2}\right)\right|^{-2d-2}.$$

Since $(\omega_j - \lambda)^{-2} n^{-2} \chi_{(\omega_{j-1}, \omega_{j+1})} c \leqslant 1/(4\pi^2)$, and

$$|\lambda|(\omega_j - \lambda)^{-2} n^{-1} \chi_{(\omega_{j-1}, \omega_{j+1})} c \leqslant \frac{|\omega_j| + |\lambda - \omega_j|}{n(\omega_j - \lambda)^2} \chi_{(\omega_{j-1}, \omega_{j+1})} c \leqslant \frac{2\pi j}{4\pi^2} + \frac{1}{2\pi},$$

we conclude that there exists a finite constant M_2 such that for all sufficiently large n,

$$|g_n(\lambda)| \leqslant M_2 \left| \sin \left(\frac{\lambda}{2} \right) \right|^{-2d-2}$$

which is integrable on $[-\pi, \pi]$ since $d < -0.5$. Since $g_n(\lambda) \to 0$ almost everywhere, it follows from Lebesgue's Dominated Convergence Theorem that $\lim_{n\to\infty} \int_{-\pi}^{\pi} g_n(\lambda) \, d\lambda = 0$. Therefore, eqn (A4) simplifies to

$$\lim_{n\to\infty} n^{2d+1} E \left(\frac{I_j}{f_j} \right) = \frac{|\pi j|^{2d}}{2\pi f^*(0)} \lim_{n\to\infty} \int_{-\pi}^{\pi} \sin^2 \left(\frac{n\lambda}{2} \right) \left| \sin \left(\frac{\lambda}{2} \right) \right|^{-2d-2} f^*(\lambda) \, d\lambda.$$
$$(A.6)$$

Now, using the fact that

$$\sin^2 \left(\frac{n\lambda}{2} \right) = \frac{1}{2} \{1 - \cos(n\lambda)\} = \frac{1}{2} \left\{ 1 - \frac{\exp(in\lambda) + \exp(-in\lambda)}{2} \right\},$$

Eqn (A.6) reduces to

$$\lim_{n\to\infty} n^{2d+1} E \left(\frac{I_j}{f_j} \right) = \frac{|\pi j|^{2d}}{4\pi f^*(0)} \int_{-\pi}^{\pi} \left| \sin \left(\frac{\lambda}{2} \right) \right|^{-2d-2} f^*(\lambda) \, d\lambda$$

$$- \frac{|\pi j|^{2d}}{8\pi f^*(0)} \lim_{n\to\infty} \int_{-\pi}^{\pi} \left| \sin \left(\frac{\lambda}{2} \right) \right|^{-2d-2} f^*(\lambda) \{\exp(in\lambda)$$

$$+ \exp(-in\lambda)\} \, d\lambda. \qquad (A.7)$$

By the Riemann–Lebesgue theorem, the second term on the right-hand side of eqn (A.7) is zero, and the proof is finished. ∎

Proof of Theorem 3 The tapered discrete Fourier transform is

$$J_{T,j} = \frac{1}{\sum w_t^2} \sum_{t=0}^{n-1} w_t x_t \exp(-i\omega_j t).$$

It follows from Bloomfield (1976: pp. 82–4) that

$$\sum_{t=0}^{n-1} w_t \exp\{-i(\omega_j - \lambda)t\} = \exp \left\{ \frac{-i(n-1)(\omega_j - \lambda)}{2} \right\} \sin \left\{ \frac{n(\omega_j - \lambda)}{2} \right\} H_{j,n}(\lambda),$$

$$(A.8)$$

where

$$H_{j,n}(\lambda) = \left[\frac{0.5}{\sin\{(\omega_j - \lambda)/2\}} - \frac{0.25}{\sin\{(\omega_{j-1} - \lambda)/2\}} - \frac{0.25}{\sin\{(\omega_{j+1} - \lambda)/2\}} \right].$$

Putting $\lambda = 0$ in eqn (A.8), we find that for $j > 1$, $\sum_{t=0}^{n-1} w_t \exp(-i\omega_j t) = 0$. Multiplying both sides of eqn (11.2) by w_t, and using an argument similar to the proofs of eqns (A.1) and (A.2), we obtain

$$J_{\text{T},j} = \int_{-\pi}^{\pi} \frac{1}{\sum w_t^2 \sin(\lambda/2)} \left[\sum_{t=0}^{n-1} w_t \exp\left(\frac{i\lambda t}{2} - i\omega_j t \right) \sin\left\{ \frac{(t+1)\lambda}{2} \right\} \right] dZ_\varepsilon(\lambda)$$

$$= \int_{-\pi}^{\pi} \frac{1}{2i \sum w_t^2 \sin(\lambda/2)} \sum_{t=0}^{n-1} w_t \exp\{-i(\omega_j - \lambda)t\} \exp\left(\frac{i\lambda}{2} \right) dZ_\varepsilon(\lambda)$$

$$= \int_{-\pi}^{\pi} \frac{1}{2i \sum w_t^2 \sin(\lambda/2)} \exp\left\{ \frac{-i(n-1)(\omega_j - \lambda)}{2} \right\} \exp\left(\frac{i\lambda}{2} \right)$$

$$\times \sin\left\{ \frac{n(\omega_j - \lambda)}{2} \right\} H_{j,n}(\lambda) \, dZ_\varepsilon(\lambda).$$

Consequently, for $j > 1$, with $\mu = n\lambda$,

$$E\left(\frac{I_{\text{T},j}}{f_j} \right) = \frac{1}{8\pi \sum w_t^2 |2 \sin(\omega_j/2)|^{-2d} f^*(\omega_j)}$$

$$\times \int_{-\pi}^{\pi} \frac{\sin^2(n\lambda/2)}{\sin^2(\lambda/2)} H_{j,n}^2(\lambda) \left| 2 \sin\left(\frac{\lambda}{2} \right) \right|^{-2(d-1)} f^*(\lambda) \, d\lambda$$

$$= \frac{1}{2\pi \sum w_t^2 |\sin(\omega_j/2)|^{-2d} f^*(\omega_j)}$$

$$\times \int_{-\pi}^{\pi} \sin^2\left(\frac{n\lambda}{2} \right) \left| \sin\left(\frac{\lambda}{2} \right) \right|^{-2d} f^*(\lambda) H_{j,n}^2(\lambda) \, d\lambda$$

$$= \frac{1}{2\pi \sum w_t^2 |\sin(\omega_j/2)|^{-2d} f^*(\omega_j)}$$

$$\times \int_{-n\pi}^{n\pi} \sin^2\left(\frac{\mu}{2} \right) \left| \sin\left(\frac{\mu}{2n} \right) \right|^{-2d} f^*\left(\frac{\mu}{n} \right) H_{j,n}^2\left(\frac{\mu}{n} \right) \frac{1}{n} \, d\mu. \quad \text{(A.9)}$$

It can be shown that Lebesgue's Dominated Convergence Theorem can be applied to eqn (A.9) when $d \in (-2.5, 1.5)$, so that

$$\lim_{n \to \infty} E\left\{ \frac{I_{\text{T}}(\omega_j)}{f(\omega_j)} \right\} = \int_{-\infty}^{\infty} \lim_{n \to \infty} \frac{1}{2\pi \sum w_t^2 |\sin(\omega_j/2)|^{-2d} f^*(\omega_j)}$$

$$\times \sin^2\left(\frac{\mu}{2} \right) \left| \sin\left(\frac{\mu}{2n} \right) \right|^{-2d} f^*\left(\frac{\mu}{n} \right) H_{j,n}^2\left(\frac{\mu}{n} \right) \frac{1}{n} \, d\mu.$$

$$\text{(A.10)}$$

Now, using the facts that

$$\frac{1}{n}\sum_{t=0}^{n-1} w_t^2 \rightarrow \frac{3}{8}$$

and $f^*(\mu/n)/f^*(\omega_j) \rightarrow 1$, a straightforward calculation shows that the right-hand side of eqn (A.10) is

$$(2\pi j)^{2d}\frac{2}{\pi}\frac{8}{3}\int_{-\infty}^{\infty}$$

$$\times \sin^2\left(\frac{\mu}{2}\right)\left\{\frac{0.5}{2\pi j - \mu} - \frac{0.25}{2\pi(j-1) - \mu} - \frac{0.25}{2\pi(j+1) - \mu}\right\}^2 |\mu|^{-2d}\,d\mu. \ \blacksquare$$

References

Agiakloglou, C., Newbold, P. and Wohar, M. (1993), 'Bias in an estimator of the fractional difference parameter', *Journal of Time Series Analysis*, 14, 235–46.

Bloomfield, P. (1976), *Fourier Analysis of Time Series: An Introduction*. New York: Wiley.

—— (1991), 'Time series methods'. In *Statistical Theory and Modeling* in D. V. Hinkley, N. Reid and E. J. Snell, eds, London: Chapman and Hall, 152–76.

Brillinger, D. R. (1981), *Time Series: Data Analysis and Theory*, expanded edn. New York: Holt, Rinhart and Winston.

Cheung, Y.-W. and Lai, K. S. (1993), 'A fractional cointegration analysis of purchasing power parity', *Journal of Business Economics and Statistics*, 11, 103–12.

Crato, N. (1992), 'Spectral analysis of nonstationary time series'. Department of Mathematical Sciences, University of Delaware.

Dickey, D. A. and Fuller, W. A. (1979), 'Distribution of the estimators for autoregressive time series with a unit root', *Journal of the American Statistical Association*, 74, 427–31.

Diebold, F. X. and Rudebusch, G. D. (1991), 'On the power of Dickey Fuller tests against fractional alternatives', *Econometrics Letters*, 35, 155–60.

Geweke, J. and Porter-Hudak, S. (1983), 'The estimation and application of long memory time series models', *Journal of Time Series Analysis*, 4, 221–38.

Granger, C. W. J. and Joyeux, R. (1980), 'An introduction to long-memory time series models and fractional differencing', *Journal of Time Series Analysis*, 1, 15–29.

Hassler, U. (1993), 'Unit root tests: the autoregressive approach in comparison with the periodogram regression', *Statistische Hefte*, 67–82.

Hosking, J. R. M. (1981), 'Fractional differencing', *Biometrika*, 68, 165–76.

—— (1984), 'Modeling persistence in hydrological time series using fractional differencing', *Water Resources Research*, 20, 1898–908.

Hurvich, C. M. and Beltrao, K. I. (1993), 'Asymptotics the low-frequency ordinates of the periodogram of a long memory time series', *Journal of Time Series Analysis*, 14, 455–72.

Robinson, P. M. (1991), 'Log-periodogram regression of time series with long range dependence', London School of Economics.

Saikkonen, P. and Luukkonen, R. (1993), 'Testing for a moving average unit root in autoregressive integrated moving average models', *Journal of the American Statistical Association*, 88, 596–601.

Solo, V. (1990), 'Intrinsic random functions and the paradox of $1/f$ noise'. Department of Electrical and Computer Engineering, Johns Hopkins University.

12

Limit Theorems for Regressions with Unequal and Dependent Errors

FRIEDJIELM EICKER

12.1 Introduction (Notations)

There exist a considerable number of publications dealing with the asymptotic normality of parameter estimates for linear regressions, many of which deal with specific cases, however, or are unnecessarily narrow in the assumptions made. The most general paper among these, and the one closest to the present note, seems to be that of Hannan (1961). In that paper vectorial regression equations are considered while we are predominantly interested in scalar ones. For that case, however, the assumptions of Hannan (1961) are more restrictive than those of the present note.

For the individual components of the vectorial LSE the asymptotic normality was already proved under general assumptions in Eicker (1963a) for the case of independent errors.

The system of (scalar) linear regression equations is denoted by

$$y_t = x_{t1}\beta_1 + \cdots + x_{tq}\beta_q + \epsilon_t, \quad t = 1, \ldots, n, \tag{12.1}$$

$n \geq q$ being the sample size. In matrix notation this becomes

$$y(n) = X_n\beta + \epsilon(n), \tag{12.2}$$

where $y(n)$ is the (column) vector of observations (n-dimensional), $X_n = (x_{tj})$, the $(n \times q)$-matrix of known regression constants assumed to be of full rank throughout, $\beta = (\beta_1, \ldots, \beta_q)'$ is the vector of unknown regression parameters ($'$ denotes the transpose), and $\epsilon(n) = (\epsilon_1, \ldots, \epsilon_n)'$ is the n-vector of error random variables (r.v.'s) about which we assume throughout that

$$E\epsilon_t = 0, \quad 0 < E\epsilon_t^2 < \infty, \quad \text{for all } t. \tag{12.3}$$

All quantities are real.

This work was first published as Eicker, F. 'Limit theorems for regressions with unequal and dependent errors' in L. Le Cam and J. Neyman (1967) *Proceedings of the Fifth Berkeley Symposium on Mathematical Statistics and Probability*. Copyright © 1967 The Regents of the University of California. Reproduced by kind permission.

Research supported in part by National Science Foundation Grant NSF-GP-3694 at Columbia University.

Let $P_n = X'_n X_n$. Then the vectorial LSE for β, denoted by $b(n) = (b_1(n), \ldots, b_q(n))'$, become

$$b(n) = P_n^{-1} X'_n y(n) = \beta - 1 - P_n^{-1} X'_n \epsilon(n). \tag{12.4}$$

The row vectors of X_n will be denoted by $r'_1, \ldots, r'_n$, and the column vectors by $x_1(n), \ldots, x_q(n)$. By F, we denote a (nonempty) set of d.f.'s whose elements G have the properties

$$\int x \, dG(x) = 0, \qquad 0 < \int x^2 \, dG(x) < \infty. \tag{12.5}$$

12.2 Independent Nonidentically Distributed Errors

12.2.1 The asymptotic normality of the $b(n)$
In order to find a limiting d.f., the vectors

$$b(n) - \beta = P_n^{-1} X'_n \epsilon(n), \tag{12.6}$$

have to be normalized by premultiplication by certain matrices B_n. Let

$$\Sigma_n = \operatorname{cov} \epsilon(n) \epsilon'(n) = \operatorname{diag}(\sigma_1^2, \ldots, \sigma_n^2), \quad \sigma_k^2 = \operatorname{var} \epsilon_k, \tag{12.7}$$

be the covariance matrix of the error vector, and write

$$B_n^2 = P_n^{-1} X'_n \Sigma_n X_n P_n^{-1}. \tag{12.8}$$

If B_n is the unique positive definite square root of B_n^2, the q-vectors

$$B_n^{-1} P_n^{-1} X'_n \epsilon(n), \tag{12.9}$$

all have expectation zero and covariance matrix I_q ($= q$-dimensional identity matrix).

Let $\mathcal{F}(F)$ be the set of all sequences $\epsilon \equiv \{\epsilon_1, \epsilon_2, \ldots\}$ of independent error r.v.'s ϵ_t (independent within each sequence) whose d.f.'s belong to some set F subject to (12.5). For a given sequence ϵ, the vectors $\epsilon(n)$ have as components the first n members of ϵ, $n = 1, 2, \ldots$.

Theorem 1 of Eicker then applies without further ado and yields the following theorem.

Theorem 1 The d.f.'s of $B_n^{-1} P_n^{-1} X'_n \epsilon(n)$ tend to the q-dimensional normal d.f. N $(0, I_q)$ and the summands of $B_n^{-1} P_n^{-1} X'_n \epsilon(n)$, are infinitesimal both uniformly for all sequences $\epsilon \equiv \{\epsilon_1, \epsilon_2, \ldots\} \in \mathcal{F}(F)$, if and only if the following three conditions are satisfied:

Condition 1 $\qquad \max_{k=1,\ldots,n} r_k' P_n^{-1} r_k \to 0,$

Condition 2 $\qquad \sup_{G \in F} \int_{|x|>c} x^2 \, dG(x) \to 0, \quad \text{as } c \to \infty,$

Condition 3 $\qquad \inf_{G \in F} \int x^2 \, dG(x) > 0.$

(All limits throughout the chapter hold for $n \to \infty$ unless otherwise stated.)

The fact that the assertion of the theorem holds for all sequences $\{\epsilon_t\} \in \mathcal{F}(F)$ makes it particularly useful in practice, since one usually does not know the error d.f.'s if they are not identical. It may also be pointed out that condition (1) on the regression matrices does not necessitate any knowledge about the error sequence present in a particular regression. Analogously, (2) and (3) concern only the set F of admissible error d.f.'s. If the ordinary clt were applied, one would obtain conditions concerning, simultaneously, the error sequence and the regression sequences. It is interesting that the consideration of the whole class $\mathcal{F}(F)$ implies the necessity of the Conditions (1)–(3). Condition (2) means *uniform integrability* of the variance integrals with respect to the class F.

As it stands, Theorem 1 is still of limited practical use since the normalizing matrices require the knowledge of the usually unknown error variances σ_k^2. Applying a law of large numbers for nonnegative random variables (Gnedenko and Kolmogorov 1954: p. 143), this defect can be removed by replacing σ_k^2 by the square of the kth residual

$$e_k(n) = y_k - r_k' b(n) = \epsilon_k - r_k' P_n^{-1} X_n' \epsilon(n), \quad k = 1, \ldots, n. \qquad (12.10)$$

The matrix $D_n^2 \equiv P_n^{1/2} B_n^2 P_n^{1/2}$ is then replaced by

$$C_n^2 = P_n^{-1/2} X_n' S_n X_n P_n^{-1/2}, \qquad (12.11)$$

with $S_n = \text{diag}\,(e_1^2(n), \ldots, e_n^2(n))$. This replacement amounts to an estimation of the matrix D_n^2 in the sense of (12.14), as will be shown below. After this substitution the estimator

$$C_n^{-1/2} P_n^{-1} X_n' y(n), \qquad (12.12)$$

no more contains any unknown quantity. Without any new assumptions we then obtain the next theorem.

Theorem 2 Under the assumptions (1)–(3) of the preceding theorem, the d.f.'s of

$$C_n^{-1} P_n^{-1/2} (b(n) - \beta) \qquad (12.13)$$

tend to $N(0, I_q)$ uniformly for all error sequences $\epsilon \in \mathcal{F}(F)$.

For the proof we need a lemma on matrices whose entries are random variables.

Lemma 1 A sequence of symmetrical random $q \times q$-matrices $A_n \to I_q$, i.p. if and only if $c'A_nc \to 1$, i.p. for all unimodular constant q-vectors c.

Proof The 'only if' part follows from Slutsky's theorem. To show the converse, first take $c = v_k$ (= kth unit vector), $k = 1, \ldots, q$, and then c proportional to $v_k + v_i$ any pair $k \neq j$. ∎

Proof of Theorem 2 We show first

$$D_n^{-1}C_n^2 D_n^{-1} \to I_q, \quad \text{i.p.} \tag{12.14}$$

We introduce the vectors

$$(u_1(n), \ldots, u_n(n))' = c'D_n^{-1}P_n^{-1/2}X_n', \quad n = q, q+1, \ldots,$$

with some unimodular constant q-vector c. Then

$$c'D_n^{-1}C_n^2 D_n^{-1}c = \sum_{k=1}^{n} u_k^2(n)e_k^2(n). \tag{12.15}$$

Since $\sum_k u_k(n)\epsilon_k = c'D_n^{-1}P_n^{-1/2}(b(n)-\beta)$ is a sum of independent infinitesimal r.v.'s whose d.f. for $n \to \infty$ tends to $N(0, 1)$ as a consequence of Theorem 1, we have

$$\sum_k u_k^2(n)\epsilon_k^2 \to 1, \quad \text{i.p.} \tag{12.16}$$

by theorem 4 of Gnedenko and Kolmogorov (1954: p. 443).

Taking account of the second terms of $e_k(n)$ as given by (12.10), we have

$$E(r_k'P_n^{-1}X_n'\epsilon(n))^2 \le Mr_k'P_n^{-1}r_k \tag{12.17}$$

where the existence of $M = \sup_G \int x^2 dG(x) < \infty$ is implied by Condition (2). Putting $m = \inf_G \int x^2 dG(x)[> 0]$ and denoting by $\|\cdot\|$ the Euclidean norm, we have

$$\sum_k u_k^2(n) = c'D_n^{-1}c \le \frac{1}{m} \tag{12.18}$$

Hence, by Condition 1,

$$E\left(\sum_k u_k^2(n)(r_k'P_n^{-1}X_n'\epsilon(n))^2\right) \le (M/m)\max_k r_k'P_k^{-1}r_k \to 0, \tag{12.19}$$

and consequently, $\sum_k u_k^2(n)(r_k'P_n^{-1}X_n'\epsilon(n))^2 \to 0$, i.p. Finally, $\sum_k u_k^2(n) \times e_k^2(n) \to 1$, i.p. for all unimodular vectors c. Because of lemma 1, this proves (12.14).

We now prove

$$C_n - D_n \to 0, \quad \text{i.p.} \tag{12.20}$$

Put $E_n = C_n^2 - D_n^2$. By (12.14) there exists a sequence of events Ω_n with $P\Omega_n \to 1$ such that $\sup_{\Omega_n} \|E_n\| \to 0$ where $\|E_n\| = \max_{i,j=1,\ldots,q} |(E_n)_{ij}|$. In the following all quantities, and equations in quantities, showing the index n are considered only on the event Ω_n, $n = q, q+1, \ldots$.

Putting $(D_n^2 + E_n)^{1/2} = D_n + E_n^*$, we then have to show $\|E_n^*\| \to 0$, that is, according to our convention, $\sup_{\Omega_n} \|E_n^*\| \to 0$. Now $E_n = D_n E_n^* + E_n^* D_n + E_n^{*2}$; E_n^*, being a real symmetric matrix, has only real characteristic values (c.r.'s), to be denoted by $\lambda_{1n}(\omega) \geq \lambda_{2n}(\omega) \geq \cdots \geq \lambda_{qn}(\omega)$, $\omega \in \Omega_n$, and possesses q orthogonal real characteristic vectors (c.v.'s). Suppose $\sup_{\omega \in \Omega_n} \lambda_{1n}(\omega) \equiv \Lambda_n > 0$ for an infinite set Γ of naturals. We assume for simplicity that there exists a matrix $E_n^*(\omega_n)$, $\omega_n \in \Omega_n$, that actually possesses Λ_n as a c.r. (otherwise we can always find an E_n^* whose maximum c.r. differs arbitrarily little from Λ_n). Let $v_n \in S_q$ (the unit sphere $\subset R_q$) be a c.v. of $E_n^*(\omega_n)$ associated with Λ_n. Then at $\omega = \omega_n$ respectively,

$$v_n E_n v_n = 2\Lambda_n v_n' D_n v_n + \Lambda_n^2 \to 0, \quad n \in \Gamma, \tag{12.21}$$

since $\|E_n\| \to 0$ and the c.v. of D_n are bounded between the finite positive constants m and M.

On the other hand, suppose $\inf_{\omega \in \Omega_{ln}} \lambda_{qn}(\omega) = \lambda_n < 0$ on some infinite set Γ' of integers, and let $u_n \in S_q$ be a c.v. associated with λ_n and a suitable matrix E_n^*. Then again $u_n' E_n u_n = \lambda_n(\lambda_n + 2u_n' D_n u_n) \to 0$, $n \in \Gamma'$, and hence either $\lambda_n \to 0$ or $\lambda_n + 2u_n' D_n u_n \to 0$ for some infinite sequence $\Gamma'' \subset \Gamma'$. But $0 < u_n'(D_n^2 + E_n)^{1/2} u_n = u_n' D_n u_n(1-2) + o(n)$ for $n \in \Gamma''$ which is impossible. Thus all c.r.'s of E_n^* tend to zero, which implies (12.20).

Finally, (12.20) implies $C_n^{-1} D_n \to I_q$ i.p., and premultiplication of

$$D_n^{-1} P_n^{-1/2} X_n' \epsilon(n)$$

yields the assertion.

Uniformity in $\epsilon \in \mathcal{F}(F)$ of (12.13) follows from the fact that the preceding proof remains valid if instead of one and the same ϵ for each n, we take for each n an arbitrary $\epsilon(n) \in \mathcal{F}(F)$. Thus, (12.13) holds for all sequences of sequences $\epsilon(n)$, and this is equivalent with uniformity in ϵ. ∎

12.2.2 Remarks

(1) In practice, for finite n, one uses Theorem 2 in the form

$$\text{d.f. } (b(n)) \sim N(\beta, C_n^2). \tag{12.22}$$

In certain situations this relation may save the trouble of computing the inverse square root of C_n^2.

(2) Theorem 2 of Eicker (1963a) states the asymptotic normality of the single components $b_j(n)$, after suitable normalization. We remark without proof that this theorem also remains valid under unchanged assumptions if, as in Section 12.3.1, the unknown variances σ_k^2 in the normalizing factor

are replaced by the squares of the residuals (12.10). Thus, the additional assumptions given in theorem 3 of Eicker (1963a) are in fact superfluous.

The progress of the present chapter over Eicker (1963a) lies essentially in the determination of the joint asymptotic d.f. of the vectorial LSE $b(n)$ which was not possible by the method used in Eicker (1963a). Besides that, Condition (1) of the present chapter is simpler than the corresponding condition in Eicker (1963a).

(3) If F contains only one element, say G, then Conditions (2) and (3) reduce to $0 < \int x^2 \, dG(x) < \infty$. The errors are, in this case, identically distributed.

(4) The assumptions (1)–(3) are no longer necessary in Theorem 2. However, since they are necessary in Theorem 1, the necessary and sufficient assumptions of Theorem 2 presumably do not differ very much from Conditions 1–3.

(5) Concerning the admissible sequences of regression matrices we prove the following lemma.

Lemma 2 Condition (1) implies

$$\lambda_{\min}(P_n) \to \infty$$

and

$$\max_{k=1,\dots,n} |x_{k,j}|/\|x_j(n)\| \to 0, \quad j = 1, \dots, q. \tag{12.23}$$

Here $\lambda_{\min}$ denotes the minimum characteristic value and $\|\cdot\|$ the Euclidean norm. Regression vectors, $x_j(n)$ satisfying (12.23) are called *slowly increasing* (compare: Grenander and Rosenblatt, 1957, p. 233).

Proof We introduce the $q \times q$ diagonal matrices

$$D_n = \operatorname{diag}(\|x_1(n)\|, \dots, \|x_q(n)\|), \quad n = 1, 2, \dots. \tag{12.24}$$

Since $\operatorname{tr}(D_n^{-1} P_n D_n^{-1}) = q$, we have

$$\sup_n \lambda_{\max}(D_n^{-1} P_n D_n^{-1}) \le q.$$

Now

$$r_k' P_n^{-1} r_k = r_k' D_n^{-1} (D_n^{-1} P_n D_n^{-1})^{-1} D_n^{-1} r_k \ge q^{-1} \|D_n^{-1} r_k\|^2. \tag{12.25}$$

By (1), $\max_k \|D_n^{-1} r_k\| \to 0$ and thus (12.23) follows. Equation (12.23) implies $\|x_j(n)\| \to \infty$, for all j, which in turn implies (12.22).

(6) Sufficient for (1) is the relation (12.23) together with

$$\inf_n \lambda_{\min}(D_n^{-1} P_n D_n^{-1}) > 0, \tag{12.26}$$

as can be seen from an inequality similar to (12.25). However, Conditions (12.26) plus (12.23) are, in general, not necessary. In particular, (12.26) is satisfied if

$$D_n^{-1} P_n D_n^{-1} \to R, \tag{12.27}$$

where R is some positive definite $q \times q$ matrix. ∎

12.2.3 Examples

We now discuss some examples of regression matrices and check whether they possess property (1) or not.

(1) *Polynominal regression*. Let $x_{k,j} = k^{cj}, c_1 > \cdots > c_q > -\frac{1}{2}; j = 1, \ldots, q, k = 1, 2, \ldots$. The c_j need not be integers. Then (compare: Eicker, 1963b, p. 469) the $x_j(n)$ are slowly increasing and $D_n^{-1} P_n D_n^{-1} \to H$ where H is a positive definite sub-matrix of the Hilbert matrix. Hence, (12.27), and consequently Condition (1), is satisfied.

(2) *Trigonometric regression*. Let

$$x_{k,2j-1} = \cos w_j k, \quad x_{k,2j} = \sin w_j k, \quad j = 1, \ldots, q, \quad k = 1, 2, \ldots \tag{12.28}$$

where the w_j are such that rank $X_n = 2q$. Then (see: Eicker, 1963b, p. 477) the $x_j(n)$ are slowly increasing and $n^{-1} P_n \to Iq^*$ where Iq^* is a diagonal matrix with diagonal elements 1 or $\frac{1}{2}$. Again (12.27), and thus Condition (1), is satisfied.

(3) *Mixed trigonometric and polynomial regression*. Let

$$\begin{cases} x_{kj} = k^{ci}, & j = 1, \ldots, q \quad \text{with} \quad c_1 > \cdots > c_q > -\frac{1}{2}, \\ x_{kj} = e^{ikw_i}, & j = q+1, \ldots, Q \quad \text{with} \quad 0 < w_j < 2\pi, \ w_j \neq w_k \text{ for } j \neq k. \end{cases} \tag{12.29}$$

Again (12.27) is satisfied with a matrix R of the form

$$R = \left(\begin{array}{c|c} H & 0 \\ \hline 0 & I_{Q-q} \end{array} \right), \tag{12.30}$$

where H is as in example 1. In order to prove this, we observe first that

$$n^{-c-1} \sum_{k=1}^{n} k^c e^{ikw} \to 0 \quad \text{for } n \to \infty \tag{12.31}$$

if $c = 0, 1, \ldots$ and $0 < w < 2\pi$. This can be seen by deriving $\sum_{k=1}^{n} e^{ikw} = e^{iw}(1 - e^{inw})/(1 - e^{iw})$ repeatedly with respect to iw. In order to prove for nonintegers c, we derive the left-hand side with respect to c and obtain

$$n^{-c-1} \sum_{k} k^c e^{ikw} \ln(k/n), \tag{12.32}$$

which remains bounded for $c > -\frac{1}{2}$ as $n \to \infty$, since

$$\int_0^1 \left(\frac{t}{n}\right)^c \ln \frac{t}{n} \mathrm{d}\left(\frac{t}{n}\right) = -(c+1)^{-2} \tag{12.33}$$

and since we have proved (12.31) already for integers c, it holds also for non-integers $c > -\frac{1}{2}$.

Finally, because the matrix R is positive definite, property (1) holds.

(4) *Analysis of variance case.* Consider a one-way classification with q classes having $N_1, \ldots, N_q$ observations respectively. The regression matrix is given by

$$X_n' = \begin{pmatrix} 1, \ldots, 1, & 0, \ldots, 0, & \cdots & 0 \\ 0, \ldots, 0, & 1, \ldots, 1, & 0, \ldots, & 0 \\ & \cdots & & \\ 0, & \cdots & 0, & 1, \ldots, 1 \end{pmatrix}_{q \times n}, \quad n = N_1 + \cdots + N_q. \tag{12.34}$$

Then $P_n = \mathrm{diag}(N_1, \ldots, N_q)$. Condition (1) is satisfied if $\min_j N_j \to \infty$. Hence, in this case the LSE of the effects are asymptotically normally distributed for every error sequence $\epsilon \in \mathcal{F}(F)$.

(5) *Exponential regression.* The regression vectors $x_j(n)$ with $x_{kj} = c_j^k$ where $c_1 > c_2 > \cdots > c_q > 1$, are not slowly increasing since

$$c_j^{2n} \Big/ \sum_{k=1}^n c_j^{2k} > \frac{c_j}{c_j^2} > 0. \tag{12.35}$$

Therefore, by Lemma 2, Condition 1 cannot be satisfied.

(6) *Mixed polynominal-exponential regression*, $x_{kj} = k^{c_j} d_j^k$, $d_j > 1$, $c_j > 0$. Again the regression vectors are not slowly increasing, since $\|x_j(n)\|^2 = O(n^{2c_j} d_j^{2n})$.

12.3 Dependent Errors (Regression for Time Series)

12.3.1 The case of one regression vector

Since for dependent errors the results are not of as closed a form as for independent errors, we consider first the case of only one regression vector ($q = 1$). This case already shows the main deviations from the previous results.

System (12.1) reduces now to

$$y_t = x_t \beta + \epsilon_t, \quad t = 1, 2, \ldots, n, \tag{12.36}$$

in vectorial notation $y(n) = x(n)\beta + \epsilon(n)$. We assume, as is typical also for the analysis of time series,

$$\epsilon_t = \sum_{j=-\infty}^{\infty} c_j \eta_{t+j} = \sum_{j=-\infty}^{\infty} c_{j-t} \eta_j, \quad t = 1, 2, \ldots, \tag{12.37}$$

where the sequence $c \equiv \{c_j\}$ of real constants is square summable, that is, $c \in l^2$ (the Hilbert space of all square summable sequences of real numbers). In all of the previous publications the stronger assumption $\sum_{j=-\infty}^{\infty} |c_j| < \infty$ has been made (e.g. see Hannan 1961).

As is well known, the condition $c \in l^2$ is necessary in order that (12.37) holds as a limit in quadratic mean. The r.v.'s η_j are assumed to be independent with expectations zero, but they need not be identically distributed. Let their d.f.'s, as in Section 12.2, lie in a set F where F satisfies (12.5) and Conditions (2) and (3) of Section 12.2.1. Then each ϵ_t is, in fact, defined as a limit in the mean of the sums $\sum_{j=-n}^{n} c_j \eta_{t+j}$ for every sequence $\{n_i\} \in \mathcal{F}(F)$ (see Section 12.2.1). Random sequences $\{\epsilon_t\}$ of this type have been called *generalized linear processes* (Eicker 1965). If the η_t are identically distributed, the ϵ_t form a strictly stationary linear stochastic process.

With $P_n = \|x(n)\|^2$, the (scalar) LSE's of β are

$$b(n) = \|x(n)\|^{-2} x'(n) y(n), \quad n = 1, 2, \dots. \tag{12.38}$$

In order to investigate the asymptotic normality of the sequence $\{b(n)\}$, put

$$\zeta_n = (\text{var } b(n))^{-1/2} (b(n) - \beta)$$

$$= (\text{var } b(n))^{-1/2} \|x(n)\|^{-2} x'(n) \epsilon(n)$$

$$= (\text{var } b(n))^{-1/2} \|x(n)\|^{-2} \sum_{j=-\infty}^{\infty} \left(\sum_{t=1}^{n} x_t c_{j-t} \right) \eta_j. \tag{12.39}$$

We have $E\zeta_n = 0$, var $\zeta_n = 1$ for all n. Put

$$A_{nj} = \sum_{t=1}^{n} x_j c_{j-t},$$

$$S_n = \sum_{j=-\infty}^{\infty} A_{nj}^2. \tag{12.40}$$

Clearly, always $S_n < \infty$. In (Eicker 1965; p. 319) the following proposition, which also holds in a more general context, has been proved.

Theorem 3 Let $S_n > 0$ for all n. In order that (Λ): d.f. $(\zeta_n) \to N(0, 1)$, and (B): the contributions of the summands of ζ_n in the last expression of (12.39) are infinitesimal, both for every $\{\eta_j\} \in \mathcal{F}(F)$, the Conditions (2) and (3) of Theorem 1 and

$$\sup_{j=-\infty,\dots,\infty} A_{nj}^2 / S_n \to 0 \tag{12.41}$$

are jointly necessary and sufficient.

We do not emphasize the validity of the theorem for the whole class $\mathcal{F}(F)$. We rather consider a sequence $\{\eta_i\} \in \mathcal{F}(F)$ to be given and shall now analyse in detail the remaining condition (12.41).

Condition (12.41) may be verified directly for a given sequence $c \in l^2$ and a given sequence $x \equiv \{x_1, x_2, \ldots\}$ of regression constants. It would be more convenient, however, if for each c the class $\mathcal{X}_c$ of all x's satisfying (12.41), or for each x the class $\mathcal{C}_x$ of all c's satisfying (12.42) were known. It then remains only to be checked whether a given x belongs to $\mathcal{X}_c$, or a given c belongs to $\mathcal{C}_x$. Since in a regression problem x is known but c usually is not, the classes $\mathcal{C}_x$ are of greater interest. We shall, therefore, direct our attention mainly on $\mathcal{C}_x$. If we are unable to determine a class $\mathcal{C}_x$ completely, we shall try to find as large a subclass as possible.

Let $c(\lambda) \in L^2$ (the space of the complex valued functions over $\Lambda = \{\lambda: -\frac{1}{2} \leq \lambda \leq \frac{1}{2}\}$ whose moduli are Lebesgue square integrable) be such that

$$c_j = \int_\Lambda c^{-2\pi i\lambda j} c(\lambda) \, d\lambda, \qquad c(\lambda) \sim \sum_j c_j e^{2\pi i j \lambda}, \qquad (12.42)$$

and put

$$x_n(\lambda) = \sum_{t=1}^n x_t e^{2\pi i t \lambda}. \qquad (12.43)$$

Then for sufficiently large n,

$$S_n = \int_\Lambda |x_n(\lambda)c(\lambda)|^2 \, d\lambda > 0 \qquad (12.44)$$

as is required in Theorem 3; null sequences x and c are of course excluded.

Because of their importance in practical applications, one will not want to exclude all finite sequences c from any $\mathcal{C}_x$. But if $\mathcal{C}_x$ contains any finite nonnull sequence, then (12.41) implies

$$\max_{k=1,\ldots,n} |x_k|/\|x(n)\| \to 0, \qquad (12.45)$$

as will be seen from Lemma 3 (x is slowly increasing). In order to investigate the behaviour of the left-hand side of (12.41) under this additional assumption, we prove the following lemma.

Lemma 3 Equation (12.45) implies

$$\sup_j |A_{n,j}|/\|x(n)\| \to 0 \qquad (12.46)$$

for all $c \in l^2$.

Presumably, (12.46) (for fixed c) also implies (12.45), but we do not prove it here, except for finite c-sequences (Lemma 5).

Proof Choose $m_n < n, m_n \to \infty$ such that

$$m_n \max_k |x_k|/\|x(n)\| \to 0. \qquad (12.47)$$

There exists always an integer j_n such that $|A_{n,j_n}| = \sup_j |A_{n,j}|$. Put $J_n = \{j_n - n, \ldots, j_n - 1\}$ and split the sum

$$A_{n,j_n} = \sum_{t=1}^{n} x_t c_{j_n-t} = \sum_{t \in J_n} x_{j_n-t} c_t \tag{12.48}$$

into the sum $\alpha_n = \sum_{t \in J_n \cap K_n} x_{j_n-t} c_t$ and into the remainder β_n; here K_n is the index set $\{t: |t| < [m_n/2]\}$.

$$|A_{n,j_n}| \le |\alpha_n| + |\beta_n| \le m_n \sup_j |c_j| \max_k |x_k|$$
$$+ \|x(n)\| \Big(\sum_{|t| \ge [m_n/2]} c_t^2 \Big)^{1/2}. \tag{12.49}$$

After division by $\|x(n)\|$, this tends to zero for $n \to \infty$. ∎

Theorem 4 Let (12.45) be true and

$$\operatorname*{ess\,inf}_{\lambda \in \Lambda} |c(\lambda)| > 0. \tag{12.50}$$

Then (12.41) holds, and consequently, statement (Λ) is valid.

Proof The proof follows from the preceding lemma and

$$S_n \ge \|x(n)\|^2 \operatorname*{ess\,inf}_{\lambda} |c(\lambda)|^2. \tag{12.51}$$

For a large class of slowly increasing regression vectors, Condition (12.50) is not necessary. Assume, besides (12.45), that

$$\lim_{n \to \infty} \sum_{t=1}^{n-h} x_{t+h} x_t / \|x(n)\|^2 = \tilde{R}_h, \quad h = 1, 2, \ldots, \tag{12.52}$$

exists. Put $\tilde{R}_{-h} = \tilde{R}_h$. Then $\{\tilde{R}_h\}$ is a positive definite sequence, and there exists a d.f. $M(\lambda), \lambda \in \Lambda$, of finite variation such that

$$\tilde{R}_h = \int_{\Lambda} e^{2\pi i h \lambda} dM(\lambda) \tag{12.53}$$

(Grenander and Rosenblatt 1957: p. 233). We have $M\left(\frac{1}{2}\right) - M\left(-\frac{1}{2}\right) = 1$ and

$$\|x(n)\|^{-2} \int_{-1/2}^{\lambda} |x_n(\mu)|^2 d\mu \equiv M_n(\lambda) \to M(\lambda) \tag{12.54}$$

at continuity points of $M(\lambda)$.

Now let $\{I_i, \ldots, I_K\}$ be any partition of Λ into disjoint intervals, whose and points are not jump points of $M(\lambda)$. Let

$$\operatorname*{ess\,inf}_{I_k} |c(\lambda)|^2 \tag{12.55}$$

denote the essential infimum of $|c(\lambda)|^2$ over I_k. Then

$$S_n/\|x(n)\|^2 \geq \sum_{k=1}^{K} \operatorname*{ess\,inf}_{I_k} |c(\lambda)|^2 \Delta M_n(I_k), \tag{12.56}$$

where $\Delta M_n(I_k)$ is the variation of $M_n(\lambda)$ over I_k. Because of (12.54),

$$\liminf_{n} S_n/\|x(n)\|^2 \geq \sum_{k=1}^{K} \operatorname*{css\,inf}_{I_k} |c(\lambda)|^2 \Delta M(I_k). \tag{12.57}$$

The relation remains valid if we take on the right-hand side the supremum with respect to all admissible partitions with arbitrary K. ∎

Definition The function $|c|$ on Λ is called essentially positive at λ if

$$\sup_{I \ni \lambda} \operatorname*{ess\,inf}_{I} |c| > 0. \tag{12.58}$$

where I denotes on interval $\subset \Lambda$.

We now deduce Theorem 5.

Theorem 5 Let (12.45) and (12.52) be true, and let $|c(\lambda)|$ be essentially positive on at least one point of the spectrum of $M(\lambda)$. Then the right-hand side of (12.57) is positive, and consequently (12.41) and statement (Λ) are valid.

For some sequences of regression constants it is possible to obtain a complete characterization of the class $\mathcal{C}_x$.

Example Let $x_t = 1$, for all t. Then $\mathcal{C}_{[1]}$ consists of all $c \in l^2$ with only a small exceptional class characterized by $\lim_{n \to \infty} \sum_{j=-n}^{n} c_j = 0$ and convergence of $\sum_{j=0}^{n} c_j$ to T, say. The c's of this subclass satisfy (12.41) if and only if $\sum_{j=-n}^{n} c_j$ does not converge too fast to zero, namely, if and only if

$$\sum_{n=1}^{N} \left(\left(\sum_{j=0}^{n} c_j - T \right)^2 + \left(\sum_{j=-n}^{-1} c_j + T \right)^2 \right) \to \infty \tag{12.59}$$

for $N \to \infty$ (Eicker 1965; p. 325). If $c_{-1} = c_{-2} = \cdots = 0$, (12.59) reduces to

$$\sum_{n=1}^{N} \left(\sum_{j=0}^{n} c_j \right)^2 \to \infty. \tag{12.60}$$

We conclude this section with a remark concerning the convergence properties of the sequence of functions

$$x_n(\lambda)c(\lambda) \sim \sum_{j=-\infty}^{\infty} A_{n,j} e^{2\pi i j\lambda}. \tag{12.61}$$

Let $d_n(\lambda), n = 1, 2, \ldots, -\frac{1}{2} \leq \lambda \leq \frac{1}{2}$, be any sequence of functions with $d_n(\lambda) \in L^2$,

$$d_n(\lambda) \sim \sum_{j=-\infty}^{\infty} d_{nj} e^{2\pi i j\lambda}. \tag{12.62}$$

Then

$$\int_\Lambda \overline{d_n(\lambda)} x_n(\lambda)c(\lambda) \, \mathrm{d}\lambda = \sum_j \overline{d_{n,j}} A_{n,j}. \tag{12.63}$$

Lemma 4 Let $\sum_j |d_{n,j}|^2 = 1$, $\sum_j |d_{n,j}|^2$ converge uniformly in n. Then (12.41) implies

$$S_n^{-1/2} \int_\Lambda \overline{d_n(\lambda)} x_n(\lambda)c(\lambda) \, \mathrm{d}\lambda \to 0 \tag{12.64}$$

uniformly with respect to the functions $d_n(\lambda)$ out of the considered class.

Proof Choose $m_n < n, m_n \to \infty$ such that

$$m_n \sup j |A_{nj}|/S_n^{1/2} \to 0. \tag{12.65}$$

Then

$$S_n^{-1/2} \left| \sum_j \overline{d_{n,j}} A_{n,j} \right| \leq 2m_n \sup_j |A_{n,j}| S_n^{-1/2} + \left(\sum_{|j| \geq m_n} |d_{n,j}|^2 \right)^{1/2} \to 0. \tag{12.66}$$

Let us now put $\gamma_n(\lambda) - S_n^{-1/2} x_n(\lambda)c(\lambda)$, and assume $\gamma_n(\lambda)$ converges boundedly in measure to an integrable limiting function $\gamma(\lambda)$ on Λ. Let Re $\gamma(\lambda)$ be of one sign in some interval $\subset \Lambda$ and take all $d_n(\lambda) = d(\lambda)$, the characteristic function of this interval. Then by the bounded convergence theorem,

$$\int_\Lambda d(\lambda)\gamma_n(\lambda) \, \mathrm{d}\lambda \to \int_\Lambda d(\lambda)\gamma(\lambda) \, \mathrm{d}\lambda. \tag{12.67}$$

By the above lemma the integrals on the left tend to zero under (12.41), so that Re $\gamma(\lambda) = 0$ [a.e.] on the considered interval. Repeating the argument for Im $\gamma(\lambda)$, we obtain $\gamma(\lambda) = 0$ [a.e.]. This, however, is in contradiction with the fact that

$$\int_\Lambda |\gamma_n(\lambda)|^2 \, \mathrm{d}\lambda = 1 \quad \text{for all } n, \tag{12.68}$$

which implies

$$\int_\Lambda |\gamma(\lambda)|^2 \, d\lambda = 1. \tag{12.69}$$

The same argument holds if any infinite subsequence of $\{\gamma_n(\lambda)\}$ is taken or any measurable subset of Λ is considered instead of Λ.

Thus we have the following: under (12.41), no subsequence of $\{\gamma_n(\lambda)\}$ converges boundedly in measure to an integrable function.

Let $c(\lambda) \equiv 1$. Then (12.41) is equivalent with (12.45). In addition, let (12.52) be true, so that (12.54) holds. Assume $M(\lambda)$ possesses a bounded derivative $M'(\lambda)$. Then the preceding proposition is somewhat surprising in view of the fact that

$$\int_{-1/2}^\lambda |\gamma_n(\mu)|^2 \, d\mu = \|x(n)\|^{-2} \int_{-1/2}^\lambda |x_n(\mu)|^2 \, d\mu \rightarrow \int_{-1/2}^\lambda M'(\mu) \, d\mu \tag{12.70}$$

for all λ. One may guess that $\gamma_n(\lambda)$ must be increasingly oscillatory for $n \rightarrow \infty$, which may be due to the fact that $\gamma_n(\lambda)$ is complex-valued.

Here the remark may be of interest that always

$$\sum_{t=1}^n |x_n| \Big/ \left(\sum_{t=1}^n x_t^2\right)^{1/2} \rightarrow \infty \tag{12.71}$$

if (12.45) holds so that, at least sometimes, the boundedness condition of the convergence will be violated. ∎

12.3.2 One regression vector (errors are finite moving averages)

For further study of Condition (12.41), we now restrict ourselves to finite c-sequences. Let $k_1 < k_2, c_{k_1}, c_{k_2} \neq 0, c_j = 0$ for $j < k_1, j > k_2$. Then

$$A_{n,j} = 0 \qquad \text{for} \quad j \le k_1 \text{ and } j > k_2 + n, \qquad n = 1, 2, \ldots, \tag{12.72}$$

$$A_{n,j} \equiv A_j = \sum_{k=k_1}^{k_2} c_k x_{j-k} \quad \text{if } k_2 < j \le k_1 + n, \quad n > k_2 - k_1, \tag{12.73}$$

$$S_n = \sum_{j=k_1+1}^{k_1+n} A_j^2 + \sum_{j=k_1+n+1}^{k_2+n} A_{n,j}^2, \quad n > k_2 - k_1, \tag{12.74}$$

with $A_j = x_1 c_{j-1} + \cdots + x_{j-k_1} c_{k_1}$ for $k_1 < j \le k_2$. Since

$$\inf_n \sup_j |A_{n,j}| > 0 \tag{12.75}$$

(12.41) implies $\lim S_n = +\infty$, we have by (12.41)

$$\sum_{j=k_1+n+1}^{k_2+n} A_{n,j}^2 / S_n \rightarrow 0. \tag{12.76}$$

Thus (12.41) also implies

$$\sum_{j=k_1+1}^{n} A_j^2 \to \infty, \tag{12.77}$$

and it is equivalent to

$$\sup_j A_{n,j}^2 \Big/ \sum_{j=k_1+1}^{n} A_j^2 \to 0. \tag{12.78}$$

We have, moreover, the following lemma.

Lemma 5 If c is a finite nonnull sequence, then (12.41) implies (12.45), and is equivalent to (12.46)

Proof We have $\max_j A_{n,j}^2 \geq c_k^2 x_n^2$. Since $c(\lambda)$ is now continuous, we have $\sup_\lambda |c(\lambda)|^2 = \gamma < \infty$. Hence by (12.44)

$$S_n \leq \gamma \|x(n)\|^2. \tag{12.79}$$

Thus

$$\sup_j A_{n,j}^2 / S_n \geq \gamma' x_n^2 / \|x(n)\|^2, \quad \gamma' > 0, \tag{12.80}$$

and (12.41) implies

$$|x_n|/\|x(n)\| \to 0. \tag{12.81}$$

As seen above, (12.41) implies $S_n \to \infty$; hence,

$$\|x(n)\| \to \infty. \tag{12.82}$$

However, this together with (12.81) is equivalent to (12.45), since otherwise, with k_n chosen such that

$$|x_{k_n}| = \max_{k=1,\dots,n} |x_k|, \tag{12.83}$$

$|x_{k_n}|/\|x(k_n)\| \nrightarrow 0$, in contradiction to (12.81).

The proof of the equivalence of (12.45) and (12.46) is similar. ∎

If $c(\lambda)$, which is now a Fourier polynomial, has no zeros, and if (12.45) is true, then we have asymptotic normality by Theorem 4.

Condition (12.77) obviously is not satisfied (and thus we have no asymptotic normality) if the sequence $\{x_t\}$ satisfies the recursive system

$$\sum_{k=k_1}^{k_2} c_k x_{j-k} = A_j = 0, \quad j > k_2. \tag{12.84}$$

If the equation

$$c_{k_1} t^{k_2-k_1} + \cdots + c_{k_2-1} t + c_{k_2} = 0 \tag{12.85}$$

possesses a root of modulus one, but no larger ones ($c(\lambda)$ then has a zero in $\left[-\frac{1}{2}, \frac{1}{2}\right]$), then $\{x_t\}$ satisfies (12.45), but (12.41) does not hold if (12.84) is true.

Thus the converse of the first statement of Lemma 5 does not hold. A criterion for the validity of (12.84) is that all the determinants

$$
\begin{vmatrix} x_1, \ldots, x_n \\ \cdots \\ x_n, \ldots, x_{2n-1} \end{vmatrix} \tag{12.86}
$$

vanish for $n > k_2 - k_t$.

There are many cases of pairs (x, c) of sequences incompatible with (12.41) that are of practical interest. For example, let $x_t = 1$, for all t, and $c_0 = -c_m = 1$, all other $c_j = 0$. Then $x_t - x_{t-m} = 0$, and $t = m + 1, m + 2, \ldots$ is a recursive system.

From this example we see that a theorem of the type 'If x is any element of a class of regression sequences independent of c, and if c is any element of a class independent of x, then asymptotic normality holds', is not desirable because it excludes too many important cases. It is for this reason that we entered into a sharper analysis and tried to obtain the classes $\mathcal{C}_x$.

For the case that $c(\lambda)$ has zeros, we can state the following: if (12.45) is true, and if the sequence of functions

$$
\left(\max_{j=1,\ldots,n} x_j^2 \right)^{-1} \int_{-1/2}^{\lambda} |x_n(\mu)| \, d\mu, \quad -\tfrac{1}{2} \le \lambda \le \tfrac{1}{2}, \tag{12.87}
$$

or any subsequence thereof does not tend to a pure jump function whose jump points are all zeros of $c(\lambda)$, then asymptotic normality holds.

About, the estimation of the normalizing factors used in (12.39) some remarks are made in Section 12.3.4(3).

12.3.3 Multiple regression (dependent errors)

We use the notation of Sections 12.1 and 12.2.1 with $q > 1$ and errors given by (12.37). Let $B_n^2 = \text{cov}(b(n)b'(n))$. This matrix has no longer the simple structure (12.8) because $\sum_n$ in general is not diagonal. In order to derive the next theorem, we introduce the q-vectors

$$
\zeta(n) = B_n^{-1} P_n^{-1} X_n' \epsilon(n) = B_n^{-1} P_n^{-1} \sum_{j=-\infty}^{\infty} \left(\sum_{t=1}^{n} r_t c_{j-t} \right) \eta_j, \tag{12.88}
$$

which all have expectation zero and covariance matrix I_q. (The vector r_k was defined as the kth column of X_n'. The matrices $P_n (n \ge q)$ are assumed nonsingular.)

Now

$$
\text{d.f. } (\zeta(n)) \to N(0, I_q) \tag{12.89}
$$

if and only if

$$
\text{d.f. } (d'(n)\zeta(n)) \to N(0, 1) \tag{12.90}
$$

for all sequences of constant q-vectors $d(n)$ with $d(n) \in S_q$ (the unit sphere in R_q). For

$$d(n) = B_n P_n f(n) \| B_n P_n f(n) \|^{-1}, \quad f(n) \in S_q, \tag{12.91}$$

we have

$$d'(n)\zeta(n) = \| B_n P_n f(n) \|^{-1} \sum_j \left(\sum_{t=1}^{n} f'(n) r_t c_{j-t} \right) \eta_j. \tag{12.92}$$

Here $\{\eta_j\} \in \mathcal{F}(F)$, F subject to Conditions (2) and (3) of Theorem 1. We now have reduced the problem of finding conditions asserting (12.89) to the one-dimensional case. By Theorem 3, (12.90) holds if

$$\sup_j A_{n,j}^2 / S_n \to 0, \quad S_n = \sum_{j=-\infty}^{\infty} A_{n,j}^2, \tag{12.93}$$

where now

$$A_{n,j} = \sum_{t=1}^{n} f'(n) r_t c_{j-t} = f'(n) X_n' \begin{pmatrix} c_{j-t} \\ \vdots \\ c_{j-n} \end{pmatrix}. \tag{12.94}$$

Relation (12.89) holds if (12.93) holds for every sequence $\{f(n)\}$, $f(n) \in S_q$. That is the case if and only if

$$\sup_{f \in S_q} \left[\sup_j (f'X_n'c(j,n))^2 \Big/ \sum_{k=-\infty}^{\infty} (f'X_n'c(k,n))^2 \right] \to 0; \tag{12.95}$$

here $c(j,n) = (c_{j-1}, \ldots, c_{j-n})'$. Putting

$$\sum_{j=-\infty}^{\infty} c_{j+k}c_j = R_k = R_{-k},$$

$$R(n') = \begin{pmatrix} R_0, & \cdots, & R_{n-1} \\ & \cdots & \\ R_{n-1}, & \cdots, & R_0 \end{pmatrix}, \tag{12.96}$$

the denominator in (12.95) becomes

$$S_n = f'X_n' R(n) X_n f. \tag{12.97}$$

In analogy to (12.95), we also have, with

$$x_n(\lambda) = \sum_{t=1}^{n} f'r_t e^{2\pi i t \lambda}, \tag{12.98}$$

$$S_n = \int_{\Lambda} |x_n(\lambda)c(\lambda)|^2 \, d\lambda. \tag{12.99}$$

Suppose

$$\operatorname{ess\,inf}_{\lambda} |c(\lambda)|^2 \equiv \gamma_0 > 0. \tag{12.100}$$

Then

$$S_n \geq \gamma_0 f' P_n f, \qquad P_n = X_n' X_n, \tag{12.101}$$

and (12.95) holds, according to Lemma 3, if

$$\max_t \sup_{f \in S_q} (f' r_t)^2 / f' P_n f \to 0. \tag{12.102}$$

This is equivalent to

$$\max_t \sup_{f \in S_q} (f' P_n^{-1/2} r_t)^2 = \max_t r_t' P_n^{-1} r_t \to 0. \tag{12.103}$$

Hence, the following theorem holds. ■

Theorem 6

$$\max_t r_t' P_n^{-1} r_t \to 0 \tag{1}$$

and $\operatorname{ess\,inf}_{\lambda} |c(\lambda)| > 0$. Then (12.89) holds.

As in Section 12.1.1, assumption (12.100) may be weakened by assuming that the regression matrices X_n allow for a harmonic analysis. Generalizing (12.102), we assume

$$\max_{t=1,\dots,n} x_{tj}^2 / \|x_j(n)\|^2 \to 0 \quad \text{for } j = 1, \dots, q, \tag{12.104}$$

$$\sum_{t=1}^{n} x_{t+k,r} x_{t,s} / \|x_t(n)\| \, \|x_n(n)\| \to R_h^{(r,s)}, \quad r, s = 1, \dots, q,$$

$$h = 0, \pm 1, \dots. \tag{12.105}$$

Then $R_{-h}^{(s,r)} = R_h^{(r,s)}$, and the $(q \times q)$-matrices $\tilde{R}_h = \{R_h^{(r,s)}\}$ admit the spectral representation

$$\tilde{R}_h = \int_\Lambda e^{2\pi i h \lambda} \, dM(\lambda), \quad h = 0, \pm 1, \dots, \tag{12.106}$$

where the elements of the $(q \times q)$-matrix $M(\lambda)$ are functions of bounded variation and $M(\lambda_2) - M(\lambda_1)$ is positive semidefinite for every $\lambda_2, \lambda_1, -\frac{1}{2} \leq \lambda_1 < \lambda_2 \leq +\frac{1}{2}$ (see Grenander and Rosenblatt 1957: p. 233). The set of all points λ with $M(\lambda_2) - M(\lambda_1)$ positive definite for all $\lambda_1 < \lambda < \lambda_2$ is called the *spectrum* of $M(\lambda)$.

We assume

$$\tilde{R}_n = \int_\Lambda dM(\lambda) = M\left(\tfrac{1}{2}\right) - M\left(-\tfrac{1}{2}\right) \equiv M \tag{12.107}$$

to be nonsingular and put

$$D_n = \text{diag}(\|x_1(n)\|, \ldots, \|x_q(n)\|). \tag{12.108}$$

Then (Grenander and Rosenblatt 1957: p. 238)

$$D_n P_n^{-1} X_n' R(n) X_n P_n^{-1} D_n \to M^{-1} \int_\Lambda |c(\lambda)|^2 \, dM(\lambda) M^{-1} \equiv \tilde{M}. \tag{12.109}$$

Now from (12.97), for $n \to \infty$,

$$S_n = f' X_n' R(n) X_n f = f' P_n D_n^{-1} \tilde{M} D_n^{-1} P_n f \geq \lambda_q \| D_n^{-1} P_n f \|^2 \tag{12.110}$$

where we have put $\lambda_p = \lambda_{\min}(\tilde{M})$. Since by (12.105) $D_n^{-1} P_n D_n^{-1} \to \tilde{R}_0$, we have in the limit $S_n \geq \gamma \| D_n f \|^2$ with some new constant γ that is positive if $\lambda_q > 0$.
We now assume $\lambda_q > 0$ and have (12.95) if

$$\sup_j \sup_{f \in S_q} (f' D_n D_n^{-1} X_n' c(j, n))^2 / \| D_n f \|^2 = \sup_j \sup_{f \in S_q} (f' D_n^{-1} X_n' c(j, n))^2 \to 0.$$
$$\tag{12.111}$$

Choosing $f = D_n^{-1} X_n' c(j, n) / \| D_n^{-1} X_n' c(j, n) \|$, we see that

$$\sup_j \| D_n^{-1} X_n' c(j, n) \| \to 0, \tag{12.112}$$

or that

$$\sup_j \sum_{k=1}^n x_{tk} c_{j-k} | / \| x_k(n) \| \to 0 \quad \text{for all } k = 1, \ldots, q \tag{12.113}$$

is sufficient for (12.95). The latter, however, is true because of (12.104) and Lemma 3.

A condition implying $\lambda_q > 0$ is that $\int_\Lambda |c(\lambda)|^2 \, dM(\lambda)$ be nonsingular in case this integral is defined and is the limit of the corresponding finite approximations. It is the higher dimensional generalization of (12.57). ∎

In summary, the following holds.

Theorem 7 Let (12.104) and (12.105) be true, and let $\tilde{R}_0$ be nonsingular. Let $c(\lambda)$ be essentially positive on at least one point of the spectrum of $M(\lambda)$. Then (12.89) holds.

Concerning the estimation of the normalizing matrices B_n, the reader is referred to Section 12.3.4(3).

Finally, it may be noticed that ess $\inf_{\lambda \in \Lambda} |c(\lambda)| > 0$ can always be achieved by simply adding to the y_t's independent random variables of an artificial sequence $\{\rho_t\}$, which are also independent of the η_j. The function $c^*(\lambda)$, associated with the new combined error sequence $\rho_t + \epsilon_t$, always satisfies ess $\inf_{\lambda \in \Lambda} |c^*(\lambda)| > 0$. This method may be considered as a particular type of prewhitening. It always implies, however, an increase in variance of the LSE $b(n)$. A similar proposal has been made by Hannan (1961).

12.3.4 Concluding remarks

(1) As is well known, the Gauss–Markov estimators

$$b_G(n) = \left(X'_n \Sigma_n^{-1} X_n \right)^{-1} X'_n \Sigma_n^{-1} y(n) \qquad (12.114)$$

are the minimum variance linear unbiased estimators for β, whether the ϵ_t are correlated or not. One may, therefore, try to use $b_G(n)$ instead of the above considered LSE $b(n)$. However, in the first place, the covariance matrix Σ_n usually is unknown, and a useful estimate for Σ_n^{-1} (or for the functions of Σ_n^{-1} that occur in (12.114)) cannot be obtained from a single sequence of observations $\{y_t\}$. Instead, it appears to be more appropriate to use a distribution free method. In the second place, there is not much point in preferring $b_G(n)$ to $b(n)$, because it is known that both are equally efficient asymptotically for a rather large class of error sequences $\{\epsilon_t\}$ and of sequences of regression matrices $\{X_n\}$ (Grenander and Rosenblatt 1957).

(2) The results of the preceding sections can be extended to the case of *vectorial or multivariate regression equations*

$$y_t = x_{t1}\beta_1 + \cdots + x_{tq}\beta_q + \epsilon_t, \qquad (12.115)$$

where now $y_t = (y_t^1, \ldots, y_t^k)'$, $x_{tj} = (x_{tj}^1, \ldots, x_{tj}^k)'$, $\epsilon_t = (\epsilon_t^1, \ldots, \epsilon_t^k)'$ are k-vectors and the β_j are scalars as before (Hannan 1961, for a review see Rosenblatt 1959). Sometimes vectorial regression equations are expressed in the form

$$y_t = B\phi_t + \epsilon_t, \qquad (12.116)$$

where the vector $\phi_t = (\phi_{t1}, \ldots, \phi_{tp})'$ contains the known regression constants, and the $(k \times p)$-matrix B contains the unknown regression constants. However, (12.116) is a special case of (12.115), as is seen by putting

$$(x_{t1}, \ldots, x_{tq}) = \begin{pmatrix} \phi'_t, & 0, & \cdots & 0 \\ 0, & \phi'_t, & \cdots & 0 \\ \vdots & \vdots & & \vdots \\ \vdots & \vdots & & \vdots \\ 0, & 0, & & \phi'_t \end{pmatrix}, \qquad (12.117)$$

where the zeros represent zero row vectors of p dimensions. With $q = kp$, both sides are $(k \times kp)$-matrices. We get the row vector β by placing the rows of B one behind the other.

If the k-vectors ϵ_t are independent for different t with possibly dependent components for each fixed t, and assuming $E\epsilon_t = 0$, we have almost the case considered in Section 2, except that the error sequence $(\epsilon_1^1, \ldots, \epsilon_1^k, \epsilon_2^1, \ldots, \epsilon_2^k, \ldots)$ now is k-dependent. We might apply a clt for k-dependent r.v.'s. However, it appears to be almost as easy to appeal directly to the clt for independent r.v.'s. For this purpose, let

(i) R be a nonempty set of strictly positive definite $(k \times k)$-covariance matrices,

(ii) F be the set of d.f.'s defined in Section 12.1,

(iii) $\mathcal{G}(F, R)$ be the set of all sequences $\{\epsilon_t\}$ of independent k-vectors ϵ_t with cov $(\epsilon_t \epsilon_t') \in R$ and d.f. $(\epsilon_t^j) \in F$ for all t, j,

(iv) $b(n)$ be the vectorial LSE for β,

$$b(n) = P_n^{-1} X_n' y(n). \tag{12.118}$$

Here now $X_n = (x_{tj}; t = 1, \ldots, n, j = 1, \ldots, q)$ is a $(kn \times q)$-matrix, $P_n = X_n', X_n$, and $y(n) = (y_1^1, \ldots, y_1^k, y_2^1, \ldots, y_2^k, \ldots, y_n^k)'$. Similarly, $\epsilon(n)$ is defined. Furthermore,

$$B_n^2 = \mathrm{cov}(b(n)b'(n)) = P_n^{-1} X_n' \Sigma_n X_n P_n^{-1} \tag{12.119}$$

$$= P_n^{-1} \sum_{j-1}^{n} x_j' \rho_j x_j P_n^{-1}$$

with

$$x_j = (x_{j_1}, \ldots, x_{jq})_{k \times q},$$
$$\rho_j = E(\epsilon_j \epsilon_j')_{k \times k}. \tag{12.120}$$

Let r_m be the mth column vector of X_n', $m = 1, \ldots, kn$. Finally, let $\lambda_k(\rho)$ be the smallest characteristic root of $\rho \in R$. Then we have

Theorem 8 The d.f. $(B_n^{-1}(b(n) - \beta) \to N(0, I_q)$ for all $\{\epsilon_t\} \in \mathcal{G}(F, R)$ if

(1) $\max_{m=1,\ldots,kn} r_m' P_n^{-1} r_m \to 0$,

(2) $\sup_{G \in F} \int_{|x| > c} x^2 \, dG(x) \to 0$ for $c \to \infty$,

(3) $\inf_{\rho \in R} \lambda_k(\rho) > 0$.

We indicate the *proof* only for $q = 1$. By (12.118) we have

$$B_n^{-1}(b(n) - \beta) = B_n^{-1} P_n^{-1} \sum_{j=1}^{n} x_j' \epsilon_j \tag{12.121}$$

where $x_j' \epsilon_j = \sum_{s=1}^{k} x_{j1}^s \epsilon_j^s$ are independent r.v.'s. Putting $\inf_{\rho \in R} \lambda_k(\rho) = \lambda$ and $\max_{s=1,\ldots,k} (x_{j1}^s)^2 P_n^{-1} = \kappa_{n,j}$, we obtain

$$B_n^2 \geq P_n^{-2} \lambda \sum_{j=1}^{n} \|x_j\|^2 = \lambda P_n^{-1}. \tag{12.122}$$

Now for any $\delta > 0$,

$$P(|B_n^{-1}P_n^{-1}x_j'\epsilon_j| > \delta) \leq P(B_n^{-1}P_n^{-1}\|x_j\|\|\epsilon_j\| > \delta)$$

$$\leq P(\|\epsilon_j\|^2 > \lambda\delta^2/(k\kappa_{n,j}))$$

$$\leq \sum_{j=1}^{k} P((\epsilon_j^s)^2 > \lambda\delta^2/(k^2\kappa_{n,j}))$$

$$\leq \sum_{n=1}^{k} \frac{k^2\kappa_{n,j}}{\lambda\delta^2} \int_{x^2 > \lambda\delta^2/(k^2\kappa_{n,j})} x^2\,dG_j^s(x)$$

$$\leq k^3\kappa_{n,j}\lambda^{-1}\delta^{-2}\phi\left(\left(\frac{\lambda\delta^2}{k^2\kappa_{n,j}}\right)^{1/2}\right), \qquad (12.123)$$

where $G_j^2 = $ d.f. (ϵ_j^s) and $\phi(c) = \sup_{G\in F}\int_{|x|>c}x^2 dG(x)$. Now $\sum_{j=1}^{n}\kappa_{n,j} \leq P_n/P_n = 1$. Putting $\kappa_n = \max'_{j=1,\ldots,n}\kappa_{n,j}$, we finally have

$$\sum_{j=1}^{n} P(|B_n^{-1}P_n^{-1}x_j\epsilon_j| > \delta) \leq k^3\lambda^{-1}\delta^{-1}\phi\left(\left(\frac{\lambda\delta^2}{k^2\kappa_n}\right)^{1/2}\right) \to 0 \qquad (12.124)$$

by (2). Hence, the clt holds.

The case where the random k-vectors ϵ_t are generated by a moving average process $\epsilon_t = \sum_{j=-\infty}^{\infty} A_j\eta_{t-j}$ with independent and identically distributed random vectors η_j of r components and with constant $(k \times r)$-matrices A_j has been considered by Hannan (1961). There it is assumed that

$$\sum_{j=-\alpha}^{\infty} (\lambda_{\max}(A_j'A_j))^{1/2} < \infty, \qquad (12.125)$$

and that the regression matrices allow for a generalized harmonic analysis in order to derive the asymptotic normal distribution of the LSE. Although the results are likely to be true under more general conditions similar to those of Sections 12.3.1–12.3.3, we do not discuss this possibility here.

(3) An estimation of the normalizing constants $B_n^2 = \text{var } b(n)$, or of the normalizing matrices B_n^2 in multiple regression, for an unknown sequence of dependent errors ϵ_t with unequally distributed residuals η_j (see (12.37)) seems to be impossible if only a single sequence $\{y_t\}$ of observations is given. By estimation of B_n^2 we mean a relation like (12.14) with a suitably adapted statistic C_n^2. The reason for the difficulty lies, first of all, in the admission of nonidentically distributed residuals η_j in (12.37) which, together with the unknown c_j's, introduces to many unknown parameters.

Therefore, we restrict ourselves in the following to strictly stationary error sequences $\{\epsilon_t\}$, that is, *we assume the η_j to be independently identically distributed with variance one* (besides $E\eta_j = 0$). In fact, it suffices to have only identical variances of the η_j not necessarily equal to one. We also restrict ourselves to simple regression ($q = 1$). Then

$$B_n^2 = \|x(n)\|^{-4} E(x'(n)\epsilon(n))^2 = \|x(n)\|^{-4} x'(n) R(n) x(n) \qquad (12.126)$$

where

$$R(n) = E(\epsilon(n)\epsilon'(n)) = \begin{pmatrix} R_0, & R_1, & \ldots, & R_{n-1} \\ R_1, & R_0, & \ldots, & R_{n-2} \\ & & \ldots, & \\ R_{n-1}, & R_{n-2}, & \ldots, & R_0 \end{pmatrix} \qquad (12.127)$$

is the covariance matrix of the ϵ_t with

$$R_k = R_{-k} = E(\epsilon_t \epsilon_{t+k}) = \sum_{j=-\infty}^{\infty} c_j c_{j+k}, \quad k = 0, 1, \ldots. \qquad (12.128)$$

Putting

$$\tilde{R}_{k,n} = \tilde{R}_{-k,n} = \|x(n)\|^{-2} \sum_{t=1}^{n-k} x_t x_{t+k}, \quad k = 0, 1, \ldots, n-1, \qquad (12.129)$$

we obtain

$$B_n^2 = \|x(n)\|^{-2} \sum_{k=-n+1}^{n-1} R_k \tilde{R}_{k,n} = \|x(n)\|^{-4} S_n \qquad (12.130)$$

where S_n was defined in Section 12.3.1.

In order to estimate B_n^2, we introduce first the sample covariances of $\{\epsilon_t\}$,

$$R_{k,n} = \frac{1}{n-|k|} \sum_{h=1}^{n-|k|} \epsilon_h \epsilon_{h+|k|}, \quad k = 0, \pm 1, \ldots, \qquad (12.131)$$

and, as an estimator for B_n^2,

$$B_{n,\alpha}^2 = \|x(n)\|^{-2} \sum_{k=-n+1}^{n-1} R_{k,n} \tilde{R}_{k,n} \qquad (12.132)$$

(later we shall replace the ϵ_h by the known residuals $e_h(n)$). Since $E R_{k,n} = R_k$, then $E B_{n,\alpha}^2 = B_n^2$.

We now introduce the *additional assumption*

$$\sum_{k=-\infty}^{\infty} R_k^2 < \infty. \qquad (12.133)$$

Then (Lomnicki and Zaremba 1956: p. 16) with some $\gamma < \infty$,

$$\text{var } R_{k,n} < \gamma/(n - |k|). \tag{12.134}$$

By Schwarz's inequality,

$$\text{var } B_{n,\alpha}^2 \leq \|x(n)\|^{-1} \sum_{k=-n+1}^{n-1} \tilde{R}_{k,n}^2 \text{var } R_{k,n} \tag{12.135}$$

$$\leq \frac{\gamma}{\|x(n)\|^4} \sum_{k=-n+1}^{n-1} \tilde{R}_{k,n}^2/(n - |k|).$$

We have to divide var $B_{n\alpha}^2$ by B_n^4 and need the ratio tending to zero. We assume

$$S_n > \gamma' \|x(n)\|^2, \quad \text{for some } \gamma' > 0. \tag{12.136}$$

This is, for instance, the case under the assumptions of Theorems 4 and 5, where the asymptotic normality of the LSE $b(n)$ for the regression parameters is proved.

Now with (12.130), (12.135), and (12.136),

$$B_n^{-4} \text{var } B_{n,\alpha}^2 = \|x(n)\|^8 S_n^{-2} \text{var } B_{n,\alpha}^2$$

$$= O\left(\sum_{k=-n+1}^{n-1} \tilde{R}_{k,n}^2/(n - |k|)\right). \tag{12.137}$$

If a regression sequence $\{x_t\}$ has the property that the last expression tends to zero, then $B_n^{-2} B_{n,\alpha}^2$ is a (strongly) consistent sequence of estimators of one. This is certainly true if

$$\sup_n \sum_{k=-n+1}^{n-1} \tilde{R}_{k,n}^4 < \infty, \tag{12.138}$$

since because of the slowly increasing character of $\{x_t\}$ there exists a sequence of integers $m_n \to \infty$ such that

$$\sum_{k=m_n}^{n-1} \tilde{R}_{k,n}^2/(n - k) \to 0, \tag{12.139}$$

and also

$$\sum_{k=n-m_n}^{n} k^{-2} \to 0. \tag{12.140}$$

Now estimate the central part of

$$\sum_{k=-n+1}^{n-1} \tilde{R}_{k,n}^2/(n - |k|) \tag{12.141}$$

for $k = -m_n, \ldots, m_n$ by Schwarz's inequality. Because of (12.138) and (12.140), it tends to zero. The rest of the sum tends to zero by (12.139).

As pointed out before $B_{n,\alpha}^2$ is not yet an estimate of B_n^2, since $B_{n,\alpha}^2$ contains the error r.v.'s ϵ_t. We now proceed to replace them by the residuals

$$c_k(n) = y_k - x_k b(n) = \epsilon_k - \|x(n)\|^{-2} x_k x'(n) \epsilon(n). \tag{12.142}$$

Let

$$\hat{R}_{k,n} = \frac{1}{n-|k|} \sum_{h=1}^{n-|k|} e_h(n) e_{h+|k|}(n). \tag{12.143}$$

Putting

$$R'_{k,n} = \frac{(x'(n)\epsilon(n))^2}{(n-|k|)\|x(n)\|^4} \sum_{h=1}^{n-|k|} x_h x_{h+|k|}$$

$$= \frac{(x'(n)\epsilon(n))^2}{(n-|k|)\|x(n)\|^2} \tilde{R}_{k,n}, \tag{12.144}$$

$$R''_{k,n} = \frac{x'(n)\epsilon(n)}{(n-|k|)\|x(n)\|^2} \sum_{h=1}^{n-|k|} (\epsilon_h x_{h+|k|} + \epsilon_{h+|k|} x_h), \tag{12.145}$$

we obtain

$$\hat{R}_{k,n} = R_{k,n} + R'_{k,n} - R''_{k,n}. \tag{12.146}$$

Consider now

$$\hat{B}_n^2 = \|x(n)\|^{-2} \sum_{k=-n+1}^{n-1} \hat{R}_{k,n} \tilde{R}_{k,n} = B_{n,\alpha}^2 + B_{n,\beta}^2 - B_{n,\gamma}, \tag{12.147}$$

where

$$B_{n,\beta}^2 = \|x(n)\|^{-2} \sum_{k=-n+1}^{n-1} R'_{k,n} \tilde{R}_{k,n}$$

$$= \|x(n)\|^{-4} (x'(n)\epsilon(n))^2 \sum_{k=-n+1}^{n-1} (n-|k|)^{-1} \tilde{R}_{k,n}^2, \tag{12.148}$$

$$B_{n,\gamma} = \|x(n)\|^{-2} \sum_{k=-n+1}^{n-1} R''_{k,n} \tilde{R}_{k,n}. \tag{12.149}$$

Now

$$B_n^{-?} E B_{n,\beta}^2 - \sum_{k=-n+1}^{n-1} (n-|k|)^{-1} \tilde{R}_{k,n}^2 \to 0, \tag{12.150}$$

making use of our assumption concerning (2.137). Since $B_{n,\beta}^2 \geq 0$, this implies

$$B_n^{-2} B_{n,\beta}^2 \to 0, \quad \text{i.p.} \tag{12.151}$$

Concerning $B_{n,\gamma}$, we proceed as follows. First,

$$R_{k,n}'' = (n - |k|)^{-1} \|x(n)\|^{-2} \sum_{j,l=-\infty}^{\infty} A_{nj}(A_{n-|k|,l-|k|} + A_{nl} - A_{|k|l})\eta_j \eta_l. \tag{12.152}$$

After some straightforward computations, we obtain

$$E \left(\sum_{j,l} A_{n,j} A_{n-|k|,1-|k|} \eta_j \eta_l \right)^2 < \text{const } S_n S_{n-|k|}, \tag{12.153}$$

and similar relations hold for the other terms of $E(R_{k,n}'')^2$. Hence,

$$E B_{n,\gamma}^2 < \text{const } \|x(n)\|^{-8} S_n \sum_{|k|<n} \tilde{R}_{k,n}^2 \frac{1}{n - |k|} \sum_{k=1}^{n}(S_k + S_n - S_{n-k})k^{-1}, \tag{12.154}$$

$$B_n^{-4} E B_{n,\gamma}^2 < \text{const } \sum_{|k|<n} R_{k,n}^2 \frac{1}{n - |k|} \|x(n)\|^{-2} \sum_{k=1}^{n} k^{-1} \|x(k)\|^2$$

$$+ \|x(n)\|^2 - \|x(n - k)\|^2).$$

This tends to zero, and hence $B_n^{-2} B_{n,\gamma} \to 0$, i.p., if we assume

$$\|x(n)\|^{-2} \max_{t=1,\ldots,n} x_t^2 = O(n), \tag{12.155}$$

besides what we have assumed previously. In summary we have the following theorem.

Theorem 9 The d.f.'s of the statistics $(\hat{B}_n^2)^{-1}(b(n) - \beta)$ tend to $N(0, 1)$ if $\{\epsilon_t\}$ is a strictly stationary linear process, if ess $\inf_\lambda |c(\lambda)| > 0$ and (12.133) holds, and if the regression sequence satisfies (12.155) and

$$\sum_{k=-n+1}^{n-1} \tilde{R}_{k,n}^2 (n - |k|)^{-1} \to 0. \tag{12.156}$$

Condition (12.155) is satisfied, for example, for polynomial regression sequences $x_t = t^c, x > -\frac{1}{2}$. However, (12.156) is not satisfied at least for some polynomial sequences. A slightly weaker assumption than (12.155) that is sufficient is

$$\|x(n)\|^{-2} \sum_{t=1}^{n} x_t^2 \ln(n/t) = O(1). \tag{12.157}$$

In this subsection it was not our aim to achieve the utmost in generality, we rather wanted to indicate one possibility of replacing the normalizing constant B_n^2 by a statistic. The assumptions on the regression sequence, in particular (12.156), can be weakened considerably, if stronger assumptions are imposed on the admissible error sequences $\{\epsilon_t\}$, such as

$$\sum_{k=-\infty}^{\infty} |R_k| < \infty, \tag{12.158}$$

and if instead of $\hat{B}_n^2$, a different estimator is used, for instance,

$$\hat{B}_n^2 = \|x(n)\|^{-2} \sum_{|k|<\sqrt{n}} R_{k,n}\tilde{R}_{k,n}. \tag{12.159}$$

Clearly, (12.158) is satisfied, for example, for finite c-sequences.

References

Eicker, F. (1963a), 'Asymptotic normality and consistency of the least squares estimators for families of linear regressions', *Annals of Mathematical Statistics*, 34, 447–56.

—— (1963b), 'Uber die Konsistenz von Parameterschätzfunktionen für ein gemischtes Zeitreihen-Regressionsmodell', *Zeitschrift für Wahrscheinlichkeitstheorie und verwandte Gebiete*, 1, 456–77.

—— (1965), 'Ein Zentraler Grenzwertsatz für Summen von Variablen einer verallgemeinerten linearen Zufallsfolge', *Zeitschrift für Wahrscheinlichkeitstheorie und verwandte Gebiete*, 3, 317–27.

——, 'A central limit theorem for linear vector forms with random coefficients', to be published.

Gnedenko, B. V. and Kolmogorov, A. N. (1951), *Limit Distributions for Sums of Independent Random Variables*, Cambridge, Addison-Wesley.

Goldman, A. J. and Zelen, M. (1964), 'Weak generalized inverses and minimum variance linear unbiased estimation', *Journal of Research*, National Bureau of Standards, Washington, D.C., 68B, 151–72.

Grenander, U. and Rosenblatt, M. (1957), *Statistical Analysis of Stationary Time Series*, New York, Wiley.

Hannan, E. J. (1960), *Time Series Analysis*, London, Methuen.

—— (1961), 'A central limit theorem for systems of regressions,' *Proceedings of the Cambridge Philosophical Society*, 57, 583–88.

Lomnicki, Z. A. and Zaremba, S. K. (1956), 'On estimating the spectral density function of a stochastic process,' *Journal of the Royal Statistical Society Series B*, 18, 13–37.

Rosenblatt, M. (1959), 'Statistical analysis of stochastic processes with stationary residuals,' *Probability and Statistics* (Cramér volume), edited by U. Grenander, New York, Wiley.

13
Time Series Regression with Long-range Dependence

P. M. ROBINSON AND F. J. HIDALGO

13.1 Introduction

This chapter derives central limit theorems for estimates of the slope coefficient vector β in the multiple linear regression.

$$y_t = \alpha + \beta' x_t + u_t, \quad t = 1, 2, \ldots, \tag{13.1}$$

where both the K-dimensional column vector of regressors x_t, and the unobservable scalar error u_t are permitted to exhibit long-range dependence, α is an unknown intercept and the prime denotes transposition. Two widely used methods of estimating β are ordinary least squares and generalized least squares. Both are known to be asymptotically normal under a wide variety of regularity conditions. However, it was pointed out by Robinson (1994a) that, when x_t and u_t collectively exhibit long-range dependence of a sufficiently high order, the least squares estimate is not asymptotically normal. We show that a class of weighted least squares estimates, which includes generalized least squares as a special case, is asymptotically normal under rather general forms of long-range dependence in both x_t and u_t.

Given observations $y_t, x_t, t = 1, \ldots, n$, consider estimates of the following type [indexed by the function $\phi(\lambda)$, which is real-valued, even, integrable and periodic of period 2π]:

$$\hat{\beta}_\phi = \hat{A}_\phi^{-1} \hat{b}_\phi, \tag{13.2}$$

This work was previously published as Robinson, P. M. and Hidalgo, F. J. (1997), 'Time series regression with long-range dependence', *The Annals of Statistics*, 25, 77–104.

This research was supported by ESRC Grant R000235892.

We are very grateful for the comments of two referees and an Associate Editor on the first version of this chapter. One referee, in particular, showed how to considerably abbreviate our original proofs. Though the final version of the chapter substantially relaxes the conditions in the original version, we have made use of this referee's techniques in doing so. We also thank Yoshihiko Nishiyama for carrying out the Monte Carlo simulations.

where, with $\sum_t$ denoting $\sum_{t=1}^n$,

$$\hat{A}_\varphi = \frac{1}{2\pi} \int_{-\pi}^{\pi} I_x(\lambda)\varphi(\lambda)\, d\lambda,$$

$$\hat{b}_\varphi = \frac{1}{2\pi} \int_{-\pi}^{\pi} I_{xy}(\lambda)\varphi(\lambda)\, d\lambda,$$

$$I_x(\lambda) = w_x(\lambda)w_x(-\lambda)',$$

$$I_{xy}(\lambda) = w_x(\lambda)w_y(-\lambda),$$

$$w_x(\lambda) = \frac{1}{(2\pi n)^{1/2}} \sum_t (x_t - \bar{x})\, e^{it\lambda}, \qquad (13.3)$$

$$w_y(\lambda) = \frac{1}{(2\pi n)^{1/2}} \sum_t (y_t - \bar{y})\, e^{it\lambda},$$

$$\bar{x} = \frac{1}{n} \sum_t x_t, \qquad \bar{y} = \frac{1}{n} \sum_t y_t,$$

and $\hat{A}_\phi$ is nonsingular. Here $\hat{A}_\phi$ and $\hat{b}_\phi$ can equivalently be written as

$$\hat{A}_\phi = \frac{1}{n} \sum_t \sum_s (x_t - \bar{x})(x_s - \bar{x})'\, \phi_{t-s},$$

$$\hat{b}_\phi = \frac{1}{n} \sum_t \sum_s (x_t - \bar{x})(y_s - \bar{y})\, \phi_{t-s}, \qquad (13.4)$$

where

$$\phi_j = \frac{1}{(2\pi)^2} \int_{-\pi}^{\pi} \phi(\lambda) \cos j\lambda\, d\lambda. \qquad (13.5)$$

In case $\phi(\lambda) \equiv 1$, so that $\phi_0 = 1/2\pi$ and $\phi_j = 0$ for $j \neq 0$, $\hat{\beta}_\phi = \hat{\beta}_1$, the least squares estimate,

$$\hat{\beta}_1 = \left\{ \sum_t (x_t - \bar{x})x_t' \right\}^{-1} \sum_t (x_t - \bar{x})\, y_t. \qquad (13.6)$$

We assume throughout that u_t is covariance stationary, having mean that is (without loss of generality) 0, and absolutely continuous spectral distribution function, so that it has spectral density, denoted $f(\lambda)$, satisfying

$$\gamma_j =_{\text{def}} E(u_1 u_{1+j}) = \int_{-\pi}^{\pi} f(\lambda) \cos j\lambda\, d\lambda, \quad j = 0, 1, \dots. \qquad (13.7)$$

Suppose $f(\lambda) > 0$, $-\pi < \lambda \leq \pi$. Then taking $\phi(\lambda) = f(\lambda)^{-1}$ gives a generalized least squares estimate $\hat{\beta}_{f^{-1}}$. In case f is not known up to scale, but only up to a finite-dimensional vector of parameters θ, so that $f(\lambda) = f(\lambda; \theta)$ is a given function of λ and θ, a feasible generalized least squares estimate is $\hat{\beta}_{\hat{f}^{-1}}$, obtained by taking $\phi(\lambda) = f^{-1}(\lambda; \hat{\theta})$, where $\hat{\theta}$ is an estimate of θ.

Now suppose that x_t is a stochastic sequence, independent of the u_t sequence, and covariance stationary with autocovariance matrix $\Gamma_j = E(x_1 - Ex_1)(x_{1+j} - Ex_1)'$. Then

$$\Gamma_j = \int_{-\pi}^{\pi} e^{ij\lambda}\, dF(\lambda), \tag{13.8}$$

where the matrix F has Hermitian nonnegative definite increments and is uniquely defined by the requirement that it is continuous from the right. Under suitable conditions we then have the central limit theorem (clt)

$$n^{1/2}(\hat{\beta}_\phi - \beta) \to_d N\left(0, \Sigma_\phi^{-1}\Sigma_\psi\Sigma_\phi^{-1}\right), \tag{13.9}$$

where $\psi(\lambda) = \phi^2(\lambda)f(\lambda)$ and

$$\Sigma_\chi = \frac{1}{2\pi}\int_{-\pi}^{\pi} \chi(\lambda)\, dF(\lambda) \tag{13.10}$$

for $\chi(\lambda)$ such that Σ_χ is finite and nonsingular. In particular,

$$n^{1/2}(\hat{\beta}_1 - \beta) \to_d N\left(0, (2\pi)^2\Gamma_0^{-1}\Sigma_f\Gamma_0^{-1}\right), \tag{13.11}$$

$$n^{1/2}(\hat{\beta}_{f^{-1}} - \beta) \to_d N\left(0, \Sigma_{f^{-1}}^{-1}\right), \tag{13.12}$$

$$n^{1/2}(\hat{\beta}_{\hat{f}^{-1}} - \beta) \to_d N\left(0, \Sigma_{f^{-1}}^{-1}\right). \tag{13.13}$$

Some conditions for (13.9) and (13.11)–(13.13) have already been laid down in the literature. In particular, the case when f is at least bounded,

$$\sup_\lambda f(\lambda) < \infty, \tag{13.14}$$

has effectively been covered in a large literature, some relatively complete results appearing in Hannan (1979). There is also interest in cases where (13.14) does not hold, when we can say that u_t has long-range dependence: $f(\lambda)$ has a singularity at one or more λ, such as $\lambda = 0$. Here the relevant literature is much less extensive, and it has stressed the least squares estimate in case of nonstochastic x_t, satisfying what have come to be called 'Grenander's conditions'. Eicker (1967) gave a clt under the assumption that

$$u_t = \sum \tau_j \varepsilon_{t-j}, \quad \sum \tau_j^2 < \infty, \tag{13.15}$$

where $\sum$ will always denote a sum over $0, \pm 1, \dots$ of an obvious index, and the ε_t are independent with finite variance, and Hannan (1979) relaxed the latter assumption to square-integrable martingale differences. The square-summability condition in (13.15) is equivalent only to covariance stationarity of u_t given the other conditions (i.e. to integrability of f). Such stronger conditions on the τ_j as absolute summability would rule out long-range dependence, and imply (13.14). Mention must also be made of important early contributions in the simple location model special case. Ibragimov and Linnik (1971), theorem 18.6.5, considered this under essentially the same conditions on u_t as Eicker (1967), while Taqqu (1975) initiated a new avenue of research by assuming u_t is a nonlinear function of a Gaussian process.

The work of Yajima (1988, 1991) contained some central limit theory (under conditions on cumulants of all orders) but stressed other aspects of regression with nonstochastic regressors and long-range-dependent errors. He assumed that

$$f(\lambda) = f^*(\lambda)|\lambda|^{-2d}, \quad -\pi < \lambda \le \pi, \ 0 < d < \tfrac{1}{2}, \tag{13.16}$$

where f^* is a continuous function, and showed the importance to the properties of least squares of the behaviour near $\lambda = 0$ of the spectral distribution function, denoted $M(\lambda)$, of the generalized harmonic analysis of his nonstochastic x_t: the estimate of the ith element of β has rate of convergence depending on d if the ith diagonal element of $M(\lambda)$ has positive increment at $\lambda = 0$, and the same rate as under (13.14) otherwise. Yajima (1988, 1991) also obtained formulas for the asymptotic covariance matrix of least squares and generalized least squares, and conditions under which they have equal asymptotic efficiency. Dahlhaus (1995) proved asymptotic normality of generalized least squares estimates, and approximations thereto, for certain forms of nonstochastic regressor such that M has mass at zero frequency and under a condition similar to (13.16). Künsch, Beran and Hampel (1993) discussed the effect of long-range-dependent errors on standard independence-based inference rules in the context of certain experimental designs. Koul (1992), Koul and Mukherjee (1993) and Giraitis, Koul and Surgailis (1996) considered the asymptotic properties of various robust estimates of β.

Our setup can be compared to those of Yajima (1991) under his condition (13.16) in that, due to our mean correction, F effectively corresponds to his M in case an element of M has a jump at frequency 0. Here Yajima (1991) imposed conditions that are not innocuous. In particular, he required (case (i) of his theorem 2.1) that, in our notation, Σ_f be finite. Yajima (1991) included also situations in which Σ_f diverges due to a jump in M at frequency 0. Robinson (1994a) indicated how Σ_f can diverge in our present context. Suppose that a certain diagonal element of $F(\lambda)$ is absolutely continuous with derivative that behaves like $C|\lambda|^{-2c}$ in a neighbourhood of $\lambda = 0$, for $c \in (0, \tfrac{1}{2})$ and $C > 0$. Then $\Sigma_f < \infty$ if and only if $c + d < \tfrac{1}{2}$, which is not true if there is collectively sufficient long-range dependence in u_t and an x_t element. This does not seem implausible in a situation in which long-range dependence is suspected in u_t, because of empirical evidence of the peakedness, near zero frequency, of spectra of observable series. In this circumstance it appears that (13.11) does not hold; $\hat{\beta}_1$ has a rate of convergence slower than $n^{1/2}$, and need not be asymptotically normal. Note that $\Sigma_f = \infty$ also if sufficiently strong singularities in the spectral densities of x_t and u_t coincide at any nonzero frequency. Note also that the obvious stochastic analog of condition (23) of Yajima (1991) (employed in his clt for least squares) does not hold even in case of white noise x_t. Our focus on stochastic x_t reflects practical situations frequently encountered (e.g. in oceanography and econometrics) where the response of one series to others is of interest, yet it is natural to regard all as stochastically generated (e.g., see Hamon and Hannan 1963).

The present work identities functions $\phi(\lambda)$ which can guarantee the clt (13.9), with its $n^{1/2}$ rate of convergence, under effectively no restrictions on the degree of stationary long-range dependence in x_t and u_t. Such ϕ include f^{-1}, so that (13.12) is included. The efficiency advantages of $\hat{\beta}_{f^{-1}}$ are well known in other circumstances, but the preceding discussion of the pitfalls of $\hat{\beta}_1$ makes $\hat{\beta}_{f^{-1}}$ seem even more attractive than usual. The conditions for the clt's (13.9) and (13.12) are introduced and described in the following section, which, under the same conditions, asserts corresponding results in case $\hat{\beta}_\phi$ is replaced by

$$\tilde{\beta}_\phi = \tilde{A}_\phi^{-1} \tilde{a}_\phi, \qquad (13.17)$$

where

$$\tilde{A}_\phi = \frac{1}{n} \sum_{j=1}^{n-1} I_x(\lambda_j)\phi(\lambda_j), \qquad \tilde{a}_\phi = \frac{1}{n} \sum_{j=1}^{n-1} I_{xy}(\lambda_j)\phi(\lambda_j), \qquad (13.18)$$

where $\lambda_j = 2\pi j/n$. Note that $\tilde{\beta}_\phi$ will sometimes be preferred on computational grounds, especially as formulas for $\phi(\lambda) = f^{-1}(\lambda)$ [and indeed for many other choices of $\phi(\lambda)$] are often simpler than the corresponding ones for ϕ_j. For $1 \leq j \leq n - 1$, $w_x(\lambda_j)$ and $w_y(\lambda_j)$ are invariant to location shift and so the mean correction in the formulas in (13.3) is vacuous; it is the omission of $j = 0$ (and n) from the sums in (13.18) which permits unknown α and $E(x_t)$. (Note that $\tilde{\beta}_1 = \hat{\beta}_1$.) The clt proofs for both $\hat{\beta}_\phi$ and $\tilde{\beta}_\phi$ are given in Section 13.3. Section 13.4 discusses the clt for the case (13.13) for a parametric f. Section 13.5 includes an extension to nonlinear regression. We report in Section 13.6 the results of a small Monte Carlo study of finite-sample behaviour.

13.2 Central Limit Theorems

We introduce the following conditions.

Condition 1

$$u_t = \sum_0^\infty \tau_j \varepsilon_{t-j}, \qquad \sum_0^\infty \tau_j^2 < \infty,$$

where

$$E(\varepsilon_t | F_{t-1}) = 0, \qquad E(\varepsilon_t^2 | F_{t-1}) = E(\varepsilon_1^2) = \sigma^2 \text{ a.s.},$$

F_{t-1} being the σ-field of events generated by $\{\varepsilon_s, s \leq t - 1\}$, and the ε_t^2 being uniformly integrable.

Condition 2

$$\sum_0^\infty \tilde\phi_j < \infty \quad \text{where} \quad \tilde\phi_a = \max_{j\geq a} |\phi_j|.$$

Condition 3 The spectral density $f(\lambda)$ exists for all λ, and

$$\left(\sum_0^n |\gamma_j| + n\tilde\gamma_n\right)\left\{\left(\sum_0^n \tilde\phi_j^{1/2}\right)^2 + n\Phi_n\right\} = O(n) \quad \text{as } n \to \infty, \quad (13.19)$$

where $\tilde\gamma_a = \max_{j\geq a}|\gamma_j|$, $\Phi_a = \sum_{|j|>a} |\phi_j|$.

Condition 4 $\psi(\lambda)$ is a continuous function.

Condition 5 Σ_ϕ and Σ_ψ are nonsingular; Σ_ψ is finite.

Condition 6 $\{x_t\}$, $\{\varepsilon_t\}$ are independent sequences.

Condition 7 $\{x_t\}$ is fourth-order stationary, and

$$\Gamma_j \to 0 \quad \text{as } j \to \infty, \quad (13.20)$$

$$\lim_{|u|\to\infty} \max_{|v|,|w|<\infty} |\kappa_{abcd}(0, u, v, w)| = 0, \quad 1 \leq a, b, c, d \leq K, \quad (13.21)$$

where $\kappa_{abcd}(0, u, v, w)$ is the fourth cumulant of $x_{a0}, x_{bu}, x_{cv}, x_{dw}$, and x_{it} is x_t's ith element.

Theorem 1 Under (13.1) and Conditions 1–7, it follows that

$$n^{1/2}(\hat\beta_\phi - \beta), n^{1/2}(\tilde\beta_\phi - \beta) \to_d N\left(0, \Sigma_\phi^{-1}\Sigma_\psi\Sigma_\phi^{-1}\right), \quad (13.22)$$

and thus, when $f(\lambda) > 0$ and $\phi(\lambda) = f(\lambda)^{-1}$ for all λ,

$$n^{1/2}(\hat\beta_{f^{-1}} - \beta), n^{1/2}(\tilde\beta_{f^{-1}} - \beta) \to_d N\left(0, \Sigma_{f^{-1}}^{-1}\right). \quad (13.23)$$

The proof is reserved for the following section, and we first discuss the conditions.

Condition 1 is a relaxation of Eicker's (1967) condition on his ε_t in (13.15); see also Hannan (1979). It is restrictive in the linearity it imposes, but not otherwise.

Condition 2 implies that $\phi_j = O(\tilde\phi_j) = o(j^{-1})$ as $j \to \infty$. Then Condition 3 is automatically satisfied if also $\gamma_j = o(j^{-1})$ as $j \to \infty$. On weakening the latter requirement to $\gamma_j = O(\log^p(2 + |j|)/(1 + |j|))$, for any $p > 0$, we can satisfy Condition 3 by $\phi_j = O(1/(1 + |j|) \log^{p+2}(2 + |j|))$. On the other hand, if only $\gamma_j = O(1/\log^p(2 + |j|))$, for any $p > 0$, then $\phi_j = O(\log^{p-2}(2 + |j|)/(1+|j|))$ suffices. The first statement of Condition 3 implies, by the Riemann–Lebesgue lemma (Zygmund 1977: p. 45], the mild type of ergodicity condition $\lim_{j\to\infty} \gamma_j = 0$ and (13.19) entails no extra condition on γ_j when $\sum_0^\infty \tilde\phi_j^{1/2} < \infty$ [which implies $\phi_j = o(j^{-2})$].

Perhaps the conditions of most interest under which Condition 3 holds lie between these extremes: when, for some $d \in (0, \frac{1}{2})$,

$$\gamma_j = O(L(|j|)^{-1}(1 + |j|)^{2d-1}), \tag{13.24}$$

$$\phi_j = O(L(|j|)(1 + |j|)^{-2d-1}), \tag{13.25}$$

where L is slowly varying at ∞ (see e.g. Bingham, Goldie, and Teugels 1987). The special case of (13.24)

$$\gamma_j \sim Dj^{2d-1}, \quad D > 0 \text{ as } j \to \infty \tag{13.26}$$

(where '$\sim$' means that the ratio of left- and right-hand sides tends to 1) holds in case of fractional autoregressive integrated moving average (FARIMA) models with differencing parameter d, where f in (13.7) is given by

$$f(\lambda) = \frac{\sigma^2}{2\pi} |1 - e^{i\lambda}|^{-2d} \left| \frac{a(e^{i\lambda})}{b(e^{i\lambda})} \right|^2, \quad -\pi < \lambda \leq \pi, \tag{13.27}$$

where a and b are polynomials of finite degree, all of whose zeros are outside the unit circle, and $\sigma^2 > 0$. The model (13.27) may have originated in Adenstedt (1974), and it also satisfies (13.16). However, we stress that 'O' in (13.24) has the usual 'upper bound' meaning, not 'exact rate', so even when $L(j) \equiv 1$ it can hold also when the γ_j not only decay slowly, but indefinitely oscillate. For example, consider the spectral density

$$f(\lambda) = \frac{\sigma^2}{2\pi} \prod_{j=1}^{h} |1 - 2 \cos \omega_j e^{i\lambda} + e^{2i\lambda}|^{-2d_j} \left| \frac{a(e^{i\lambda})}{b(e^{i\lambda})} \right|^2, \quad -\pi < \lambda \leq \pi, \tag{13.28}$$

where a, b and σ^2 are as before, and the ω_j are distinct numbers in $[0, \pi]$. In case $h = 1$ and $\omega_1 = 0$, (13.28) reduces to (13.27) with $d_1 = d/2$, so $0 < d_1 < \frac{1}{4}$ is required for (13.24) to hold. When $h = 1$ but $0 < \omega_1 \leq \pi$, we have a 'cyclic' FARIMA, f having singularity at the nonzero frequency ω_1. In case $h = 1$ and $\omega_1 = \pi$, Gray, Zhang, and Woodward (1989) showed that $\gamma_j \sim D(-1)^j j^{4d_1-1}$, so again we need $0 < d_1 < \frac{1}{4}$ for (13.24) to hold, but now (13.26) does not hold. In case $h = 1$ and $0 < \omega_1 < \pi$, Gray, Zhang, and Woodward (1989, 1994) showed that $\gamma_j \sim Dj^{2d_1-1} \cos(j\omega_1)$, so (13.24) holds when $0 < d_1 < \frac{1}{2}$, but not (13.26). For $h > 1$ there is more than one spectral singularity in $[0, \pi]$, as may be a reasonable model for seasonal processes (e.g. for monthly data take $h = 7$ and $\omega_j = (j - 1)\pi/6$, $j = 1, \ldots, 7$). We conjecture that (13.24) will hold, with $d = \max(a_1, \ldots, a_h)$ and $0 < a_j/2 = d_j < \frac{1}{4}$ if $\omega_j = 0$ or π, $0 < a_j = d_j < \frac{1}{2}$ otherwise. The problem of testing hypotheses on the d_j in (13.27) and (13.28) has been studied by Robinson (1994b). Conditions that link (13.16) and (13.26) outside the classes (13.27) and (13.28) can be inferred from Yong (1974), and sufficient conditions for (13.24) with $L(j) \equiv 1$ in terms of upper bounds on

f and its derivative are in lemma 4 of Fox and Taqqu (1986), which can also be generalized to the case of singularities at nonzero frequencies.

As d decreases from $\frac{1}{2}$ to 0, (13.24) becomes stronger but (13.25) becomes weaker, and while 'O' in the latter again means 'upper bound', we have

$$\phi_j \sim D' j^{-2d-1}, \quad D' > 0 \text{ as } j \to \infty, \tag{13.29}$$

along with (13.26) in case $\phi = f^{-1}$ and u_t is a noncyclic FARIMA as in (13.27). For this ϕ, the ϕ_j are proportional to the inverse autocorrelations of u_t. Condition 4 is implied by Condition 2, if $\phi = f^{-1}$, and indicates that in order to choose ϕ merely to guarantee an $n^{1/2}$-consistent $\hat{\beta}_\phi$ we have to know the location of the singularity or singularities of f, and design ϕ to have a zero or zeros of sufficient order to cancel them. Suppose that, for some $\omega \in [0, \pi)$, $f(\lambda) \sim C|\lambda - \omega|^{-2d}/L(1/|\lambda - \omega|)$ as $\lambda \to \omega$ for $d \in (0, \frac{1}{2})$, f is continuous for $\lambda \neq \pm\omega$ and (13.24) holds (as is true in case of (13.27)). Then, for any such d,

$$\phi(\lambda) = |\lambda - \omega|, \quad 0 \leq \lambda < \pi, \tag{13.30}$$

satisfies Conditions 2, 3 (because $\psi_j = O(j^{-2})$), and 4. When f has several singularities we can extend (13.30) to have zeros of degree 1 at each.

Nonsingularity in Condition 5 is likely to hold in case of no multicollinearity in x_t. Finiteness of Σ_ϕ is a consequence of Parseval's equality and Condition 2.

Condition 6 is very restrictive. We believe it could be relaxed to independence up to some moment order, but at the cost of greater structure on x_t than required in Condition 7. Such structure (e.g. a linear process similar to that assumed for u_t) could also lead to a reduction in the fourth moment condition on x_t, but we find the extremely mild condition (13.20) aesthetically so appealing as to offset both considerations. The fourth cumulant condition (13.21) is vacuous in case of Gaussianity, and milder than the summability conditions on cumulants frequently employed. Finally, the covariance stationarity condition on x_t is very strong. There may be scope for replacing it by a stochastic version of 'Grenander's conditions'. Neither Condition 7 nor 'Grenander's conditions' are satisfied by unit root behaviour, a popular recent assumption. While there are certainly many series lending empirical support for this sort of assumption, not only is it strictly not 'weaker' than Condition 7, but it actually describes a very specialized form of nonstationarity, and covers a far narrower spectrum of the '$I(d)$' processes than does Condition 7. Note that the model (13.1) permits the vector x_t to consist of lags or leads of a basic time series (see, e.g. Hannan (1967)); parsimonious parameterizations have been proposed in this case, entailing linear restrictions on β, and it is easy to extend our results to cover these.

It is our insistence on allowing for any degree of fractional differencing, at any frequencies, in x_t which prevents the least squares choice of $\phi \equiv 1$ (see (13.6)] from being covered by the clt, and we have preferred not to investigate in detail the trade-offs between F and ϕ which the discussion of Section 13.1 indicates is possible. However, the following result will be of some use in Section 13.4. Its proof is a simple consequence of some of the properties derived in the following section, and the Toeplitz lemma (see, e.g. Stout 1974: pp. 120–1), and is omitted.

Theorem 2 Under (13.1) and Conditions 6 and 7 and the nonsingularity of Γ_0,

$$\hat{\beta}_1 = \beta + o_p \left[n^{-1} \sum_0^n |\gamma_j| \right]^{1/2} \quad \text{as } n \to \infty. \tag{13.31}$$

As an application of Theorem 2, consider a problem of relatively minor interest, estimating α in (13.1). An obvious estimate is

$$\hat{\alpha}_\phi = \bar{y} - \hat{\beta}'_\phi \bar{x} = \alpha + \bar{u} - (\hat{\beta}_\phi - \beta)'\bar{x}, \tag{13.32}$$

where $\bar{u} = n^{-1} \sum_t u_t$. If $\hat{\beta}_\phi = \beta + o_p(\{n^{-1} \sum_0^n |\gamma_j|\}^{1/2})$, as in Theorem 2, and if (13.24) holds, then $n^{1/2-d}(\hat{\alpha}_\phi - \alpha)$ has the same limit distribution as $n^{1/2-d}\bar{u}$, and if Condition 1 holds this is $N(0, D/d(2d+1))$ (cf. Taqqu 1975). On the other hand, if $f(\lambda)$ is continuous and positive at $\lambda = 0$ and $\hat{\beta}_\phi$ is $n^{1/2}$-consistent (as in Theorem 1), then the last term on the rightmost expression in (13.32) also contributes, unless $E(x_1) = 0$, while it dominates if $\hat{\beta}_\phi$ is less-than-$n^{1/2}$-consistent.

13.3 Proof of Theorem 1

Dropping the ϕ subscripts from (13.2), (13.3), (13.17) and (13.18), write $\hat{\beta} - \beta = \hat{A}^{-1}\hat{a}$, $\tilde{\beta} - \beta = \tilde{A}^{-1}\tilde{a}$, where

$$\hat{a} = n^{-1} \sum_s \sum_t (x_t - \bar{x})(u_s - \bar{u})\phi_{t-s}, \tilde{a} = n^{-1} \sum_{j=1}^{n-1} I_{xu}(\lambda_j)\phi(\lambda_j),$$

for

$$I_{xu}(\lambda) = (2\pi n)^{-1} \left(\sum_t x_t e^{it\lambda} \right) \left(\sum_t u_t e^{-it\lambda} \right).$$

With $Ex_1 = 0$, without loss of generality, write $A = n^{-1} \sum_s \sum_t x_t x'_s \phi_{t-s}$, $a = n^{-1} \sum_s \sum_t x_t u'_s \phi_{t-s}$. Define $r_t = \sum_s x_s \phi_{t-s}$, $s_u = \sum_t r_t \tau_{t-u}$, with $\tau_j = 0$, $j < 0$, and $a_1 = n^{-1} \sum_{-N}^n s_u \varepsilon_u$, $a_2 = a - a_1$, with $N = N_n$ yet to be chosen. Write

$$D_0 = E(naa'|\{x_t\}), \quad D_1 = E(na_1 a'_1|\{x_t\}), \quad D_2 = D_0 - D_1,$$

$$E = (2\pi)^{-1} \int_{-\pi}^{\pi} I_x(\lambda)\psi(\lambda) \, d\lambda.$$

Write

$$n^{1/2}(\hat{\beta} - \beta) = A^{-1}(A\hat{A}^{-1})\Sigma_\psi^{1/2} \left(\Sigma_\psi^{-1/2} E^{1/2} \right) \left(E^{-1/2} D_0^{1/2} \right)$$

$$\times (D_0^{-1/2} D_1^{1/2}) D_1^{-1/2} n^{1/2}(a_1 + a_2),$$

$$n^{1/2}(\tilde{\beta} - \hat{\beta}) = (\hat{A}^{-1} - \tilde{A}^{-1}) n^{1/2} \hat{a} + \tilde{A}^{-1} n^{1/2} (\hat{a} - \tilde{a}),$$

where $X^{1/2}$ satisfies $(X^{1/2})^2 = X$, and noting that $E\|D_2\| \leq E\|n^{1/2}a_2\|^2$, where $\|X\| = \mathrm{tr}^{1/2}(X'X)$, the proof for $\hat{\beta}$ follows immediately from Propositions 1–6, and that for $\tilde{\beta}$ from Propositions 1–5 and 7.

Proposition 1 For N increasing suitably with n,

$$\lim_{n \to \infty} E\|n^{1/2}a_2\|^2 = 0.$$

Proof With C denoting a generic constant,

$$E(n\|a_2\|^2) = \frac{\sigma^2}{n} \sum_{u=-\infty}^{-N-1} \sum_{t} \sum_{s} \sum_{r} \sum_{q} E(x_r x_q')\phi_{t-r}\phi_{s-q}\tau_{t-u}\tau_{s-u}$$

$$\leq \frac{C}{n}\left(\sum_{t}\sum_{s}|\phi_{t-s}|\right)^2 \sum_{N}^{\infty}\tau_u^2 \leq Cn\sum_{N}^{\infty}\tau_u^2,$$

which tends to 0 as $n \to \infty$ for a suitable sequence N_n, by Condition 1. ∎

Proposition 2 As $n \to \infty$,

$$D_1^{-1/2}n^{1/2}a_1 \to_d N(0, I), \tag{13.33}$$

where I is the identity matrix.

Proof Equation (13.33) is implied if the convergence holds conditional on $\{x_t\}$ and we establish the latter. For any vector z such that $\|z\| = 1$ and any N, n, define $d_u = n^{-1/2}z'D_1^{-1/2}s_u$. Then from Scott (1973), it suffices to show that, as $n, N \to \infty$,

$$\sum_{-N}^{n} d_u^2 E(\varepsilon_u^2|F_{u-1}) \to_p 1, \tag{13.34}$$

$$E\left(\sum_{-N}^{n} d_u^2 E[\varepsilon_u^2 I(|d_u\varepsilon_u| > \eta)|\{x_t\}]\right) \to 0 \quad \text{for all } \eta > 0. \tag{13.35}$$

Trivially, (13.34) follows from Condition 1 and $\sum_{-N}^{n} d_u^2 = 1/\sigma^2$. For $\delta > 0$ the left-hand side of (13.35) is bounded by

$$E\left[\sum_{-N}^{n} d_u^2 E[\varepsilon_u^2 I(|\varepsilon_u| > \eta/\delta)]\right] + P\left(\max_u |d_u| > \delta\right). \tag{13.36}$$

The first term can be made arbitrarily small by choosing δ small enough, from uniform integrability. We now modify an argument of Eicker (1967). Propositions 1,

4, and 5 and Condition 5 imply that $D_1^{-1} = O_p(1)$, so we can consider the set $\|D_1^{-1/2}\| \le C$, on which, for any $\varepsilon > 0$,

$$\max_u |d_u| \le \max_u Cn^{-1/2} \sum_t \|r_t\| |\tau_{t-u}|$$

$$\le C \left\{ \frac{\varepsilon}{n} \sum_t \|r_t\|^2 \right\}^{1/2} + CLn^{-1/2} \max_{1 \le t \le n} \|r_t\| \max_u |\tau_u|,$$

where $L = L(\varepsilon)$ is chosen such that $\sum_{u \ge L} \tau_u^2 < \varepsilon$. Then, the proof of (13.35) is completed by

$$E\|r_t\|^2 \le \sum_s \sum_u E\|x_s\|^2 |\phi_{t-s} \phi_{t-u}| \le C < \infty,$$

$$\max_{1 \le t \le n} \|r_t\| \le \max_{1 \le t \le n} \|x_t\| \sum |\phi_u| \le C \left\{ \sum_t \|x_t\|^4 \right\}^{1/4} = O_p(n^{1/4}),$$

using Markov's inequality and Conditions 2 and 7. ∎

Proposition 3 $A \to_p \Sigma_\phi$ as $n \to \infty$.

Proof Write $C_j = n^{-1} \sum_{1 \le t, t+j \le n} x_t x'_{t+j}$. By easy use of formula (30) of Anderson (1971: p. 464), it is seen that Condition 7 implies $E\|C_j - \Gamma_j\|^2 \to 0$ for all fixed j. With a simple truncation argument and Condition 2, it follows that $\sum_{1-n}^{n-1} \phi_j(C_j - \Gamma_j) \to_p 0$. But Condition 2 also implies that $\sum_{|j| \ge n} \phi_j \Gamma_j \to 0$. ∎

Proposition 4 $E \to_p \Sigma_\psi$ as $n \to \infty$.

Proof Let $\psi_J(\lambda) = 2\pi \sum_{j=-J}^{J} (1 - |j|/J) \psi_j e^{-ij\lambda}$ be the Cesaro sum to J terms of the Fourier series of $\psi(\lambda)$, where $\psi_j = (2\pi)^{-2} \int_{-\pi}^{\pi} \psi(\lambda) \cos j\lambda d\lambda$, and write

$$E - \Sigma_\psi = \frac{1}{2\pi} \int_{-\pi}^{\pi} I_x(\lambda)\{\psi(\lambda) - \psi_J(\lambda)\} d\lambda$$

$$+ \frac{1}{2\pi} \int_{-\pi}^{\pi} \psi_J(\lambda)\{I_x(\lambda) d\lambda - dF(\lambda)\}$$

$$+ \frac{1}{2\pi} \int_{-\pi}^{\pi} \{\psi_J(\lambda) - \psi(\lambda)\} dF(\lambda).$$

The second term on the right-hand side is

$$\sum_{j=-J}^{J} \psi_j \left(1 - \frac{|j|}{J}\right) (C_j - \Gamma_j) \to_p 0$$

for any fixed J, as in the proof of Proposition 3, while Condition 4 implies that for any $\varepsilon > 0$ we can choose J so large that $\sup_\lambda |\psi(\lambda) - \psi_J(\lambda)| < \varepsilon$ from Fejér's theorem (Zygmund 1977: p. 89) so that the norms of the first and last terms on the right-hand side are less than $(\varepsilon/2\pi)\mathrm{tr}(C_0)$ and $(\varepsilon/2\pi)\mathrm{tr}(\Gamma_0)$. ∎

Proposition 5 $E - D \to_p 0$ as $n \to \infty$.

Proof We have

$$\psi_m = \sum\sum \phi_j \phi_k \gamma_{m-j-k},$$

so that

$$E - D = \frac{1}{n}\sum_t\sum_u x_t x_u' \sum_{1-n}^{n-1} \gamma_j \sum_q{}' \phi_{t-j-q}\,\phi_{u-q}$$

$$+ \frac{1}{n}\sum_t\sum_u x_t x_u' \sum_{|j|\geq n} \gamma_j \Delta_{t-j-u}$$

$$= R_1 + R_2,$$

where $\sum_q'$ is a sum over $\{q > \min(n, n - j)\} \cup \{q < \max(1, 1 - j)\}$ and $\Delta_j = \sum \phi_t \phi_{t-j}$. To prove $R_1 \to_p 0$ and $R_2 \to_p 0$, it is convenient to introduce first a series of lemmas, some of which will also be useful in proving subsequent properties. Let $a_{jtu} = \sum_{q>n-j} \phi_{t-j-q}\phi_{u-q}$. ∎

Lemma 1 For $|r| < n$,

$$\sum_t \Phi_{|r-t|/2} \leq 4\sum_{t=0}^{n-1} \Phi_t.$$

Proof $\Phi_{|r-t|/2} = \Phi_{[|r-t|/2]}$, and for $1 \leq t \leq n$, $0 \leq r < n$ we have $0 \leq [\frac{1}{2}|r - t|] \leq n - 1$, with duplications due to division by 2 and taking integer parts, and to positive and negative values of $r - t$. ∎

Lemma 2

$$\sum_{t=0}^{n-1} \Phi_t = \sum_t t|\phi_t| + n\Phi_n.$$

Proof Elementary. ∎

Lemma 3

$$\sum_j t\tilde{\phi}_t \leq \left(\sum_t \tilde{\phi}_t^{1/2}\right)^2.$$

Proof $\sum_t t\tilde{\phi}_t \leq \sum_t \tilde{\phi}_t^{1/2} \sum_{u=1}^t \tilde{\phi}_u^{1/2} \leq \left(\sum_t \tilde{\phi}_t^{1/2}\right)^2.$ ∎

Lemma 4 Under Condition 2, for all integers j and all $t, u \in [1, n]$,

$$|a_{jtu}| \leq C \tilde{\phi}_{n-t}^{1/2} \tilde{\phi}_{|t-j-u|/2}^{1/2} + \tilde{\phi}_{n-t} \Phi_{|t-j-u|/2}.$$

Proof The left-hand side is bounded by

$$\sum_{i \in A} |\phi_{t-j-u-i} \phi_{-i}| + \sum_{i \in B} |\phi_{t-j-u-i} \phi_{-i}|,$$

where $A = \{i : i > n - j - u, |i| \leq \frac{1}{2}|t - j - u|\}$, $B = \{i : i > n - j - u, |i| > \frac{1}{2}|t - j - u|\}$. Because $i \in A$ or $i \in B$ imply $t - j - u - i < t - n \leq 0$ while $i \in A$ implies $t - j - u - i \leq \frac{1}{2}(t - j - u) \leq 0$, we can bound the two sums respectively by $\tilde{\phi}_{n-t}^{1/2} \tilde{\phi}_{|t-j-u|/2}^{1/2} \sum |\phi_i|$ and $\tilde{\phi}_{n-t} \Phi_{|t-j-u|/2}$, whence the result follows from Condition 2. ∎

Lemma 5 For $0 \leq j < n$,

$$\sum_t \sum_u |a_{jtu}| \leq C \left\{ \left(\sum_0^n \tilde{\phi}_t^{1/2} \right)^2 + n \Phi_n \right\}. \tag{13.37}$$

Proof Applying Lemma 4,

$$\sum_t \sum_u |a_{jtu}| \leq C \sum_t \tilde{\phi}_{n-t}^{1/2} \sum_u \tilde{\phi}_{|t-j-u|/2}^{1/2} + \sum_t \tilde{\phi}_{n-t} \sum_u \Phi_{|t-j-u|/2}$$

$$\leq C \left\{ \left(\sum_0^{n-1} \tilde{\phi}_t^{1/2} \right)^2 + \sum_t t |\phi_t| + n \Phi_n \right\}$$

by Lemmas 1 and 2, where (13.37) follows from Lemma 3. ∎

Lemma 6 $\sum_{j > 2n} |\Delta_j| \leq C \Phi_n$.

Proof We have

$$\sum_{j > 2n} |\Delta_j| \leq \sum_{|q| \leq n} |\phi_q| \sum_{j > 2n} |\phi_{q-j}| + \sum_{|q| > n} |\phi_q| \sum_{j > 2n} |\phi_{q-j}|$$

$$\leq \sum_{|q| \leq n} |\phi_q| \sum_{j > n} |\phi_j| + 2 \sum_{q > n} |\phi_q| \sum |\phi_j|$$

$$\leq C \Phi_n. ∎$$

Lemma 7 For $j > 0$,

$$|\Delta_j| \leq C \tilde{\phi}_{j/2}.$$

Proof

$$|\Delta_j| \le \sum_{|q| \le j/2} |\phi_q \phi_{q-j}| + \sum_{|q| > j/2} |\phi_q \phi_{q-j}|$$

$$\le C\tilde{\phi}_{j/2}. \quad \blacksquare$$

Consider now

$$R_1 = \sum_t \sum_u x_t x_u' b_{tu} + \sum_t \sum_u x_{n-t+1} x_{n-u+1}' b_{tu}$$

$$+ \sum_t \sum_u x_u x_t' \tilde{b}_{tu} + \sum_t \sum_u x_{n-t+1} x_{n-u+1}' \tilde{b}_{tu}$$

$$= R_{11} + R_{12} + R_{13} + R_{14},$$

where $b_{tu} = n^{-1} \sum_{j=0}^{n-1} \gamma_j a_{jtu}$, $\tilde{b}_{tu} = n^{-1} \sum_{j=1}^{n-1} \gamma_j a_{jtu}$. Then

$$|ER_1| \le \left| \sum_t \sum_u (\Gamma_{t-u} + \Gamma_{u-t}) b_{tu} \right| + \left| \sum_t \sum_u (\Gamma_{t-u} + \Gamma_{u-t}) \tilde{b}_{tu} \right|$$

$$\le \frac{4}{n} \sum_t \sum_u c_{t-u} \bar{b}_{tu},$$

where $c_j = \|\Gamma_j\|$, $\bar{b}_{tu} = \sum_{j=0}^{n-1} |\gamma_j a_{jtu}|$. By Lemma 5 and Condition 3,

$$\sum_t \sum_u \bar{b}_{tu} = O(n), \tag{13.38}$$

so that $ER_1 \to 0$ by Condition 7 and the Toeplitz lemma. Next

$$E\|R_1 - ER_1\|^2 \le 4 \sum_{j=1}^{4} E\|R_{1j} - ER_{1j}\|^2$$

$$\le \sum_t \sum_u \sum_v \sum_w h_{tuvw} \bar{b}_{tu} \bar{b}_{vw}, \tag{13.39}$$

where

$$h_{tuvw} = c_{t-v} c_{u-w} + c_{t-w} c_{u-v} + \sum_{a,b,c,d=1}^{K} |\kappa_{abcd}(t, u, v.w)|$$

$$= h_{tuvw}^{(1)} + h_{tuvw}^{(2)} + h_{tuvw}^{(3)}.$$

By Condition 7, for any $\varepsilon > 0$ there exists $L = L(\varepsilon)$ such that $h_{tuvw}^{(1)} < \varepsilon$ if $\max(|t-v|, |u-w|) \ge L$, $h_{tuvw}^{(2)} < \varepsilon$ if $\max(|t-w|, |u-v|) \ge L$ and

$h_{tuvw}^{(3)} < \varepsilon$ if $\max(|t - u|, |t - v|, |t - w|, |u - v|, |u - w|, |v - w|) \geq L$. Thus, because $\sum_t \sum_u \sum_v \sum_w \bar{b}_{tu} \bar{b}_{vw} = O(n^2)$ from (13.38), it follows that $E\|R_1 - ER_1\|^2 \to 0$, and then $R_1 \to_p 0$. Now consider R_2. We have

$$|ER_2| \leq \frac{1}{n} \sum_t \sum_u c_{t-u} d_{tu},$$

where $d_{tu} = \sum_{|j|>n} |\gamma_j \Delta_{t-j-u}|$. Applying Lemmas 3, 6, and 7 and Condition 3,

$$\frac{1}{n} \sum_t \sum_u d_{tu} \leq \tilde{\gamma}_n \sum_{1-n}^{n-1} \sum_{|j|>n} |\Delta_{r-j}|$$

$$\leq \tilde{\gamma}_n \left(\sum_1^{2n} |j\Delta_j| + n \sum_{j>2n} |\Delta_j| \right)$$

$$\leq C\tilde{\gamma}_n \left(\sum_1^{2n} j\tilde{\phi}_j + n\Phi_n \right) = O(1). \tag{13.40}$$

Thus $ER_2 \to 0$ by the Toeplitz lemma. Next

$$E\|R_2 - ER_2\|^2 \leq \frac{1}{n^2} \sum_t \sum_u \sum_v \sum_w h_{tuvw} |d_{tu} d_{vw}| \to 0$$

from (13.40) and the properties of h_{tuvw} previously indicated and the Toeplitz lemma. ∎

Proposition 6 As $n \to \infty$,

$$n^{1/2}(\hat{a} - a) \to_p 0, \quad \hat{A} - A \to_p 0.$$

Proof Put $n^{1/2}(\hat{a} - a) = b_1 + b_2 + b_3$, where

$$b_1 = -n^{-1/2} \bar{x} \sum_t \sum_s u_s \phi_{t-s}, \quad b_2 = -n^{-1/2} \bar{u} \sum_t \sum_s x_t \phi_{t-s}$$

and

$$b_3 = \bar{u}\bar{x}n^{-1/2} \sum_t \sum_s \phi_{t-s}.$$

Note first that

$$\bar{u} = O_p \left(\left\{ n^{-1} \sum_{1-n}^{n-1} (1 - |j|/n)\gamma_j \right\}^{1/2} \right), \quad \bar{x} \to_p 0, \tag{13.41}$$

where the second result uses Condition 7 and the Toeplitz lemma. First suppose $\phi(0) = 0$. Because $\phi(\lambda) = 2\pi \sum \phi_j \cos j\lambda$,

$$\sum_t \sum_s \phi_{t-s} = \sum_{1-n}^{n-1} (n - |j|)\phi_j$$

$$= -2nq_n - 2\sum_t t\phi_t$$

$$= O\left(\left\{\sum_0^n \tilde{\phi}_t^{1/2}\right\}^2 + n\Phi_n\right)$$

by Lemma 3, where $q_n = \sum_{t>n} \phi_t$. Thus $b_3 \to_p 0$ by (13.41) and Condition 3. Next, $\sum_t \sum_s x_t \phi_{t-s} = -\sum_t x_t(q_{t+1} + q_{n-t})$, so that

$$E\left\|\sum_t \sum_s x_t \phi_{t-s}\right\|^2 = \sum_{1-n}^{n-1} \text{tr}(\Gamma_j) P_j,$$

where

$$P_j = \sum_{\max(1, 1-j)}^{\min(n, n-j)} (q_{t+1} + q_{n-t})(q_{t+j+1} + q_{n-t-j}).$$

Because $P_j = O\left(\sum_0^{n+1} \Phi_t\right)$ uniformly in j, it follows from Lemmas 2 and 3, (13.41), Condition 3 and the Toeplitz lemma that $b_2 \to_p 0$. Similarly,

$$E\left(\sum_t \sum_s u_s \phi_{t-s}\right)^2 = \sum_{1-n}^{n-1} \gamma_j P_j = O(n),$$

and then $b_1 \to_p 0$ by (13.41). Now suppose $\phi(0) \neq 0$, when Condition 4 implies $f(\lambda)$ is continuous at $\lambda = 0$. Thus, from (13.41) and Fejér's theorem, $\bar{u} = O_p(n^{-1/2})$, and because Condition 2 implies $\sum_t \sum_s \phi_{t-s} = O(n)$, it follows that $b_3 \to_p 0$. Because

$$E\left\|\sum_t \sum_s x_t \phi_{t-s}\right\|^2 \leq \left(\sum |\phi_j|\right)^2 \sum_t \sum_s c_{t-s} = o(n^2),$$

we then have $b_2 \to_p 0$. To deal with b_1,

$$E\left(\sum_t \sum_s u_t \phi_{t-s}\right)^2 = \sum_t \sum_s \phi_{t-s} \sum_v \sum_{j=t-n}^{t-1} \gamma_j \phi_{t-v-j} = \sum_{i=1}^5 d_i,$$

where

$$d_1 = \sum_t \sum_s \psi_{t-s},$$

$$d_2 = -\sum_{t>n} \sum_s \phi_{t-s} \sum_v \zeta_{t-v},$$

$$d_3 = -\sum_{t\leq 0} \sum_s \phi_{t-s} \sum_t \zeta_{s-v},$$

$$d_4 = -\sum_t \sum_s \phi_{t-s} \sum_{j\geq t} \gamma_j \sum_v \phi_{t-v-j},$$

$$d_5 = -\sum_t \sum_s \phi_{t-s} \sum_{j<t-n} \gamma_j \sum_v \phi_{t-v-j},$$

where $\zeta_k = \sum \gamma_j \phi_{k-j}$. Now $d_1 = (n/2\pi)\psi_n(0) = O(n)$ by Condition 4, as in the proof of Proposition 4. Uniformly in t,

$$\sum_v |\zeta_{t-v}| \leq \sum_v \left\{ \sum_{-n}^n |\gamma_j \phi_{t-v-j}| + \sum_{j>n} |\gamma_j \phi_{t-v-j}| + \sum_{j<-n} |\gamma_j \phi_{t-v-j}| \right\}$$

$$= O\left(\sum_0^n |\gamma_j| + n\bar{\gamma}_n \right),$$

so that d_2 and d_3 are $O(n)$ by Lemma 3 and Condition 3. It is easily seen that

$$|d_4| \leq C \sum_t \sum_{j\geq t} |\gamma_j| \sum_v |\phi_{t-v-j}|$$

$$\leq \left(\sum_j |\gamma_j| + n\bar{\gamma}_n \right) \left\{ \sum_j j|\phi_j| + n\Phi_n \right\} = O(n)$$

by Condition 3. In the same way $d_5 = O(n)$. Thus $b_1 \to_p 0$ by (13.41). The proof that $\hat{A} - A \to_p 0$ is omitted because it follows from similar but more straightforward calculations to those above after writing

$$\hat{A} = A = \bar{x}\bar{x}' \frac{1}{n} \sum_t \sum_s \phi_{t-s} - \bar{x}\frac{1}{n} \sum_t \sum_s x'_s \phi_{t-s} - \frac{1}{n} \sum_t \sum_s x_t \phi_{t-s}\bar{x}'. \; \blacksquare$$

Proposition 7 As $n \to \infty$,

$$n^{1/2}(\tilde{a} - a) \to_p 0, \quad \tilde{A} - A \to_p 0.$$

Proof Because

$$\sum_{j=0}^{n-1} \exp(i\ell\lambda_j) = \begin{cases} n, & \ell = 0, \mod n, \\ 0, & \text{otherwise,} \end{cases}$$

we have

$$n^{1/2}(\tilde{a} - a) = n^{-1/2} \sum_t \sum_s x_t u_s \omega_{t-s} - n^{1/2} \bar{x} \bar{u} \sum \phi_j,$$

where $\omega_j = \sum_{|i| \geq 1} \phi_{j+in}$. The second term on the right-hand side is identically 0 if $\phi(0$ is 0); otherwise, because Condition 4 implies $f(0) < \infty$, and thus $\bar{u} = O_p(n^{-1/2})$, it is $o_p(1)$. The first term has mean 0 and variance bounded by

$$\frac{1}{n} \sum_s \sum_t \sum_u \sum_v c_{t-u} |\gamma_{s-v} \omega_{t-s} \omega_{u-v}|.$$

By the Toeplitz lemma it suffices to show that

$$\sum_s \sum_t \sum_u \sum_v |\gamma_{s-v} \omega_{t-s} \omega_{u-v}| = O(n).$$

The left-hand side is $O\left(\sum_{j=0}^{n} |\gamma_j| \sum_{1-n}^{n-1} |\omega_j| \sum_t \sum_s |\omega_{t-s}|\right)$. In view of Condition 3 and Lemma 3, the proof that $n^{1/2}(\tilde{a} - a) \rightarrow_p 0$ is completed by the following lemmas.

Lemma 8

$$\sum_{j=1-n}^{n-1} |\omega_j| < \infty.$$

Proof The left-hand side is bounded by $2 \sum |\phi_j| < \infty$. ∎

Lemma 9

$$\sum_t \sum_s |\omega_{t-s}| \leq C \left(\sum_j j|\phi_j| + n\Phi_n \right).$$

Proof The left-hand side is bounded by

$$2 \sum_{j=1}^{n-1} j|\phi_j| + 4 \sum_{\ell \geq 1} \sum_{\ell n}^{(\ell+1)n-1} |\phi_j|(n - |j - \ell n|).$$

The second term is bounded by $4n\Phi_n$. ∎

Finally, the proof that $\tilde{A} - A \rightarrow_p 0$ follows from

$$\tilde{A} - A = \frac{1}{n} \sum_t \sum_s x_t x_s' \omega_{t-s} - \bar{x} \bar{x}' \sum \phi_j.$$

The second term is clearly $o_p(1)$. The first has mean $O(n^{-1} \sum_t \sum_s c_{t-s} |\omega_{t-s}|) \rightarrow 0$ from the Toeplitz lemma and Lemma 8. The latter property is also used to show that the variance of the first term tends to 0, on dealing with this term as we did with R_1 and R_2 in the proof of Proposition 5. ∎

13.4 Feasible Generalized Least Squares

We now wish to estimate β efficiently in the absence of knowledge of f up to scale. We write $f(\lambda) = f(\lambda; \theta_0)$, where $f(\lambda; \theta)$ is a known function of λ and the p-dimensional vector θ, and θ_0 is unknown. For an estimate $\hat{\theta}$ of θ_0, define $\hat{\beta}_{\hat{f}-1}, \tilde{\beta}_{\hat{f}-1}$ by putting $\phi(\lambda) = f^{-1}(\lambda; \hat{\theta})$ in $\hat{\beta}_\phi$ and $\tilde{\beta}_\phi$. If θ includes the scale factor of f, and this is functionally unrelated to the remaining elements of θ, then these estimates of β are invariant to it, but the scale factor has to be estimated in order to estimate the limiting covariance matrix $\Sigma_{f^{-1}}^{-1}$.

Now introduce the following condition.

Condition 8

$$\hat{\theta} - \theta_0 = O_p\left(\left\{\sum_0^n |\gamma_j|\right\}^{-1/2}\right) \tag{13.42}$$

and, defining $N_\delta = \{\theta: \|\theta - \theta_0\| < \delta\}$, there exists $\delta > 0$ such that

$$\phi_j(\theta) = \frac{1}{(2\pi)^2} \int_{-\pi}^{\pi} f(\lambda; \theta)^{-1} \cos j\lambda \, d\lambda$$

is differentiable in N_δ, for each j, and

$$\sum \sup_{\theta \in N_\delta} \left\|\frac{\partial \phi_j(\theta)}{\partial \theta}\right\| < \infty. \tag{13.43}$$

Theorem 3 Under (13.1) and Conditions 1–8, it follows that

$$n^{1/2}(\hat{\beta}_{\hat{f}-1} - \beta), \quad n^{1/2}(\tilde{\beta}_{\hat{f}-1} - \beta) \to_d N\left(0, \Sigma_{f^{-1}}^{-1}\right), \tag{13.44}$$

and $\Sigma_{f^{-1}}$ is consistently estimated by both

$$\frac{1}{n}\sum_t \sum_s (x_s - \bar{x})(x_t - \bar{x})' \phi_{t-s}(\hat{\theta}) \quad \text{and} \quad \frac{1}{n}\sum_{j=1}^{n-1} I_x(\lambda_j)\phi(\lambda_j; \hat{\theta}). \tag{13.45}$$

The proof is omitted because it makes straightforward use of Theorem 1 and its proof, the mean value theorem and the fact that

$$E\left\|n^{-1}\sum_{1 \le t, t+j \le n} (x_t - Ex_1)u_t\right\|^2 = o\left(n^{-1}\sum_{j=0}^{n-1} |\gamma_j|\right)$$

as $n \to \infty$ uniformly in j, to show that $\hat{\beta}_{\hat{f}-1} - \hat{\beta}_{f^{-1}}$ and $\tilde{\beta}_{\hat{f}-1} - \tilde{\beta}_{f^{-1}}$ are $o_p(n^{1/2})$. We discuss Condition 8, however.

In the simple model

$$\phi_j(\theta) = \theta_1 j^{-2\theta_2 - 1}, \quad j = 1, 2, \ldots, \quad \theta_1 > 0, \quad 0 < \theta_2 < \tfrac{1}{2},$$

(13.43) clearly holds. More generally, lemma 5 of Fox and Taqqu (1986) establishes (13.43) under certain regularity conditions. These conditions entail a singularity in $f(\lambda; \theta)$ at $\lambda = 0$ only, and are satisfied in the FARIMA case (13.27). It seems likely that Fox and Taqqu's (1986) result can be extended to enable (13.43) to be checked in case of singularities at nonzero frequencies.

Condition (13.42) is clearly milder than the requirement

$$\hat{\theta} - \theta_0 = O_p(n^{-1/2}). \tag{13.46}$$

It is necessary to say something about how $\hat{\theta}$ is obtained. First, pretend that the u_t are observable, and denote by $\tilde{\theta}$ an estimate of θ_0 based on them. In any given model any number of $\tilde{\theta}$ is, in principle, available, some of which have been explicitly discussed in the literature, and for some of these, rates of convergence have been obtained. For example, defining $I(\lambda) = (2\pi n)^{-1} \left| \sum_{t=1}^{n} (u_t - \bar{u}) e^{it\lambda} \right|^2$, consider $\tilde{\theta}$ minimizing

$$\int_{-\pi}^{\pi} \left\{ \log f(\lambda; \theta) + \frac{I(\lambda)}{f(\lambda; \theta)} \right\} d\lambda. \tag{13.47}$$

Conditions for $\tilde{\theta} - \theta_0 = O_p(n^{-1/2})$ have here been given by Fox and Taqqu (1986), Dahlhaus (1989) and Giraitis and Surgailis (1991) in case f has a singularity at $\lambda = 0$ only, while Dahlhaus (1989) also considered approximations to (13.47) which lead to the same rate of convergence. Hosoya (1993) has similar results in case of singularities at known, nonzero frequencies; see (13.28). It appears that $n^{1/2}$-consistency can be achieved by many other types of estimates of parametric models for long-range dependence. While no rigorous asymptotics exist for it, this seems true of the explicitly defined estimate (or a modification thereof) of Kashyap and Eom (1988) for the FARIMA(0, d, 0) model given by (13.27) with $a = b \equiv 1$, and an extension of this type of estimate to the model

$$f(\lambda; \theta) = \exp\left[\sum_{j=1}^{p-1} \theta_j \cos\{(j-1)\lambda\} \right] |1 - e^{i\lambda}|^{-2\theta_p}$$

proposed by Robinson (1994a).

Now suppose that for some such $\tilde{\theta}$ we define $\hat{\theta}$ correspondingly after replacing u_t in $\tilde{\theta}$ by

$$\hat{u}_t = y_t - \bar{y} - \hat{\beta}_\phi'(x_t - \bar{x}) \tag{13.48}$$

for some ϕ. (Of course, $\tilde{\beta}_\phi$ can be used in place of $\hat{\beta}_\phi$.) Then, given that $\tilde{\theta} - \theta_0 = O_p(\{\sum_0^n |\gamma_j|\}^{-1/2})$, it suffices to show that $\hat{\theta} - \tilde{\theta} = O_p(\{\sum_0^n |\gamma_j|\}^{-1/2})$. Unfortunately, this requires a somewhat case-by-case treatment, and in a given case a reasonably detailed proof may be lengthy, especially for estimates that are only implicitly defined, where a preliminary consistency proof for $\hat{\theta}$ may be needed; this is true of estimates minimizing (13.47) with $I(\lambda)$ replaced by $\hat{I}(\lambda) = (2\pi n)^{-1} |\sum_t \hat{u}_t e^{it\lambda}|^2$, which are among those of most interest because of their efficiency under Gaussianity. There seems little interest in presenting the details in even a single case, especially as the statistical literature contains many demonstrations that errors can be replaced by suitable residuals without affecting first-order asymptotics, and which suggest that the same is likely to be true in our problem, at least if $\hat{\beta}_\phi$ is $n^{1/2}$-consistent, because $\hat{u}_t = u_t - \bar{u} - (\hat{\beta}_\phi - \beta)'(x_t - \bar{x})$.

It appears that (13.42) is capable of significant relaxation, subject to additional regularity conditions (such as on higher-order derivatives of the $\phi_j(\theta)$) and that a slower rate of convergence of $\hat{\beta}_\phi$ would suffice, such that possibly least squares $\hat{\beta}_1$ (see Theorem 2) could be used in (13.48). However, verification of these conjectures would not only be lengthy but of limited practical value because it seems desirable in finite-sample practice to use estimates of θ_0 and β in $\hat{\beta}_{\hat{f}-1}$ or $\tilde{\beta}_{\hat{f}-1}$ with the maximum, $n^{1/2}$, rate of convergence, and Theorem 1 and our recent discussion indicate that this is likely to be possible. In practice, we may also wish to iterate the above practice, employing $\hat{\beta}_{\hat{f}-1}$ or $\tilde{\beta}_{\hat{f}-1}$ to recalculate residuals (13.48) and thence new estimates of β which would have no greater asymptotic efficiency but might have better finite-sample properties.

13.5 Nonlinear Regression Models

An important extension of (13.1) consists of regression models with nonlinearity in parameters and possibly in regressors also. Consider

$$y_t = \alpha + z_t(\beta_0) + u_t, \quad t = 1, 2, \ldots, \tag{13.49}$$

where $z_t(\beta)$ is a given function of the vector β and of a stochastic regression vector observed at time t, whose presence is indicated only by the t-subscript. Note that β is zero-subscripted, as was θ in the previous section, because of the need to explicitly define an objective function for estimation in nonlinear problems. As an example of $z_t(\beta)$, consider the multiplicative form

$$z_t(\beta) = \beta_1 \prod_{i=2}^{K} x_{it}^{\beta_i}. \tag{13.50}$$

Following the work of Jennrich (1969) and Malinvaud (1970), on nonlinear least squares in case of independent u_t, authors such as Gallant and Goebel (1975), Hannan (1971) and Robinson (1972) studied asymptotic theory for estimates of models such as (13.49) in case of weakly dependent u_t. We allow u_t to have long-range dependence. Nonlinear extremum estimates in regression models with long-range-dependent errors have been discussed by Koul (1992), Koul and Muhkerjee (1993) and Robinson (1994a). However, the regression models considered in the former two references are linear ones, the nonlinearity coming from the nonquadratic objective functions used, and the conditions and theoretical treatment are very different from ours, while the latter reference provides only a heuristic account of broad issues.

Define

$$\hat{\beta}_\phi = \arg\min_B \hat{Q}_\phi(\beta),$$

$$\hat{Q}_\phi(\beta) = \int_{-\pi}^{\pi} \{|w_y(\lambda) - w_z(\lambda; \beta)|^2 - I_y(\lambda)\}\phi(\lambda)\,d\lambda$$

$$= \int_{-\pi}^{\pi} [I_z(\lambda; \beta) - 2R\{I_{zy}(\lambda; \beta)\}]\phi(\lambda)\,d\lambda,$$

$$I_z(\lambda; \beta) = |w_z(\lambda; \beta)|^2, \quad I_{zy}(\lambda; \beta) = w_z(\lambda; \beta)w_y(-\lambda),$$

$$w(\lambda; \beta) = \frac{1}{(2\pi n)^{1/2}} \sum_t \{z_t(\beta) - \bar{z}(\beta)\}e^{it\lambda}, \quad \bar{z}(\beta) = \frac{1}{n} \sum_t z_t(\beta),$$

where B is a compact subset of $\Re^K$. As in Section 13.1 it can be argued that the clt with $n^{1/2}$ convergence rate can fail for the nonlinear least squares choice $\phi(\lambda) \equiv 1$, whereas $\phi = f^{-1}$ is an efficient choice among those ϕ which do lead to $n^{1/2}$-consistency.

This setup does not quite cover models with infinite-dimensional leads or lags, such as where

$$z_t(\beta) = \sum \chi_j(\beta)x_{t-j}, \tag{13.51}$$

where the regression vector x_t is observed only at $t = 1, \ldots, n$ and the row vector $\chi_j(\beta)$ can be nonlinear in β; for example, a popular choice has been the geometric weights

$$\chi_j(\beta) = \begin{cases} \beta_1\beta_2^j, & j \geq 0, \\ 0, & j < 0, \end{cases} \quad |\beta_2| < 1, \ K = 1. \tag{13.52}$$

However, we can replace $z_t(\beta)$ in (13.51) by

$$\tilde{z}_t(\beta) = \sum_{t-n}^{t-1} \chi_j(\beta)x_{t-j}, \tag{13.53}$$

and it is easily shown that the error in $\hat{\beta}_\phi$ is no more than

$$O_p\left(n^{-1} \sum_{j=-n}^{n} |j|\|\chi_j(\beta_0)\| + \sum_{|j|>n} \|\chi_j(\beta_0)\|\right),$$

and by the Kronecker lemma we can apply (13.58) of Condition 14 below to show that this is $o_p(n^{-1/2})$, as desired.

We define

$$\tilde{\beta}_\phi = \arg\min_B \tilde{Q}(\beta),$$

$$\tilde{Q}_\phi(\beta) = \frac{1}{n} \sum_{j=1}^{n-1} [I_z(\lambda_j; \beta) - 2R\{I_{zy}(\lambda_j; \beta)\}]\phi(\lambda_j).$$

In case (13.51) we may replace $\tilde{Q}_\phi(\beta)$ by

$$\frac{1}{n} \sum_{j=1}^{n-1} [\chi(\lambda_j; \beta) I_x(\lambda_j) \chi'(-\lambda_j; \beta) - 2R\{\chi(\lambda_j; \beta) I_{xy}(\lambda_j)\}] \phi(\lambda_j), \quad (13.54)$$

where

$$\chi(\lambda; \beta) = \sum \chi_j(\beta) e^{ij\lambda},$$

the frequency response function $\chi(\lambda; \beta)$ typically being of simpler form than the $\chi_j(\beta)$ and written down by inspection, for example $\chi(\lambda; \beta) = \beta_1(1 - \beta_2 e^{i\lambda})^{-1}$ in case (13.52).

We now introduce some additional conditions. Define

$$N_\delta(\beta) = \{b: b \in B, \|b - \beta\| < \delta\}.$$

Condition 9 β_0 is an interior point of the compact set B.

Condition 10 For all $\beta \in B$, $\{z_t(\beta)\}$ is strictly stationary, $z_1(\beta)$ is continuous, there exists $\delta > 0$ such that

$$E\left(\sup_{b \in N_\delta(\beta)} z_1^2(b)\right) < \infty, \quad (13.55)$$

and, for all $b, \beta \in B$,

$$\lim_{n \to \infty} \frac{1}{n} \sum_{j=1}^{n} \gamma_j^2(b, \beta) = 0, \quad (13.56)$$

where $\gamma_j(b, \beta) = \text{cov}\{z_1(b), z_{1+j}(\beta)\}$, and, for all k,

$$\lim_{n \to \infty} \frac{1}{n} \sum_{j=1}^{n} |\text{cum}\{z_1(b), z_{1+j}(b), z_{1+k}(\beta), z_{1+j+k}(\beta)\}| = 0.$$

Condition 11 $\{z_t(\beta)\}$ and $\{\varepsilon_t\}$ are independent for all $\beta \in B$.

Condition 12 For all $\beta \in B \backslash \{\beta_0\}$,

$$\int_{-\pi}^{\pi} \phi(\lambda) \, d\{G(\lambda; \beta, \beta) - G(\lambda; \beta, \beta_0) \quad (13.57)$$

$$- G(\lambda; \beta_0, \beta) + G(\lambda; \beta_0, \beta_0)\} > 0,$$

where, for all $b, \beta \in B$, G is given by

$$\gamma_j(b, \beta) = \int_{-\pi}^{\pi} e^{ij\lambda} \, dG(\lambda; b, \beta),$$

and the matrix

$$\begin{bmatrix} G(\lambda; b, b) & G(\lambda; \beta, b) \\ G(\lambda; \beta, b) & G(\lambda; \beta, \beta) \end{bmatrix}$$

has Hermitian nonnegative definite increments and is continuous from the right.

Condition 13 $z_t(\beta)$ is twice differentiable in β, $(\partial/\partial\beta)z_1(\beta)$ and $(\partial^2/\partial\beta\,\partial\beta')z_1(\beta)$ are continuous at β_0 and there exists $\delta > 0$ such that

$$E\left\{\sup_{\beta\in N_\delta(\beta_0)}\left\|\frac{\partial}{\partial\beta}z_1(\beta)\right\|^2 + \sup_{\beta\in N_\delta(\beta_0)}\left\|\frac{\partial^2 z_1(\beta)}{\partial\beta\,\partial\beta'}\right\|^2\right\} < \infty.$$

Condition 14 In case $z_t(\beta)$ is given by (13.51), and $z_t(\beta)$ is replaced by (13.53) in $\hat{\beta}_\phi$, or $\tilde{Q}_\phi(\beta)$ is replaced by (13.54), we have $\max_t E\|x_t\|^2 < \infty$ and

$$\sum |j|^{1/2}\|\chi_j(\beta_0)\| < \infty. \tag{13.58}$$

We will apply Conditions 5 and 7 and the notation in (13.8) with x_t there replaced by

$$\frac{\partial z_t(\beta_0)}{\partial\beta}. \tag{13.59}$$

Conditions 9–11 are introduced partly for the consistency proof which is needed as a preliminary to the clt's in Theorems 4 and 5 below. Condition 9 is fairly standard in asymptotic theory for extremum estimates, and can be removed with respect to a multiplicative scale parameter in $z_t(\beta)$. Conditions 10 and 13 can be checked for (13.50) if the x_{it} are positive, bounded away from 0, depend only on a stationary and ergodic Gaussian vector and satisfy a suitable moment condition. The components (13.56) of Condition 10, and (13.20) of Condition 7 with (13.59) substituted, can more readily be explained in terms of conditions on the observables on which $z_t(\beta)$ (in general nonlinearly) depends than can 'Grenander's conditions', linear processes or unit roots; (13.56) is equivalent to the absence of jumps in $G(\lambda; b, \beta)$, and is implied if $\gamma_j(b, \beta) \to 0$ as $j \to \infty$. Conditions 10 and 13 are of a different character from the corresponding ones of Hannan (1971) and Robinson (1972), which are expressed in terms of limits. Condition 11 holds in (13.50) under Condition 6, and, in general, implies that the derivatives in Condition 13 will also be independent of the ε_t. Condition 12 is an inevitable identifiability condition which, like Condition 7 with (13.59), is violated in (13.50) if, for example, $x_{i1} = x_{j1}$ a.s., for some $i \neq j$.

Theorem 4 Under (13.49) and Conditions 1–5, 7 and 9–14, it follows that (13.22) holds, and thus (13.23) holds in case $f(\lambda) > 0$ and $\phi(\lambda) = f(\lambda)^{-1}$.

Theorem 5 Under (13.49) and Conditions 1–5 and 7–14 and $f(\lambda) > 0$ for all λ, it follows that (13.44) holds and $\Sigma_{f^{-1}}$ is consistently estimated by (13.45) with x_t replaced by $(\partial/\partial\beta)z_t(\hat{\beta}_{f^{-1}})$.

It suffices to indicate how the proof of Theorem 4 differs from the many other proofs for extremum estimates, and how it uses Theorem 1. We discuss only $\hat{\beta}_\phi$. The first step is to show that $\hat{\beta}_\phi \to_p \beta_0$, where the broad argument is similar to that of, for example, Malinvaud (1970); see also the proofs of a.s. convergence of Jennrich (1969) and others. Writing $Q_\phi(\beta)$ in time domain form using (13.5), because of Conditions 3 and 13 it suffices to show that, for all fixed j,

$$\frac{1}{n}\sum_t{}' z_t(\beta)u_{t+j} \to_p 0, \qquad \frac{1}{n}\sum_t{}' z_t(b)z_{t+j}(\beta) \to_p \gamma_j(b,\beta) \qquad (13.60)$$

uniformly in $b, \beta \in B$, where $\sum_t'$ is a sum over $1 \le t, t+j \le n$. Pointwise mean square convergence is an easy consequence of Conditions 2 and 10, and uniformity follows from this, Condition 9 and equicontinuity resulting from the Cauchy inequality and the fact the $\sup_{b\in N_\delta(\beta)} |z_1(b) - z_1(\beta)|^2$ for all $\beta \in B$ as $\delta \to 0$, which is due to continuity, the domination condition (13.53) and a routine extension of DeGroot (1970: p. 206). Next we have

$$0 = \frac{\partial Q_\phi(\hat\beta_\phi)}{\partial\beta} = \frac{\partial Q_\phi(\beta_0)}{\partial\beta} + \tilde{H}(\hat\beta_\phi - \beta_0),$$

with probability approaching 1 as $n \to \infty$, where $\tilde{H}$ is a matrix with ith row $(\partial^2/\partial\beta_i\partial\beta')Q_\phi(\beta^{(i)})$ such that $\|\beta^{(i)} - \beta_0\| \le \|\hat\beta_\phi - \beta_0\|, i = 1, \ldots, K$. In view of (13.59) we can rewrite $(\partial/\partial\beta)Q_\phi(\beta)$ as $-2\hat{a}$ (where $\hat{a}$ is defined at the start of Section 13.3) so after applying Theorem 1 it suffices to show that $2\hat{A} - \tilde{H} \to_p 0$. This follows straightforwardly from consistency of $\hat\beta_\phi$ and Condition 13 using techniques in the proof of (13.60).

13.6 Monte Carlo Simulations

Finite-sample performance of estimates was investigated in a small Monte Carlo study. In the linear model (13.1) we took $K = 1$ and $\alpha = 0, \beta = 1$; our results are invariant to the choice of α and β. The scalar processes x_t and u_t were both Gaussian FARIMAs with spectra (see (13.27)) $dF(\lambda)/d\lambda = (2\pi)^{-1}|1 - e^{i\lambda}|^{-2c}$ and $f(\lambda) = (2\pi)^{-1}|1 - e^{i\lambda}|^{-2d}$, for the grid of values $c, d = 0.05(0.1)0.45$. In $\hat\beta_\phi$ and $\tilde\beta_\phi$ we took $\phi(\lambda) = |2\sin\frac{1}{2}\lambda|$, which satisfies our conditions for all the above c, d, noting that $\phi_j = 2/\{\pi^2(1 - 4j^2)\}, j = 0, \pm 1, \ldots$. The asymptotic relative efficiency is

$$\frac{\{\Gamma(2-2c)\Gamma(1-c+d)\Gamma(2-c-d)\}^2}{\Gamma(3/2-c)^4\Gamma(1-2c+2d)\Gamma(3-2c-2d)},$$

which is given in Table 13.1. The efficiency increases in d and decreases in c, such that it is effectively 100 per cent for all c when $d = 0.45$, and is generally very satisfactory for the other larger values of d, and falls below 50 per cent only when $c = 0.45, d = 0.05$. Notice that least squares is asymptotically normal for all cases above the south-west/north-east diagonal.

We generated 1000 replications of series x_t and u_t, of lengths $n = 64, 128$ and 256, via the algorithm of Davies and Harte (1987), noting that x_t and u_t have

Table 13.1. Asymptotic relative efficiency of $\hat{\beta}_\phi$ or $\tilde{\beta}_\phi$ to
$\hat{\beta}_{f^{-1}}$ or $\tilde{\beta}_{f^{-1}}$ with $\phi(\lambda) = |2\sin\lambda/2|$

			d		
c	0.05	0.15	0.25	0.35	0.45
0.05	0.8295	0.8992	0.9494	0.9820	0.9980
0.15	0.7792	0.8701	0.9351	0.9769	0.9975
0.25	0.7039	0.8270	0.9139	0.9695	0.9966
0.35	0.5850	0.7597	0.8811	0.9580	0.9954
0.45	0.3823	0.6470	0.8268	0.9392	0.9933

Table 13.2. Ratios of $n\Sigma_{f^{-1}}^{-1}$ to Monte Carlo MSE's of $\hat{\beta}_\phi$ and $\tilde{\beta}_\phi$ with
$\phi(\lambda) = |2\sin\lambda/2|$

c/d	$\hat{\beta}_\phi$					$\tilde{\beta}_\phi$				
	0.05	0.15	0.25	0.35	0.45	0.05	0.15	0.25	0.35	0.45
$n = 64$										
0.05	0.829	0.889	0.918	0.964	0.963	0.821	0.881	0.918	0.959	0.962
0.15	0.768	0.875	0.821	0.945	1.042	0.760	0.869	0.809	0.938	1.008
0.25	0.709	0.756	0.922	0.905	0.920	0.700	0.747	0.916	0.886	0.901
0.35	0.584	0.744	0.890	0.898	1.018	0.567	0.730	0.870	0.870	0.979
0.45	0.395	0.640	0.777	0.930	0.930	0.385	0.629	0.765	0.916	0.897
$n = 128$										
0.05	0.813	0.894	0.880	0.988	0.973	0.809	0.891	0.882	0.981	0.973
0.15	0.813	0.818	0.970	0.990	0.989	0.809	0.808	0.964	0.990	0.976
0.25	0.668	0.796	1.006	0.955	0.998	0.661	0.790	1.007	0.939	0.980
0.35	0.607	0.766	0.849	1.049	0.971	0.601	0.750	0.838	1.029	0.939
0.45	0.406	0.662	0.769	0.971	0.979	0.403	0.647	0.762	0.956	0.955
$n = 256$										
0.05	0.837	0.894	0.915	0.937	0.995	0.833	0.895	0.910	0.934	0.994
0.15	0.758	0.909	0.965	0.950	0.984	0.755	0.906	0.959	0.945	0.969
0.25	0.699	0.833	0.898	0.941	1.003	0.697	0.834	0.892	0.936	0.997
0.35	0.611	0.758	0.927	1.018	0.975	0.607	0.749	0.922	0.998	0.955
0.45	0.409	0.620	0.821	0.980	1.027	0.406	0.615	0.806	0.969	1.001

autocovariances

$$\Gamma_j = \frac{(-1)^j \Gamma(1-2c)}{\Gamma(1+j-c)\Gamma(1-j-c)},$$

$$\gamma_j = \frac{(-1)^j \Gamma(1-2d)}{\Gamma(1+j-d)\Gamma(1-j-d)}, \quad j = 0, \pm 1, \ldots$$

Table 13.3. Ratios of $n\Sigma_{\hat{f}-1}^{-1}$ to Monte Carlo MSE's of $\hat{\beta}_{\hat{f}-1}$ and $\tilde{\beta}_{\hat{f}-1}$

c/d	$\hat{\beta}_{\hat{f}-1}$					$\tilde{\beta}_{\hat{f}-1}$				
	0.05	0.15	0.25	0.35	0.45	0.05	0.15	0.25	0.35	0.45
$n = 64$										
0.05	0.964	0.935	0.933	0.955	0.955	0.964	0.934	0.936	0.951	0.954
0.15	0.916	0.931	0.848	0.947	0.955	0.916	0.931	0.843	0.942	0.992
0.25	0.906	0.874	0.901	0.912	1.020	0.905	0.869	0.903	0.898	0.893
0.35	0.737	0.882	0.911	0.907	0.907	0.734	0.878	0.903	0.885	0.941
0.45	0.583	0.792	0.809	0.933	0.906	0.584	0.788	0.802	0.922	0.878
$n = 128$										
0.05	0.981	0.980	0.926	1.003	0.962	0.981	0.981	0.928	0.998	0.962
0.15	0.997	0.908	1.004	0.984	0.978	0.997	0.905	1.004	0.986	0.966
0.25	0.891	0.899	1.028	0.956	1.001	0.891	0.898	1.033	0.942	0.984
0.35	0.846	0.951	0.936	1.048	0.960	0.847	0.945	0.925	1.036	0.931
0.45	0.614	0.905	0.854	0.962	0.960	0.616	0.907	0.852	0.954	0.939
$n = 256$										
0.05	1.023	0.966	1.001	0.916	0.899	1.021	0.968	0.998	0.941	0.989
0.15	0.899	1.021	0.999	0.911	0.923	0.898	1.023	1.013	0.957	0.971
0.25	1.034	0.990	0.927	0.846	0.808	1.032	0.990	0.951	0.933	0.988
0.35	0.940	0.953	0.963	0.909	0.727	0.937	0.950	1.002	1.009	0.946
0.45	0.654	0.794	0.892	0.858	0.613	0.653	0.797	0.932	1.014	0.986

(see Adenstedt 1974). Deriving the y_t from (13.1), we then computed $\hat{\beta}_\phi$, $\hat{\beta}_{\hat{f}-1}$, $\tilde{\beta}_\phi$ and $\tilde{\beta}_{\hat{f}-1}$ in each case, where

$$\hat{d} = \arg \min_{\varepsilon \le \delta \le 1/2 - \varepsilon} \sum_{j=1}^{n-1} \frac{\left| \sum_{t=1}^{n} (y_t - \hat{\beta}_\phi x_t) e^{it\lambda_j} \right|^2}{f(\lambda_j; \delta)}$$

for $f(\lambda; \delta) = (2\pi)^{-1} |1 - e^{i\lambda}|^{-2\delta}$ and $\varepsilon = 0.001$.

Table 13.2 displays the ratio of the asymptotic variance of $\hat{\beta}_{\hat{f}-1}$, namely,

$$n^{-1} \Sigma_{\hat{f}-1}^{-1} = \Gamma(1 - c + d)^2 / n\Gamma(1 - 2c + 2d),$$

to the Monte Carlo mean-squared errors of $\hat{\beta}_\phi$ and $\tilde{\beta}_\phi$. The differences between the $\hat{\beta}_\phi$ and $\tilde{\beta}_\phi$ values decrease as n increases, but even for $n = 64$ they seem rather slight so we detect no finite-sample grounds for preferring one of these estimates over the other. Throughout Table 13.2 there are noticeable discrepancies relative to Table 13.1 (mostly the Table 13.2 values are smaller), with some departures from monotonicity in c and d, and little evidence of convergence over the range of n considered; indeed, in some cases the ratios drift in the wrong direction. However, nowhere is the asymptotic theory grossly misleading, and generally it seems to have performed fairly well.

Table 13.3 contains ratios of $\Sigma_{\hat{f}-1}^{-1}$ to Monte Carlo mean-squared errors of $\hat{\beta}_{\hat{f}-1}$ and $\tilde{\beta}_{\hat{f}-1}$. They are predominantly less than 1, and for the smaller values of c and d substantially so. In some cases there are significant, even large, differences between the $\hat{\beta}_{\hat{f}-1}$ and $\tilde{\beta}_{\hat{f}-1}$ values, though often they are very slight. The $\tilde{\beta}_{\hat{f}-1}$ demonstrate the greater degree of convergence to 1 with n. One expects that a major effect on the results is the estimation of d and that they could deteriorate if a richer model of $f(\lambda)$ were estimated. Overall, while there is clearly a degree of sensitivity to c and d, the estimates considered appear to cope adequately with long-range dependence in both x_t and u_t.

References

Adenstedt, R. K. (1974), 'On large-sample estimation of the mean of a stationary random sequence', *Annals of Statistics* 2, 1095–107.

Anderson, T. W. (1971), *The Statistical Analysis of Time Series*. Wiley, New York.

Bingham, N. H., Goldie, C. M. and Teugels, J. L. (1987), *Regular Variation*. Cambridge University Press.

Dahlhaus, R. (1989), 'Efficient parameter estimation for self-similar processes', *Annals of Statistics* 17, 1749–66.

—— (1995), 'Efficient location and regression estimation for long range dependent regression models', *Annals of Statistics* 23, 1029–47.

Davies, R. B. and Harte, D. S. (1987), 'Tests for Hurst effect', *Biometrika* 74, 95–101.

DeGroot, M. H. (1970), *Optimal Statistical Decisions*. McGraw-Hill, New York.

Eicker, F. (1967), 'Limit theorems for regressions with unequal and dependent errors', *Proceedings of Fifth Berkeley Symposium on Mathematical Statistics and Probability* 1, 59–82. University of California Press, Berkeley.

Fox, R. and Taqqu, M. S. (1986), 'Large-sample properties of parameter estimates for strongly dependent stationary Gaussian time series', *Annals of Statistics* 14, 517–32.

Gallant, A. R. and Goebel, J. J. (1976), 'Nonlinear regression with autocorrelated errors', *Journal of the American Statistical Association* 71, 961–7.

Giraitis, L., Koul, H. L. and Surgailis, D. (1996), 'Asymptotic normality of regression estimators with long memory errors', *Statistics and Probability Letters*, 29, 317–35.

—— and Surgailis, D. (1991), 'A central limit theorem for quadratic forms in strongly dependent linear variables and its application to asymptotic normality of Whittle's estimate', *Probability Theory and Related Fields* 86, 87–104.

Gray, H. L., Zhang, N.-F. and Woodward, W. A. (1989), 'On generalized fractional processes', *Journal of Time Series Analysis* 10, 233–57.

——, —— and —— (1994), 'On generalized fractional processes—a correction', *Journal of Time Series Analysis* 15, 561–2.

Hamon, B. V. and Hannan, E. J. (1963), 'Estimating relations between time series', *Journal of Geophysical Research* 68, 1033–41.

Hannan, E. J. (1967), 'The estimation of a lagged regression relation', *Biometrika* 54, 409–18.

—— (1971), 'Non-linear time series regression', *Journal of Applied Probability* 8, 767–80.

—— (1979), 'The central limit theorem for time series regression', *Stochastic Processes and Applications* 9, 281–9.

Hosoya, Y. (1993), A limit theory in stationary processes with long-range dependence and related statistical models. Unpublished manuscript.

Ibragimov, I. A. and Linnik, Y. V. (1971), *Independent and Stationary Sequences of Random Variables*. Wolters-Noordhoff, Groningen.

Jennrich, R. I. (1969), 'Non-linear least squares estimators', *Annals of Mathematical Statistics* 40, 633–43.

Kashyap, R. L. and Eom, K.-B. (1988), 'Estimation in long-memory time series model', *Journal of Time Series Analysis* 9, 35–41.

Koul, H. L. (1992), '*M*-estimators in linear models with long range dependent errors', *Statistics and Probability Letters* 14, 153–64.

—— and Mukherjee, K. (1993), 'Asymptotics of R-, MD-, and LAD estimators in linear regression models with long range dependent errors', *Probability Theory and Related Fields* 95, 533–53.

Künsch, H., Beran, J. and Hampel, R. (1993), 'Contrasts under long-range correlations', *Annals of Statistics* 21, 943–64.

Malinvaud, E. (1970), 'The consistency of nonlinear regression', *Annals of Mathematical Statistics* 41, 959–69.

Robinson, P. M. (1972), 'Non-linear regression for multiple time series', *Journal of Applied Probability* 9, 758–68. (Reprinted in *Nonlinear Models I* (H. Bierens and A. R. Gallant, eds), 192–202. International Library of Critical Writings in Econometrics, Edward Elgar.)

—— (1994a), 'Time series with strong dependence', in *Advances in Econometrics: Sixth World Congress* (C. A. Sims, ed.) 1, 47–95. Cambridge University Press.

—— (1994b), 'Efficient tests of nonstationary hypotheses', *Journal of the American Statistical Association* 89, 1420–37.

Scott, D. J. (1973), 'Central limit theorems for martingales and for processes with stationary increments using a Skorokhod representation approach', *Advances in Applied Probability* 5, 119–37.

Stout, W. F. (1974), *Almost Sure Convergence*. Academic Press, New York.

Taqqu, M. S. (1975), 'Weak convergence to fractional Brownian motion and to the Rosenblatt process', *Zeitschrift für Wahrscheinlichkeitstheorie und Verwandte Gebiete* 31, 287–302.

Yajima, Y. (1988), 'On estimation of a regression model with long-memory stationary errors', *Annals of Statistics* 16, 791–807.

—— (1991), 'Asymptotic properties of the LSE in a regression model with long-memory stationary errors', *Annals of Statistics* 19, 158–77.

Yong, C. H. (1974), *Asymptotic Behaviour of Trigonometric Series*. Chinese University Hong Kong.

Zygmund, A. (1977), *Trigonometric Series*. Cambridge University Press.

14

Semiparametric Frequency Domain Analysis of Fractional Cointegration

P. M. ROBINSON AND D. MARINUCCI

14.1 Introduction

Cointegration analysis has developed as a major theme of time series econometrics since the article of Engle and Granger (1987), much applied interest prompting considerable methodological and theoretical development during the past decade. Numerous empirical studies have investigated the possibility of cointegration in areas of economics in which a long-run relationship can be conjectured between nonstationary variables, including stock prices and dividends, consumption and income, wages and prices, short- and long-run interest rates, monetary aggregates and nominal GNP, exchange rates and prices, GNP and public debt, and many others.

The bulk of theoretical and applied work has focused on what might be called the '$I(1)/I(0)$' paradigm. We say a scalar process $u_t, t = 0, 1, \ldots,$ is $I(0)(u_t \equiv I(0))$ if it is covariance stationary and has spectral density that is finite and positive at zero frequency, and, for a scalar process v_t, that $v_t \equiv I(d), d > 0$, if

$$(1 - L)^d v_t = u_t 1(t > 0), \quad t > 0, \tag{14.1}$$

where $1(.)$ is the indicator function and formally

$$(1 - L)^d = \sum_{j=0}^{\infty} \frac{\Gamma(j - d)}{\Gamma(-d)\Gamma(j + 1)} L^j, \quad \Gamma(a) = \int_0^{\infty} x^{a-1} e^{-x} \, dx,$$

using, for $d = 0, 1, \ldots, \Gamma(0)/\Gamma(0) = 1, \Gamma(-d) = \infty$. For $d > 0, v_t$ generated by (14.1) is said to have long memory, d measuring the extent of the 'memory'. The initial condition that the innovation in (14.1) is zero for $t \leq 0$ implies that v_t is nonstationary for all $d > 0$. However, for $0 < d < 1/2, v_t$ is asymptotically stationary, while for $d \geq \frac{1}{2}$ it is not. In the former case we might replace the right hand side of (14.1) by u_t, to make v_t covariance stationary (and for convenience we do so in Section 14.3). In much of the existing literature the definition of $I(0)$ effectively employed relaxes covariance stationarity to a form of asymptotic stationarity or stable heterogeneity because limit theorems for relevant statistics are

This research was supported by ESRC grant R000235892. © P. M. Robinson and D. Marinucci 2003.

available in such settings (though the mixing conditions employed are typically in other respects stronger than ours, implying that the limiting analogue of the spectral density is bounded at all frequencies). The $I(1)/I(0)$ paradigm envisages a vector of economic variables z_t which are all $I(1)$ and are cointegrated if there exists a linear combination $e_t = \alpha' z_t$ which is $I(0)$, the prime denoting transposition. Some approaches employ parametric time series models, such as autoregression for z_t and white noise for e_t (e.g. Johansen 1988, 1991), while others adopt a nonparametric characterization of $I(0)$ to cover a wide range of stationary behaviour (e.g. Phillips 1991a). In both approaches, it is standard to test such assumptions on z_t and e_t. At the same time, the $I(1)$ and $I(0)$ classes are clearly highly specialized forms of, respectively, nonstationary and stationary processes, for example when nested in the $I(d)$ processes, for real-valued d.

To define fractional cointegration for a $p \times 1$ vector z_t whose ith element $z_{it} \equiv I(d_i), d_i > 0, i = 1, \ldots, p$, we say $z_t \equiv FCI(d_1, \ldots, d_p; d_e)$ if there exists a $p \times 1$ vector $\alpha \neq 0$ such that $e_t = \alpha' z_t \equiv I(d_e)$ where $0 \leq d_e < \min_{1 \leq i \leq p} d_i$. This property is possible and meaningful if and only if $d_i = d_j$, some $i \neq j$; moreover, a necessary condition for α to be a cointegrating vector is that its ith component be equal to zero if $d_i > d_j$ for all $j \neq i$.

In case $d_1 = \cdots = d_p = d$ it is usual to write $z_t \equiv CI(d, b)$, where $b = d - d_e$ measures the strength of the cointegrating relationship. Here the possibility of fractional (i.e. non-integer) d or d_e was mentioned in the original paper of Engle and Granger (1987). That paper focused, however, on the $FCI(1, \ldots, 1; 0)$ or $CI(1, 1)$ case, which we hereafter abbreviate to $CI(1)$, and several explanations can be suggested for the overwhelming interest of the subsequent literature in this case. First, unit roots can sometimes be viewed as a consequence of economic theory, for example the efficient markets hypothesis and the random walk hypothesis for consumption. Second, standard tests have failed to reject the unit root hypothesis in very many time series. Third, the computational implications of the unit root hypothesis, that simple differencing removes nonstationarity, are attractive. Fourth, asymptotic theory for statistics based on $I(0)$ sequences was much better developed than that for stationary $I(d)$ series with $d \neq 0$. Finally, rules of inference relating to stationary and nonstationary fractional $I(d)$ processes were not well developed.

On the other hand, as argued by Sims (1988), unit root theory in economics typically rests on strong assumptions, or provides only very approximate justification. The power of unit root tests has often been criticized, and the bulk of these have been directed against $I(0)$ alternatives (for example, stationary autoregressive ones), while data that seem consistent with the $I(1)$ hypothesis might well also be consistent with $I(d)$ behaviour on some interval of d values. Fractional differencing is not a prohibitive computational drawback nowadays, and in recent years progress has been made on rigorously justifying large sample inference for stationary $I(d)$ processes, in both parametric and nonparametric directions. Indeed the relative 'smoothness' of the $I(d)$ class for real d can lead to standard limit distribution theory and optimality theory in situations where the $I(1)/I(0)$ approach entails nonstandard and discontinuous asymptotics; for example, tests of a unit root null against fractional alternatives were shown by Robinson (1994b) to have a standard null distribution and optimal asymptotic local power (and to extend immediately to test the $I(d)$ hypothesis for any other value of d) whereas statistics for testing the value of an autoregressive parameter have nonstandard asymptotics at the unit root, as evaluated by Dickey and Fuller (1979) and others, but standard asymptotics at stationary points arbitrarily close to the unit circle.

It is important to recognize that the $CI(1)$ setting has entailed challenges which have been surmounted with ingenuity and difficulty. It is also important to appreciate that $CI(1)$-based inference rules are largely invalidated when in fact $(d_1, \ldots, d_p; d_e) \neq (1, \ldots, 1; 0)$. It is possible to imagine how aspects of the $CI(1)$ methodology and theory can be extended to more general $FCI(d_1, \ldots, d_p; d_e)$ situations where the d_i and d_e are, while not necessarily one and zero, known values, especially in view of limit theory for nonstationary fractional processes of Akonom and Gourieroux (1987), Silveira (1991) and Chan and Terrin (1995). Indeed Dolado and Marmol (1998) have recently pursued this line of study. However, when non-integral d_i and d_e are envisaged, assuming their values seems somehow more arbitrary than stressing the $CI(1)$ case in an autoregressive setting. Moreover for $\frac{1}{2} < d_i < \frac{3}{2}$ Dolado and Marmol's (1998) definition of a nonstationary fractionally integrated process differs from ours, being the partial sum of stationary, long-memory innovations, and leading to 'cointegration with fractionally integrated errors' (as in Jeganathan (1999)). The parametrization we adopt for the nonstationary case (which corresponds to theirs for $d_i > \frac{3}{2}$) relies more directly on the linear expansion of the fractional differencing operator; see Section 14.4. With an empirical focus, the issue of fractional cointegration is also dealt with by Cheung and Lai (1993) and others. It seems of greatest interest to study the problem in the context of unknown orders of integration d_i and d_e in the observed and cointegrated processes, possibly less or greater than unity. For example, in circumstances where $CI(1)$ cointegration has been rejected it may be possible to find evidence of $FCI(d_1, \ldots, d_p; d_e)$ cointegration for some $(d_1, \ldots, d_p; d_e) \neq (1, \ldots, 1; 0)$. Our allowance for some 'memory' remaining in the cointegrating residual e_t (i.e. $d_e > 0$), is appealing, especially recalling how d_e can be linked to the speed of convergence to long run equilibrium (compare for instance Diebold and Rudebusch (1989)).

Cointegration is commonly thought of as a stationary relation between nonstationary variables (so that $d_i \geq \frac{1}{2}$, for all $i, d_e < \frac{1}{2}$). Other circumstances covered by our definition of cointegration are also worth entertaining. One case is $d_e \geq \frac{1}{2}$, when both z_t and e_t are nonstationary. Another is $0 \leq d_i < \frac{1}{2}$, for all i, when both z_t and e_t are stationary.

The latter situation was considered by Robinson (1994a), as an application of limit theory for averages of periodogram ordinates on a degenerating frequency band in stationary long-memory series. Ordinary least squares estimates (OLS) (and other 'full-band' estimates such as generalized least squares) are inconsistent due to the usual simultaneous equation bias. Robinson (1994a) showed, in case of bivariate z_t, that a narrow-band frequency domain least squares (FDLS) estimate of (a normalized) α can be consistent. It is possible that some macroeconomic time series that have been modelled as nonstationary with a unit root could arise from stationary $I(d)$ processes with d near $\frac{1}{2}$, say, and interest in the phenomenon of cointegration of stationary variates has recently emerged in a finance context. Moreover, it is likely to be extremely difficult in practice to distinguish a z_t with unit root from one, say, composed additively of a stationary autoregression with a root near the unit circle, and a stationary long-memory process.

FDLS is defined in the following section, after which in Section 14.3 we extend Robinson's (1994a) results to a more general stationary vector setting, with rates of convergence. We go on in Section 14.4 to establish results of some independent interest on the approximation of sample moments of nonstationary sequences by narrow-band periodogram averages. These results, which constitute the main technical innovation of the paper, are exploited in Section 14.5 to demonstrate the usefulness of FDLS for nonstationary z_t: correlation between z_t and e_t does not prevent consistency of OLS, but it produces a larger second order bias relative to FDLS in the $CI(1)$ case, and a slower rate of convergence in many circumstances in which z_t exhibits less-than-$I(1)$ nonstationarity. When e_t is itself nonstationary, the two estimates share a common limit distribution. Our theoretical result for the $CI(1)$ case is supported in finite samples by Monte Carlo simulations in Section 14.6. Section 14.7 describes a semiparametric methodology for investigating the question of cointegration in possibly fractional conditions, and applies it to series that were studied in the early papers of Engle and Granger (1987) and Campbell and Shiller (1987). Section 14.8 mentions possibilities for further work. Proofs are collected in an Appendix.

OLS by no means represents the state of the art in $CI(1)$ analysis. A number of more elaborate estimates have been proposed and shown to have advantages over OLS, such as Engle and Granger's (1987) two step regression; Johansen's (1988, 1991) maximum likelihood estimate for the error-correction mechanism (ECM); Phillips and Hansen's (1990) fully modified least squares; Phillips' (1991b) spectral regression for the ECM; Stock's (1987) nonlinear least squares; and Bossaerts' (1988) canonical correlation approach. Many of these methods make use of OLS at an early stage, so one implication of our results for the $CI(1)$ case is that FDLS be substituted here. These methods are all specifically designed for the $CI(1)$ case, and in more general settings the validity and optimality of the associated inference procedures will be lost, and they may have no obvious advantage over OLS. Moreover, like OLS, they will not even be consistent in the stationary case. For the computationally simple FDLS procedure, our paper demonstrates a consistency-robustness not achieved by OLS and other procedures, a matching of limit distributional properties in some cases, and superiority in others, including the standard $CI(1)$ case.

14.2 Frequency Domain Least Squares

For a sequence of column vectors $a_t, t = 1, \ldots, n$, define the discrete Fourier transform

$$w_a(\lambda) = \frac{1}{\sqrt{2\pi n}} \sum_t a_t \, \mathrm{e}^{\mathrm{i}t\lambda},$$

where the sample mean of the a_t is given by $\bar{a} = \sqrt{(2\pi/n)}w_a(0)$ and $\sum_t$ will throughout denote $\sum_{t=1}^n$. If b_t, $t = 1, \ldots, n$, is also a sequence of column vectors, define the (cross-) periodogram matrix

$$I_{ab}(\lambda) = w_a(\lambda)w_b^*(\lambda),$$

where * indicates transposition combined with complex conjugation. Further, for $\lambda_j = 2\pi j/n$, define the (real part of the) averaged periodogram

$$\hat{F}_{ab}(k, \ell) = \frac{2\pi}{n} \sum_{j=k}^{\ell} \text{Re}\{I_{ab}(\lambda_j)\}, \quad 1 \le k \le \ell \le n - 1.$$

In case $\hat{F}_{ab}(k, \ell)$ is a vector we shall denote its ith element $\hat{F}_{ab}^{(i)}(k, \ell)$, and in case it is a matrix we shall denote its (i, j)th element $\hat{F}_{ab}^{(i,j)}(k, \ell)$; analogously we will use $I_{ab}^{(i)}(\lambda)$ and $I_{ab}^{(i,j)}(\lambda)$ to denote respectively the ith element of the vector $I_{ab}(\lambda)$ and the (i, j)th element of the matrix $I_{ab}(\lambda)$.

Now suppose we observe vectors $z_t = (x_t', y_t)'$, $t = 1, \ldots, n$, where y_t is real-valued and x_t is a $(p - 1) \times 1$ vector with real-valued elements. Consider, for various m, the statistic

$$\hat{\beta}_m = \hat{F}_{xx}(1, m)^{-1}\hat{F}_{xy}(1, m),\tag{14.2}$$

assuming the inverse exists. We can interpret $\hat{\beta}_m$ as estimating the unknown β in the 'regression model'

$$y_t = \beta' x_t + e_t, \quad t = 1, 2, \ldots.\tag{14.3}$$

Notice that

$$\hat{F}_{xx}(1, n - 1) = \frac{1}{n} \sum_t (x_t - \bar{x})(x_t - \bar{x})', \; \hat{F}_{xy}(1, n - 1) = \frac{1}{n} \sum_t (x_t - \bar{x})(y_t - \bar{y}).$$

$$\tag{14.4}$$

Thus $\hat{\beta}_{n-1}$ is the OLS estimate of β with allowance for a non-zero mean in the unobservable e_t. Our main interest is in cases $1 < m < n - 1$, where, because $w_a(\lambda)$ has complex conjugate $w_a(2\pi - \lambda)$, we restrict further to $1 < m < n/2$. Then we call $\hat{\beta}_m$ an FDLS estimate. Properties of e_t will be discussed subsequently, but these permit it to be correlated with x_t as well as y_t, $\hat{\beta}_m$ being consistent for β due to $\hat{F}_{ee}(1, m)$ being dominated by $\hat{F}_{xx}(1, m)$ in a sense to be indicated. This can happen when z_t is stationary with long memory and e_t is stationary with less memory, if

$$\frac{1}{m} + \frac{m}{n} \to 0, \quad \text{as } n \to \infty,\tag{14.5}$$

which rules out OLS. Under (14.5) $\hat{\beta}_m$ can be termed a 'narrow-band' FDLS estimator. It can also happen when x_t is nonstationary while e_t is stationary or nonstationary with less memory, if only

$$m < n/2, \quad m \to \infty, \quad \text{as } n \to \infty,\tag{14.6}$$

which includes OLS. In both situations $z_t \equiv FCI(d_1, \ldots, d_p; d_e)$ and the focus on low frequencies is thus natural. Notice that when $\lim(m/n) = \theta \in \left(0, \frac{1}{2}\right)$ (so that $\hat{\beta}_m$ is not narrow-band), $\hat{\beta}_m$ is a special case of the estimate introduced by Hannan (1963) and developed by Engle (1974) and others. However, while such m satisfy (14.6), our primary interest is in the narrow-band case (14.5) where $\hat{\beta}_m$ is based on a degenerating band of frequencies and its superiority over OLS can be established under wider circumstances. It is the stationary case which we first discuss.

14.3 Stationary Cointegration

The covariance stationary processes with which we shall be concerned will always be assumed to have absolutely continuous spectral distribution function. Thus, for jointly covariance stationary column vector processes $a_t, b_t, t = 1, 2, \ldots$, define the (cross) spectral density matrix $f_{ab}(\lambda)$ to satisfy

$$E(a_0 - E(a_0))(b_j - E(b_0))' = \int_{-\pi}^{\pi} f_{ab}(\lambda) e^{ij\lambda} \, d\lambda, \quad j = 0, \pm 1, \ldots.$$

We impose the following condition on z_t introduced earlier. For two matrices A and B, of equal dimension and possibly complex-valued elements, we say that $A \sim B$ if, for each (i, j), the ratio of the (i, j)th elements of A and B tends to unity.

Assumption 1 The vector process z_t is covariance stationary with

$$f_{zz}(\lambda) \sim \Lambda G \Lambda, \quad \text{as } \lambda \to 0^+,$$

where G is a Hermitian nonnegative definite matrix whose leading $(p-1) \times (p-1)$ submatrix has full rank and

$$\Lambda = \text{diag}\{\lambda^{-d_1}, \ldots, \lambda^{-d_p}\},$$

for $0 < d_i < \frac{1}{2}, 1 \leq i \leq p$, and there exists a $p \times 1$ vector $\alpha \neq 0$, and a $c \in (0, \infty)$ and $d_e \in [0, d_{\min}), d_{\min} = \min_i d_i$, such that

$$\alpha' f_{zz}(\lambda)\alpha \sim c\lambda^{-2d_e}, \quad \text{as } \lambda \to 0^+.$$

Assumption 1 is similar to that introduced by Robinson (1995a), where it is shown to hold for vector stationary and invertible fractional ARIMA processes (we could allow here, as there, for negative orders of integration greater than $-\frac{1}{2}$). However, there G was positive definite, whereas if Assumption 1 is imposed it has reduced rank, because otherwise

$$\alpha' f_{zz}(\lambda)\alpha \sim (\alpha' \Lambda)G(\Lambda \alpha) \geq \tilde{c}\lambda^{-2d_{\min}} \quad \text{as } \lambda \to 0^+,$$

for $0 < \tilde{c} < \infty$. Nevertheless G must be nonnegative definite because $f_{zz}(\lambda)$ is, for all λ. Notice that $z_{it} \equiv I(d_i), i = 1, \ldots, p$ and $e_t \equiv I(d_e)$ if we adopt the stationary definition (replacing $u_t 1(t > 0)$ by u_t in (14.1)) of an $I(d)$ process. Notice then that Assumption 1 follows if $z_t \equiv FCI(d_1, \ldots, d_p; d_e)$ with $d_e < \tilde{d}$. We adopt the normalization given by $\alpha = (-\beta', 1)'$, so the cointegrating relation is given by (14.3) if $e_t \equiv I(d_e)$. We stress that Assumption 1 does not restrict the spectrum of e_t away from frequency zero, because it is only local properties that matter since here we consider $\hat{\beta}_m$ under (14.5). Asymptotic properties of $\hat{\beta}_m$ require an additional regularity condition, such as

Assumption 2

$$z_t = \mu_z + \sum_{j=0}^{\infty} A_j \varepsilon_{t-j}, \quad \sum_{j=0}^{\infty} \|A_j\|^2 < \infty, \tag{14.7}$$

where $\mu_z = E(z_0)$, and the $p \times 1$ vectors ε_t satisfy

$$E(\varepsilon_t \mid \Im_{t-1}) = 0, \quad E(\varepsilon_t \varepsilon_t' \mid \Im_{t-1}) = \Sigma, \text{ a.s.,}$$

for a constant, full rank matrix Σ, $\Im_t$ being the σ-field of events generated by $\varepsilon_s, s \le t$, and $\|.\|$ denoting Euclidean norm, and the $\varepsilon_t \varepsilon_t'$ are uniformly integrable.

This assumption is a generalization of that of Robinson (1994a), the square summability of the A_j only confirming, in view of the other assumptions, the finite variance of z_t implied by Assumption 1. We could replace the martingale difference assumption on ε_t and $\varepsilon_t \varepsilon_t' - \Sigma$ by fourth moment conditions, as in Robinson (1994c). Notice that it would be equivalent to replace z_t by $(x_t', e_t)'$ with e_t given by (14.3). When z_t satisfies both Assumptions 1 and 2, the A_j are restricted by the requirement that $\|\alpha' A(e^{i\lambda})\| \sim c^* \lambda^{-d_e}$ as $\lambda \to 0^+$, for $0 < c^* < \infty$. A very simple model covered by Assumptions 1 and 2 is (14.3) and $x_t = \varphi x_{t-1} + u_t$, with $p = 2, 0 < \varphi < 1, u_t \equiv I(d_1)$ (implying $x_t \equiv I(d_1)$), and $e_t \equiv I(d_e), 0 \le d_e < d_1 < \frac{1}{2}$. When u_t and e_t are not orthogonal OLS is of course inconsistent for β, as indeed is any other standard cointegration estimator, notwithstanding the fact that for φ close enough to unity x_t is indistinguishable for any practical purpose from a unit root process. The following theorem extends Robinson's (1994a) consistency result for the scalar x_t case.

Theorem 1 Under Assumption 1 with $\alpha = (-\beta', 1)'$, Assumption 2 and (14.5), as $n \to \infty$

$$\hat{\beta}_{im} - \beta_i = O_p \left(\left(\frac{n}{m} \right)^{d_e - d_i} \right), \quad i = 1, \ldots, p - 1,$$

where $\hat{\beta}_{im}$ and β_i are the ith elements of, respectively, $\hat{\beta}_m$ and β.

It follows that if there is cointegration, so $d_{\min} > d_e$, $\hat{\beta}_m$ is consistent for β. In case the d_i are identical there is a common stochastic order $O_p((n/m)^{-b})$, varying inversely with the strength $b = d_i - d_e$ of the cointegrating relation. We conjecture that Theorem 1 is sharp and that under suitable additional conditions the

$(n/m)^{d_i - d_e}(\hat{\beta}_{im} - \beta_i)$ will jointly converge in probability to a non-null constant vector. We conjecture also that after bias-correction and with a different normalization the limit distribution will be normal in some proper subset of stationary $(d_1, \ldots, d_{p-1}, d_e)$-space, and non-normal elsewhere (cf. the derivation of Lobato and Robinson (1996) of the limit distribution of the scalar averaged periodogram). A proper study of this issue would take up considerable space, however, whereas our principal purpose here is to establish consistency, with rates, as an introduction to a study of $\hat{\beta}_m$ in nonstationary environments.

14.4 The Averaged Periodogram in Nonstationary Environments

In order to analyse FDLS in case x_t, and possibly e_t, is nonstationary, we provide some basic properties of the averaged (cross-) periodogram that are of more general interest. For these, it suffices to consider a bivariate sequence $(a_{1t}, a_{2t}), t = 1, 2, \ldots$, given by

$$a_{it} = \sum_{j=1}^{t} \varphi_{i,t-j} \eta_{ij}, \quad i = 1, 2, \tag{14.8}$$

where we impose the following assumptions.

Assumption 3 $(\eta_{1t}, \eta_{2t}), t = 0, \pm 1, \ldots$, is a jointly covariance stationary process with zero mean and bounded spectral density matrix.

Assumption 4 For $i = 1, 2, 0 \leq \gamma_2 \leq \gamma_1, \gamma_1 > \frac{1}{2}, \gamma_2 \neq \frac{1}{2}$, the sequences φ_{it} satisfy $\varphi_{it} = \varphi_t(\gamma_i)$, where for $t \geq 0$,

$$\varphi_t(\gamma) = 1(t = 0), \quad \gamma = 0,$$
$$= O((1+t)^{\gamma - 1}), \quad \gamma > 0,$$
$$= 1, \quad \gamma = 1,$$
$$|\varphi_t(\gamma) - \varphi_{t+1}(\gamma)| = O\left(\frac{|\varphi_t(\gamma)|}{t}\right), \quad \gamma > 0.$$

As we shall see in the next section, Assumption 4 covers cases where a_{1t} is nonstationary whereas a_{2t} is either $I(0)$ ($\gamma_2 = 0$), has asymptotically stationary long memory ($0 < \gamma_2 < \frac{1}{2}$), or is nonstationary ($\gamma_2 > \frac{1}{2}$).

We consider here not only the statistic $\hat{F}_{aa}^{(1,2)}(1, m)$, but also $\hat{F}_{aa}^{(1,2)}(m + 1, M)$, where $0 < m < M \leq n/2$. The latter arises as follows. We have

$$\hat{F}_{aa}^{(1,2)}(1, m) = \frac{\pi}{n} \sum_{j=1}^{m} \left\{ I_{aa}^{(1,2)}(\lambda_j) + I_{aa}^{(1,2)}(\lambda_{n-j}) \right\}$$

$$= \frac{1}{2} \left\{ \hat{F}_{aa}^{(1,2)}(1, m) + \hat{F}_{aa}^{(1,2)}(n - m, n - 1) \right\}$$

$$= \frac{1}{2} \hat{F}_{aa}^{(1,2)}(1, n - 1) - \frac{1}{2} \hat{F}_{aa}^{(1,2)}(m + 1, n - m - 1). \tag{14.9}$$

For n odd (14.9) is

$$\tfrac{1}{2} \hat{F}_{aa}^{(1,2)}(1, n - 1) - \hat{F}_{aa}^{(1,2)}(m + 1, (n - 1)/2),$$

and for n even it is

$$\tfrac{1}{2} \hat{F}_{aa}^{(1,2)}(1, n - 1) - \tfrac{1}{2} \hat{F}_{aa}^{(1,2)}(m + 1, n/2) - \tfrac{1}{2} \hat{F}_{aa}^{(1,2)}(m + 1, n/2 - 1),$$

where

$$\hat{F}_{aa}^{(1,2)}(1, n - 1) = \frac{1}{n} \sum_t (a_{1t} - \bar{a}_1)(a_{2t} - \bar{a}_2).$$

This development follows Robinson (1994c).

When $\gamma_1 + \gamma_2 > 1$ we will deduce that

$$\hat{F}_{aa}^{(1,2)}(m + 1, M) = o_p(n^{\gamma_1 + \gamma_2 - 1})$$

by showing that both the mean and the standard deviation of the left side are $o(n^{\gamma_1 + \gamma_2 - 1})$. Thus if $\frac{1}{2} n^{1 - \gamma_1 - \gamma_2} \hat{F}_{aa}^{(1,2)}(1, n - 1)$ has a nondegenerate limit distribution, $n^{1 - \gamma_1 - \gamma_2} \hat{F}_{aa}^{(1,2)}(1, m)$ shares it. For $\gamma_1 + \gamma_2 \leq 1$, only the standard deviation of $\hat{F}_{aa}^{(1,2)}(m + 1, M)$ is $o(n^{\gamma_1 + \gamma_2 - 1})$ and it is necessary to estimate $E(\hat{F}_{aa}^{(1,2)}(1, m))$, which differs non-negligibly from $E(\hat{F}_{aa}^{(1,2)}(1, [(n - 1)/2]))$. We first consider the means.

Proposition 1 Under (14.8), Assumptions 3 and 4 and (14.6), for $0 < m < M \leq n/2$

$$E\left(\hat{F}_{aa}^{(1,2)}(m + 1, M) \right) = o(n^{\gamma_1 + \gamma_2 - 1}), \quad \gamma_1 + \gamma_2 > 1, \tag{14.10}$$

and

$$E\left(\hat{F}_{aa}^{(1,2)}(1, m) \right) = O(n^{\gamma_1 + \gamma_2 - 1}), \quad \gamma_1 + \gamma_2 > 1, \tag{14.11}$$

$$= O\left(\left(\frac{n}{m} \right)^{\gamma_1 + \gamma_2 - 1} \right), \quad \gamma_1 + \gamma_2 < 1. \tag{14.12}$$

The proof of this, and of Proposition 2 below, is contained in Appendix 14.1. To consider the variances we impose the additional:

Assumption 5 (η_{1t}, η_{2t}) is fourth order stationary with bounded fourth-order cross-cumulant spectrum $f(\mu_1, \mu_2, \mu_3)$ satisfying

$$k(j, k, l) = \int_{-\pi}^{\pi} \int_{-\pi}^{\pi} \int_{-\pi}^{\pi} f(\mu_1, \mu_2, \mu_3) \exp(ij\mu_1 + ik\mu_2 + il\mu_3) \prod_{k=1}^{3} d\mu_k,$$

where $k(j, k, l)$ is the fourth order cumulant of $\eta_{10}, \eta_{2j}, \eta_{1, j+k}, \eta_{2, j+k+l}$ for $j, k, l = 0, \pm 1, \ldots$.

Proposition 2 Under (14.8), Assumptions 3–5 and (14.6), for $0 < m < M \leq n/2$, as $n \to \infty$

$$\text{Var}(\hat{F}_{aa}^{(1,2)}(m+1, M)) = o(n^{2(\gamma_1 + \gamma_2 - 1)}) \tag{14.13}$$

and

$$\text{Var}(\hat{F}_{aa}^{(1,2)}(1, M)) = O(n^{2(\gamma_1 + \gamma_2 - 1)}). \tag{14.14}$$

The proof of Proposition 2 is rather lengthy and as it is essentially contained in that of Theorem 5.1 of Robinson and Marinucci (2001), it is omitted. In view of (14.4), $I_{aa}^{(1,2)}(\lambda_j)$ distributes the sample covariance $\hat{F}_{aa}^{(1,2)}(1, n-1)$ across the Fourier frequencies λ_j, $j = 1, \ldots, n-1$. Propositions 1 and 2 suggest that $\hat{F}_{aa}^{(1,2)}(1, n-1)$ is dominated by the contributions from a possibly degenerating frequency band $(0, \lambda_m)$ when the collective memory in a_{1t}, a_{2t} is sufficiently strong $(\gamma_1 + \gamma_2 > 1)$ while otherwise $\hat{F}_{aa}^{(1,2)}(1, m) - \frac{1}{2}\hat{F}_{aa}^{(1,2)}(1, n-1)$ is estimated by its mean, in view of (14.12). These results are crucial to the derivation of the asymptotic behaviour of $\hat{F}_{xx}(1, m)$ and $\hat{\beta}_m$ introduced in Section 14.2.

14.5 Nonstationary Fractional Cointegration

For z_t nonstationary $(d_i > \frac{1}{2}, i = 1, 2, \ldots, p)$, we find it convenient to stress a linear representation for $w_t = (x_t', e_t)'$ in place of that for $z_t = (x_t', y_t)'$ in Assumption 2. Introduce the diagonal fractional operator

$$D(L) = \text{diag}\{(1-L)^{-d_1}, \ldots, (1-L)^{-d_{p-1}}, (1-L)^{-d_e}\}$$

and

Assumption 6 The vector sequence w_t is given by

$$D^{-1}(L)(w_t - \mu_w) = u_t 1(t > 0) \tag{14.15}$$

for a fixed vector μ_w with pth element zero, where

$$d_i > \tfrac{1}{2}, \quad i = 1, \ldots, p-1, \quad d_e \geq 0,$$

$$u_t = C(L)\varepsilon_t, \quad C(L) = \sum_{j=0}^{\infty} C_j L^j, \tag{14.16}$$

$$\det\{C(1)\} \neq 0, \tag{14.17}$$

$$\sum_{j=0}^{\infty} \left(\sum_{k=j}^{\infty} \|C_k\|^2\right)^{1/2} < \infty, \tag{14.18}$$

and the ε_t are independent and identically distributed $p \times 1$ vectors such that

$$E(\varepsilon_t) = 0, \quad E(\varepsilon_t \varepsilon_t') = \Sigma, \qquad \text{rank}\,(\Sigma) = p, \tag{14.19}$$

$$E\|\varepsilon_t\|^\theta < \infty, \qquad \theta > \max\left(4, \frac{2}{2d_{\min} - 1}\right). \tag{14.20}$$

Assumption 6 strengthens the requirements on ε_t of Assumption 2. Under (14.18)

$$\sum_{j=0}^\infty \|C_j\| \le \sum_{j=0}^\infty (\sum_{k=j}^\infty \|C_k\|^2)^{1/2} < \infty, \tag{14.21}$$

so (14.17) and (14.18) imply that all elements of u_t are in $I(0)$, whereas with reference to (14.1), for $i = 1, \ldots, p - 1$ the ith element of w_t (and thus of x_t) is in $I(d_i)$, while its pth element, e_t, is in $I(d_e)$, so in particular w_t could be a vector fractional ARIMA process. We have allowed for an unknown intercept, μ_w, in w_t.

From (14.15) and (14.16), we can write

$$w_t = \mu_w + \sum_{j=0}^\infty B_{jt}\varepsilon_{t-j}, \tag{14.22}$$

where

$$B_{jt} = \sum_{i=0}^{\min(j,t-1)} D_i C_{j-i}, \tag{14.23}$$

with D_j given by the formal (binomial) expansion

$$D(L) = \sum_{j=0}^\infty D_j L^j.$$

Defining the nonsingular matrix

$$P = \begin{bmatrix} I_{p-1} & O_{p-1} \\ -\beta' & 1 \end{bmatrix},$$

where I_j and O_j are respectively the j-rowed identity matrix and the $j \times 1$ vector of zeroes, we find that (14.15) is equivalent to

$$z_t = \mu_z + \sum_{j=0}^\infty A_{jt}\varepsilon_{t-j}, \tag{14.24}$$

where $\mu_z = P^{-1}\mu_w$, $A_{jt} = P^{-1}B_{jt}$. The representation (14.24) can be compared with the time-invariant one (14.7) for the stationary case (in which $A_j = P^{-1}B_{j\infty}$).

Define $d = (d_1, \ldots, d_{p-1})'$ and

$$\Delta(d) = \text{diag}\{n^{(1/2)-d_1}, \ldots, n^{(1/2)-d_{p-1}}\}, \quad G(r, d) = \text{diag}\{r^{d_1-1}, \ldots, r^{d_{p-1}-1}\}.$$

Let Ω be the leading $(p-1) \times (p-1)$ submatrix of $C(1) \sum C(1)'$, and $B(r, \Omega)$ be $(p-1)$-dimensional Brownian motion with covariance matrix Ω (which has full rank under Assumption 6). Let

$$W(r; d, \Omega) = \int_0^r G(r - s; d) \, dB(s; \Omega), \quad W(d, \Omega) = \int_0^1 W(r; d, \Omega) \, dr,$$

$$V(d, \Omega) = \int_0^1 \{W(r; d, \Omega)W'(r; d, \Omega) - W(d, \Omega)W(d, \Omega)'\} \, dr.$$

We call $W(r; d, \Omega)$ a multivariate fractional Brownian motion, following Akonom and Gourieroux (1987) and Gourieroux, Maurel, and Monfort (1989). A somewhat different definition of (scalar) fractional Brownian motion was proposed by Mandelbrot and Van Ness (1968), and has since prevailed in the probability literature (though the latter authors also mentioned $W(r; d, \Omega)$). Note that the components of $W(r; d, \Omega)$ are continuous Gaussian processes with zero means and variances that grow like r^{2d_i-1}, $i = 1, \ldots, p-1$. Let $\Rightarrow$ denote weak convergence and $\bar{x} = n^{-(1)} \sum_t x_t$.

Theorem 2 Under Assumption 6, as $n \to \infty$

$$\Delta(d)x_{[nr]} \Rightarrow W(r; d, \Omega), \quad 0 < r \leq 1, \tag{14.25}$$

$$\Delta(d)\bar{x} \Rightarrow W(d, \Omega), \tag{14.26}$$

$$\Delta(d)\hat{F}_{xx}(1, n-1)\Delta(d) \Rightarrow V(d, \Omega). \tag{14.27}$$

The proof of (14.25), which relates to results of Akonom and Gourieroux (1987), Gourieroux, Maurel and Monfort (1989) and Silveira (1991), is given by Marinucci and Robinson (2000), whence (14.26) and (14.27) follow from the continuous mapping theorem. For $d_1 = \cdots = d_{p-1} = 1$, fractional Brownian motion reduces to classical Brownian motion and so (14.25) includes a multivariate invariance principle for $I(1)$ processes, as can be found for instance in Phillips and Durlauf (1986). (14.27) provides an invariance principle for the sample covariance matrix of x_t (see (14.4)), and due to the following lemma Propositions 1 and 2 can be applied to deduce one for $\hat{F}_{xx}(1, m)$.

Lemma 1 Let Assumption 6 hold. Then with the choices

$$(a_{1t}, a_{2t}) = (x_{it}, e_t), \quad \gamma_1 = d_i, \quad \gamma_2 = d_e, \quad i = 1, \ldots, p-1, \tag{14.28}$$

or

$$(a_{1t}, a_{2t}) = (x_{it}, x_{jt}), \quad \gamma_1 = d_i, \quad \gamma_2 = d_j, \quad i, j = 1, \ldots, p-1, \tag{14.29}$$

it follows that Assumptions 3–5 are satisfied.

Lemma 2 Under Assumption 6 and (14.6), as $n \to \infty$

$$\Delta(d)\hat{F}_{xx}(1, m)\Delta(d) \implies V(d, \Omega).$$

We can now proceed to investigate asymptotic behaviour of OLS and FDLS in a variety of the cases that arise when x_t is nonstationary, and e_t has short memory or stationary or nonstationary long memory.

Case I: $d_i + d_e < 1, i = 1, \dots, p - 1.$

Here not only does x_t possess less-than-unit-root nonstationarity, but the collective memory in x_t and e_t is more limited than in the $CI(1)$ case. It corresponds to $\gamma_1 + \gamma_2 < 1$ of Section 14.4, and we require first a more precise result than (14.12) in case $m = n - 1$. Let b'_{ij} be the ith row of $B_j = B_{j\infty} = \sum_{i=0}^{j} D_i C_{j-i}$ given by (14.23). Define

$$\xi_i = \sum_{j=0}^{\infty} b'_{ij} \Sigma b_{pj}, \quad i = 1, \dots, p - 1.$$

Lemma 3 Under Assumption 6 with $d_i + d_e < 1, d_i > \frac{1}{2}, i = 1, \dots, p - 1,$

$$\lim_{n\to\infty} E\{\hat{F}_{xx}^{(i)}(1, n - 1)\} = \xi_i, \quad i = 1, \dots, p - 1,$$

where the right-hand side is finite.

Lemma 3 (and (14.12) of Proposition 1) are of some independent interest in that they indicate how sample covariances between a nonstationary and a stationary sequence can be stochastically bounded and have the same structure as when both sequences are stationary, so long as the memory parameters sum to less than 1, as automatically applies in the fully stationary case.

Define ξ to be the $(p - 1) \times 1$ vector with ith element ξ_i if $d_i = d_{\min}$ and zero if $d_i > d_{\min}$, where $d_{\min} = \min_{1 \le i \le p-1} d_i$.

Theorem 3 Let Assumption 6 hold with $d_i + d_e < 1, d_e < d_i, i = 1, \dots, p - 1,$ and

$$\text{rank}\{V(d, \Omega)\} = p - 1, \text{ a.s.} \tag{14.30}$$

Then as $n \to \infty$

$$n^{d_{\min}-(1/2)}\Delta(d)^{-1}(\hat{\beta}_{n-1} - \beta) \Rightarrow V(d, \Omega)^{-1}\xi, \tag{14.31}$$

and under (14.5)

$$n^{d_{\min}-(1/2)}\Delta(d)^{-1}(\hat{\beta}_m - \beta) = O_p\left(\left(\frac{n}{m}\right)^{d_{\min}+d_e-1}\right) = o_p(1). \tag{14.32}$$

Theorem 2 indicates that so long as ξ is non-null, the $n^{d_{\min}+d_i-1}(\hat{\beta}_{i,n-1}-\beta_i)$ have a nondegenerate limit distribution, whereas when the interval $(0, \lambda_m)$ degenerates $\hat{\beta}_{im} - \beta_i = o_p(n^{1-d_{\min}-d_i})$, $i = 1, \ldots, p-1$, so that FDLS converges faster than OLS. In view of the 'global' nature of $\hat{\beta}_{n-1}$ and the 'local' nature of $\hat{\beta}_m$ this outcome is at first sight surprising, but it is due to the bias of $\hat{\beta}_m$ becoming negligible relative to that of $\hat{\beta}_{n-1}$. Notice that the rate of convergence of $\hat{\beta}_{n-1}$ is independent of d_e.

Case II: The $CI(1)$ case ($d_i = 1, i = 1, \ldots, p-1, d_e = 0$).

Now we consider the case considered in the bulk of the cointegration literature, where z_t has a unit root and the cointegrating error is $I(0)$. Write ι for the $(p-1)\times 1$ vector of units, so that in the present case $d = \iota$ and $W(r; d, \Omega) = B(r; \Omega)$. Let $B_*(r; \Omega_*)$ be p-dimensional Brownian motion with covariance matrix $\Omega_* = C(1)\sum C(1)'$, and thus write

$$B_*(r; \Omega_*) = \begin{pmatrix} B(r; \Omega) \\ b(r; \sigma^2) \end{pmatrix},$$

where σ^2 is the (p, p)th element of Ω_*, and in general $B(r; \Omega)$ and $b(r; \sigma^2)$ are correlated and thus in effect depend not only on Ω and σ^2 but on the other elements of Ω_* also. Write

$$U(\Omega_*) = \int_0^1 B(r; \Omega) \, db(r; \sigma^2).$$

Denote by γ_j the $(p-1) \times 1$ vector with ith element $E(u_{it}e_{t+j})$, recalling that $d_e = 0$ implies $e_t = u_{pt}$. Now define

$$\Gamma_j = \sum_{\ell=|j|}^{\infty} \gamma_\ell \operatorname{sign}(\ell), \quad j = 0, \pm 1, \ldots, \tag{14.33}$$

with the convention that $\operatorname{sign}(0)$ is negative, so that $\Gamma_j = \sum_{\ell=j}^{\infty} \gamma_\ell$ for $j > 0$ and $\Gamma_j = \sum_{\ell=-\infty}^{j} \gamma_\ell$ for $j \leq 0$, and the sum (14.33) converges absolutely for all j under Assumption 6. Let $h(\lambda)$ be the vector function with Fourier coefficients given by

$$\Gamma_{|j|} - \Gamma_{-|j|-1} = \int_{-\pi}^{\pi} h(\lambda)e^{ij\lambda}d\lambda, \quad j = 0, \pm 1, \ldots.$$

Assumption 7 $h(\lambda)$ is continuous at $\lambda = 0$, and integrable.

Assumption 7 is implied by $\sum_{j=0}^{\infty} \|\Gamma_{|j|} - \Gamma_{-|j|-1}\| < \infty$, which is in turn implied by

$$\sum_{j=0}^{\infty}(j + 1)\|\gamma_j - \gamma_{-j-1}\| < \infty, \tag{14.34}$$

in which case we may write

$$h(0) = \frac{1}{2\pi} \sum_{j=0}^{\infty} (2j + 1)(\gamma_j - \gamma_{-j-1}).$$

Of course (14.34) is itself true if $\sum_{j=-\infty}^{\infty} (|j| + 1)\|\gamma_j\| < \infty$ for which a sufficient condition in terms of (14.16) is

$$\sum_{j=0}^{\infty} (j + 1)\|C_j\| < \infty, \tag{14.35}$$

which is stronger than (14.18), while holding when u_t is a stationary ARMA process.

Lemma 4 Under Assumption 7 and (14.5)

$$\lim_{n \to \infty} E\left(\frac{n}{m} \hat{F}_{xe}(1, m)\right) = \frac{1}{2} h(0).$$

Theorem 4 Let Assumption 6 hold with $d = \iota$, $d_e = 0$. Then as $n \to \infty$

$$n(\hat{\beta}_{n-1} - \beta) \Rightarrow V(\iota, \Omega)^{-1}\{U(\Omega) + \Gamma_0\} \tag{14.36}$$

and if also Assumption 7 and (14.5) hold

$$n(\hat{\beta}_m - \beta) \Rightarrow V(\iota, \Omega)^{-1} U(\Omega). \tag{14.37}$$

Thus in the $CI(1)$ case $\hat{\beta}_{n-1}$ and $\hat{\beta}_m$ have the same rate of convergence but under (14.5) $\hat{\beta}_m$ does not suffer from the 'second-order bias' term Γ_0 incurred by $\hat{\beta}_{n-1}$. More precisely, as the proof of Theorem 4 indicates, there is a second-order bias of order $O(m/n^2)$ in $\hat{\beta}_m$ which is thus too small to contribute to (14.37), by comparison with the $O(n^{-1})$ second-order bias in (14.36). Phillips (1991b) considered a form of narrow-band spectral regression in the $CI(1)$ case, albeit stressing a system type of estimate which has superior limit distributional properties to $\hat{\beta}_m$, assuming the $CI(1)$ hypothesis is correct. However his proof is based on weighted autocovariance spectrum estimates, rather than our averaged periodogram ones. As is well known, in many stationary environments these two types of estimate are very close asymptotically, but in the $CI(1)$ case the weighted autocovariance version of $\hat{\beta}_m$ turns out to exhibit second-order bias due to correlation between u_t and e_t (specifically, to their cross-spectrum at zero frequency).

.**Case III:** The case $d_i + d_e > 1$, $i = 1, \ldots, p - 1$, $d_e < \frac{1}{2}$.

We now look at the case where the collective memory in each (x_{it}, u_t) combination exceeds that of the previous two cases, yet e_t is still stationary. Thus x_t could have less than unit root nonstationarity but in that case the memory in e_t must compensate suitably. On the other hand x_t could exhibit nonstationarity of arbitrarily high degree.

Theorem 5 Let Assumption 6 hold with $d_i + d_e > 1, 0 \leq d_e < \frac{1}{2} < d_i, i = 1, \ldots, p - 1$, and let (14.30) hold. Then for $i = 1, \ldots, p - 1$, as $n \to \infty$

$$\hat{\beta}_{i,n-1} - \beta_i = O_p(n^{d_e - d_i}), \tag{14.38}$$

and if also (14.6) holds

$$\hat{\beta}_{im} - \hat{\beta}_{i,n-1} = O_p(n^{d_e - d_i}), \tag{14.39}$$

$$\hat{\beta}_{im} - \beta_i = O_p(n^{d_e - d_i}). \tag{14.40}$$

The results (14.38) and (14.40) only bound the rates of convergence of OLS and FDLS, and we have been unable to characterize even the exact rate of convergence of OLS in the present case, due to the fact that on the one hand e_t is stationary so that the continuous mapping theorem does not suffice, whereas on the other hand e_t cannot be approximated by a semi-martingale, unlike in the short-memory case $d_e = 0$ (where in fact an exact rate and limit distribution can be derived, as it was in the $CI(1)$ case). We conjecture, however, that at least under some additional conditions the rate in (14.38) is exact, whereupon (14.39) implies immediately that $\hat{\beta}_m$ shares the same rate and limit distribution as $\hat{\beta}_{n-1}$.

Case IV: The case $d_e > \frac{1}{2}$.

Now we suppose that cointegration does not account for all the nonstationarity in z_t, so that e_t is nonstationary, as is motivated by some of the empirical experience to be described in Section 14.7. Write $d_* = (d', d_e)'$ and

$$\Delta_*(d_*) = \text{diag} \left\{ n^{(1/2) - d_1}, \ldots, n^{(1/2) - d_{p-1}}, n^{(1/2) - d_e} \right\},$$

$$G_*(r; d_*) = \text{diag}\{r^{d_1 - 1}, \ldots, r^{d_{p-1} - 1}, r^{d_e - 1}\},$$

$$W_*(r; d_*, \Omega_*) = \int_0^r G_*(r - s; d_*) dB_*(s; \Omega_*) = \begin{bmatrix} W(r; d, \Omega) \\ w(r; d_e, \sigma^2) \end{bmatrix},$$

$$W_*(d_*, \Omega_*) = \int_0^1 W_*(r; d_*, \Omega_*) dr,$$

$$U(d_*, \Omega_*) = \int_0^1 \{W(r; d, \Omega) - W(d, \Omega)\} w(r; d_e, \sigma^2) dr.$$

Let $\bar{w} = n^{-1} \sum_t w_t$. The following theorem is analogous to Theorem 2 and needs no additional explanation.

Theorem 6 Under Assumption 6 and $\frac{1}{2} < d_e < d_i, i = 1, \ldots, p - 1$, as $n \to \infty$

$$\Delta_*(d_*) w_{[nr]} \Rightarrow W_*(r; d_*, \Omega_*),$$

$$\Delta_*(d_*) \bar{w} \Rightarrow W(d_*, \Omega_*),$$

$$n^{(1/2) - d_e} \Delta(d) \hat{F}_{xe}(1, n - 1) \Rightarrow U(d_*, \Omega_*). \tag{14.41}$$

Theorem 7 Under Assumption 6, $\frac{1}{2} < d_e < d_i, i = 1, \ldots, p - 1$, and (14.30), as $n \to \infty$

$$n^{(1/2)-d_e} \Delta(d)^{-1} (\hat{\beta}_{n-1} - \beta) \Rightarrow V(d, \Omega)^{-1} U(d_*, \Omega_*), \qquad (14.42)$$

and if also (14.6) holds

$$n^{(1/2)-d_e} \Delta(d)^{-1} (\hat{\beta}_m - \beta) \Rightarrow V(d, \Omega)^{-1} U(d_*, \Omega_*). \qquad (14.43)$$

Now so long as m is regarded as increasing with n, the limit distribution is unaffected however many frequencies we omit from $\hat{\beta}_m$. Notice that in case the d_i are all equal the rate of convergence reflects the cointegrating strength b defined in Section 14.1, such that $\hat{\beta}_m$ is n^b-consistent.

14.6 Monte Carlo Evidence

Because OLS is often used as a preliminary step in $CI(1)$ analysis, the previous section suggests that even if fractional possibilities are to be ignored, FDLS might be substituted at this stage. To compare the performance of FDLS with OLS in moderate sample sizes a small Monte Carlo study in the $CI(1)$ case was conducted. The models we employed are as follows. For $i = 1, 2$, let u_{it} be a sequence of $N(0, i)$ random variables, independent across t.

Model A: AR(1) cointegrating error, $p = 2$, in (14.3) with

$$(1 - L)x_t = u_{1t},$$

$$(1 - \varphi L)e_t = u_{2t},$$

$$E(u_{1t}u_{2t}) = 1, \quad \varphi = 0.8, 0.6, 0.4, 0.2.$$

Model B: AR(2) cointegrating error, $p = 2$, in (14.3) with

$$(1 - L)x_t = u_{1t},$$

$$(1 - \varphi_1 L - \varphi_2 L^2)e_t = u_{2t},$$

$$E(u_{1t}u_{2t}) = 1, \quad \varphi_2 = -0.9; \quad \varphi_1 = 0.947, 0.34, -0.34, -0.947.$$

We fix $\varphi_2 = -0.9$ in Model B to obtain a spectral peak for e_t in the interior of $(0, \pi)$, in particular at $\lambda^* = \arccos(-\varphi_1(1 + \varphi_2)/4\varphi_2)$, that is at $\lambda^* = \pi/3, 4\pi/9, 5\pi/9$ and $2\pi/3$, respectively, for the four φ_1. On the other hand in Model A e_t always has a spectral peak at zero frequency.

Series of lengths $n = 64, 128$, and 256 were generated. $\hat{\beta}_{n-1}$ and $\hat{\beta}_m$, for $m = 3, 4, 5$ were computed, as were an estimate superior to OLS in the $CI(1)$ case, the fully-modified least squares estimate (FM-OLS, denoted $\tilde{\beta}_{FM}$) of Phillips and Hansen (1990) which uses OLS residuals at a first step, and also a modified version of this (denoted $\tilde{\beta}_{FM}^*$), using FDLS residuals. Bartlett nonparametric spectral estimation was used in $\tilde{\beta}_{FM}$ and $\tilde{\beta}_{FM}^*$, with lag numbers $\tilde{m} = 4, 6, 8$, for $n = 64, 128, 256$ respectively.

Monte Carlo bias and mean squared error (MSE), based on 5000 replications, are reported in Tables 14.1 and 14.2 for Models A and B, respectively. For each m, FDLS is superior to OLS in terms of both bias and MSE in every single case, often significantly. In fact $\hat{\beta}_m$ is best for the smallest m, 3. Our modified version $\widetilde{\beta}^*_{FM}$ of FM-OLS improves on the standard one $\widetilde{\beta}_{FM}$ in 23 out of 24 cases in terms of bias, and 16 out of 24 in terms of MSE, with 8 ties. The intuition underlying FDLS is that on the smallest frequencies cointegration implies a high signal-to-noise ratio, so it is not surprising that FDLS performs better for AR(2) e_t than AR(1) e_t, especially as λ^* increases. It is possible to devise an e_t with such power around $\lambda = 0$ that, in finite samples, FDLS performs worse than OLS, for example when $\varphi_1 \simeq 2$, $\varphi_2 \simeq -1$, $\varphi_2 + \varphi_1 < 1$, so e_t is 'near-$I(2)$' and in small samples e_t dominates x_t. However, given the intuition underlying the concept of cointegration, we believe this could be described as a 'pathological' case.

14.7 Empirical Examples

Our empirical work employs the data of Engle and Granger (1987) and Campbell and Shiller (1987). We consider seven bivariate series, denoting by y the variable chosen to be 'dependent' and by x the 'independent' one in (14.3), and by d_y, d_x integration orders. We describe the methodology used in three steps.

1. *Memory of raw data*
A necessary condition for cointegration is $d_x = d_y$, which can be tested using estimates of d_x and d_y. Three types of estimate were computed, and one test statistic. The estimates are all 'semiparametric', based only on a degenerating band of frequencies around zero frequency and assuming only a local-to-zero model for the spectral density (cf. Assumption 1) rather than a parametric model over all frequencies. The semiparametric estimates are inefficient when the parametric model is correct, but are consistent more generally and seem natural in the context of the present paper. Their asymptotic properties were established by Robinson (1995a,b) under the assumption of stationarity and invertibility (having integration order between $-\frac{1}{2}$ and $\frac{1}{2}$) and so because our raw series seem likely to be nonstationary, and quite possibly with integration orders between $\frac{1}{2}$ and $\frac{3}{2}$, we first-differenced them prior to d estimation, and then added unity. The stationarity assumption is natural in view of the motivation of these estimates, but recently Hurvich and Ray (1995), Velasco (1997a,b) have shown that they can still be consistent and have the same limit distribution as in Robinson (1995a,b) under nonstationarity (although with a different definition of $I(d)$ nonstationarity from ours), at least if a data taper is used. We thus estimated d_x and d_y directly from the raw data also, but as the results were similar they are not reported.

Denote by Δz_t either Δx_t or Δy_t where Δ is the difference operator. We describe the estimation and testing procedures as follows

(i) Log-periodogram regression. For $z = x, y$ we report in the tables $\hat{d}_z = 1 + \hat{\delta}_z$, where $\hat{\delta}_z$ is the slope estimate obtained by regressing $\log(I_{\Delta z \Delta z}(\lambda_j))$ on $-2\log(\lambda_j)$ and an intercept, for $j = 1, \ldots, \ell$, where ℓ is a bandwidth

Table 14.1. Monte Carlo bias and MSE for model A

n = 64

	Bias						MSE					
φ_1	$\hat\beta_3$	$\hat\beta_4$	$\hat\beta_5$	$\hat\beta_{n-1}$	$\tilde\beta_{FM}$	$\tilde\beta^*_{FM}$	$\hat\beta_3$	$\hat\beta_4$	$\hat\beta_5$	$\hat\beta_{n-1}$	$\tilde\beta_{FM}$	$\tilde\beta^*_{FM}$
0.8	0.194	0.210	0.229	0.295	0.171	0.154	0.128	0.123	0.130	0.154	0.094	0.091
0.6	0.068	0.083	0.096	0.177	0.062	0.041	0.033	0.033	0.034	0.059	0.027	0.024
0.4	0.031	0.037	0.046	0.125	0.036	0.014	0.015	0.014	0.014	0.031	0.011	0.009
0.2	0.017	0.020	0.026	0.097	0.024	0.003	0.009	0.007	0.008	0.019	0.007	0.006

n = 128

	Bias						MSE					
φ_1	$\hat\beta_3$	$\hat\beta_4$	$\hat\beta_6$	$\hat\beta_{n-1}$	$\tilde\beta_{FM}$	$\tilde\beta^*_{FM}$	$\hat\beta_3$	$\hat\beta_4$	$\hat\beta_6$	$\hat\beta_{n-1}$	$\tilde\beta_{FM}$	$\tilde\beta^*_{FM}$
0.8	0.074	0.087	0.110	0.175	0.075	0.062	0.034	0.034	0.037	0.057	0.026	0.024
0.6	0.020	0.023	0.035	0.096	0.018	0.006	0.008	0.008	0.008	0.018	0.005	0.006
0.4	0.008	0.010	0.015	0.066	0.010	$-1e-4$	0.004	0.003	0.003	0.009	0.002	0.002
0.2	0.003	0.005	0.007	0.051	0.008	$-4e-4$	0.001	0.001	0.001	0.003	0.001	0.001

n = 256

	Bias						MSE					
φ_1	$\hat\beta_6$	$\hat\beta_8$	$\hat\beta_{10}$	$\hat\beta_{n-1}$	$\tilde\beta_{FM}$	$\tilde\beta^*_{FM}$	$\hat\beta_6$	$\hat\beta_8$	$\hat\beta_{10}$	$\hat\beta_{n-1}$	$\tilde\beta_{FM}$	$\tilde\beta^*_{FM}$
0.8	0.038	0.048	0.053	0.097	0.026	0.019	0.008	0.009	0.009	0.018	0.006	0.005
0.6	0.008	0.013	0.015	0.050	0.003	-0.002	0.002	0.002	0.002	0.005	0.001	0.001
0.4	0.004	0.005	0.006	0.034	0.003	-0.001	$7e-4$	$7e-4$	$7e-4$	0.002	$5e-4$	$5e-4$
0.2	0.001	0.003	0.004	0.026	0.002	-0.001	$4e-4$	$3e-4$	$4e-4$	0.001	$3e-4$	$3e-4$

Table 14.2. Monte Carlo bias and MSE for model B

	Bias						MSE					
n = 64												
λ^*	$\hat{\beta}_3$	$\hat{\beta}_4$	$\hat{\beta}_5$	$\hat{\beta}_{n-1}$	$\tilde{\beta}_{FM}$	$\tilde{\beta}^*_{FM}$	$\hat{\beta}_3$	$\hat{\beta}_4$	$\hat{\beta}_5$	$\hat{\beta}_{n-1}$	$\tilde{\beta}_{FM}$	$\tilde{\beta}^*_{FM}$
$\frac{\pi}{3}$	−0.008	−0.010	−0.010	0.089	0.025	0.007	0.007	0.007	0.006	0.025	0.010	0.007
$\frac{4\pi}{9}$	−0.005	−0.006	−0.007	0.057	0.032	0.022	0.003	0.002	0.002	0.011	0.006	0.004
$\frac{5\pi}{9}$	−0.002	−0.005	−0.006	0.040	0.011	0.003	0.001	0.001	0.001	0.007	0.002	0.002
$\frac{2\pi}{3}$	−0.003	−0.003	−0.004	0.030	0.031	0.026	0.001	0.001	7e − 4	0.005	0.004	0.003
n = 128												
λ^*	$\hat{\beta}_3$	$\hat{\beta}_4$	$\hat{\beta}_6$	$\hat{\beta}_{n-1}$	$\tilde{\beta}_{FM}$	$\tilde{\beta}^*_{FM}$	$\hat{\beta}_3$	$\hat{\beta}_4$	$\hat{\beta}_6$	$\hat{\beta}_{n-1}$	$\tilde{\beta}_{FM}$	$\tilde{\beta}^*_{FM}$
$\frac{\pi}{3}$	−0.001	−0.003	−0.004	0.044	0.007	6e − 4	0.002	0.001	0.001	0.005	0.002	0.001
$\frac{4\pi}{9}$	−0.001	−0.001	−0.002	0.026	0.005	5e − 4	5e − 4	5e − 4	4e − 4	0.002	5e − 4	4e − 4
$\frac{5\pi}{9}$	−0.001	−0.001	−0.001	0.020	0.014	0.011	3e − 4	3e − 4	2e − 4	0.001	7e − 4	5e − 4
$\frac{2\pi}{3}$	−0.001	−0.001	−0.001	0.015	0.011	0.009	2e − 4	2e − 4	2e − 4	9e − 4	5e − 4	4e − 4
n = 256												
λ^*	$\hat{\beta}_6$	$\hat{\beta}_8$	$\hat{\beta}_{10}$	$\hat{\beta}_{n-1}$	$\tilde{\beta}_{FM}$	$\tilde{\beta}^*_{FM}$	$\hat{\beta}_6$	$\hat{\beta}_8$	$\hat{\beta}_{10}$	$\hat{\beta}_{n-1}$	$\tilde{\beta}_{FM}$	$\tilde{\beta}^*_{FM}$
$\frac{\pi}{3}$	−5e-4	−0.001	−0.002	0.022	−0.002	−0.004	3e − 4	3e − 4	3e − 4	0.001	3e − 4	3e − 4
$\frac{4\pi}{9}$	−0.001	−7e − 4	−7e − 4	0.013	0.005	0.004	1e − 4	1e − 4	1e − 4	4e − 4	2e − 4	1e − 4
$\frac{5\pi}{9}$	−3e − 4	−6e − 4	−0.001	0.009	0.005	0.003	5e − 5	5e − 5	5e − 5	2e − 4	1e − 4	7e − 5
$\frac{2\pi}{3}$	−3e − 4	−5e − 4	−0.001	0.007	0.007	0.006	3e − 5	3e − 5	3e − 5	2e − 4	1e − 4	9e − 5

number, tending to infinity slower than n. This is the version proposed by Robinson (1995*a*) rather than the original one of Geweke and Porter-Hudak (1983), except that we do not trim out any frequencies; recent evidence of Hurvich, Deo and Brodsky (1998) suggests that this is not necessary for nice asymptotic properties.

(ii) Test of $d_x = d_y$. We report the Wald statistic, denoted W in the tables, of Robinson (1995*a,b*), based on the difference $\hat{d}_x - \hat{d}_y = \hat{\delta}_x - \hat{\delta}_y$. The significance of W is judged by comparison with the upper tail of the χ_1^2 distribution, the 5% and 1% points being respectively, 3.78 and 5.5.

(iii) GLS log-periodogram regression. Given that $d_x = d_y$ we estimate the common value by $\hat{d}_{x=y} = 1 + \hat{\delta}_{x=y}$ where $\hat{\delta}_{x=y}$ is the generalized least squares (GLS) log-periodogram estimate of Robinson (1995*a*) based on the bivariate series $(\Delta x_t, \Delta y_t)$, using residuals from the regression in (i), $\hat{\delta}_{x=y}$ is asymptotically more efficient than $\hat{\delta}_x$ and $\hat{\delta}_y$ when $d_x = d_y$.

(iv) Gaussian estimation. For $z = x$, y we report $\tilde{d}_z = 1 + \tilde{\delta}_z$ where $\tilde{\delta}_z$ minimizes

$$\log\left(\sum_{j=1}^{\ell} \lambda_j^{2d} I_{\Delta z \Delta z}(\lambda_j)\right) - \frac{2d}{\ell} \sum_{j=1}^{\ell} \log(\lambda_j),$$

which is a concentrated narrow-band Gaussian pseudo-likelihood; see Künsch (1987), Robinson (1995*b*). As shown by Robinson (1995*b*), $\tilde{\delta}_z$ is asymptotically more efficient than $\hat{\delta}_z$.

For the estimates in (iii) and (iv) we report also approximate 95 per cent confidence intervals (denoted *CI* in the tables) based on the (normal) asymptotic distribution theory developed by Robinson (1995*a,b*). Robinson (1995*a*) assumed Gaussianity in establishing consistency and asymptotic normality of the estimates in (i) and (iii), but recent work of Velasco (1997*b*) suggests that this can be relaxed. For the estimates in (iv), Robinson (1995*b*) assumed a linear filter of martingale differences satisfying mild moment conditions. Although progress is currently being made on the choice of bandwidth ℓ in log-periodogram and Gaussian estimation, we have chosen a grid of three arbitrary values for each data set analysed in order to judge sensitivity to ℓ. Note that the estimates are $\ell^{1/2}$-consistent.

2. Cointegration analysis

We report $\hat{\beta}_m$ and also a 'high-frequency' estimate

$$\hat{\beta}_{-m} = \frac{\hat{F}_{xy}(m+1, [(n-1)/2])}{\hat{F}_{xx}(m+1, [(n-1)/2])}$$

based on the remaining frequencies, substantial deviations between $\hat{\beta}_m$ and $\hat{\beta}_{-m}$ suggesting that a full-band estimate such as OLS could be distorted by misspecification at high frequencies which is irrelevant to the essentially low-frequency concept of cointegration.

The tables include results for three values of m for each data set. These are much smaller than the bandwidths ℓ used in inference on d_x and d_y due to the anticipation of nonstationarity in the raw data; for stationary x_t, y_t optimal rules of bandwidth choice would lead to m that are more comparable with the ℓ we have used. After computing residuals $\hat{e}_t = y_t - \hat{\beta}_m x_t$, we obtained the low- and high-frequency R^2 quantities

$$R^2_m = 1 - \frac{\hat{F}_{\hat{e}\hat{e}}(1, m)}{\hat{F}_{yy}(1, m)}, \qquad R^2_{-m} = 1 - \frac{\hat{F}_{\hat{e}\hat{e}}(m + 1, [(n - 1)/2])}{\hat{F}_{yy}(m + 1, [(n - 1)/2])}.$$

We can judge the fit of a narrow-band regression by R^2_m and by comparing this with R^2_{-m} see to what extent this semiparametric fit compares with a parametric one.

For each m we report also the fractions

$$r_{xx,m} = \frac{\hat{F}_{xx}(1, m)}{\hat{F}_{xx}(1, [(n - 2)/2])}, \qquad r_{xy,m} = \frac{\hat{F}_{xy}(1, m)}{\hat{F}_{xy}(1, [(n - 2)/2])},$$

their closeness to unity indicating directly the empirical, finite sample relevance of Propositions 1 and 2 (though note that $r_{xy,m}$ need not lie in $[0, 1]$.)

3. Memory of cointegrating error
We estimated d_e first by $\hat{d}_e$ and $\tilde{d}_e$, which are respectively the log-periodogram and Gaussian estimates of (i) and (iv) above, based on first differences of the $\hat{e}_t$ and then adding unity. We also report $\hat{d}^*_e$ and $\tilde{d}^*_e$ which use the raw $\hat{e}_t$ and do not add unity, because in general we have little prior reason for believing e_t is either stationary or nonstationary. In addition we report 95 per cent confidence intervals based on the asymptotic theory of Robinson (1995a,b), though strictly this has not been justified in case of the residuals $\hat{e}_t$.

Tables 14.3–14.9 report empirical results based on several data sets.

(a) Consumption (y) and income (x) (quarterly data), 1947Q1–1981Q2

Engle and Granger (1987) found evidence of $CI(1)$ cointegration in these data. Table 14.3 tends to suggest an integration order very close to one for both variables, the estimates ranging from 0.89 to 1.08 for income and from 1.04 to 1.13 for consumption. The Wald statistic is at most 1.06, so we can safely not reject the $d_x = d_y$ null. Exploiting this information, one obtains GLS estimates ranging from 0.953 to 1.02; but with confidence intervals all so narrow as to exclude unity. The $\hat{\beta}_m$ are about 0.232, which is close to OLS (0.229), but the high frequency estimates $\hat{\beta}_{-m}$ are closer to 0.20. The unexplained variability is four times smaller around frequency zero $(1 - R^2_m)$ than at short run frequencies $(1 - R^2_{-m})$. Variability concentrates rapidly around frequency zero, 85.1 per cent of the variance of income being accounted for by the three smallest periodogram ordinates, less than 5 per cent of the total. This proportion rises to 92.6 per cent for six frequencies, and is even greater for the cross-periodogram, confirming the high

Table 14.3. Consumption (y) and income (x) ($n = 138$, $\hat{\beta}_{n-1} = 0.229$, $1 - R^2 = 0.009$)

1 *Memory of raw data*

ℓ	$\hat{d}_x$	$\hat{d}_y$	W	$\hat{d}_{x=y}$	CI	$\tilde{d}_x$	CI	$\tilde{d}_y$	CI
22	0.89	1.13	1.06	0.95	0.94, 0.97	0.99	0.78, 1.20	1.13	0.92, 1.34
30	0.95	1.04	0.02	0.98	0.97, 0.99	1.03	0.84, 1.21	1.10	0.93, 1.29
40	1.02	1.04	0.02	1.02	1.02, 1.03	1.08	0.92, 1.24	1.12	0.96, 1.28

2 *Cointegration analysis*

m	$\hat{\beta}_m$	$\hat{\beta}_{-m}$	$r_{xx,m}$	$r_{xy,m}$	$1 - R^2_m$	$1 - R^2_{-m}$
3	0.231	0.219	0.85	0.86	0.003	0.013
4	0.232	0.210	0.88	0.89	0.004	0.013
6	0.232	0.201	0.93	0.93	0.004	0.013

3 *Memory of cointegrating error*

ℓ	$\hat{d}_e^*$	CI	$\hat{d}_e$	CI	$\tilde{d}_e^*$	CI	$\tilde{d}_e$	CI
22	0.20	$-0.05, 0.46$	0.56	0.29, 0.84	0.44	0.22, 0.65	0.62	0.41, 0.83
30	0.57	0.27, 0.87	0.84	0.60, 1.07	0.68	0.49, 0.86	0.78	0.60, 0.96
40	0.61	0.38, 0.84	0.86	0.66, 1.06	0.76	0.60, 0.92	0.87	0.71, 1.02

Table 14.4. Stock prices (y) and dividends (x) ($n = 116$, $\hat{\beta}_{n-1} = 30.99$, $1 - R^2 = 0.15$)

1 *Memory of raw data*

ℓ	$\hat{d}_x$	$\hat{d}_y$	W	$\hat{d}_{x=y}$	CI	$\tilde{d}_x$	CI	$\tilde{d}_y$	CI
22	0.91	0.96	0.07	0.94	0.92, 0.96	0.36	0.15, 0.57	1.04	0.83, 1.25
30	0.86	0.83	0.04	0.84	0.83, 0.85	0.48	0.30, 0.66	0.91	0.73, 1.09
40	0.91	0.84	0.36	0.87	0.86, 0.87	0.70	0.54, 0.86	0.90	0.74, 1.06

2 *Cointegration analysis*

m	$\hat{\beta}_m$	$\hat{\beta}_{-m}$	$r_{xx,m}$	$r_{xy,m}$	$1 - R^2_m$	$1 - R^2_{-m}$
3	33.16	23.24	0.78	0.84	0.076	0.215
4	33.55	21.49	0.79	0.85	0.093	0.210
6	32.47	22.81	0.85	0.89	0.114	0.190

3 *Memory of cointegrating error*

ℓ	$\hat{d}_{e^*}$	$\hat{d}_e$	CI	$\tilde{d}_{e^*}$	$\tilde{d}_e$	CI
22	0.73	0.74	0.47, 1.01	0.95	1.04	0.83, 1.26
30	0.60	0.60	0.36, 0.83	0.85	0.91	0.73, 1.09
40	0.64	0.66	0.46, 0.86	0.84	0.90	0.74, 1.06

Table 14.5. Log prices (y) and log wages (x)
($n = 360$, $\hat{\beta}_{n-1} = 0.706$, $1 - R^2 = 0.033$)

Memory of raw data

ℓ	$\hat{d}_x$	$\hat{d}_y$	W	$\tilde{d}_x$	CI	$\tilde{d}_y$	CI
30	1.16	1.60	5.84	1.07	0.89, 1.25	1.24	1.06, 1.42
40	1.03	1.54	11.1	1.07	0.92, 1.23	1.25	1.09, 1.41
60	0.99	1.54	19.9	1.07	0.94, 1.20	1.27	1.14, 1.40

Table 14.6. Log L (y) and log nominal GNP (x)
($n = 90$, $\hat{\beta}_{n-1} = 1.039$, $1 - R^2 = 0.00085$)

Memory of raw data

ℓ	$\hat{d}_x$	$\hat{d}_y$	W	$\tilde{d}_x$	CI	$\tilde{d}_y$	CI
16	1.29	1.61	5.51	1.23	0.98, 1.48	1.46	1.21, 1.71
22	1.36	1.68	6.30	1.25	1.03, 1.46	1.56	1.35, 1.77
30	1.29	1.68	9.99	1.22	1.04, 1.40	1.60	1.42, 1.68

Table 14.7. Log M2 (y) and log nominal
GNP (x) ($n = 90$, $\hat{\beta}_{n-1} = 0.99$, $1 - R^2 = 0.0026$)

1 *Memory of raw data*

ℓ	$\hat{d}_x$	$\hat{d}_y$	W	$\hat{d}_{x=y}$	CI	$\tilde{d}_x$	CI	$\tilde{d}_y$	CI
16	1.29	1.53	0.69	1.29	1.28, 1.30	1.23	0.98, 1.48	1.35	1.10, 1.60
22	1.36	1.56	0.83	1.38	1.37, 1.39	1.25	1.03, 1.46	1.47	1.25, 1.69
30	1.29	1.64	3.67	1.33	1.32, 1.34	1.22	1.04, 1.40	1.59	1.41, 1.78

2 *Cointegration analysis*

m	$\hat{\beta}_m$	$\hat{\beta}_{-m}$	$r_{xx,m}$	$r_{xy,m}$	$1 - R_m^2$	$1 - R_{-m}^2$
3	0.99	0.98	0.83	0.84	0.002	0.003
4	0.99	0.99	0.87	0.87	0.002	0.003
6	0.99	0.99	0.91	0.91	0.003	0.003

3 *Memory of cointegrating error*

ℓ	$\hat{d}_x^*$	$\hat{d}_e$	CI	$\tilde{d}_e^*$	$\tilde{d}_e$	CI
16	1.15	1.19	0.88, 1.52	1.20	1.23	0.98, 1.48
22	1.10	1.16	0.89, 1.43	1.04	1.10	0.89, 1.31
30	1.10	1.15	0.92, 1.38	1.02	1.09	0.91, 1.27

Table 14.8. Log M1 (y) and log nominal GNP (x) ($n = 90$, $\hat{\beta}_{n-1} = 0.643$, $1 - R^2 = 0.00309$)

1 *Memory of raw data*

ℓ	$\hat{d}_x$	$\hat{d}_y$	W	$\hat{d}_{x=y}$	CI	$\tilde{d}_x$	CI	$\tilde{d}_y$	CI
16	1.29	1.53	7.70	1.31	1.30, 1.32	1.23	0.98, 1.48	1.39	1.14, 1.64
22	1.36	1.42	0.346	1.39	1.38, 1.40	1.25	0.97, 1.46	1.33	1.12, 1.55
30	1.29	1.29	0.001	1.29	1.28, 1.30	1.22	1.04, 1.40	1.27	1.09, 1.45

2 *Cointegration analysis*

m	$\hat{\beta}_m$	$\hat{\beta}_{-m}$	$r_{xx,m}$	$r_{xy,m}$	$1 - R_m^2$	$1 - R_{-m}^2$
3	0.645	0.632	0.83	0.84	0.003	0.003
4	0.645	0.630	0.87	0.88	0.003	0.003
6	0.645	0.629	0.91	0.91	0.003	0.003

3 *Memory of cointegrating error*

ℓ	$\hat{d}_e^*$	$\hat{d}_e$	CI	$\tilde{d}_e^*$	$\tilde{d}_e$	CI
16	0.99	1.20	0.88, 1.52	0.97	1.06	0.81, 1.31
22	0.76	1.07	0.80, 1.34	0.77	0.92	0.71, 1.13
30	0.78	0.88	0.64, 1.11	0.76	0.85	0.67, 1.03

Table 14.9. Log M3 (y) and log nominal GNP (x) ($n = 90$, $\hat{\beta}_{n-1} = 1.0997$, $1 - R^2 = .0023$)

1 *Memory of raw data*

ℓ	$\hat{d}_x$	$\hat{d}_y$	W	$\hat{d}_{x=y}$	CI	$\tilde{d}_x$	CI	$\tilde{d}_y$	CI
16	1.29	1.45	0.79	1.33	1.33, 1.34	1.23	0.98, 1.48	1.34	1.08, 1.60
22	1.36	1.62	2.01	1.44	1.43, 1.44	1.25	1.03, 1.46	1.50	1.28, 1.71
30	1.29	1.71	7.04	1.42	1.41, 1.43	1.22	1.04, 1.40	1.65	1.47, 1.83

2 *Cointegration analysis*

m	$\hat{\beta}_m$	$\hat{\beta}_{-m}$	$r_{xx,m}$	$r_{xy,m}$	$1 - R_m^2$	$1 - R_{-m}^2$
3	1.10	1.10	0.83	0.83	0.002	0.002
4	1.10	1.10	0.87	0.87	0.002	0.002
6	1.10	1.10	0.91	0.91	0.002	0.002

3 *Memory of cointegrating error*

ℓ	$\hat{d}_e^*$	$\hat{d}_e$	CI	$\tilde{d}_e^*$	$\tilde{d}_e$	CI
16	0.88	0.89	0.57, 1.21	1.02	1.02	0.76, 1.28
22	0.97	1.00	0.73, 1.27	1.05	1.08	0.87, 1.29
30	0.96	1.01	0.78, 1.21	0.98	1.04	0.86, 1.22

coherency of the two series at low frequencies. The residual diagnostics are less clear-cut, but in only one case out of 12 does the confidence interval for d_e include zero, providing strong evidence against weak dependence. The estimates of d_e vary quite noticeably with ℓ and the procedure adopted, ranging from 0.2 to 0.87.

(b) Stock prices (y) and dividends (x) (annual data), 1871–1986

The idea that these might be cointegrated follows mainly from a present value model, which asserts that an asset price is linear in the present discounted value of future dividends, $y_t = \theta(1-\delta) \sum_{i=0}^{\infty} \delta^i E_t(x_{t+i}) + c$, where δ is the discount factor; see Campbell and Shiller (1987). In Table 14.4, the estimates of d_x, d_y appear close to unity, although now the hypothesis that dividends are mean-reverting ($d_x < 1$) appears to be supported. The Wald statistics for testing $d_x = d_y$ are always manifestly insignificant. A marked difference between $\hat{\beta}_m$ and $\hat{\beta}_{-m}$ is found, the former oscillating around 33 and the latter below 24. The spectral R^2 still indicate a much better fit at low frequencies, but empirical evidence of cointegration is extremely weak. Notice in particular that if y and x are not cointegrated, $d_e = \max(d_x, d_y)$, as is amply confirmed by the Gaussian estimates, where one gets identical estimates of d_x and $d_e = (1.04, 0.91, 0.90)$ for $\ell = 22, 30, 40$. The results of Campbell and Shiller on this data set were, in their own words, inconclusive; our findings confirm those of Phillips and Ouliaris (1988), who were unable to reject the null of no cointegration at the 10 per cent level.

(c) Log prices (y) and wages (x) (monthly data), 1960M1–1979M12

The results in Table 14.5 tend to develop those of Engle and Granger (1987) by supporting an absence of a cointegrating relationship of any order. Where our conclusions differ is in the integration orders of x and y, in particular of log prices, which appear not to be unity, ranging from 1.54 to 1.60, while confidence intervals never include unity. This is not very surprising in that the inflation rate might plausibly be characterized as a stationary long-memory process. W is always above 5.8, so we reject also at the 1 per cent level the hypothesis that $d_x = d_y$, and so because this necessary condition for cointegration is not satisfied the analysis is taken no further.

(d) Quantity theory of money (quarterly data): log M1, M2, M3 or L (y) and log GNP (x), where L denotes total liquid assets, 1959Q1–1981Q2

Engle and Granger (1987) found the classical equation $MV = PY$ of the quantity theory of money to hold for $M = M2$, but not $M1, M3, L$. This is somewhat unsatisfactory since the latter monetary aggregates are linked with $M2$ in the long run, so that there might exist cointegration (albeit of different orders) between more than one of these aggregates and GNP. For log L in Table 14.6 we reject at the 1 per cent level the hypothesis that GNP shares the same integration order. For $M2$, in Table 14.7, the necessary condition for cointegration is met, GLS confidence intervals tending to suggest integration orders around 1.3, which seems unsurprising since both aggregates are nominal. The $\hat{\beta}_m$ are not noticeably influenced by m and are indeed the same as OLS (0.99). Estimates of the d_e are strongly inconsistent with stationarity, ranging from 1.02 to 1.23, the confidence intervals excluding values below 0.88. Overall, it seems very difficult to draw reliable conclusions about the existence of fractional cointegration between these variables given such a small sample. The relationship of nominal GNP with $M1$ and $M3$, in Tables 14.8 and 14.9, appears much closer to that with $M2$ than Engle and Granger concluded, exploiting the greater flexibility of our framework. In particular, the common integration order for the bivariate raw data is again estimated via GLS to be 1.31, 1.39, 1.29 (for $m = 16, 22, 30$) for nominal GNP and $M1$ and 1.33, 1.44, 1.42 for nominal GNP and $M3$; estimates of d_e range from 0.76 to 1.20 for the former case and from 0.88 to 1.08 for the latter.

14.8 Final Comments

The chapter demonstrates that OLS estimates of a cointegrating vector are asymptotically matched or bettered in a variety of stationary and nonstationary cases by a narrow-band frequency domain estimate, FDLS. The overall superiority of FDLS relies on correlation between the cointegrating errors and regressors; in the absence of such correlation FDLS is inferior to OLS for stationary data, and comparable for nonstationary data. The finite-sample advantages of FDLS in correlated situations are observed in a small Monte Carlo study. FDLS is incorporated in a semiparametric methodology for investigating the possibility of fractional cointegration, which is applied to bivariate macroeconomic series.

The chapter leaves open numerous avenues for further research. It is possible that the whole of x_t does not satisfy the conditions of either Section 14.3 or one of the cases I, II or III/IV of Section 14.5, but rather that subsets of x_t are classified differently. It is straightforward to extend our results to cover such situations, and we have not done so for the sake of simplicity, and because the case $p = 2$ is itself of practical importance. A more challenging development would cover such omitted cases as when x_t has integration order $\frac{1}{2}$, on the boundary between stationarity and nonstationarity, though this can be thought of as occupying a measure-zero subset of the parameter space. From a practical viewpoint a significant deficiency of our treatment of nonstationarity is the lack of allowance for deterministic trends, such as (possibly nonintegral) powers of t, but if these are suitably dominated by the stochastic trends it appears that the results of Sections 14.4 and 14.5 continue to hold.

A more challenging area for study is the extent to which we can improve on FDLS in our semiparametric context, with unknown integration orders, to mirror the improvement of OLS by various estimates in the $CI(1)$ case. There is also a need, as in the $CI(1)$ case, to allow for the possibility of more than one cointegrating relation, where we might wish to permit these to have different integration orders. Certainly it seems clear that results such as Propositions 1 and 2 and Lemma 1 can be established for more general quadratic forms, and so the extensive asymptotic theory for quadratic forms of stationary long-memory series can be significantly extended in a nonstationary direction. The choice of bandwidth m in $\hat{\beta}_m$ seems less crucial under nonstationarity than under stationarity, but nevertheless some criterion must be given to practitioners. For the stationary case, which seems of interest in financial applications, bandwidth theory of Robinson (1994c) can be developed, but there is a need also to develop asymptotic distribution theory for FDLS, useful application of which is likely to require bias-correction due to correlation between x and e. For the relatively short macroeconomic series analyzed in the present chapter the semiparametric approach employed, while based on very mild assumptions, will not produce as reliable estimates of integration orders as correctly specified parametric time series models, and it is possible to analyse narrow-band β estimates in such a parametric framework also. In this connection a recent treatment of parametric inference in multivariate stationary long memory is given by Hosoya (1997).

Appendix 14.1

Proof of Theorem 1 From (14.2), (14.3) we have

$$\hat{\beta}_m - \beta = \hat{F}_{xx}(1, m)^{-1} \hat{F}_{xe}(1, m). \tag{A.1}$$

By the Cauchy inequality, as in Robinson (1994a)

$$|\hat{F}_{xe}^{(i)}(1, m)| \le \left\{ \hat{F}_{xx}^{(i,i)}(1, m) \hat{F}_{ee}(1, m) \right\}^{1/2}. \tag{A.2}$$

For any non-null $p \times 1$ vectors γ and δ, by Assumption 1

$$\gamma' \{ \hat{F}_{zz}(1, m) - F_{zz}(\lambda_m) \} \delta = o_p(\{ \gamma' F_{zz}(\lambda_m) \gamma \delta' F_{zz}(\lambda_m) \delta \}^{1/2}), \quad \text{as } n \to \infty \tag{A.3}$$

by a straightforward multivariate extension of theorem 1 of Robinson (1994a) (see also Lobato 1997), with

$$F_{zz}(\lambda) = \int_0^\lambda \text{Re} \{ f_{zz}(\mu) \} \, d\mu \sim G(\lambda), \quad \text{as } \lambda \to 0^+, \tag{A.4}$$

where $G(\lambda)$ has (i, j)th element

$$G_{ij}(\lambda) = \frac{G_{ij} \lambda^{1-d_i-d_j}}{1 - d_i - d_j},$$

G_{ij} being the (i, j)th element of G. Applying (A.3), (A.4) and (14.5),

$$\Lambda_m^{-1} \{ \hat{F}_{zz}(1, m) - F_{zz}(\lambda_m) \} \Lambda_m^{-1} = o_p(\lambda_m),$$

where $\Lambda_m = \text{diag}\{\lambda_m^{-d_1}, \ldots, \lambda_m^{-d_p}\}$, so

$$\hat{F}_{xx}^{(i,j)}(1, m) = G_{ij}(\lambda_m) + o_p(\lambda_m^{1-d_i-d_j}),$$

$$\hat{F}_{ee}(1, m) = \alpha' \hat{F}_{zz}(1, m)\alpha = \alpha' F_{zz}(\lambda_m)\alpha + o_p(\alpha' F_{zz}(\lambda_m)\alpha)$$
$$= O_p(\lambda_m^{1-2d_e}),$$

because

$$\alpha' F_{zz}(\lambda)\alpha = \int_0^\lambda \alpha' f_{zz}(\mu)\alpha \, d\mu \sim c \int_0^\lambda \mu^{-2d_e} \, d\mu = c\frac{\lambda^{1-2d_e}}{1 - 2d_e}, \quad \text{as } \lambda \to 0^+.$$

Denote by $\tilde{\Lambda}_m, \tilde{G}(\lambda)$ the leading $(p-1) \times (p-1)$ submatrices of $\Lambda_m, G(\lambda)$. For any $(p-1) \times 1$ non-null vector $v = (v_1, \ldots, v_{p-1})'$

$$v'\tilde{\Lambda}_m^{-1} F_{xx}(\lambda_m)\tilde{\Lambda}_m^{-1}v \sim v'\tilde{\Lambda}_m^{-1}\tilde{G}(\lambda_m)\tilde{\Lambda}_m^{-1}v \geq \omega \int_0^{\lambda_m} \sum_{j=1}^{p-1} \left\{ (\lambda_m/\lambda)^{d_j} v_j \right\}^2 d\lambda$$

$$= \omega\lambda_m \sum_{j=1}^{p-1} \frac{v_j^2}{1 - 2d_j},$$

where ω is the smallest eigenvalue of the leading $(p-1) \times (p-1)$ submatrix of G, which is positive definite by Assumption 1. It follows that $\tilde{\Lambda}_m \hat{F}_{xx}^{-1}(\lambda_m)\tilde{\Lambda}_m = O_p(n/m)$, whence the proof is completed by elementary manipulation. ∎

To assist proof of Propositions 1 and 2, we introduce the following Lemma. In the sequel, C denotes a generic positive constant.

Lemma 5 Under Assumption 4,

$$S_{uv}(\lambda, \gamma) \stackrel{\text{def}}{=} \sum_{t=u}^{v} e^{it\lambda}\varphi_t(\gamma)$$

satisfies, for $0 \leq u < v, 0 \leq |\lambda| \leq \pi$,

$$S_{uv}(\lambda, 0) = 1, \quad u = 0,$$
$$= 0, \quad u > 0,$$

and for $\gamma > 0$

$$|S_{uv}(\lambda, \gamma)| \leq C \min\left(v^\gamma, \frac{(u+1)^{\gamma-1}}{|\lambda|}, \frac{1}{|\lambda|^\gamma} \right), \quad 0 < \gamma \leq 1,$$

$$|S_{uv}(\lambda, \gamma)| \leq C \min\left(v^\gamma, \frac{v^{\gamma-1}}{|\lambda|} \right), \quad \gamma > 1.$$

Proof The proof for $\gamma = 0$ is trivial, so we consider $\gamma > 0$. Drop the argument γ from $S_{uv}(\lambda, \gamma)$ and $\varphi_t(\gamma)$. Obviously $|S_{uv}(\lambda)| \leq Cv^\gamma$. For $0 < \gamma \leq 1$ we can write, for $u < s < v$,

$$S_{uv}(\lambda) = \sum_{t=u}^{s-1} \varphi_t e^{it\lambda} + \sum_{t=s}^{v-1}(\varphi_t - \varphi_{t+1})\sum_{v=s}^{t} e^{iv\lambda} + \varphi_v \sum_{t=s}^{v} e^{it\lambda}$$

by summation-by-parts. Thus because

$$\left|\sum_{v=s}^{t} e^{iv\lambda}\right| \le \frac{C(t-s)}{1+(t-s)|\lambda|}, \qquad |\lambda| < \pi, \tag{A.5}$$

Assumption 4 implies that

$$|S_{uv}(\lambda)| \le C\left((s-1)^{\gamma} + \frac{s^{\gamma-1}}{|\lambda|}\right). \tag{A.6}$$

For $1/|\lambda| \le Cv$ we may choose $s \sim |\lambda|^{-1}$ so that (A.6) is $O(|\lambda|^{-\gamma})$. On the other hand we also have

$$S_{uv}(\lambda) = \sum_{t=u}^{v-1}(\varphi_t - \varphi_{t+1})\sum_{s=u}^{t} e^{is\lambda} + \varphi_v\sum_{t=u}^{v} e^{it\lambda}, \tag{A.7}$$

to give $|S_{uv}(\lambda)| \le C(u+1)^{\gamma-1}/|\lambda|$. For $\gamma > 1$, (A.7) gives instead

$$|S_{uv}(\lambda)| \le \frac{Cv^{\gamma-1}}{|\lambda|}. \qquad \blacksquare$$

Proof of Proposition 1 The discrete Fourier transform of a_{jt} is, from (14.8),

$$w_j(\lambda) = \frac{1}{\sqrt{2\pi n}}\sum_t a_{jt}e^{it\lambda} = \frac{1}{\sqrt{2\pi n}}\sum_t \varphi_{j,n-t}(\lambda)\,e^{it\lambda}\eta_{jt}, \qquad j = 1, 2,$$

where

$$\varphi_{jt}(\lambda) \equiv \sum_{s=0}^{t} \varphi_{js}\,e^{is\lambda}, \qquad j = 1, 2.$$

Thus by Assumption 3

$$E\left(I_{aa}^{(1,2)}(\lambda)\right) = \frac{1}{2\pi n}\int_{-\pi}^{\pi} \Phi_1(\lambda, -\mu)\Phi_2(-\lambda, \mu)f_{12}(\mu)\,d\mu, \tag{A.8}$$

where $f_{ij}(\mu)$ is the cross spectral density of η_{it}, η_{jt} and

$$\Phi_j(\lambda, \mu) = \sum_t \varphi_{j,n-t}(\lambda)\,e^{it(\lambda+\mu)}, \qquad j = 1, 2.$$

The modulus of (A.8) is bounded by

$$\frac{C}{n}\sup_{\mu}|f_{12}(\mu)|\left\{\int_{-\pi}^{\pi}|\Phi_1(\lambda, -\mu)|^2 d\mu \int_{-\pi}^{\pi}|\Phi_2(-\lambda, \mu)|^2 d\mu\right\}^{1/2}$$

$$\le \frac{C}{n}\left\{\prod_{j=1}^{2}\sup_{\mu} f_{jj}(\mu)\sum_t |\varphi_{j,n-t}(\lambda)|^2\right\}^{1/2}, \tag{A.9}$$

from Assumption 3. From Lemma A.1, for $0 < |\lambda| < \pi, t = 1, \ldots, n$ and $i = 1, 2$

$$|\varphi_{it}(\lambda)| = O\left(\frac{n^{\gamma_i - 1}}{|\lambda|}1(\gamma_i > 1) + \frac{1}{|\lambda|^{\gamma_i}}1(\gamma_i \le 1)\right)$$

$$= O\left(\frac{n^{\max(\gamma_i - 1, 0)}}{|\lambda|^{\min(\gamma_i, 1)}}\right) \qquad (A.10)$$

when $\gamma_i > 0$. The latter bound also applies for $\gamma_i = 0$, when it is $O(1)$. Thus (A.9) is, for $0 < |\lambda| < \pi$,

$$O\left(\frac{n^{\max(\gamma_1 - 1, 0) + \max(\gamma_2 - 1, 0)}}{|\lambda|^{\min(\gamma_1, 1) + \min(\gamma_2, 1)}}\right).$$

For $\lambda = \lambda_j, j = 1, \ldots, M$, it is

$$O\left(\frac{n^{\gamma_1 + \gamma_2}}{j^{\min(\gamma_1, 1) + \min(\gamma_2, 1)}}\right).$$

Hence, when $\gamma_1 + \gamma_2 > 1$, by (14.6)

$$|E(F_{aa}^{(1,2)}(m + 1, M))| \le \frac{2\pi}{n} \sum_{j=m+1}^{M} |E(I_{aa}^{(1,2)}(\lambda_j))|$$

$$\le Cn^{\gamma_1 + \gamma_2 - 1} \sum_{j=m}^{\infty} j^{-\min(\gamma_1, 1) - \min(\gamma_2, 1)}$$

$$= o(n^{\gamma_1 + \gamma_2 - 1}),$$

$$|E(F_{22}^{(1,2)}(1, m))| \le Cn^{\gamma_1 + \gamma_2 - 1} \sum_{j=1}^{m} j^{-\min(\gamma_1, 1) - \min(\gamma_2, 1)}$$

$$= O(n^{\gamma_1 + \gamma_2 - 1}).$$

Likewise, when $\gamma_1 + \gamma_2 < 1$,

$$|E(F_{aa}^{(1,2)}(1, M))| \le Cn^{\gamma_1 + \gamma_2 - 1} \sum_{j=1}^{m} \frac{1}{j^{\gamma_1 + \gamma_2}}$$

$$= O\left(\left(\frac{n}{m}\right)^{\gamma_1 + \gamma_2 - 1}\right). \blacksquare$$

Proof of Lemma 1 From (14.15)

$$x_{it} - (1 - L)^{-d_i}\{u_{it}1(t > 0)\}, \quad i = 1, \ldots, p - 1,$$

$$e_t = (1 - L)^{-d_e}\{u_{pt}1(t > 0)\},$$

where u_{it} is the ith element of u_t. We take

$$\varphi_k(\gamma) = \frac{\Gamma(k+\gamma)}{\Gamma(\gamma)\Gamma(k+1)}, \tag{A.11}$$

since this is the coefficient of L^k in the Taylor expansion of $(1-L)^{-\gamma}$, and choose $\varphi_{1k} = \varphi_k(d_i)$ in both cases (14.28) and (14.29), with $\varphi_{2k} = \varphi_k(d_e)$ in (14.28) and $\varphi_{2k} = \varphi_k(d_j)$ in (14.29). Now

$$\varphi_k(\gamma) = O((1+k)^{\gamma-1}) \tag{A.12}$$

from Abramowitz and Stegun (1970). Also

$$\varphi_k(\gamma) - \varphi_{k+1}(\gamma) = -\varphi_{k+1}(\gamma-1) = O((1+k)^{\gamma-2}),$$

to check Assumption 4. Next we take $\eta_{1t} = u_{it}$ in each case and $\eta_{2t} = u_{pt}$ and $\eta_{2t} = u_{jt}$ in (14.28) and (14.29) respectively. Now the spectral density matrix of u_t is $(2\pi)^{-1}C(e^{i\lambda})\sum C(e^{i\lambda})^*$ whose modulus is bounded by $C(\sum_{j=0}^{\infty}\|C_j\|)^2 < \infty$ from (14.21). Thus Assumption 3 is satisfied. Finally the fourth cumulant of $u_{i0}, u_{ia}, u_{j,a+b}, u_{j,a+b+c}$ is, for $a, b, c \geq 0$

$$\text{cum}\left(\sum_{-\infty}^{\infty} c'_{i,-d}\varepsilon_d, \sum_{-\infty}^{\infty} c'_{i,a-e}\varepsilon_e, \sum_{-\infty}^{\infty} c'_{j,a+b-f}\varepsilon_f, \sum_{-\infty}^{\infty} c'_{j,a+b+c-g}\varepsilon_g\right),$$

where c'_{ij} is the ith row of C_j. This is bounded in absolute value by

$$C\sum_{d=-\infty}^{\infty} \|c_{i,-d}\| \|c_{i,a-d}\| \|c_{i,a+b-d}\| \|c_{i,a+b+c-d}\|.$$

Because the sum of this, over all a, b, c, is finite due to (14.21), it follows that the Fourier coefficients of the fourth cumulant spectrum of $u_{it}, u_{it}, u_{jt}, u_{jt}$ are absolute summable, so that their spectrum is indeed bounded and Assumption 5 is satisfied. ∎

Proof of Lemma 2 For brevity we shall write $\Delta = \Delta(d)$ in the sequel.

$$\Delta\hat{F}_{xx}(1,m)\Delta = \Delta\hat{F}_{xx}(1,n-1)\Delta$$
$$- \Delta\left\{\hat{F}_{xx}(m+1,n-1) - E\left(\hat{F}_{xx}(m+1,n-1)\right)\right\}\Delta$$
$$- \Delta E\left(\hat{F}_{xx}(m+1,n-1)\right)\Delta.$$

In view of Lemma 1 the last two components are $o_p(1)$ and $o(1)$ from Propositions 2 and 1, respectively. The proof is completed by appealing to Theorem 2. ∎

Proof of Lemma 3 We begin by estimating the b_{ijt}. First

$$\|C_j\| \leq \left(\sum_{\ell=j}^{\infty} \|C_\ell\|^2 \right)^{1/2} \tag{A.13}$$

$$\leq \frac{2}{j} \sum_{k=[\frac{j}{2}]}^{j} \left(\sum_{\ell=k}^{\infty} \|C_\ell\|^2 \right)^{1/2} \tag{A.14}$$

$$= o(j^{-1}) \quad \text{as } j \to \infty, \tag{A.15}$$

where (A.14) is due to monotonic decay of the right-hand side of (A.13), and (A.15) follows from (14.18). The ith diagonal element of D_k is $\varphi_k(d_i)$ for $i = 1, \ldots, p-1$ and $\varphi_k(d_e)$ for $i = p$, where $\varphi_k(\gamma)$ is given by (A.11). Now from (14.26), for $i = 1, \ldots, p-1$

$$\|b_{ijt}\| \leq \sum_{\ell=0}^{r} \varphi_\ell(d_i) \|C_{j-\ell}\| + \sum_{\ell=r+1}^{\min(j,t-1)} \varphi_\ell(d_i) \|C_{j-\ell}\|$$

$$\leq C \max_{j-r \leq \ell \leq j} \|C_\ell\| \sum_{\ell=0}^{r} \ell^{d_i-1} + C r^{d_i-1} \sum_{0}^{\infty} \|C_\ell\|$$

$$\leq C r^{d_i} (j-r)^{-1} + C r^{d_i-1}$$

for $1 \leq r < \min(j, t-1)$. It follows that for $j \leq 2t$

$$\|b_{ijt}\| \leq C j^{d_i-1}, \quad i = 1, \ldots, p-1, \tag{A.16}$$

on taking $r \sim j/2$. For $j > 2t$ we have more immediately

$$\|b_{ijt}\| \leq \sum_{\ell=0}^{t-1} |\varphi_\ell(d_i)| \|C_{j-\ell}\| \leq C(j-t)^{-1} t^{d_i}, \quad i = 1, \ldots, p-1.$$

Similarly

$$\|b_{pjt}\| \leq O(j^{d_e-1}), \quad j \leq 2t,$$

$$= O\left((j-t)^{-1} t^{d_e} \right), \quad j > 2t. \tag{A.17}$$

Next notice that $b_{ij} = b_{ijt}$ for $0 \le j < t$, so from (14.22)

$$
E(x_{it}e_t) = \sum_{j=0}^{t-1} b'_{ij} \sum b_{pj} + \sum_{j=t}^{\infty} b'_{ijt} \sum b_{pjt}
$$

$$
= \xi_i - \sum_{j=t}^{\infty} b'_{ij} \sum b_{pj} + \sum_{j=t}^{2t-1} b'_{ijt} \sum b_{pjt} + \sum_{j=2t}^{\infty} b'_{ijt} \sum b_{pjt}
$$

$$
= \xi_i + O\left(\sum_{j=t}^{\infty} j^{d_i+d_e-2} + t.t^{d_i+d_e-2} + t^{d_i+d_e} \sum_{j=t}^{\infty} j^{-2} \right)
$$

$$
= \xi_i + O(t^{d_i+d_e-1})
$$

because $d_i + d_e < 1$, where this and (A.16), (A.17) imply that $|\xi_i| < \infty$. On the other hand

$$
|E(\bar{x}_i \bar{e})| \le \frac{C}{n^2} \sum_s \sum_t \sum_{j=0}^{\infty} \sum_{k=0}^{\infty} \|b_{ijs}\| \|b_{pkt}\|
$$

$$
\le \frac{C}{n^2} \left\{ \sum_s \sum_t \sum_{\max(0,s-t)}^{\min(2s,s+t)} j^{d_i-1}(j+t-s)^{d_e-1} \right.
$$

$$
+ \sum_s \sum_t t^{2d_e} \sum_{s+t}^{2s} j^{d_i-1}(j-s)^{-1}
$$

$$
+ \sum_s \sum_t s^{d_i} \sum_{2s}^{s+t} (j+t-s)^{d_e-1}(j-s)^{-1}
$$

$$
\left. + \sum_s \sum_t s^{d_i} t^{d_e} \sum_{\min(2s,s+t)}^{\infty} (j-s)^{-2} \right\}
$$

$$
= O(n^{d_i-d_e-1}).
$$

The proof is routinely completed in view of (14.4). ■

Proof of Theorem 3 It is convenient to introduce the abbreviating notation

$$
\tilde{A} = \hat{F}_{xx}(1, n-1), \quad \tilde{a} = \hat{F}_{xe}(1, n-1), \quad \hat{A} = \hat{F}_{xx}(1, m), \quad \hat{a} = \hat{F}_{xe}(1, m).
$$

Thus

$$
\hat{\beta}_{n-1} - \beta = \Delta(\Delta\tilde{A}\Delta)^{-1}\Delta\tilde{a}, \tag{A.18}
$$

$$
\hat{\beta}_m - \beta = \Delta(\Delta\hat{A}\Delta)^{-1}\Delta\hat{a}. \tag{A.19}
$$

Now

$$\Delta(\widetilde{A} - \hat{A})\Delta = \Delta\{(\widetilde{A} - \hat{A}) - E(\widetilde{A} - \hat{A})\}\Delta + \Delta E(\widetilde{A} - \hat{A})\Delta$$
$$\rightarrow_p 0 \tag{A.20}$$

from Propositions 2 and 1, Assumption 6 and Lemma 1, so that $\Delta\widetilde{A}\Delta$, $\Delta\hat{A}\Delta \Rightarrow V(d, \Omega)$ by Theorem 2. Now denote by $\widetilde{a}_i$, $\hat{a}_i$ the ith elements of $\widetilde{a}$, $\hat{a}$. From Proposition 2, Assumption 6 and Lemma 1

$$\widetilde{a}_i = E(\widetilde{a}_i) + (\widetilde{a}_i - E(\widetilde{a}_i)) = E(\widetilde{a}_i) + O_p(n^{d_i + d_e - 1}), \quad i = 1, \dots, p - 1,$$

whereas from Lemma 3

$$\lim_{n \to \infty} n^{d_{\min} - (1/2)} \Delta E(\widetilde{a}) = \xi.$$

Then (14.31) follows from (14.23). Finally

$$\hat{a}_i = E(\hat{a}_i) + \{\hat{a}_i - E(\hat{a}_i)\} = O\left(\left(\frac{n}{m}\right)^{d_i + d_e - 1}\right) + O_p(n^{d_i + d_e - 1}),$$

from Proposition 2 and 1, Assumption 6 and Lemma 1 so the ith element of (A.19) is

$$O_p\left(n^{(1/2) - d_i} \max_{1 \le j < p} n^{(1/2) - d_j} \left(\frac{n}{m}\right)^{d_j + d_e - 1}\right) = O_p\left(n^{1 - d_i - d_{\min}} \left(\frac{n}{m}\right)^{d_{\min} + d_e - 1}\right)$$
$$= o_p\left(n^{1 - d_i - d_{\min}}\right),$$

since $d_{\min} + d_e < 1$, to complete the proof of (14.32). ∎

Proof of Lemma 4 For $1 \le j \le m$, writing $\widetilde{\Gamma}_j = \sum_{\ell \ge j}^{\infty} \gamma_\ell$,

$$E\{I_{xe}(\lambda_j)\} = \frac{1}{2\pi n} \sum_s \sum_t (\gamma_{s-1} + \cdots + \gamma_{s-t}) e^{i(t-s)\lambda_j}$$

$$= \frac{1}{2\pi n} \sum_s \sum_t \left(\widetilde{\Gamma}_{s-t} - \widetilde{\Gamma}_{s-1}\right) e^{i(t-s)\lambda_j}$$

$$= \frac{1}{2\pi} \sum_{1-n}^{n-1} \left(1 - \frac{|\ell|}{n}\right) \widetilde{\Gamma}_\ell e^{-i\ell\lambda_j}, \tag{A.21}$$

because

$$D_n(\lambda_k - \lambda_j) = n, \quad j = k \bmod n,$$
$$= 0. \quad j \ne k \bmod n.$$

Now for $\ell \geq 0$ $\tilde{\Gamma}_\ell = \Gamma_\ell$, whereas for $\ell < 0$, $\tilde{\Gamma}_\ell = \Gamma_0 + \Gamma_{-1} - \Gamma_{\ell-1}$, so (A.35) has real part

$$\frac{1}{2\pi} \sum_{0}^{n-1} \left(1 - \frac{\ell}{n}\right) \Gamma_\ell \cos \ell\lambda_j + \frac{1}{2\pi} \sum_{1-n}^{-1} \left(1 + \frac{\ell}{n}\right) (\Gamma_0 + \Gamma_{-1} - \Gamma_{\ell-1}) \cos \ell\lambda_j.$$

(A.22)

The first term can be written

$$\frac{1}{4\pi} \sum_{1-n}^{n-1} \left(1 - \frac{|\ell|}{n}\right) \Gamma_{|\ell|} \cos \ell\lambda_j + \frac{\Gamma_0}{4\pi}.$$

To deal with the second term of (A.22) note that for $1 \leq j \leq n-1$

$$\sum_{\ell=0}^{n-1} \ell e^{i\ell\lambda_j} = \frac{e^{i\lambda_j} - 1}{(1 - e^{i\lambda_j})^2} - \frac{(n-1)}{1 - e^{i\lambda_j}} = \frac{-n}{1 - e^{i\lambda_j}},$$

which has real part

$$-\frac{n}{2} \left(\frac{1}{1 - e^{i\lambda_j}} + \frac{1}{1 - e^{-i\lambda_j}}\right) = -\frac{n}{2} \left(\frac{2 - 2\cos\lambda_j}{2 - 2\cos\lambda_j}\right) = -\frac{n}{2}.$$

Thus, the second term in (A.22) is

$$\frac{(\Gamma_0 + \Gamma_{-1})}{2\pi} \sum_{0}^{n-1} \left(1 - \frac{\ell}{n}\right) \cos \ell\lambda_j - \frac{\Gamma_0 + \Gamma_{-1}}{2\pi}$$

$$- \frac{1}{4\pi} \sum_{1-n}^{n-1} \left(1 - \frac{|\ell|}{n}\right) \Gamma_{-|\ell|-1} \cos \ell\lambda_j + \frac{\Gamma_{-1}}{4\pi}$$

$$= -\frac{1}{4\pi} \sum_{1-n}^{n-1} \left(1 - \frac{|\ell|}{n}\right) \Gamma_{-|\ell|-1} \cos \ell\lambda_j + \frac{\Gamma_0}{4\pi}.$$

It follows that (A.21) has real part

$$\frac{1}{4\pi} \sum_{1-n}^{n-1} \left(1 - \frac{|\ell|}{n}\right) (\Gamma_{-|\ell|} - \Gamma_{-|\ell|-1}) \cos \ell\lambda_j,$$

which is the Cesaro sum, to $n-1$ terms, of the Fourier series of $h(\lambda_j)/2$. Equivalently we can write

$$E\left\{\frac{n}{m} \hat{F}_{xe}(1, m)\right\} = \frac{1}{4\pi nm} \sum_{j=1}^{m} \int_{-\pi}^{\pi} |D_n(\lambda - \lambda_j)|^2 h(\lambda) \, d\lambda.$$

(A.23)

Fix $\varepsilon > 0$. There exists $\delta > 0$ such that $\|h(\lambda) - h(0)\| < \varepsilon$ for $0 < |\lambda| \le \delta$. Let n be large enough that $2\lambda_m < \delta$. Then the difference between the right-hand side of (A.23) and $h(0)/2$ is bounded in absolute value by

$$\frac{1}{4\pi nm} \sum_{j=1}^{m} \int_{-\pi}^{\pi} |D_n(\lambda - \lambda_j)|^2 \|h(\lambda) - h(0)\| d\lambda$$

$$\le \frac{1}{4\pi nm} \left\{ \varepsilon \max_{1 \le j \le m} \int_{-\delta}^{\delta} |D_n(\lambda - \lambda_j)|^2 d\lambda \right.$$

$$\left. + \sup_{\frac{\delta}{2} < |\lambda| < \pi} |D_n(\lambda)|^2 \left(\int_{-\pi}^{\pi} \|h(\lambda)\| d\lambda + 2\pi \|h(0)\| \right) \right\}$$

$$= O\left(\varepsilon + \frac{1}{n} \right),$$

because of Assumption 7, (A.5) and

$$\int_{-\pi}^{\pi} |D_n(\lambda)|^2 \, d\lambda = 2\pi n.$$

Because ε is arbitrary, the proof is complete. ∎

Proof of Theorem 4 (14.36) is familiar under somewhat different conditions from ours (e.g. see Stock 1987; Phillips 1988), but we briefly describe its proof in order to indicate how the outcome differs from (14.37). We have

$$n(\hat{\beta}_{n-1} - \beta) = (n^{-1}\widetilde{A})^{-1} \{(\widetilde{a} - E(\widetilde{a})) + E(\widetilde{a})\}.$$

Now

$$n^{-1}\widetilde{A} \Rightarrow V(\iota, \Omega), \quad \widetilde{a} - E(\widetilde{a}) \Rightarrow U(\Omega), \qquad \text{as } n \to \infty \qquad (A.24)$$

from Assumption 6, Theorem 2 and the continuous mapping theorem. Because $E(\widetilde{a}) \to \Gamma_0$ by elementary calculations and $V(\iota, \Omega)$ is a.s. of full rank by Phillips and Hansen (1990), (14.36) is proved. Next

$$n(\hat{\beta}_m - \beta) = [n^{-1}\widetilde{A} + n^{-1}\{(\hat{A} - \widetilde{A}) - E(\hat{A} - \widetilde{A})\} + n^{-1}E(\hat{A} - \widetilde{A})]^{-1}$$
$$\times [\widetilde{a} - E(\widetilde{a}) + \{(\hat{a} - \widetilde{a}) - E(\hat{a} - \widetilde{a})\} + E(\hat{a})].$$

Since $n^{-1}\{\hat{A} - \widetilde{A} - E(\hat{A} - \widetilde{A})\} \to_p 0$, $n^{-1}E(\hat{A} - \widetilde{A}) \to 0$ and $\hat{a} - \widetilde{a} - E(\hat{a} - \widetilde{a}) \to_p 0$ from Propositions 1 and 2, and Lemma 1, the proof of (14.36) is completed by invoking (A.24), Lemma 4 and (14.5). ∎

Proof of Theorem 5 From (A.18), Theorem 2, (14.30), (14.11) of Proposition 1, and (14.14) of Proposition 2 we deduce (14.38). Next, using (A.19),

$$\hat{\beta}_{n-1} - \hat{\beta}_m = \Delta(\Delta\hat{A}\Delta)^{-1}\{\Delta(\hat{A} - \tilde{A})\Delta\}(\Delta\tilde{A}\Delta)^{-1}\Delta\tilde{a} \tag{A.25}$$

$$- (\Delta\hat{A}\Delta)^{-1}\Delta\left\{(\hat{a} - \tilde{a}) - E(\hat{a} - \tilde{a})\right\}. \tag{A.26}$$

The ith element of the right side of (A.25) is $o_p(n^{d_i + d_e})$ by arguments used in the previous proof and (A.20), while the ith element of (A.26) is also $o_p(n^{d_i + d_e})$ on applying also (14.10) of Proposition 1 and (14.13) of Proposition 2, to prove (14.39). Then (14.40) is a consequence of (14.38) and (14.39). ■

Proof of Theorem 7 The proof of (14.42) follows routinely from (A.18), (14.27), (14.30) and (14.41). Then (14.43) is a consequence of (14.39) and (14.42), because in view of (A.26) it is clear that (14.39) holds for all $d_e < d_{min}$. ■

References

Abramovitz, M. and Stegun, I. (1970), *Handbook of Mathematical Functions*, New York: Dover.

Akonom, J. and Gourieroux, C. (1987), 'A functional central limit theorem for fractional processes', preprint, CEPREMAP.

Bossaerts, P. (1988), 'Common nonstationary components of asset prices', *Journal of Economic Dynamics and Control*, 12, 347–64.

Campbell, J. Y. and Shiller, R. J. (1987), 'Cointegration and tests of present value models', *Journal of Political Economy*, 95, 1062–89.

Chan, N. H. and Terrin, N. (1995), 'Inference for unstable long-memory processes with applications to fractional unit root autoregressions', *Annals of Statistics*, 23, 1662–83.

Cheung, Y. and Lai, K. (1993), 'A fractional cointegration analysis of purchasing power parity', *Journal of Business and Economic Statistics*, 11, 103–12.

Dickey, D. A. and Fuller, W. A. (1979), 'Distribution of the estimators for autoregressive time series with a unit root', *Journal of the American Statistical Association*, 74, 427–31.

Diebold, F. X. and Rudebusch, G. (1989), 'Long memory and persistence in aggregate output', *Journal of Monetary Economics*, 24, 189–209.

Dolado, J. and Marmol, F. (1998), 'Efficient estimation of cointegrating relationships among higher order and fractionally integrated processes'. Preprint.

Engle, R. (1974), 'Band spectrum regression', *International Economic Review*, 15, 1–11.

—— and Granger, C. (1987), 'Cointegration and error correction: representation, estimation and testing', *Econometrica*, 55, 251–76.

Geweke, J. and Porter-Hudak, S. (1983), 'The estimation and application of long memory time series models', *Journal of Time Series Analysis*, 4, 221–38.

Gourieroux, C., Maurel, F., and Monfort, A. (1989), 'Least squares and fractionally integrated regressors', *Working Paper #8913*, INSEE.

Hannan, E. J. (1963), 'Regression for time series with errors of measurement', *Biometrika*, 50, 293–302.

Hosoya, Y. (1997), 'A limit theory of long-range dependence and statistical inference in related models', *Annals of Statistics*, 25, 105–37.

Hurvich, C. M., Deo, R., and Brodsky, J. (1998), 'The mean squared error of Geweke and Porter-Hudak's estimates of the memory parameter of a long memory time series', *Journal of Time Series Analysis*, 19, 19–46.

—— and Ray, B. K. (1995), 'Estimation of the memory parameter for nonstationary or noninvertible fractionally integrated processes', *Journal of Time Series Analysis*, 16, 17–41.

Jeganathan, P. (1999), 'On asymptotic inference in cointegrated time series with fractionally integrated errors', *Econometric Theory*, 15, 583–621.

Johansen, S. (1988), 'Statistical analysis of cointegration vectors', *Journal of Economic Dynamics and Control*, 12, 231–54.

—— (1991), 'Estimation and hypothesis testing of cointegration vectors in Gaussian vector autoregression models', *Econometrica*, 59, 1551–80.

Künsch, H. R. (1987), 'Statistical aspects of self-similar processes', *Proceedings of the First World Congress of the Bernoulli Society*, VNU Science Press, 1, 67–74.

Lobato, I. (1997), 'Consistency of the averaged cross-periodogram in long memory series', *Journal of Time Series Analysis*, 18, 137–56.

—— and Robinson, P. M. (1996), 'Averaged periodogram estimation of long memory', *Journal of Econometrics*, 73, 303–24.

Mandelbrot, B. B. and Van Ness, J. R. (1968), 'Fractional Brownian motions, fractional noises and applications', *SIAM Review*, 10, 422–37.

Marinucci, D. and Robinson, P. M. (2000), 'Weak convergence of multivariate fractional processes', *Stochastic Processes and their Applications*, 86, 103–20.

Phillips, P. C. B. (1988), 'Weak convergence of sample covariance matrix to stochastic integrals via martingale approximation', *Econometric Theory*, 4, 528–33.

—— (1991*a*), 'Optimal inference in cointegrated systems', *Econometrica*, 59, 283–306.

—— (1991*b*), 'Spectral regression for cointegrated time series', *Nonparametric and Semiparametric Methods in Econometrics and Statistics* (W. A. Barnett, J. Powell, and G. Tauchen, eds), Cambridge: Cambridge University Press, 413–35.

—— and Durlauf, N. (1986), 'Multiple time series regression with integrated processes', *Review of Economic Studies*, 53, 473–95.

—— and Hansen, B. E. (1990), 'Statistical inference in instrumental variable regression with $I(1)$ variables', *Review of Economic Studies*, 57, 99–125.

—— and Ouliaris, S. (1988), 'Testing for co-integration using principal components methods', *Journal of Economic Dynamics and Control*, 12, 205–30.

Robinson, P. M. (1994*a*), 'Semiparametric analysis of long-memory time series', *Annals of Statistics*, 22, 515–39.

—— (1994*b*), 'Efficient tests of nonstationary hypotheses', *Journal of the American Statistical Association*, 89, 1420–37.

—— (1994*c*), 'Rates of convergence and optimal spectral bandwidth for long range dependence', *Probability Theory and Related Fields*, 99, 443–73.

—— (1995*a*), 'Log-periodogram regression of time series with long range dependence', *Annals of Statistics*, 23, 1048–72.

——(1995*b*), 'Gaussian semiparametric estimation of long-range dependence', *Annals of Statistics*, 23, 1630–61.

—— and Marinucci, D. (2001) 'Narrow-band analysis of nonstationary processes', *Annals of Statistics*, 29, 947–86.

Silveira, G. (1991), *Contributions to Strong Approximations in Time Series with Applications in Nonparametric Statistics and Functional Central Limit Theorems*, Ph.D. Thesis, University of London.

Sims, C. A. (1988), 'Bayesian skepticism on unit root econometrics', *Journal of Economic Dynamics and Control*, 12, 463–74.

Stock, J. H. (1987), 'Asymptotic properties of least squares estimators of cointegrating vectors', *Econometrica*, 55, 1035–56.

Velasco, C. (1999*a*), 'Non-stationary log-periodogram regression', *Journal of Econometrics*, 91, 325–71.

—— (1999*b*), 'Gaussian semiparametric estimation of non-stationary time series', *Journal of Time Series Analysis*, 20, 87–127.

Index